D1198972

# STUDIES IN THE VEGETATIONAL HISTORY OF THE BRITISH ISLES

# STUDIES IN THE VEGETATIONAL HISTORY OF THE BRITISH ISLES



ESSAYS IN HONOUR OF
HARRY GODWIN

EDITED BY
D. WALKER & R. G. WEST

CAMBRIDGE
AT THE UNIVERSITY PRESS
1970

Published by the Syndics of the Cambridge University Press
Bentley House, 200 Euston Road, London N.W.1
American Branch: 32 East 57th Street, New York, N.Y.10022


Library of Congress Catalogue Card Number: 69–75829
Standard Book Number: 521 07565 3

Printed in Great Britain
at the University Printing House, Cambridge
(Brooke Crutchley, University Printer)

(*facing p.* v)

Harry Godwin entered Cambridge University as an undergraduate in 1919 and remained there until his retirement from the Chair of Botany in 1968. In the intervening years his enthusiasm, industry and humanity have been inexhaustible sources of stimulus and guidance to innumerable students and colleagues. The authors of these essays offer them in gratitude for having had an opportunity to share in the excitement.

# CONTENTS



*The photograph facing p.v is by F. T. N. Elborn*

# PLEISTOCENE HISTORY OF THE BRITISH FLORA

by R. G. West

*Subdepartment of Quaternary Research, Botany School, University of Cambridge*

## INTRODUCTION

In dealing with the vegetational history and environmental changes of the period since the last glaciation, we have a straightforward connection between past and present. Landscapes have changed little; basins of accumulation are recognizable and often still existent, and the vegetational changes are continuous and directly relative to present British vegetation and phytogeography. Naturally, therefore, it is this recent part of Pleistocene history which bears most interest for the ecologist and phytogeographer. Farther back in geological time, the changes in landscape consequent upon multiple glaciation, and the repeated and unknown number of climatic fluctuations in the Pleistocene, make it difficult to see any significance in the vegetational changes of those remoter times for the present British flora. What interest these periods lose for the ecologist, they gain for the geologist, for in recent years it has become obvious in north-west Europe, at least, that vegetational history can be the key for the unravelling of the Pleistocene sequence. But, on closer consideration, and in the wider context of the evolution of the flora and its response to environmental changes over a long period, the study of these older floras and their relation to the present flora on the one hand, and to the Pliocene flora on the other hand, can reveal overall patterns of migration and extinction of species, and patterns of aggregation and dispersion of species within vegetation types, which have an important bearing on the view we form of the interrelations of species of present plant communities, as well as of course providing the substantive evidence for the past history of the species.

Edward Forbes (1846) recognized at an early date the importance of climatic change, large enough to induce glaciation, for the history of the British flora. Thus he said that the glacial (arctic-alpine) element in the present flora appeared in Britain during the glacial epoch and had survived in favourable situations since that time. The biological importance of glaciation was recognized by Clement and Eleanor Reid (1915), who explained the depauperate nature of the Pleistocene flora, as compared with the Pliocene flora, as a result of climatic refrigeration and enforced migration acting in conjunction with barriers to migration formed, in Europe, by the predominantly west–east mountain chains—the Reids' ideas were based on the study of fossil floras of macroscopic plant remains. More recently, many more details of the Pleistocene history of the flora have been gained by studies of the microfossils, and the outline of the Pleistocene history of the flora is rather clear. But such studies have not only revealed the history of the flora. They have made possible a reconstruction in considerable detail of the sequence of climatic episodes, and of the environments against which the history of the flora must be viewed. It is the purpose of this essay to present such reconstructions that have been made, and to discuss what interest they hold for the botanist and geologist, and what may be expected in the future exploration of this field of study.

## ENVIRONMENTAL DIVERSITY AND THE RELATION BETWEEN CLIMATE AND FLORA IN THE PLEISTOCENE

We can usefully distinguish in the Pleistocene three major divisions of environment—the glacial environment, the periglacial environment and the non-glacial environment. With the first we are not concerned, even though it is the environment in which is produced the thick glacial deposits which characterize the Pleistocene over much of north-west Europe. The periglacial environments can be considered as the dominant form of environment during the glaciations in non-glacierized areas. As far as can be seen it is a tundra-like environment of open vegetation and low mean temperatures, though the details vary according to climatic variation with time and with latitude and longitude. At some times and in some places permafrost may develop with indications of considerable continentality of climate; at other times and places more oceanic conditions may ameliorate the climate.

The non-glacial environment, transitional from and to the periglacial environment, shows the development of more temperate floras in the intervals between glacial advances. Physical evidence suggests a mean temperature difference of some 5–7 °C between the cold and temperate extremes. The type of vegetation developed depends on the length of the non-glacial period and the degree of climatic amelioration manifested. If, in a cycle of climatic change within a non-glacial period, the time or the degree of climatic amelioration only permits the development of boreal forest, the period is termed interstadial. If time and climatic amelioration allows the development of temperate deciduous forest, of a kind characterizing the Flandrian (Post-glacial) climatic optimum, then the period is termed interglacial. These definitions are not expected to be clear cut, for both latitude and longitude will play a part in determining non-glacial vegetational sequences, as will the relation of the periglacial to the non-glacial floras in particular areas. For example, we might envisage as a possibility the situation where a flora described as interglacial in an area of continental climate, where a rich periglacial flora survived, might be of the same age as a flora described as interstadial in an oceanic area distant from the glacial refuges of the thermophilous species required to transform it to an interglacial.

However, the concept of interstadial and interglacial ameliorations is useful in our area of the British Isles in categorizing the type of vegetational history to be found in non-glacial intervals.

These categories of climatic change, defined by vegetational change, are of course very broad. They concern on the one hand regional aspects of vegetation, and on the other hand the gross climatic effects on this vegetation. The details of seasonal distribution of temperature and rainfall and the extremes of variation in microhabitat tend to escape us. Yet from the point of view of vegetational history it is just these facts that we should wish to know in assessing the importance of climatic variations for plant distribution in the past. We should, for example, like to know the mosaic of communities which characterized the periglacial flora at any one time. Such investigations require the study of fossil floras known to be synchronous, and living under different microhabitats and differing edaphic conditions. The possibilities of diversity will be seen to be enormous and are well illustrated by the great diversity of treeless vegetation types in northern Eurasia at the present time. This complicated nature of the full-glacial flora, with its many phytogeographical elements, contrasts with the more easily explicable and uniform floras of the warmer episodes, with their variation in microhabitat muted by a more uniform forest cover.

The history of particular species will depend on climatic change, microhabitat diversity, reproductive ability and chance. In many examples we can trace the history of the species throughout the Pleistocene, noting their disappearance or appearance in cold or temperate conditions, as will be shown subsequently. It would be satisfactory if we could use the present distribution of such species as closer indicators of climate and environment in the past. But usually too little is known of the autecology of the species concerned or of their behaviour in communities, so that only broad generalizations about the environment can be made. It might be hoped that the history of the flora in the last few centuries, combined with our knowledge of climatic change within that time, might be used as a basis for extrapolating backwards in time on the relation between vegetation and climate. But the anthropogenic factor here confuses the relation.

The importance of relating known climatic change to known floristic change cannot be overstressed; only in this way will we be able to interpret the details of past climatic change and the effects of the present climatic tendencies.

## VEGETATIONAL HISTORY OF THE PLEISTOCENE

### Units of vegetational history

The biostratigraphical units which comprise vegetational history are pollen zones. These are assemblage zones, typified by dominant regional genera of the pollen spectrum. The pollen zones of the cold or glacial stages are divisible into full-glacial and interstadial types. The former show the prevalence of open vegetation, the latter the presence of trees of present boreal distribution. There is yet insufficient evidence for the erection of long sequences of pollen zones in the cold stages, though variations are seen in the full-glacial spectra, in particular in the diversity of taxa recorded. On the other hand, during the temperate interglacial stages, the marked vegetational changes give rise to sequences of pollen zones (Turner & West, 1968). In general, during these temperate stages, four major assemblage zones may be discerned, typified as follows (the youngest first):

ZONE IV (post-temperate zone). Amongst the trees, the dominant genera are boreal—*Betula*, *Pinus*, *Picea*. There is a thinning of the forest and non-tree pollen types are frequent, particular pollen of Ericales, associated with damp heathland.

ZONE III (late-temperate zone). The zone is characterized by the expansion of forest trees not abundant earlier in the stage—*Carpinus*, *Abies*, *Picea*, and perhaps *Tsuga*—at the expense of mixed oak forest genera already present.

ZONE II (early temperate zone). This zone is dominated by forest trees of the mixed oak forest—*Quercus*, *Ulmus*, *Fraxinus*, *Corylus*.

ZONE I (pre-temperate zone). This zone is characterized by the presence of boreal trees, *Betula* and *Pinus*, accompanied by significant quantities of pollen-types of light-demanding shrubs and herbs.

This sequence of zones is found in the temperate Pastonian, Cromerian, Hoxnian and Ipswichian stages, though in each sequence there are minor differences which permit distinctions to be drawn between each of them. The zones of the different stages can be conveniently signified by pre-fixing the initial letter of the stage to the zone (table 1).

In building up our sequence of assemblage zones, we have finally to draw a boundary between the cold or glacial stages and the temperate stages. The lower boundary of a temperate stage may be placed at the point where the tree-pollen percentages consistently exceed those of non-tree pollen, signifying the change from open vegetation to (boreal) forest. A similar definition, but reversed, can be used for the temperate/cold stage boundary, though this is more difficult to apply as the change is a more gradual one.

## THE PLEISTOCENE SEQUENCE

Figure 1 summarizes the sequence of stages now known from the British Pleistocene. The extent of the fossil record in relation to these stages is also shown in the figure. It will be seen that there is a record, at one place or another in Britain, over much of the Pleistocene. Much of the record comes from sites in East Anglia. Regional variation at one time, however, is hardly known, because of the low total number of sites studied, except of course in the Flandrian. The curve shown in the figure is not to be taken as a complete expression of climatic change, but is inserted merely to give some kind of scale to the type of flora recorded and the approximate extent in relative time covered by the fossil floras.

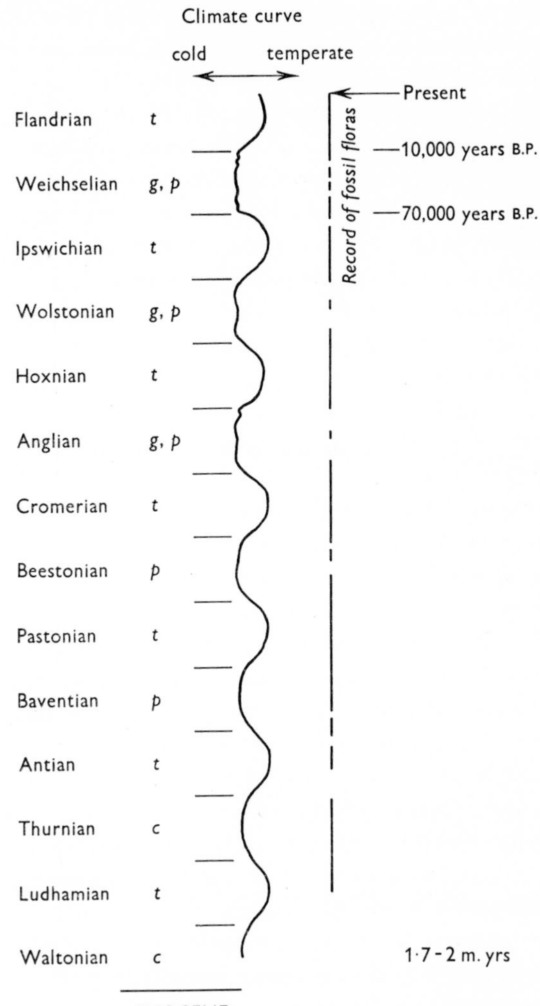


Figure 1. The sequence of Pleistocene stages in relation to climatic change and the fossil record. *t*, temperate; *g*, glacial; *p*, permafrost; *c*, other evidence of cold conditions.

## PREGLACIAL VEGETATIONAL HISTORY

We consider first the vegetational history of pre-Anglian times, that is before the first great glaciation of midland Britain. Here the record is confined to East Anglia. Unfortunately there are not yet any certain records of the Pliocene flora in Britain, though on the continent there are rich fossil floras of Late Tertiary age. These show that the Pliocene flora of continental Europe was characterized by a considerable number of exotic genera, many now native in eastern Asia and north America. Such genera as *Sequoia*, *Tsuga*, *Carya*, and *Nyssa* are commonly found. This exotic element forms part of the circumboreal so-called Arcto–Tertiary flora of the northern Hemisphere. At the onset of the Pleistocene climatic deterioration, the exotic temperate genera must have been forced out from north-west Europe, and on the return of a temperate climate after the first cold stage (Pretiglian of the Netherlands) many did not reappear, and thus became extinct in north-west Europe as a result of climatic change. C. and E. M. Reid (1915) spoke of barriers to migration in the form of west–east oriented mountain ranges, and the Mediterranean, in western Eurasia, whereas in eastern Asia and north America the diversity of habitats given by the altitudinal range of the mountains and their alignment allowed the survival of the Arcto–Tertiary flora during the cold stages. The explanation is a very reasonable one. It must then be presumed that the surviving genera in western Europe were particularly hardy or were rich enough in biotypes to find suitable perglacial refugia.

In Britain few sites of Lower Pleistocene age have been studied, but there is a satisfactory long record from one or two boreholes. The pollen diagrams from these boreholes show fluctuations in vegetation, with mixed oak forest genera and *Tsuga* characterizing the temperate stages (Ludhamian, Antian) and higher percentages of non-tree pollen, principally Ericales, and boreal trees characterizing the cooler stages (Thurnian, Baventian) (West, 1961).

As there is no record of Pliocene flora we cannot say what effect the earliest cold period (Waltonian?) had on the Pliocene flora. But if the Ludhamian is equivalent in age to the Tiglian flora of the Netherlands, as seems likely, it follows that the flora of East Anglia was far poorer in exotic species than the continental flora, for there is a considerable percentage of exotic pollen types in the Tiglian flora, whereas only *Tsuga* is represented abundantly as a widespread exotic genus in East Anglia. Perhaps some of this difference may be a consequence of the fact that the Tiglian flora is a freshwater one whereas the pre-Baventian pollen floras of East Anglia are found only in marine deposits; but the magnitude of the contrast suggests that there is indeed a real difference, and that west European vegetation was clearly differentiated across the lines of longitude, more so perhaps than today.

The marine deposits of the pre-Pastonian cool stages (Baventian, Antian) are characterized by high percentages of Ericales pollen, much of it of *Empetrum*-type, while Baventian freshwater sediments show a contrast, with high percentages of non-tree pollen giving a true full-glacial aspect to the flora. This contrast between the two types of pollen flora results from differential transport according to environment, at present being investigated, and the meaning of the difference in regional vegetational terms is not clear. At any rate, between the forested stages there are clearly cooler and perhaps wetter intervals characterized by open vegetation and heath.

The Pastonian temperate stage, in contrast to the earlier temperate stages, shows a characteristic interglacial-type sequence of pollen zones, with very low frequencies of *Tsuga* confined to Zone P III. It is remarkable also in showing *Carpinus* present in Zone P II, in the mixed oak forest pollen flora. The full-glacial floras of the Beestonian are similar to those of the later glaciations. The Cromerian vegetational history is again a typical interglacial sequence. *Tsuga* is absent. *Carpinus* and *Abies* appear in Zone C III, as in the Hoxnian temperate stage. The flora of the Cromerian is remarkable in showing so few non-British species, even though it predates the earliest glacial deposits of East Anglia. In fact all Cromerian species known are either British or occur in later interglacial floras (e.g. *Salvinia natans*, *Azolla filiculoides*, *Trapa natans*).

What evidence we have, therefore, suggests that even before the time of the earliest known glacial advance into East Anglia a flora resembling the modern British flora was present in East Anglia, that modification of the flora from a possibly rich Pliocene flora occurred at the beginning of the Pleistocene, with *Tsuga* the only characteristic survivor into the Lower Pleistocene, and that a difference existed between the British Lower Pleistocene flora and the synchronous European flora.

### The glacial stages

The floras of the glacial stages are divisible into full-glacial floras and interstadial floras. The full-glacial floras are characterized by high proportions of non-tree pollen, with a varying diversity of pollen types. The vegetation indicated by the plant remains is open and a mixture of plants of many ecological and geographical categories is usually present, e.g. northern and montane plants, halophytes, and other maritime plants, many plants of wide distribution, often 'weeds', and a few species of more southern distribution, often aquatics and marsh plants. Species in these groups were listed by Godwin (1956), but many more records have accumulated since then, and we are in need of a detailed re-assessment of the full-glacial flora.

A few of the more characteristic species may be mentioned:

Northern and montane plants: *Salix herbacea*, *Dryas octopetala*, *Draba incana*, *Thalictrum alpinum*, *Saxifraga oppositifolia*.

TABLE 1. *Zone characters of temperate stages (the frequency references ase to pollen percentages)*

| | Pastonian | | Cromerian | | Hoxnian | | Ipswichian | | Flandrian | |
|---|---|---|---|---|---|---|---|---|---|---|
| | e Be | | e A | | e Wo | | e We | | | |
| IV | P IV | *Pinus*, *Picea*, *Betula*, *Alnus*, Ericales | C IV | *Pinus*, *Picea*, *Betula*, *Alnus*, Ericales | H IV | *Pinus*, *Betula* N.A.P. higher | I IV | *Pinus*, N.A.P. higher | | — |
| III | P III | M.o.f., *Carpinus*, *Picea*, *Tsuga* | C III | M.o.f., *Abies*, *Carpinus* | H III | M.o.f., *Abies*, *Carpinus* | I III | *Carpinus* | | |
| II | P II | M.o.f., low *Carpinus*, *Picea*, and *Corylus* | C II | M.o.f., high *Ulmus*, low *Corylus* | H II | M.o.f., *Taxus*, *Corylus* | I II | M.o.f., *Pinus*, *Acer*, high *Corylus* | F II | M.o.f. |
| I | P I | *Pinus*, *Betula* | C I | *Pinus*, *Betula* | H I | *Betula*, *Pinus* | I I | *Betula*, *Pinus* | F I | *Betula*, *Pinus*, *Corylus* |
| | l Ba | — | l Be | — | l A | (*Hippophaë*) | l Wo | — | l We | — |

Ba, Baventian; e, early; l, late; C. Cromerian; Be, Beestonian; P, Pastonian; Wo, Wolstonian; H, Hoxnian; A, Anglian; F, Flandrian; We, Weichselian; I, Ipswichian

Maritime plants: *Armeria maritima*, *Atriplex hastata*, *Plantago maritima*, *Sueda maritima*.

Weeds: *Polygonum aviculare*, *Potentilla anserina*, *Ranunculus repens*.

Plants of a more southern distribution: *Potamogeton crispus*, *P. densus*, *Ranunculus sceleratus*.

This type of plant list is recorded from floras of Beestonian, Anglian, Wolstonian and Weichselian age, and the assemblage recurs in Pleistocene time. The mixed assemblages must derive from the diversity of habitat and microclimate of the periglacial area, with its possibilities for permafrost, waterlogged soil, sunny banks, solifluction slopes and so on; and perhaps from minor climatic fluctuations in the periglacial area which affected the constitution of the assemblages. The considerable increase in knowledge of full-glacial floras in recent years must lead to a much more detailed knowledge of the species concerned and how they combined into communities. Studies of the macroscopic plant remains in conjunction with studies of pollen, local sediment and surrounding soil types should lead to a greater understanding of the peculiar full-glacial flora.

Interstadial floras, showing evidence for boreal forested conditions during glacial stages, are very few. The best known is the Chelford interstadial (Simpson & West, 1958) from the last (Weichselian) glaciation, where *Betula–Pinus–Picea* forest was the dominant local vegetation. If the Chelford interstadial is the same age as the Brørup interstadial of continental north-west Europe, it appears that the *Picea omoricoides* recorded from the continent at this time did not reach this country. On the other hand, *Picea abies* is represented. This species did not reach Britain in the Ipswichian, although characteristic of the later part of the equivalent Eemian interglacial on the continent.

There is some evidence for interstadial conditions at some time during the Wolstonian glacial stage at Mildenhall, with boreal forest present. On the other hand the only flora known from between the two ice advances of the Anglian glaciation, at Lowestoft, is of a glacial type, and was probably formed during an adjustment of ice sheets rather than during a definite climatic amelioration (West & Wilson, 1968).

### Interglacial and other temperate stages

The interglacial deposits found in Britain may be referred to the Hoxnian or Ipswichian interglacial stages. The outline of forest history is shown in table 1, which tabulates the zones. The Flandrian (Postglacial) is also shown in this table, as well as two earlier temperate stages which have clear zonations.

We may briefly give the salient points which appear to characterize the different histories of the temperate stages as follows:

Flandrian: High *Corylus* frequencies in Zone F I.

Ipswichian: High *Corylus* early in Zone I II. Well-marked *Carpinus* zone (I III).

Hoxnian: High frequencies of *Hippophaë* pollen in the late glacial (l A).
Considerable *Tilia* frequencies in late Zone H II. *Corylus* maximum later in Zone H II. *Abies*, *Picea*, *Carpinus* and *Pterocarya* in Zone H III.

Cromerian: Low frequencies of *Corylus*. High frequencies of *Ulmus* in Zone C II, but *Tilia* low or absent. *Abies*, *Picea* and *Carpinus* occur in Zone C III.

Pastonian: *Carpinus* in Zones P II and P III. *Tsuga* in Zone P III. No or very low *Abies* in Zone P III.

### The zonal sequence of the temperate stages

The sequence of four zones already outlined for the temperate stages must now be related to environmental changes forming the cycle of change during the stages. The fact that the four-zone system appears applicable to a number of temperate stages does support the idea that we are dealing with similar effects in each temperate stage. Zones I and II are clearly an expression of the early expansion of forest genera in the order boreal to temperate. Zone I is characterized by light-demanding genera, Zone II by the spreading of the temperate shade-giving genera, *Ulmus*, *Quercus*, *Alnus* and *Corylus*. This change must partly be a result of succession on soils improved from the raw immediately post-glacial state to richer mull soils, and partly a result of the times of immigration and expansion of genera.

The change from Zone II to Zone III is given by the expansion of genera, in particular *Carpinus*, *Abies*, and *Picea* at the expense of the mixed oak-forest genera. This expansion may be accompanied by an increase in Ericales pollen. Basically this change results from the expansion of genera either already present in low frequencies or because of late immigration (e.g. like that of *Picea* in north-west Europe during the late Flandrian). It is likely that soil development from a mull to a mor state is related to the expansion of certain of the genera (e.g. *Picea*) which are known to cause soil deterioration after their introduction (Andersen, 1966). The increased acidification of soils lead to expansion of conifers and often heathland. In Zone IV this expansion continued and is accompanied by a reduction in the number of thermophilous forest genera. It is probable here that climatic deterioration played a part in the restriction of the flora, as well as the increased soil deterioration.

It will be clear that the interactions of climate, soil and plants make it exceedingly difficult to disentangle the role of each in determining vegetational change. A possible scheme of such changes, outlined above, may be summarized as follows:

1 Climatic amelioration.
2 Expansion of light-demanding genera.
3 Soil improvement to mull condition and expansion of shade-giving forest genera.
4 Soil deterioration to mor condition and/or expansion of late-arriving genera.
5 Climatic deterioration, restriction of thermophilous genera, expansion of heathland.

This last episode gives way to the periglacial environment with its open vegetation and soils enriched by solifluction and freeze/thaw processes.

### Distinctions between the zones of the different temperate stages

Though the zonal sequence of each temperate stage is similar, there are considerable differences between the zones of the several stages. Many factors can be involved in such differences. They include climatic differences between the stages, different barriers to migration during the stages, especially that formed by the present English Channel, differing distances of glacial refuges from which genera expanded, changes in ecological tolerance and variability within genera, and other changes consequent upon evolution or extinction. Again, it is most difficult to disentangle these factors. The difficulty of inferring climate from vegetation or flora records has been referred to many times. Perhaps a general trend of oceanicity or continentality may be discerned. For example, in the differences between the Hoxnian and Ipswichian interglacials: the former has an abundance of *Taxus*, *Ilex* and *Alnus* pollen less well represented in the latter, while the latter contains records of many continental thermophilous genera indicating higher summer temperatures than at present. Again this particular element in the Ipswichian flora is lacking in the Flandrian, which suggests that the Ipswichian climatic optimum may have been warmer than that of the Flandrian. But apart from these indications, it is difficult to draw further or close conclusions from analyses of fossil floras.

Some conclusions may perhaps be drawn regarding the persistence of the Channel barrier during the temperate stages. The resemblance of the Hoxnian of East Anglia and Holsteinian interglacial of continental Europe suggests no barrier between them in the early part of the interglacial. On the other hand, the considerable differences between the Ipswichian and Eemian (notably the lack of development in the Ipswichian of zones with *Tilia* and *Picea*) suggest a barrier early in the interglacial, perhaps formed by the connection through the Channel of the Eem Sea and the Atlantic. Such contrasts in the last two interglacials are in accord with marine mollusc faunas. The Eemian fauna has connections with the

Lusitanian fauna, while the Hoxnian fauna appears to have no such connection.

Differing distances of glacial refuges may account for the different behaviour of *Corylus* in the Hoxnian, Ipswichian and Flandrian stages. The notable difference is the progressively earlier time of expansion in England, late in the Hoxnian, at the time of the mixed oak forest in the Ipswichian, and before this in the Flandrian. This may be related to the increasing nearness of the glacial refuges to the ice-freed areas. If so, it may suggest a change in the ecological tolerances of *Corylus*, allowing it to survive in progressively nearer refuges, assuming that the intervening glacial climates were of a similar type. Here we come up against the question of changing biotypes or ecological tolerances within species and genera, which can more conveniently be dealt with separately below.

A further detailed discussion of the differences between the zones of the successive temperate stages would be out of place here. It will be apparent that much further information on the zones and their regional differentiation must be obtained before we can make more progress in disentangling the causes of the differences.

## The behaviour of certain forest genera

The interpretation of past vegetation and its relation to climate is based on the present behaviour of genera or species. Convincing generalizations can be drawn because the parallel behaviour of different taxa in fossil sequences leads to reasonable environmental conclusions. If we examine the zones of the temperate stages we find little or no evidence to refute this view. The genera behave together as would be expected in a climatic cycle. But if we examine the history of each genus in more detail, we may discern that certain genera have a changing behaviour within climatic cycles assumed to be similar. Of course such changes may be the result of chance or other effects on distribution, and differences in the intensity of glaciation and climate of the cool stages. But because certain trends do emerge, it is possible that such effects do not have an overall importance, and we may then consider whether there is any evidence for evolutionary change.

Let us take certain temperate genera, and consider their behaviour in East Anglia during the temperate stages, summarized in figure 2. Some, like *Quercus* and *Ulmus*, occur consistently in each temperate stage. They are the earliest temperate forest genera to occur in each temperate cycle and expand as part of the temperate forest flora in Zones II and III. *Tilia* is not so consistently a part of the mixed oak forest. It is rare or absent in the Pastonian, Cromerian and Ipswichian, only forming an obvious forest component in the Hoxnian and the Flandrian. When present it is characteristic of Zone II. The behaviour of *Corylus* in the Flandrian, Ipswichian and Hoxnian has been mentioned already. In the Pastonian and Cromerian this genus plays a very minor part in the mixed oak forest. Figure 3 compares pollen curves for *Corylus* which have been found in East Anglia for the temperate stages. The change in time of the behaviour of this taxon shows that during the earlier stages its expansion corresponded with that of the mixed oak forest; but in the later temperate stages it may be suggested that there were changes in the biotypes which allowed it to survive in periglacial refuges in western Europe and to expand rapidly in the Flandrian before the expansion of the mixed oak forest and compete more effectively in the mixed oak forest.

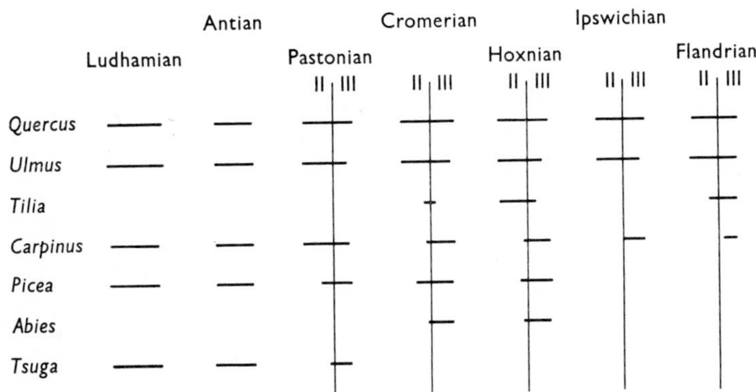


Figure 2. The occurrence of certain forest genera in East Anglia during the temperate stages of the Pleistocene.

Alternatively, chance survival or the differing extent of glaciation may have had similar effects on the perglacial survival of *Corylus*. A regional study of the behaviour of *Corylus* in the temperate stages over north-west Europe would be likely to lead to a solution of this problem.

*Carpinus* occurs in Zones II and III of the Pastonian, but only in Zone III of the Cromerian, Hoxnian and Ipswichian. Thus in these latter stages it shows a late expansion, small in the Cromerian and Hoxnian and Flandrian, but massive in the Ipswichian. This massive expansion contrasts with the less marked (Zone g) expansion on the continent, where it is accompanied by a rise in *Picea* pollen frequencies. We may suggest that the absence of *Picea* from the Ipswichian, perhaps a result of the absence of a land-bridge to the continent in the interglacial, allowed the unhindered expansion of *Carpinus*, while on the continent the spread of *Picea* favoured the podsolization of soils and limited conditions suitable for *Carpinus*. Likewise, the absence or rarity of heathland in Zone IV of the Ipswichian, compared with its presence in Zone IV of the previous temperate stages, might partly be related to a much reduced tendency to soil deterioration. In contrast, on the continent acidification of the soils towards the end of this interglacial does seem to have lead to the development of heathland.

The restriction of *Carpinus* to Zone III in the four latest stages (but not in the Pastonian) may suggest this genus has undergone a reduction in biotypic variation which only allowed it to survive in later glaciations at a distance which gave rise to late expansion in the temperate stages.

*Picea*, in the earlier temperate stages (Pastonian to Hoxnian), occurs in Zones II and III, though there is a tendency for expansion in Zone III. In the Ipswichian, no marked pollen frequencies occur; neither do they in the Flandrian. On the continent, in these two most recent stages, *Picea* shows late expansion. Thus the late immigration in these two stages did not allow spread to Britain, contrasted with its earlier appearance in the older stages. We may perhaps conclude that the successive glaciations impoverished the biotypes of *Picea* in such a way that refugia in western Europe were not available in the last two glaciations, though they were in earlier cold stages.

*Abies* is a characteristic tree of Zone III in the Cromerian and Hoxnian, but not in the other temperate stages. On the continent in north-west Europe it is a late immigrant, occurring in Zone III in the Holsteinian (Hoxnian) and Eemian (Ipswichian). The sporadic appearance of this genus in our interglacials is probably related to its rather slow spread from continental refuges during the temperate stages. Thus in the Eemian it failed to reach Denmark and the present western seaboard.

*Tsuga* is a characteristic tree only of the Ludhamian and Antian temperate stages, though it also occurs in very low pollen frequencies in Zone III of the Pastonian. In later temperate stages it is absent. This behaviour is most easily explicable by considering

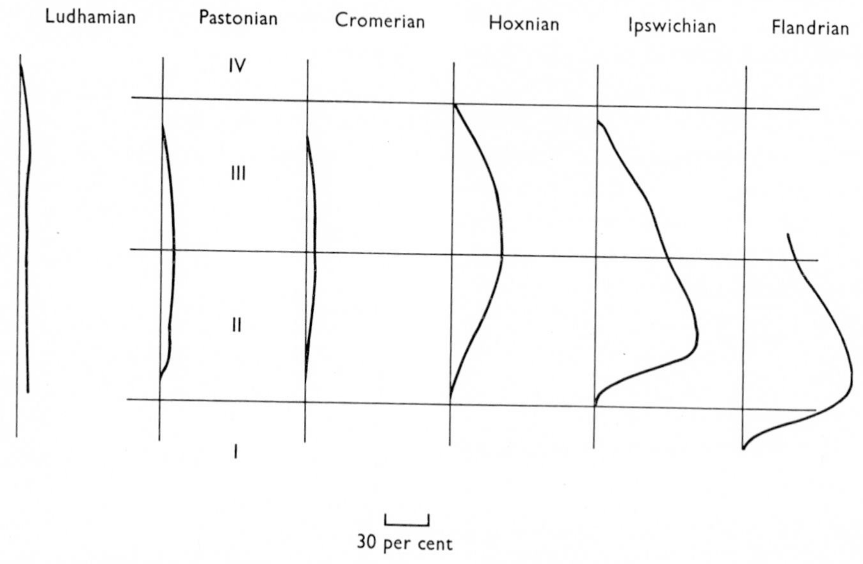


Figure 3. Schematic *Corylus* pollen curves for temperate stages in East Anglia, presented as percentages of total tree pollen and related to the zone system described in the text.

*Tsuga* as a late immigrant in the Pastonian, having suffered severe restriction in western Europe during the previous (Baventian) cold stage, then extinction in western Europe during the succeeding (Beestonian) cold stage. The genus was thus an early loss in the west European forest flora, though in southern Europe it may have survived into later temperate stages.

These remarks on the history of particular genera show how it may be possible in future to decipher in more detail the changing status of particular genera, and how each responded to environmental change. It would be of interest to relate the behaviour of these genera with their known taxonomic diversity. The Arcto-Tertiary flora will have contained genera in various states of evolution and biotypic variety, and it might be expected that extinction by environmental change in western Europe would affect most of those genera with minimum variation. Unfortunately the state of taxonomy of the major forest trees does not permit any correlation at present. In summary, figure 2 underlines the fact that those genera (*Quercus*, *Ulmus* and *Corylus*) which are most successful are those with a widespread distribution in western Europe, while others have suffered restriction in the course of time, perhaps caused by loss of biotypes (*Carpinus*, *Picea*, *Tsuga*), and yet others (*Tilia*, *Abies*) show an inconsistent behaviour more probably related to a combination of environmental effects and change in biotype content.

## HISTORY OF PHYTOGEOGRAPHICAL ELEMENTS

The British flora has been divided into a number of phytogeographical elements from the point of view of revealing the relations of the British flora to the European flora as a whole. Such geographical elements within the British flora were considered by Edward Forbes (1846) to relate to the origin of the flora in the sense that certain elements were of pre-glacial origin (Lusitanian, atlantic, southern continental), the arctic-alpine element appeared during the glacial period and the widespread 'Germanic' element was post-glacial in origin. Though each element is heterogeneous in that each species within it is likely to have had a different history, Forbes's deductions regarding the glacial and Lusitanian elements have support from fossil evidence. The earliest full-glacial floras (Baventian) are like those of the last glaciation and it is clear that in each of five cold stages a similar full-glacial flora was present.

TABLE 2. *Some fossil occurrence of species in different phytogeographical elements of Matthews (1955)*

| | | |
|---|---|---|
| Oceanic southern | *Damasonium alisma* | Weichselian; periglacial |
| | *Ranunculus parviflorus* | Ipswichian; temperate, Zone I II |
| Oceanic west European | *Daboecia cantabrica* | Wolstonian; periglacial |
| | *Erica mackaiana* | Hoxnian (Gortian); temperate, Zone H III |
| Continental southern | *Buxus sempervirens* | Hoxnian; temperate, Zone H III |
| | *Helianthemum canum* | Weichselian; periglacial |
| Continental | *Carpinus betulus* | Ipswichian; temperate, Zone I III |
| | *Ranunculus lingua* | Weichselian; periglacial |
| | *Stratiotes aloides* | Beestonian; periglacial |
| Continental northern | *Cicuta virosa* | Cromerian; temperate, Zone C II |
| | *Potamogeton praelongus* | Weichselian; periglacial |
| Oceanic northern | *Armeria maritima* | Weichselian; periglacial |
| | *Naias flexilis* | Flandrian; temperate, Zone VI<br>Weichselian; periglacial |
| North American | *Eriocaulon septangulare* | Flandrian; temperate, Zone VII a<br>Wolstonian; periglacial |

Presumably in each of the temperate stages this was dispersed to refugia as it is now. There is also fossil evidence for considering the Lusitanian element to contain species relict from Lower Pleistocene times in oceanic western Europe, though much more work needs to be done on the phytogeography of the Late Tertiary and Lower Pleistocene. Pliocene and Lower Pleistocene floras suggest considerable oceanicity in western Europe in those times, and it is probable that Lusitanian species were more widespread then.

It is more useful to discuss, though, the fossil history of particular species rather than of phytogeographical elements in the flora. There is a substantial record of species from nearly all the geographical categories named by Matthews (1955). Even a short list, such as that in table 2, shows that species in each category can have very different environments of fossil occurrence. We cannot therefore generalize from the history of particular species to the history of phytogeographical elements. Some species occur in both periglacial and temperate environments, others in periglacial environments only or temperate environments only. The reconstruction of the past history by the assembling of fossil records is the most important key to present distribution. A most striking example has been the finds of species (e.g. *Erica mackaiana*, *Daboecia cantabrica*) of the so-called 'Lusitanian' element in the Hoxnian (Gortian) interglacial of Ireland (Watts, 1959). Evidently these species have a long history in Ireland and further evidence may well demonstrate that some of these species were able to survive the periglacial environment.

## CONCLUSIONS

We have considered in general terms the changing flora of Britain during the Pleistocene; such a general treatment is partly a result of the lack of knowledge of regional diversity of floras of the British Isles. While much is known of regional diversity, in terms of altitude, latitude and longitude and to some extent surface rock type, in the Flandrian and Late-Weichselian, hardly anything of this sort is known for early periods. Only in the Hoxnian interglacial (Gortian of Ireland) can it be said that regional variation is known to some extent, and that only in terms of longitudinal variation resulting from increased oceanicity in the west. Much more study is required of sites in Britain and Ireland, to detect regional trends and their relation to the continent. The relation of the Irish Pleistocene flora in other interglacials than the Hoxnian will be of particular interest, in view of the possibility that the Irish flora has occupied a very isolated position throughout the Pleistocene and perhaps Pliocene.

The regional variation of full-glacial floras, of obvious importance to problems of perglacial survival, is again unknown. One of the difficulties here is the necessity for comparison of floras known to be synchronous. To take full advantage of sites it is clearly necessary to carry out combined studies of macroscopic plant remains with pollen studies, to obtain evidence of regional and local vegetation. We may expect improved interpretation of fossil assemblages from the currently developing work and interest in recent pollen rain and sedimentation of macroscopic remains, and their relation to present plant communities. With this information available we may be able to say more of the process of build-up of communities in the temperate stages, of retrogressive succession in the latter parts of these stages, and of the selection and spread of the periglacial floras.

Since the publication of Professor Godwin's book on *The History of the British Flora* in 1956, fossil plant records have accumulated more and more rapidly, with the increased number of researchers in the field, both in respect of Flandrian times and the older Pleistocene stages. So much so that to make full use of the data for reconstruction of the history of the flora, it will be necessary to have some sort of data retrieval system whereby records for locality, age, and local environment for each species can be grouped, of the kind already started by Professor Godwin. The accumulation of records in the British Isles, where the variety of regional climates and rock types within a relatively small area, isolated from the continent, and with a satisfactory stratigraphical record in the Pleistocene, will be very great value in providing the data for interpretation of the effects of environmental changes on the evolution of floras, a subject which becomes increasingly more important as man's pressure on natural vegetation systems increases.

## SUMMARY

The major Pleistocene environments, glacial, periglacial and non-glacial, are discussed with reference to their characteristic flora and vegetation. The extent of the Pleistocene fossil record is demonstrated. Preglacial, periglacial and temperate floras are described, and a general zonation applicable to temperate stages is described and its meaning discussed. Analogous zones in the successive temperate stages are compared with especial reference to the history of forest tree genera in East Anglia. The history of phytogeographical elements in the British flora is discussed.

I am pleased to acknowledge helpful discussions with Dr Charles Turner during the preparation of this essay.

## REFERENCES

Andersen, S. Th. (1966). Interglacial vegetational succession and lake development in Denmark. *Palaeobotanist* **15**, 117–27.

Forbes, E. (1846). Of the connection between the distribution of the existing fauna and flora of the British Isles, and the geological changes which have affected their area, especially during the epoch of the Northern Drift. *Mem. Geol. Survey Gt. Britain* **1**, 336–432.

Godwin, H. (1956). *The History of the British Flora.* Cambridge University Press.

Matthews, J. R. (1955). *Origin and distribution of the British Flora.* London: Hutchinson.

Reid, C. & Reid, E. M. (1915). The Pliocene floras of the Dutch–Prussian Border. *Med. Rijksops. Delfstoffen* no. 6.

Simpson, I. M. & West, R. G. (1958). On the stratigraphy and palaeobotany of a late-Pleistocene organic deposit at Chelford, Cheshire. *New Phytol.* **57**, 239–50.

Turner, C. & West, R. G. (1968). The subdivision and zonation of interglacial periods. *Eiszeit. u. Gegenwart*, **19**, 93–101.

Watts, W. A. (1959). Interglacial deposits at Kilbeg and Newtown, Co. Waterford. *Proc. R. Ir. Acad.* **60** B, 79–134.

West, R. G. (1961). Vegetational history of the Early Pleistocene of the Royal Society borehole at Ludham, Norfolk. *Proc. R. Soc.* B **155**, 437–53.

West, R. G. & Wilson, D. G. (1968). Plant remains from the Corton Beds at Lowestoft, Suffolk. *Geol. Mag.* **105**, 116–23.

# THE HISTORY OF THE ERICACEAE IN IRELAND DURING THE QUATERNARY EPOCH

by G. F. Mitchell and W. A. Watts

*Trinity College, Dublin*

There is no direct evidence of what the vegetation of Ireland was like at the end of the Tertiary Epoch, as no Pliocene deposits are known. In the early Tertiary (Eocene/Oligocene) there are in the north-east the inter-basaltic lignites and the Lough Neagh Clays, and in the south the clay at Ballymacadam, Tipperary (Watts, 1962). The Lough Neagh Clays have Taxodiaceae, and there is no reason to think that the later Tertiary vegetational history of Ireland was very different from that of western Europe. Ericaceous pollen is recorded at Ballymacadam. The residual ore deposit, probably Tertiary in age, at Tynagh, Galway, contains altered wood, determined as *Cupressus* (M. Scannell, unpublished).

Our knowledge of the Pliocene vegetation of western Europe is based on the Reuverian lignites of Holland (Zagwijn, 1960), with added information from the black clay at La Londe in Normandy (Elhaï, 1963, pp. 50–61), from the shelly marine clay at Bosq d'Aubigny also in Normandy (see Roger & Freneix, 1946, p. 103; Andrew & Mitchell, unpublished), from the lowest brown clay at Gurp, near the mouth of the Gironde (Paquereau & Schoeller, 1959, p. 82), and from the clay at Coulgens, Charente (Depape, Florschütz & Guillien, 1954, p. 198). Reuver has pollen of Ericales, and La Londe pollen of Ericaceae (unfortunately not further subdivided). At Bosq d'Aubigny pollen of Ericaceae are common. One grain was of *E. terminalis* type, and there was also a leaf fragment of *E. tetralix* type. Ericaceae are not recorded at the other sites.

It is now known that in the Pleistocene there were many more episodes of cold than the four postulated by Penck and Brückner. Recent work in East Anglia and Holland suggests that there were at least seven major periods of cold and of warmth. Within each such major period minor swings between warmer and colder conditions took place. The very tentative table (p. 15) of possible Pleistocene correlations in western Europe is included only for the assistance of readers of this paper. The first three columns (reading from the left) are largely based on the work of R. G. West, both published (1961) and unpublished; the fourth column is largely based on Mitchell (1960); the fifth column is taken from Woldstedt (1965); the sixth column is highly speculative. The locations of the sites are shown on the map (figure 1); the map also shows a number of Late- and Post-glacial sites which are not discussed in the text: they are included to indicate the considerable amount of information that is now available.

The transition from the Pliocene to the Pleistocene is perhaps seen at Ste-Reine in the Cantal where pollen-counts dominated by trees (including Tertiary elements) give way to counts with higher values for herbs, suggesting a Pre-Ludhamian age (Durand & Rey, 1964). The marine clay at *c.* 30 m at St-Jean-la-Poterie in Brittany may also lie near the transition. Durand (1960, p. 256) gives a list of pollen dominated by Tertiary trees, but another count from a different sample showed dominant *Pinus* of *sylvestris* type, some *Abies* and *Tsuga*, and only one grain of *Liquidambar*; several types of Ericaceous pollen were also present (Andrew & Mitchell, unpublished). At a site, Bidart A, near Biarritz, which Oldfield (1968, Fig. 23) considers to lie near the Pliocene/Pleistocene boundary, a mor soil yielded abundant Ericaceous macrofossils, including seeds of *Erica ciliaris*, *E. mackaiana*, *E. scoparia*, *E. tetralix* and *E. vagans*.

The base of the Ludham borehole in East Anglia (West, 1961) shows a warm period, the Ludhamian, with Ericaceous pollen present. Pollen of Ericales was common at the corresponding site at Tegelen in Holland (Zagwijn, 1960). This stage is probably represented at Bidart (Site Bidart B) where seeds of *Erica ciliaris* and pollen of *E. vagans* type are recorded (Oldfield, 1968, Fig. 23). Ericaceous pollen is also recorded in the upper deposit at Gurp (Elhaï & Prenant, 1963), perhaps of the same age, and at Joursac in the Cantal (Durand & Rey, 1963), broadly described as Villafranchian. Ericaceae are not recorded in the Villafranchian deposits at Senèze, Haute-Loire (Elhaï & Grangeon, 1963), where some oscillations in climate are suggested.

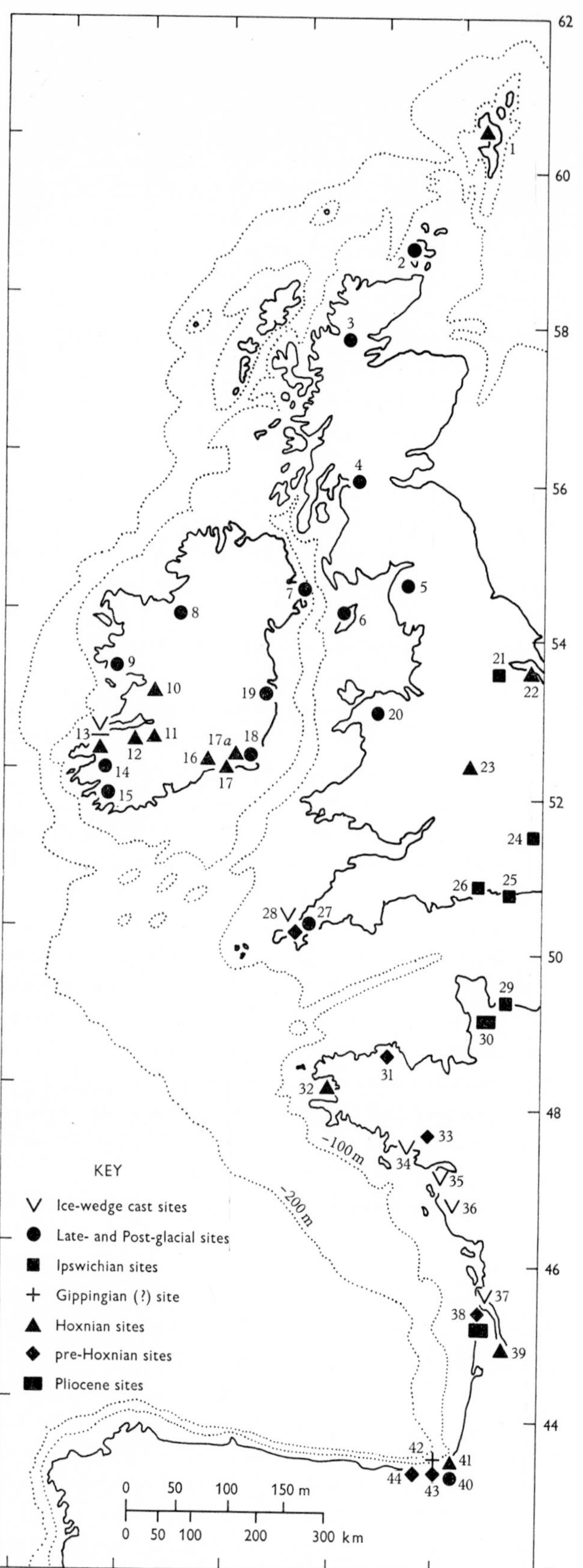


Figure 1. Some sites of deposits with plant fossils and of ice-wedge casts in western Europe.

KEY TO NUMBERS

1 Fugla Ness, Shetland (Moar, unpublished) Erratic of peat (probably Hoxnian in age) in till.
2 Yesnaby, Orkney (Moar, unpublished) Late- and Post-glacial deposit.
3 Lough Droma, Ross and Cromarty (Kirk & Godwin, 1963) Late- and Post-glacial deposit.
4 Garscadden Mains, Glasgow (Mitchell, 1952) Late- and Post-glacial deposit.
5 Cumberland Lowlands (Walker, 1966) Late- and Post-glacial deposits.
6 Kirkmichael, Isle of Man (Dickson, Dickson & Mitchell, unpublished) Late-glacial deposits.
7 Roddans Port, Down (Morrison & Stephens, 1965) Late- and Post-glacial deposits.
8 Treanscrabbagh, Sligo (Mitchell, 1953) Late- and Post-glacial deposits.
9 Roundstone, Galway (Jessen, 1949) Late- and Post-glacial deposits.
10 Gort, Galway (Jessen, Andersen & Farrington, 1959) Hoxnian deposit.
11 Kildromin, Limerick (Watts, 1967) Hoxnian deposit.
12 Baggotstown, Limerick (Watts, 1964) Hoxnian deposit.
13 Spa, Kerry (Mitchell, unpublished) Ice-wedge casts in head overlying Hoxnian deposit.
14 Killarney, Kerry (Jessen, 1949) Late- and Post-glacial deposits.
15 Whiddy Island, Cork (Stillman, 1968) Late-glacial freshwater deposit at −50 m.
16 Kilbeg, Waterford (Watts, 1959) Hoxnian deposit.
17 Newtown, Waterford (Watts, 1959) Hoxnian deposit.
17*a* Ballykeerogemore, Wexford (Mitchell, unpublished) Hoxnian deposit.
18 Kilmacoe, Wexford (Mitchell, 1951) Late- and Post-glacial deposits.
19 Ballybetagh, Dublin (Jessen & Farrington, 1938) Late- and Post-glacial deposits.
20 Nant Ffrancon, Caerns. (Seddon, 1962) Late-glacial deposits.
21 Austerfield Low Common, Yorks. (Rep. of Director, Inst. Geol. Sci. Lond. 1966) Ipswichian deposit.
22 Kirmington, Lincs. (Watts, 1959) Hoxnian deposit.
23 Nechells, Warwicks. (Kelly, 1964) Hoxnian deposit.
24 Trafalgar Square, London (Franks, 1960) Ipswichian deposit.
25 Selsey, Sussex (West & Sparks, 1960) Ipswichian deposit.
26 Stone, Hants. (West & Sparks, 1960) Ipswichian deposit.
27 Hawk's Tor, Cornwall (Conolly, Godwin & Megaw, 1950) Late-glacial deposit.
28 St Erth, Cornwall (Mitchell, unpublished) Ice-wedge cast and Pre-Hoxnian deposit.
29 St Côme de Fresné, Calvados (West & Sparks, 1960) Ipswichian deposit.
30 Bosq d'Aubigny, Manche (Andrew & Mitchell, unpublished) Pliocene deposit.
31 Quemperven, Côtes-du-Nord (Kerfourn, 1965; Pinot, 1966) Pre-Hoxnian deposit.
32 Trez Rouz, Finistère (Kerfourn, 1965) Hoxnian deposit.
33 St-Jean-la-Poterie, Ille-et-Vilaine (Durand, 1960; Mitchell, unpublished) Late Pliocene deposit.
34 Penerf, Ille-et-Vilaine (Rivière *et al.* 1966) Ice-wedge casts.
35 La Sennetiere, Loire-Atlantique (Rivière *et al.* 1966) Ice-wedge casts.
36 Château de Vie, Vendée (Ters, 1961) Ice-wedge casts.
37 Royan, Charente-Maritime (Mitchell, unpublished) Ice-wedge casts.
38 Gurp, Gironde (Paquereau & Scholler, 1959) Pliocene deposits, (Elhaï & Prenant, 1963) Pre-Hoxnian deposits.
39 Bruges, Gironde (Elhaï, 1966) Hoxnian deposit.
40 Le Moura, Basses Pyrénées (Oldfield, 1964) Late- and Post-glacial deposits.
41 Marbella, Basses Pyrénées (Oldfield, 1962) Hoxnian deposits.
42 Chabiague, Basses Pyrénées (Oldfield, 1968) Gippingian (?) deposit.
43 Mouligna-cliff, Basses Pyrénées (Oldfield, 1968) Pre-Hoxnian deposits.
44 Bidart, Basses Pyrénées (Oldfield, 1968) Pre-Hoxnian deposits.

TABLE 1. *Possible correlations in western Europe*

| Index fossils | Climate | East Anglia | Ireland | General | Possible stratigraphical position of West European sites |
|---|---|---|---|---|---|
| | Warm | Flandrian | Post-glacial | Holozän | Roundstone: Le Moura |
| | Cold | Weichselian | Tipperarian | Würm | Le Moura |
| | Warm | Ipswichian | | Eem | Selsey |
| | Cold | Gippingian | Ballycroneenian | Riss | Chabiague? |
| *Abies* disappears | Warm | Hoxnian | Gortian | Holstein | Shetland: Gort: Trez Rouz: Bruges: Marbella |
| | Cold | Lowestoftian | Pre-Gortian | Mindel | |
| | Warm | Cromerian *s.s.* | | Cromer *s.l.* | Mouligna-cliff? |
| | Cold | Beestonian | | | |
| *Tsuga* disappears | Warm | Pastonian | | | Quemperven |
| | Cold | Baventian | | Gunz und älterer kalt- und warm-zeiten etwa = Villafranchian | |
| | Warm | Antian | | | St Erth |
| | Cold | Thurnian | | | |
| | Warm | Ludhamian | | | Bidart B: Gurp: Joursac: Senèze: Ste Reine |
| | Cold | Pre-Ludhamian | | | |
| *Nyssa, Taxodium* & *Sciadopitys* virtually disappear | Warm | Late Pliocene | | | Bidart A: St-Jean-la-Poterie: Ste Reine |
| | Warm | Pliocene *s.l.* | | Pliozän | La Londe: Bosq d'Aubigny: Gurp: Coulgens |

The Ludhamian stage is probably the equivalent of the lower Crags of East Anglia, which include the Oakley horizon of the Waltonian (Harmer, 1900, p. 717). The Irish Sea must have existed at the Oakley stage, for glacial gravels at Killincarrig, Wicklow, have yielded a rich collection of shells with Oakley affinities (McMillan, 1938), presumably pushed up from the seafloor by later ice. Much smaller numbers of Crag shells are known from glacial deposits in Kintyre in Scotland, in the Isle of Man and in Wexford (McMillan, 1964), and it can be assumed that an Irish Sea, probably very similar in outline to that of today, was in existence during the warm periods of the Quaternary Epoch. During the cold periods its floor, when not covered by ice, was probably dry.

On both sides of the Irish Sea a wave-cut shore-platform, about 6 m above modern sea-level, is to be seen. It follows every detail of the modern shoreline, and indeed indicates a coastline even more indented than that of today. In many cases narrow coastal fissures were later filled either with head or with boulder clay, and the modern waves have not yet succeeded in cutting away the unconsolidated material. A similar shore-platform can be traced from the Hebrides to the mouth of the Gironde, but the possibility that the platform is a composite feature cannot be overlooked. The date of the cutting of the platform or platforms is obviously of great interest for the phytogeography of Ireland. In western France the platform is cut in Eocene limestone (Durand, 1960, p. 98); in Ireland late Hoxnian deposits rest on it at Spa, Kerry (Mitchell, unpublished). Large erratics rest on the platform in Cornwall (Flett, 1946, p. 108) and in Normandy (Elhaï, 1963, Pl. VII), and their transport may perhaps be due to pack-ice or ice-bergs of Lowestoft age.

The shores of the Irish Sea may thus be of considerable antiquity. What of its depths? Each successive ice-sheet that ploughed down the floor of the Irish Sea—and there have been at least three of them—caught up debris from the sea-floor and deposited it either on the south-east shores of Ireland or on the continental shelf south of Wexford. The Irish Sea may have become progressively deeper, and hence a more substantial barrier to migration, as the Ice Age proceeded. On the other hand the retreating ice may have built up morainic ridges, and these may have served to establish temporary links between Ireland and Britain (Mitchell, 1963).

In East Anglia the Thurnian Cold Period succeeds the Ludhamian, and Ericales pollen, chiefly of *Empetrum*-type, becomes more common. At the end of the Last Cold Period *Empetrum* heaths were present in western Ireland (Jessen, 1949), Cumberland (Walker, 1966) and Orkney (Moar, unpublished). Such a development was probably repeated in each cold period of the Pleistocene.

With climatic amelioration in the Antian Warm Period the forests expanded again, and Ericales, though remaining in evidence, made a smaller contribution to the pollen-rain. There will have been a marine transgression towards the end of this period, and the marine clay at +40 m at St Erth in Cornwall (Mitchell, unpublished) may perhaps belong here. *Tsuga* is still present, and the foraminifera have an archaic appearance, similar to those at Bosq d'Aubigny (Funnell in Mitchell, 1965). Pollen and leaf fragments of conifers and Ericales are present, and a heath with coniferous trees and a mor soil was probably overrun by rising sea-level. *Erica tetralix*, *Calluna* and *Vaccinium* are represented by macrofossils, and there is in addition pollen of *Empetrum* and other ericaceous types, though no grains of a size which would suggest *Arbutus* or *Rhododendron* were noted. *Tsuga* and *Osmunda* suggest oceanic conditions.

Cold conditions returned to East Anglia, and the Baventian Cold Period was established. Pollen of Ericales (mostly *Empetrum nigrum*-type) became common again.

We now enter the phase of the Pleistocene associated with the historic Cromer Forest Bed, and West (personal communication) is of opinion that there are three stages involved: a warm Pastonian with *Tsuga* still just surviving, a cold Beestonian associated with ice-wedge casts and other frost features, and a warm Cromerian *s.s.* from which *Tsuga* is absent. A similar sub-division will probably follow in Holland.

Sea-level will have had a high stage in the Pastonian, and there is at Quemperven in Brittany a marine clay which may be of this age, lying at an altitude of about 24 m (Pinot, 1966). The content of Foraminifera is very different from that at Bosq d'Aubigny and at St Erth—though all three lie open to influence from the Atlantic—and suggests a younger age. *Tsuga* is still present, and the deposit is unlikely to be as young as Cromerian *s.s.*; the absence of *Abies* seems to rule out a still younger Hoxnian. *Ilex* and *Hedera* suggest oceanic conditions, and there is a small amount of Ericales pollen, both *Erica* and *Empetrum* (Kerfourn, 1965*a*).

Little is as yet known of the Beestonian Cold Period, but the presence of cold in East Anglia is shown by ice-wedge casts which pierce the Pastonian sediments, but are themselves truncated or overlain by the temperate Cromerian *s.s.* deposits.

In the Cromerian *s.s.* Warm Period *Tsuga* has disappeared, and Ericales values are lower; West suggests that highly oceanic conditions no longer penetrated into East Anglia. In the fluctuations of Early Pleistocene climate there may have been a trend towards larger ice-masses during the cold periods and more continental climates during the warm periods, so that the moisture-loving elements in the Early Pleistocene flora tended to survive only farther and farther towards the west. At Mouligna-cliff in the vicinity of Biarritz, Oldfield (1968) has recorded a pollen flora which appears to lie between that at Bidart B for which a Tiglian age has been suggested and a younger one of Hoxnian age at Marbella (Oldfield, 1962). *Arbutus*, *Daboecia*, *Erica* cf. *vagans*, and *Rhododendron* were present and also *Empetrum* and *Vaccinium*. A rich ericaceous flora must have flourished in the vicinity of Biarritz in all the warm periods of the early Pleistocene.

During the next cold period, the Lowestoftian, parts of the British Isles are known to have been buried under ice. In eastern England ice advanced as far as the Thames valley, and ice from Scotland may have established a floating ice-shelf in the Irish Sea. Only two agencies can have transported the far-travelled stones that are found in later beach deposits as far south as Noirmoutier in western France—an ice-sheet or floating ice. Few would extend the ice-sheets into the Bay of Bisquay, and though the concept of a high glacial sea-level with pack-ice in the English Channel and the Bay of Bisquay presents its own difficulties, it is perhaps the more acceptable alternative. We have no knowledge of the vegetation of north-west Europe during the Lowestoft Cold Period.

With the waning of that cold phase there begins a remarkable record of vegetational development. The record is most complete at Marks Tey in Essex (C. Turner, in preparation) where the open vegetation of the final Lowestoft phase gives way to the forests of the Hoxnian Warm Period, and these in turn give way to the open vegetation of the first phase of the following Gippingian Cold Period. In the later stages of woodland development *Abies* is prominent, and *Calluna*, *Erica scoparia* and *E.* cf. *terminalis* are present. In Ireland part of the record was first seen by Kinahan (1865) near Gort in Galway, though it was not more fully revealed until 1959 through the work of Jessen, Andersen and Farrington. In England the record was first seen by Pike & Godwin (1953) at

Clacton, and was substantially expanded by West (1956) at Hoxne.

The chain of records now runs from the foothills of the Pyrenees (Oldfield, 1962), past the Gironde (Elhaï, 1966), through the westernmost tip of Brittany (Kerfourn, 1965*b*), through south and west Ireland (Watts, 1967), and on to the Shetland Islands where a block of peat, probably lying as an erratic in till, and containing *Erica* cf. *scoparia* and *E. tetralix*, almost certainly belongs to the same period (Moar, unpublished). We can clearly see a ribbon of Ericaceae extending from the coasts of Portugal to the Shetland Islands.

TABLE 2. *Quaternary Ericaceae of Western Europe*

| | Indigenous today | | | Hoxnian fossils | |
|---|---|---|---|---|---|
| | Western Europe | Great Britain | Ireland | Great Britain | Ireland |
| *Arbutus unedo* | + | . | + | . | . |
| *Calluna vulgaris* | + | + | + | + | + |
| *Daboecia cantabrica* | + | . | + | . | + |
| *Erica arborea* | + | . | . | . | . |
| *E. australis* | + | . | . | . | . |
| *E. ciliaris* | + | + | + | . | + |
| *E. cinerea* | + | + | + | . | + |
| *E. hibernica* | + | . | + | . | . |
| *E. lusitanica* | + | . | . | . | . |
| *E. mackaiana* | + | . | + | . | + |
| *E. scoparia* | + | . | . | + | + |
| *E. terminalis* | + | . | . | + (type) | . |
| *E. tetralix* | + | + | + | . | + |
| *E. umbellata* | + | . | . | . | . |
| *E. vagans* | + | + | + | . | . |
| *Rhododendron ponticum* | + | . | . | . | + |
| | 16 | 5 | 9 | 3 | 8 |
| | | 9 | | 9 | |

What were the components of this ribbon? Which are recorded in the Hoxnian Warm Period in the British Isles? Which are still to be found there?

With two exceptions, *Erica scoparia* var. *macrosperma* (whose fossil seeds cannot be matched by modern material) recorded from Gort, Kilbeg, Baggotstown and probably Shetland, and *Rhododendron ferrugineum*, because pollen of this type was found in the Biarritz area (Oldfield, unpublished), no fossil forms are known which do not grow in lowland western Europe today. The sixteen forms in the list may give a good indication of the strength of the west European Ericaceae at the opening of the Pleistocene. Of the sixteen forms, twelve are known from the British Isles, either living or fossil. Nine still grow in western Ireland, and eight are known as Hoxnian fossils from Ireland. *Arbutus* and *E. vagans*, living in Ireland today, are not known as interglacial fossils; *E. scoparia* and *Rhododendron ponticum*, not indigenous in Ireland today, are known as fossils.

There must have been a mobile population of Ericaceae in western Europe throughout the Pleistocene, and various members of it re-appeared in the warm periods depending on their manner of dispersal in the preceding cold period. As the warm periods proceeded, climate deteriorated, soils became leached, and deciduous woodland gave way to open coniferous woodlands with heaths, and perhaps bogs (see Iversen, 1958; Andersen, 1964). Towards the end of the Hoxnian Warm Period *Calluna*, *Daboecia*, *Erica ciliaris*, *E. cinerea*, *E. mackaiana*, *E. scoparia* and *E. tetralix* were all growing freely in the south and west of Ireland.

When the next cold period, the Gippingian, set in, *Pinus* survived the longest of the trees, but gradually even the pine trees disappeared and tundra developed. As the sea-level fell, the shore-line migrated westwards, drawing beach ridges, dunes, and intervening wetter 'slacks' in its wake. Plants of these habitats had a good chance of survival. The former shore-line (which is essentially the same as that of today) eventually found itself some 100 m above the glacial sea-level enduring conditions of severe permafrost. Ice-wedge casts are now known from Spa, Kerry (Mitchell, in preparation), from St Erth in Cornwall (Mitchell, in preparation), and from the west of France (Ters, 1961; Rivière *et al.* 1966; Mitchell, unpublished). Their general distribution in Europe is discussed by Dylik & Maarleveld (1967). It cannot yet be proved that these ice-wedge casts are of Gippingian age; they may belong to the Last Cold Period, but whatever their age the Gippingian climate is unlikely to have been less severe than that of the Last Cold Period. Modern ice-wedges have been intensively studied in Alaska (Péwé, 1963), where they do not develop actively south of the mean January −12 °C isotherm. They form where thin snow cover permits loss of ground heat by radiation.

In late Hoxnian time a broad ribbon of Ericaceae stretched along the western shores of Europe from the coasts of Portugal to the Shetland Islands. If no northern refugia existed, then only that portion of the ribbon which lay to the south of the Pyrenees can have survived the rigours of the Gippingian Cold Period. As sea-level fell, it exposed, except in three areas, broad expanses of the continental shelf, and here severe climatic conditions would have obtained. But off the mountain ranges of Connemara, Kerry and the Pyrenees steep slopes continue seawards, slopes that even with a lower sea-level would continue to receive the full force of the Gulf Stream. Here ridges and valleys provided shelter, slopes drained the soils, and oceanic waters provided a moderating influence. Possible refugia are to be found here.

In Ireland today most of the localities for the rarer Ericaceae lie west of the mean January 5 °C isotherm. Their range may not be limited by temperature, as Perring (1962) has pointed out. But nonetheless, if effective refugia were to exist on the Atlantic coast at the height of the glaciations, there must have been a remarkably steep temperature gradient. The glacial coast was perhaps 25 km west of, and 100 m lower than, the modern coast where the ice-wedge casts are now seen. In that short descent January mean temperature must have risen from −12 °C to a level that the Ericaceae could tolerate. On the west coast of Norway today the January gradient is about 1 °C in 15 km, and the gradient off south-west Ireland must have been steeper. Granted the Norwegian gradient, the coastal temperature would have been −10·5 °C. At Kilbeg (Watts, 1959*a*) *Daboecia* survived until temperatures had fallen markedly late in the interglacial, for it was associated with plants that would have been at home in central Sweden today where January mean temperatures are below −10 °C. At Chabiague in the Biarritz area pollen of *Daboecia* and of *Erica vagans* type was found in a relatively cold deposit, perhaps of Gippingian age. Even if we double the Norwegian gradient, the temperature off the Irish coast would have been perhaps −9 °C. Could the other Ericaceae have survived such a temperature? In support of the concept of a refuge off the present coast of Connemara, we may note that Donegal, similar in terrain and vegetation to Connemara, which has not got steep offshore slopes, is today much poorer in Ericaceae.

Of the vegetational history of the Last Warm Period, the Ipswichian (West, 1957), we know much less than that of the Hoxnian. The site at Ardcavan, Wexford (Mitchell, 1948), is of Post-glacial not interglacial age. A disturbed marine clay, perhaps of Ipswichian age, has recently been found in Wexford (Colhoun & Mitchell, in preparation). The English sites at Ipswich, and also at Selsey and Stone (West & Sparks, 1960), have no macrofossils and very little pollen of Ericaceae. But at Weeze on the lower Rhine (von der Brelie & Rein, 1956) Ericaceae have high values in the later stages of the interglacial, and a similar vegetational development probably took place. Until sites of the Last Warm Period have been discovered and studied in the west of Ireland, there must be a lacuna in the Pleistocene history of the Ericaceae.

The Last Cold Period then developed, and though the south-east of Ireland was not covered by ice, those parts of Ireland that are the main headquarters of the Ericaceae today were covered. Ice covered most of Connemara, there was a local ice-cap in Kerry, and there were severe frost conditions over most of north-west Europe. Van der Hammen *et al.* (1967) have shown that at times there were cold polar desert conditions in the Netherlands, as also must have been the case in western France at the time of the formation of the ice-wedges there. Again there are two possibilities. Either the Irish Ericaceae were wiped out, and there again had to be Post-glacial immigration from south of the Pyrenees, because Oldfield (1964), who has studied plant-fossils from the Last Cold Period at Le Moura between Biarritz and the Pyrenees, considers that Ericaceae could not have survived there. Or they again survived in the same refugia off the coasts of Kerry and Galway.

The current warm period opened about 12,500 B.C. when limited plant growth became possible (Zone I). A period of marked amelioration (Zone II) from 10,000 to 8800 B.C. was followed by a return of cold (Zone III) for about 500 years. At Whiddy Island, Cork, a boring at the site of a new pier passed through a silty freshwater lake mud 54 m below modern sea-level. The mud contained pollen which suggested a Zone II age (Stillman, 1968). Thus even at the end of the Last Cold Period sea-level was still very low, and large areas of the floor of the Irish Sea and the adjoining continental shelf were still dry. In the Pays Basque *Erica tetralix*, *Calluna* and *Empetrum nigrum* are recorded from Zone III (Oldfield, 1964).

Full warmth was restored about 8300 B.C., and the Early Post-glacial period in Ireland is characterized in pollen-diagrams first by a very brief well-marked peak in Gramineae, and then by successive peaks of *Juniperus* and *Betula* (Watts, 1963). This period corresponds approximately to Zone IV and Zone V of Jessen (1949) which are not satisfactorily defined, largely because *Juniperus* pollen was not identified at the time of Jessen's work. With the exception of isolated occurrences of pollen of *Arctostaphylos uva-ursi* in the *Juniperus* peak zone of the Burren region of County Clare, no ericaceous pollen is recorded from either the Late-glacial period or the beginning of Early Post-glacial time (Watts, unpublished). One leaf of *Erica tetralix* was recorded from a mud of Zone II age in Sligo (Mitchell, 1953), but as the mud was sampled by using a Hiller drill through many metres of *Sphagnum* peat, the possibility of contamination cannot be ruled out.

Jessen (1949, p. 222) emphasized the importance of *Empetrum nigrum* in the Late-glacial and earliest Post-glacial of the Roundstone area of County Galway and elsewhere in western Ireland. Recent investigations (Watts, 1963) show *Empetrum nigrum* to have been rare or absent over most of Ireland in Zone III, but at Roundstone a new study of Jessen's Site 37 shows that it was present throughout the whole Late-glacial and earliest Post-glacial in quantity, though rarer in Zone III than earlier or later (Watts, unpublished). At Site 37 *Empetrum nigrum* is the only Ericales-type pollen present until the end of the *Juniperus* peak zone when it suddenly becomes rare and disappears, giving place to pollen of *Calluna* which appears in the *Betula* peak zone accompanied by *Cladium*. Jessen found macro-fossils of *Erica cinerea* and *E. tetralix* in the *Betula* peak zone, and of the eu-oceanic *E. mackaiana* in the subsequent Boreal period. *E. mackaiana* occurs in quantity at Site 37 today, and, of other eu-oceanic species, *E. ciliaris* occurs only a few hundred metres away (Webb, 1966), while *E. hibernica* (*E. mediterranea* of the older literature) has one stand nearby and *Daboecia* is locally frequent.

Jessen supposed that the occurrence of *Empetrum* provided evidence for the occurrence of oceanic heaths in western Ireland in his Zones I to IV, and referred to the occurrence of *Empetrum*-dominated heaths in northern oceanic regions such as northern Norway and Iceland today. Clearly, however, such heaths have only a broad physiognomic similarity to the heaths in which eu-oceanic species occur today. There appears to be no fossil evidence for the occurrence of the type of heath and bogland communities in which eu-oceanic Ericaceae now occur before the *Betula* peak at the end of the Early Post-glacial. It must be concluded that for some 3,000 years after the melting of the ice of the Last Cold Period the Ericaceae were not present in the Roundstone area, their present main centre of distribution in Ireland, or were present in such sites or such small quantities as to escape palaeontological detection.

Jessen comments on the evidence from Ireland for rather late development (his Zone VIII) of extensive typical blanket bog vegetation. E. Vokes (unpublished) has shown that in the Killarney area the main development of blanket bog took place as recently as 3,000 years ago. At Roundstone, *Erica mackaiana* occurs fossil long before the main bog development, and it is probable, as Jessen suggests, that the Ericaceae of the Mid Post-glacial were growing in the field layer of woods around the Roundstone lakes. It is interesting to note that the habitats in which the eu-oceanic Ericaceae occur in Spain are heaths with thin peat or mineral soil or in the field layer of wet, open oak-woods (Webb, 1955; Woodell, 1958). In the Cantabrian mountains several of these species may experience snow cover for relatively long periods in winter, and generally more continental conditions than ever occur in western Ireland. It is not generally appreciated that the wet blanket bogs, which in Ireland are regarded as a specially characteristic habitat for the eu-oceanic species, were not in existence when these species first re-appeared, and that the same species in their southern localities occur in different types of habitat. It may be that they should be thought of rather as relict species now found in a habitat that has developed as a response to climatic change in the last 3,000 years, and is not especially congenial to them. Some of their peculiarities, such as the sterility of *Erica mackaiana* in its Irish stations, and the extraordinarily small single stands of *E. ciliaris* and *E. vagans*, may perhaps be explained as the responses of species occurring in essentially relict localities to habitats which are not entirely suitable for them. Comparative autecological studies of the environment and performance of selected eu-oceanic species in their Irish and Spanish localities would be helpful in understanding these problems.

There are no Post-glacial fossil records for *Daboecia*, *Erica ciliaris*, *E. hibernica* or *E. vagans*, though *Daboecia* and *E. hibernica* are sufficiently abundant in the modern flora to make a search for fossils in suitable localities worthwhile. *Arbutus unedo* now occurs in several places in western Ireland, but it is much commoner at Killarney than elsewhere (Sealy, 1949). It appears to have a very low pollen production because difficulty is experienced in finding any pollen of the species, even in *Sphagnum* patches growing directly beneath well-grown small trees (Watts, unpublished). Recently Vokes (unpublished) has shown that *Arbutus* was present at Killarney perhaps as early as 4,000 years ago. Isolated pollen tetrads are found after the *Alnus* curve is fully established, and about the time that *Fraxinus* was first beginning to appear in quantity. *Arbutus* was established at Killarney while *Pinus sylvestris* was still abundant prior to its extinction, and before blanket bog began to develop. Thus the circumstances of its appearance in the fossil record resemble the history of *Erica mackaiana* at Roundstone.

Both their record as fossils, and the habitats in which we find them today, suggest that the Ericaceae have been a remarkably resilient group in western Ireland. Unlike other groups, they do not seem to have been substantially reduced in numbers as the Pleistocene Epoch proceeded. Their variety in the Roundstone area, coupled with steep slopes offshore from western Connemara, suggests survival in nearby refuges. But if the refuges were close at hand, the gap of three thousand years between the end of the Last Cold Period and the first record of Ericaceae in Post-glacial Ireland is not easy to understand. Their re-appearance in the Biarritz area was more rapid, and their main area of refuge may have lain far to the south of Ireland. Despite the important amount of new information that has come to light in the last thirty years, largely under the inspiration of Jessen and Godwin, the full history of the Ericaceae in Ireland cannot yet be written.

The authors wish to thank Professor Frank Oldfield and all those other colleagues who made the results of their work, both published and unpublished, so freely available. For any conclusions drawn from the work the authors alone are responsible.

## REFERENCES

Andersen, S. Th. (1964). Interglacial plant successions in the light of environmental changes. *Rep. 6th INQUA Cong. Warsaw 1961* **2**, 359–68.

Von der Brelie, G. & Rein, U. (1956). Pollenanalytische Untersuchungen zur Gliederung des Pleistozäns am linken Niederrhein. *Geologie Mijnb.* **16**, 423–5.

Conolly, A. P., Godwin, H. & Megaw, E. M. (1950). Studies in the Post-glacial history of British vegetation. XI. Late-glacial deposits in Cornwall. *Phil. Trans. R. Soc.* B **234**, 397–469.

Depape, G., Florschütz, F. & Guillien, Y. (1954). La Flore des Argiles de Coulgens (Charente). *Bull. Soc. Géol. Fr.* Ser. 6, **4**, 193–201.

Durand, S. (1960). Le Tertiare de Bretagne. *Mem. Soc. Géol. Miner. Bretagne* **12**, 1–384.

Durand, S. & Rey, R. (1963). Les formations à végétaux de Joursac (Cantal) peuvent être datées du Villafranchian par l'analyse pollinique. *C. R. hebd. Séanc. Acad. Sci. Paris* **257**, 2692.

Durand, S. & Rey, R. (1964). Le dépôt de la diatomite de Sainte-Reine (Cantal) débute au Pliocene supérieur et permet de déceler les traces du refroidissement prétiglian. *C. R. hebd. Séanc. Acad. Sci. Paris* **259**, 1978–80.

Dylik, J. & Maarleveld, G. C. (1967). Frost cracks, frost fissures and related polygons. *Meded. Geol. Sticht.* N.S. **18**, 7–22.

Elhaï, H. (1963). *La Normandie Occidentale.* Bordeaux: Bière.

Elhaï, H. (1966). Deux Gisements Du Quaternaire Moyen. *Bull. Assoc. Franc. Quaternaire* **1**, 69–78.

Elhaï, H. & Grangeon, P. (1963). Nouvelles recherches sur le gisement villafranchian de Senèze (Haute-Loire, France). *Bull. Soc. Géol. Fr.* Ser. 7, **5**, 483–8.

Elhaï, H. & Prenant, A. (1963). Présence et Extension d'un Niveau Marin Littoral Interglaciaire sur la Côte du Medoc. *Bull. Soc. Géol. Fr.* Ser. 7, **5**, 495–507.

Flett, J. S. (1946). The geology of the Lizard and Meneage. *Mem. Geol. Surv. U.K.*

Florschütz, F. & Amor, J. M. (1963). Analyse Palynologique d'un Gisement de Diatomite au Cantal (Massif Central). *Grana Palynol.* **4**, 452–8.

Franks, J. W. (1960). Interglacial deposits at Trafalgar Square, London. *New Phytol.* **59**, 145–52.

Van der Hammen, T., Vogel, J. C., Maarleveld, G. C. & Zagwijn, W. H. (1967). Stratigraphy, climatic succession and radiocarbon dating of the Last glacial in the Netherlands. *Geologie Mijnb.* **46**, 79–95.

Harmer, F. W. (1900). The Pliocene Deposits of the East of England. *Quart. Jl. Geol. Soc. Lond.* **56**, 705–44.

Iversen, J. (1958). The bearing of glacial and interglacial epochs on the formation and extinction of plant taxa. *Uppsala Univ. Årsskt.* 1958, 210–15.

Jessen, K. (1949). Studies in Late Quaternary deposits and flora-history of Ireland. *Proc. R. Ir. Acad.* **52** B, 88–279.

Jessen, K., Andersen, S. T. & Farrington, A. (1959). The Interglacial deposit near Gort, Co. Galway, Ireland. *Proc. R. Ir. Acad.* **60** B, 1–77.

Jessen, K. & Farrington, A. (1938). The Bogs at Ballybetagh, near Dublin, with remarks on late-glacial conditions in Ireland. *Proc. R. Ir. Acad.* **44** B, 205–60.

Kelly, M. R. (1964). The Middle Pleistocene of North Birmingham. *Phil. Trans. R. Soc.* B **247**, 533–92.

Kerfourn, M.-T. (1965*a*). L'analyse pollinique permet de rapporter au Pléistocène inférieur un dépôt littoral découvert à Lanmerin (Côtes-du-Nord). *C. R. hebd. Séanc. Acad. Sci. Paris* **260**, 254–5.

Kerfourn, M.-T. (1965*b*). Le dépôt tourbeux de l'anse de Trez-Rouz à Camaret (Finistère) peut-être rapporté à l'Interglaciaire Mindel-Riss. *C. R. hebd. Séanc. Acad. Sci. Paris* **260**, 2024–6.

Kinahan, G. H. (1865). Explanation to accompany sheets 115 and 116. *Geol. Surv. Ireland.*

Kirk, W. & Godwin, H. (1963). A Late-glacial site at Loch Droma, Ross and Cromarty. *Trans. R. Soc. Edinb.* **65**, 225–49.

McMillan, N. F. (1938). On an occurrence of Pliocene shells in Co. Wicklow. *Proc. Lpool. Geol. Soc.* **17**, 255–66.

McMillan, N. F. (1964). The Mollusca of the Wexford Gravels (Pleistocene), south-east Ireland. *Proc. R. Ir. Acad.* **63** B, 265–89.

Mitchell, G. F. (1948). Two Interglacial deposits in south-east Ireland. *Proc. R. Ir. Acad.* **52** B, 1–14.

Mitchell, G. F. (1951). Studies in Irish Quaternary deposits, no. 7. *Proc. R. Ir. Acad.* **53** B, 111–206.

Mitchell, G. F. (1952). Late-glacial deposits at Garscadden Mains, Glasgow. *New Phytol.* **50**, 277–86.

Mitchell, G. F. (1953). Further identifications of macroscopic plant fossils from Irish Quaternary deposits, especially from a Late-glacial deposit at Mapastown, Co. Louth. *Proc. R. Ir. Acad.* **55** B, 225–81.

Mitchell, G. F. (1960). The Pleistocene history of the Irish Sea. *Advmt Sci., Lond.* **17**, 313–25.

Mitchell, G. F. (1963). Morainic ridges on the floor of the Irish Sea. *Ir. Geogr.* **5**, 335–44.

Mitchell, G. F. (1965). The St Erth Beds—An alternative explanation. *Proc. Geol. Ass.* **76**, 345–66.

Morrison, M. E. S. & Stephens, N. (1965). A submerged Late-Quaternary deposit at Roddans Port on the north-east coast of Ireland. *Phil. Trans. R. Soc.* B **249**, 221–55.

Oldfield, F. (1962). Quaternary Plant Records from the Pays Basque. I. Le Moura, Mouligna, Marbella. *Bull. Cent. Étud. Rech. scient., Biarritz* **2**, 211–17.

Oldfield, F. (1964). Late-Quaternary deposits at Le Moura, Biarritz, south-west France. *New Phytol.* **63**, 374–409.

Oldfield, F. (1968). The Quaternary succession of the French Pays Basques. I. Stratigraphy and pollen analysis. *New Phytol.* **67**, 677–731.

Paquereau, M.-M. & Schoeller, M. (1959). Quaternaire et Pliocène du Gurp (Gironde). *Bull. Soc. Géol. Fr.* Ser. 7, **5**, 79–83.

Perring, F. H. (1962). The Irish problem. *Proc. Bournemouth Nat. Sci. Soc.* **52**, 1–13.

Péwé, T. (1963). Ice wedges in Alaska—classification, distribution and climatic significance. *Rep. Int. Congr. Permafrost, Sect. 2b.*

Pike, K. & Godwin, H. (1953). The Interglacial at Clacton-on-Sea, Essex. *Quart. Jl. Geol. Soc. Lond.* **108**, 261–72.

Pinot, J.-P. (1966). Quelques Hauts Niveaux Marins Quaternaires de la Côte du Tregor Central. *Bull. Assoc. Franc. Quaternaire* **2**, 139–51.

Report of Director of 1965 (1966). *Inst. Geol. Sci. Lond.* 4.

Rivière, A., Vernhet, S., Arbey, F. & Rivière, M. (1966). Sur les terrains récents des côtes atlantiques. *C. R. hebd. Séanc. Acad. Sci. Paris* **262**, D, 5–7.

Roger, J. & Freneix, S. (1946). Remarques sur les Faunes de Foraminiféres du Redonien. *Bull. Soc. Géol. Fr.* Ser. 5, **16**, 103–33.

Sealy, J. R. (1949). *Arbutus unedo. J. Ecol.* **37**, 365–88.

Seddon, B. (1962). Late-glacial deposits at Llyn Dwythwch and Nant Ffrancon, Caernarvonshire. *Phil. Trans. R. Soc.* B **244**, 459–81.

Stillman, C. J. (1968). The post-glacial change in sea level in south-western Ireland: new evidence from fresh-water deposits on the floor of Bantry Bay. *Sci. Proc. R. Dubl. Soc.* Ser. A, **3**, 125–7.

Ters, M. (1961). *La Vendée Littorale.* Paris: C.N.R.S.

Walker, D. (1966). The Late Quaternary history of the Cumberland Lowland. *Phil. Trans. R. Soc.* B **251**, 1–210.

Watts, W. A. (1959*a*). Interglacial Deposits at Kilbeg and Newtown, Co. Waterford. *Proc. R. Ir. Acad.* **60**, B, 79–134.

Watts, W. A. (1959*b*). Pollen spectra from the interglacial deposits at Kirmington, Lincolnshire. *Proc. Yorks. Geol. Soc.* **32**, 145–51.

Watts, W. A. (1962). Early Tertiary pollen deposits in Ireland. *Nature, Lond.* **193**, 600.

Watts, W. A. (1963). Late glacial pollen zones in Western Ireland. *Ir. Geogr.* **4**, 367–76.

Watts, W. A. (1967). Interglacial deposits in Kildromin Townland, near Herbertstown, Co. Limerick. *Proc. R. Ir. Acad.* **65** B. 339–48.

Webb, D. A. (1955). *Erica mackaiana. J. Ecol.* **43**, 319–30.

Webb, D. A. (1966). *Erica ciliaris* L. in Ireland. *Proc. Bot. Soc. Br. Isl.* **6**, 221–5.

West, R. G. (1956). The Quaternary deposits at Hoxne, Suffolk. *Phil. Trans. R. Soc.* B **239**, 265–356.

West, R. G. (1957). Interglacial deposits at Bobbitshole, Ipswich. *Phil. Trans. R. Soc.* B **241**, 1–31.

West, R. G. (1961). Vegetational history of the early Pleistocene of the Royal Society borehole at Ludham, Norfolk. *Proc. R. Soc.* B **155**, 437–53.

West, R. G. & Sparks, B. W. (1960). Coastal interglacial deposits of the English Channel. *Phil. Trans. R. Soc.* B **243**, 95–133.

West, R. G. & Wilson, D. G. (1966). Cromer Forest Bed Series. *Nature, Lond.* **209**, 497–8.

Woldstedt, P. (1965). *Das Eiszeitalter, III.* Stuttgart: Enke.

Woodell, S. R. J. (1958). *Daboecia cantabrica. J. Ecol.* **46**, 205–16.

Zagwijn, W. H. (1960). Aspects of the Pliocene and early Pleistocene vegetation in the Netherlands. *Meded. Geol. Sticht.* Series C3, no. 5, 27–42.

# LAND/SEA LEVEL CHANGES IN SCOTLAND

by J. J. Donner

*Department of Geology and Palaeontology, University of Helsinki*

## INTRODUCTION

The melting of the glaciers of the last, Weichselian, glaciation resulted in a eustatic rise of the ocean level of at least 100 m, with a more rapid rise between 14,000 B.P. and 6000 B.P., after which the rate of rising slowed down (Godwin, Suggate & Willis, 1958; Jelgersma, 1966). In areas with a stable earth's crust large parts of the coasts became submerged, but in those areas which during the glaciation had been covered by extensive ice sheets, there was an isostatic recovery of the earth's crust. This resulted in an uplift of the coasts occurring simultaneously with the eustatic rise of sea level. Where the isostatic rise was greater than the eustatic rise, formerly submerged coastal areas emerged above the present sea level. Scotland is such an area, but the isostatic rise of the earth's crust was here relatively slow and small so that at times when the eustatic rise of sea level was rapid, it overtook the isostatic rise, which resulted in a temporary transgression of the sea level interrupting the emergence of the coasts. As a result of this, marine and freshwater sediments alternate with each other in Scotland, in contrast to areas which had a greater isostatic adjustment, like the central parts of Fennoscandia, where there has been a continuous emergence after the time of the withdrawal of the ice.

Consequently the land/sea level changes in Scotland, following the deglaciation, can be studied in the coastal areas now above sea level. In this formerly submerged area there are a number of raised beaches, mostly depositional gravel terraces or terraces consisting of re-worked drift. These conspicuous morphological features clearly show at what altitudes the sea level has stood at various times in relation to its present position. The raised beaches can sometimes be used in demonstrating variations in uplift between different areas, if particular raised beaches can be followed in the field from one area to another. The raised beaches cannot, however, be used in dating former land/sea level changes. For this purpose the dating methods available in Quaternary stratigraphical investigations have to be employed. On, and in some sites below, the marine clays and silts deposited in the lower parts of the coasts in Scotland, there are peats and freshwater muds. These alternating deposits of marine and freshwater sediments comprise the most valuable record of land/sea level changes, and it is on the basis of these deposits that any conclusions about these changes have to be based. In dating and correlating the above-mentioned sediments the relative chronology of pollen stratigraphy combined with radiocarbon determinations has been used. The conclusions first drawn by Fraser & Godwin (1955) that there is a parallelism in the vegetation history between Scotland and other parts of the British Isles and that, as a result of this, the same division of the pollen diagrams into zones as in England can be used in Scotland, were confirmed in later pollen analytical work in various parts of Scotland, as, for instance, by Durno (1956, 1957), Durno & McVean (1959), Donner (1957, 1962) and Turner (1965). Before these investigations very few pollen diagrams had been analysed from Scotland after the first preliminary work by Erdtman (1923, 1924, 1928).

As the same pollen zones could be distinguished in Scotland, at least in its central and southern parts, as in England, it was assumed that they represent synchronous periods. Therefore the results of the radiocarbon dating of English pollen diagrams, particularly the detailed dating of Scaleby Moss, Cumberland (Godwin, Walker & Willis, 1957; Godwin & Willis, 1959*b*; Walker, 1966), were used for the pollen diagrams from Scotland (Donner, 1962). The radiocarbon dates for Late-Weichselian (Late-glacial) deposits (Godwin & Willis, 1959*a*, 1959*b*, 1960) and for Flandrian (Post-glacial) deposits in Scotland, some of which will be discussed later, confirm the view that the main zones are likely to be broadly of the same age as those further south in Britain, taking into account the accuracy which radiocarbon dating permits (for use of the term Late-Weichselian and Flandrian see West, 1963).

In the present account the available results of pollen analyses and radiocarbon dating of deposits from sites in some way of use in dating the Late-Weichselian and Flandrian land/sea level changes in

Scotland will be summarized and viewed in relation to the occurrence of the raised beaches. Finally the land/sea level changes in Scotland will be compared with those in Norway, which forms the western part of the large Fennoscandian area of uplift lying immediately east and north-east of Scotland.

## STRATIGRAPHICAL RESULTS CONCERNING LAND/SEA LEVEL CHANGES

Already before pollen analysis was used the main outlines of land/sea level changes were known in Scotland on the basis of stratigraphical studies. Following the high position of the sea level at the end of the Weichselian glaciation of Scotland, when the ice withdrew and deposited outwash gravels and sands outside the receding ice margin, the sea level was below or at the altitude of the present sea level. Some of the organic deposits from that time were subsequently covered by the Carse Clays, deposited during the transgression which submerged the lower parts of central Scotland. These changes were described already in detail by Jamieson (1865); later, Praeger, in his studies of the stratigraphy and the molluscan fauna of the clays in Northern Ireland, concluded that the transgression covering the submerged organic deposits occurred during the Flandrian (Post-glacial) climatic optimum first discovered by him (Praeger, 1888, 1892, 1897; Coffey & Praeger, 1904; see also Wright, 1937). The submerged peats covered by the deposits of the Flandrian transgression, now in Scotland at or above the present sea level because of the subsequent land uplift, also occur in other parts of Britain but are still submerged, as seen from the studies by Godwin (1943, 1945). These studies, in which pollen analyses were used, also showed that the last rise of sea level took place during the later part of Zone VI of the Boreal period (see also Godwin, 1956), a result later confirmed by radiocarbon determinations of the rise of the ocean level after the Weichselian glaciation, of which the dates were already mentioned in the introduction (Godwin, Suggate & Willis, 1958; Jelgersma, 1966). From further radiocarbon determinations from Scotland the Flandrian transgression, which deposited the Carse Clays and beach material on top of the submerged organic deposits, was dated to have taken place here between 8000 B.P. and 5000 B.P. (Godwin & Willis, 1961, 1962).

The land/sea level changes in Scotland, already broadly known on the basis of stratigraphical changes, have been dated in more detail with the help of pollen analyses and radiocarbon determinations. In a more detailed analysis these results become more complex and the shortcomings of the dating methods used become more obvious. Many sites which are important in dating the land/sea level changes have thin peat beds lying underneath or between marine sands or silts. In sections of this kind only the peat beds are suitable for pollen analysis whereas the marine sediments seldom contain enough pollen to allow counting, and if they do the pollen flora is often not representative. The short pollen diagrams from the peat beds, which are often also affected by the local habitats at the shore, are frequently difficult to place in a particular zone in the Late-Weichselian or Flandrian vegetation history. Further, the regional differences between the pollen diagrams in Scotland are not yet known well enough to allow a detailed zoning everywhere (for problems related to regional differences in pollen diagrams see Smith, 1965).

Because of the above-mentioned difficulties in the use of pollen diagrams for dating sections important in the study of land/sea level changes, dating with the help of radiocarbon determinations is particularly important; but even this method, especially when a single date from a section is used, is not without its errors. The possibility that wrong dates may be obtained with this method calls for caution in using the determinations. The best results are obtained by combining pollen analytical investigations with radiocarbon datings, a requirement which, unfortunately, is not often fulfilled.

Before any detailed conclusions are drawn about the land/sea level changes in Scotland, those sites will be listed in which the stratigraphical record together with pollen analyses and/or radiocarbon determinations can be used for dating these changes and possibly in giving information about the range of them. The sites listed below are all seen in figure 1. For each site the references will be given to papers in which the site is discussed together with the available details about datings and altitude of site and thickness of strata. The sites will be listed by area, starting with the sites in the lowlands of central Scotland. The numbers in brackets refer to site numbers used in the figures.

(1) *Eastfield of Dunbarney* (*Bridge of Earn*), Perthshire. Peat bed below Carse Clay formed in pollen Zone VI a, with a radiocarbon age of 8421 ± 157 B.P., 6461 B.C. (Q-421, Godwin & Willis, 1961; Godwin, 1960). Altitude of site not given.

(2) *Broombarns*, near *Forgandenny*, Perthshire. Peat

Figure 1. Stratigraphically studied sites in Scotland with numbers used in the text, and base line for diagram in figure 5.

below Carse Clay formed in pollen Zone VIa or VIb, with a radiocarbon age of 8354 ± 143 B.P., 6394 B.C. (Q-422, Godwin & Willis, 1961; Godwin, 1960). The site is presumed (Godwin & Willis, 1961) to be identical with the site studied by Erdtman (1928) and referred to as Forgandenny and dated to belong to the Boreal period. Altitude of site not given.

(3) *Airth Colliery*, Stirlingshire. *c.* 10 cm thick peat layer on top of beach material and underneath Carse Clay, the base of which is at *c.* 775 cm O.D. Pollen analyses show that the lower part of the peat may represent Zone V but the upper part formed in Zone VI. The radiocarbon age of this part is 8421 ± 157 B.P., 6461 B.C. (Q-280), but for the lower part 11,024 ± 199 B.P., 9074 B.C. (Q-281), a date which is likely to be erroneous (Godwin & Willis, 1961).

(4) *Flanders Moss*, Perthshire. *c.* 400 cm of *Sphagnum*-peat, the lowermost part formed in Zone VIIa,

on top of Carse Clay (Turner, 1965). A radiocarbon determination of a sample 10–12 cm above the surface of the clay gave an age of 5492 ± 130 B.P., 3542 B.C. (Q-533, Godwin & Willis, 1962). Altitude of site is not given, but on the basis of details given by Durno (1956) and Sissons (1966) in their investigations of the same area, and particularly on the basis of the Ordnance Survey One Inch Map, it is estimated to be *c.* 18 m O.D. The surface of the Carse Clay would then be at *c.* 14 m O.D.

(5) *Kippen*, Stirlingshire. *c.* 50 cm thick peat bed, the base of which is at *c.* 715 cm O.D., on top of clay-silt and below Carse Clay, formed in Zone V and VIa (Newey, 1966). The radiocarbon age for the top of the peat is 8270 ± 160 B.P., 6220 B.C., and for the base 8690 ± 140 B.P., 6740 B.C. (I-1838 and I-1839, Sissons, 1966).

(6) *West Flanders Moss*, Stirlingshire. *c.* 90 cm of *Sphagnum*-peat, the base of which is at *c.* 10 m O.D., on top of clay-silt and below Carse Clay. The lowermost part of the peat was formed in Zone V, or possibly in Zone IV, and the top in Zone VIa (Newey, 1966). At East Flanders Moss the peat formation has earlier been shown to have started in Zone V (Durno, 1956).

(7) *Garscadden Mains*, Dunbartonshire. Marine clay, formed in the end of the Late-Weichselian Zone I and reaching up to an altitude of 25 m O.D., underneath Zone II (Alleröd) freshwater muds up to *c.* 2 m thick, underlying Zone III freshwater clay (Mitchell, 1952; Donner, 1957).

(8) *Irvine*, Ayrshire. *c.* 60 cm thick peat bed, at *c.* 610 cm O.D., overlying boulder clay and underlying beach sands and gravels (Jardine, 1964). Two radiocarbon determinations gave the ages 9620 ± 150 B.P. and 9530 ± 150 B.P., the average being 9575 B.P., 7625 B.C. (Q-642, Godwin & Willis, 1962).

(9) *Girvan, Enoch*, Ayrshire. *c.* 30 cm thick peat at *c.* 515 cm O.D., on clay and underlying beach sands and clays, with a radiocarbon age of 9362 ± 150 B.P., 7312 B.C. (Q-641, Godwin & Willis, 1962; Jardine, 1964). The stratigraphy of the sediments at sites 9 and 10 at Girvan were studied in detail by Jardine (1962, 1963).

(10) *Girvan, railway station*, Ayrshire. *c.* 30 cm thick layer with wood fragments, at *c.* 190 cm O.D., between sands and gravels, with a radiocarbon age of 9020 ± 1 B.P., 7070 B.C. (Q-640, Godwin & Willis, 1962; Jardine, 1964).

(11) *Newton Stewart*, Wigtownshire. Wood fragments at *c.* 450 cm O.D. in gravels overlying clay and underlying beach deposits consisting of banded clays and sands. The wood gave a radiocarbon age of 6159 ± 120 B.P., 4209 B.C. (Q-639, Godwin & Willis, 1962; Jardine, 1964).

(12) *Gatehouse of Fleet*, Kirkudbrightshire. *c.* 60 cm thick lens of peat, at *c.* 750 cm O.D., underlain by gravel and clay and overlain by silts. Wood from the lens of peat gave a radiocarbon age of 6244 ± 140 B.P., 4294 B.C. (Q-818, Godwin, Willis & Switsur, 1965).

(13) *Lochar Moss*, Dumfriesshire. *c.* 2 m of peat, the base of which is at *c.* 7 m O.D., on top of fine sands. The radiocarbon age of the basal peat is 6645 ± 120 B.P., 4695 B.C. (Q-638, Godwin & Willis, 1962; Jardine, 1964; Churchill, 1965).

(14) *Redkirk Point*, Dumfriesshire. *c.* 23 cm thick peat bed at *c.* 275 cm O.D., about 60–90 cm below normal HWL, overlain by clays and underlain by sands. Wood from a tree stump in the peat gave a radiocarbon age of 8135 ± 150 B.P., 6185 B.C. (Q-637, Godwin & Willis, 1962; Jardine, 1964). Another bed with laminated peat is found *c.* 120 cm below the upper peat. The radiocarbon age for this peat is 10,300 ± 185 B.P., 8350 B.C. (Q-815, Godwin, Willis & Switsur, 1965). Another sample from the same peat, which varies in thickness, gave an age of 12,290 ± 250 B.P., 10,340 B.C. (Q-816, Godwin & Switsur, 1966).

(15) *Loch Mòr, Soay*, near Skye. A bed of *Sphagnum*-peat, formed in Zone VI, overlain by marine clay and marine diatom mud, formed in Zones VI and VII, and peats, the total thickness of the layers being 320 cm (Harrison, 1948; pollen diagram by Blackburn). Taking into account later pollen analytical work in Scotland, the marine deposits were most likely formed at the end of Zone VI and in the beginning of Zone VIIa. No details about altitude were given, but on the basis of the Ordnance Survey One Inch Map the surface of the bog was estimated to be at *c.* 4 m O.D. The peat under the marine deposits would then lie immediately above O.D. The age of the peat (Zone VII) was questioned by Sissons (1965).

There are two sites from which there are radiocarbon dates for wood samples below sediments described as Carse Clays which were not included in the list above, because the dates are clearly at variance with the other dates from the area. They are, for instance, younger than the date of 3542 B.C. for peat above Carse Clay in Flanders Moss (site 4). One of the samples, Q-666, was from Heathershot, Bridge of Allan, Perthshire, and gave an age of 3656 ± 150 B.P.,

1706 B.C., and the other, Q-667, near Flanders Moss, was from Littleward, Kippen, Perthshire, and gave an age of 3249 ± 160 B.P., 1299 B.C. (Godwin & Switsur, 1966). In addition to these samples, three radiocarbon dates for samples from near Paisley, Renfrewshire (Shotton, Blundell & Williams, 1967), are related to land/sea level changes but were not included in the list because fuller details about their stratigraphical position have not yet been given. These dates are not, however, at variance with the dates listed from other sites. One sample, Birm-2, gave an age of 3572 ± 64 B.P., 1622 B.C., for wood fragments in the basal peat, at *c.* 8 m O.D., on top of silts, in Linwood Moss, north-west of Paisley. Another sample from Linwood Moss, Birm-3, gave an age of 9231 ± 96 B.P., 7281 B.C., for the basal peat where it is higher above O.D. A sample, Birm-4, from a bed, at *c.* 335 cm O.D., of sands and gravels with organic material, from the floodplain at Wester Fulwood, north-west of Paisley, gave an age of 8039 ± 128 B.P., 6089 B.C. The bed overlies marine clay and is covered with sands and gravels.

In addition to the sites listed above there are some sites where peats found at or below the present sea level show that it has risen to its present position relatively late and that it was never higher in Flandrian time. At the following sites from the Outer Hebrides and Shetland Islands the land/sea level changes have been investigated in detail:

(16) *Isle of Calvay*, South Uist, Outer Hebrides. On the shore of Isle of Calvay a 200 cm thick peat bed, only completely exposed at low tide. The base of the peat was formed in Zone VI (Harrison & Blackburn, 1946).

(17) *Borve*, The Uists, Outer Hebrides. Intertidal organic deposits reaching down to a depth of *c.* −180 cm O.D. A wood sample from *c.* −60 cm O.D. gave a radiocarbon age of 5700 ± 170 B.P., 3750 B.C. (Ritchie, 1966; radiocarbon determination by Isotopes Inc., New Jersey).

(18) *Symbister, Whalsay*, Shetland. A peat bed, the base of which is just over −8 m O.D., resting on bedrock and underlying *c.* 2 m of sand. Five radiocarbon determinations of wood or peat have been obtained from three cores; from the first an age of 6030 ± 80 B.P., 4080 B.C. (St-1552), from the second 5455 ± 170 B.P., 3505 B.C. (St-1811), from the top of a 20 cm thick peat layer and 5945 ± 230 B.P., 3795 B.C. (St-1812), from the base, and from the third core an age of 6970 ± 100 B.P., 5020 B.C. (St-1809), for wood and 6670 ± 100 B.P., 4720 B.C. (St-1925), for peat, both from the same depth (Hoppe, 1965; Engstrand, 1967). There are also radiocarbon determinations from Shetland from organic sediments of tidal lakes and lakes above the present sea level (Engstrand, 1967; Olsson, Stenberg & Göksu, 1967).

On the basis of the sites for which details have been given above, curves can be constructed showing the Late-Weichselian and Flandrian land/sea level changes in Scotland. As Scotland is an area which has risen isostatically after the Weichselian glaciation, there are differences in the total amount of uplift between different areas. Thus, the amount of the relative sea level changes has varied from place to place, depending on the position of the site in relation to the centre of the area of uplift. To lessen the effect of these differences in uplift when drawing the curves for the land/sea level changes, the sites were divided into three groups, which will be treated separately. The division into these three groups on a regional basis helps to demonstrate how great the differences of the land/sea level changes were between central Scotland and the marginal parts of Scotland.

In the diagrams in figures 2, 3 and 4 the altitudes in relation to O.D. are given on the ordinate and the time scale in thousands of years on the abscissa, which also gives the pollen zoning. On the left the stratigraphy of the sites used in each diagram is drawn in relation to the scale of the ordinate. In addition, the pollen zones are given in those profiles in which pollen analyses have been made. The sites were used in drawing the curves of land/sea level changes in the following way. If a radiocarbon age was obtained from a profile it was marked with a dot in the diagram in the place where the horizontal line, showing the altitude of the dated sample in the profile, meets the vertical line giving the age (and the sample number for each radiocarbon determination). If the pollen were studied in a profile the pollen analytical dating of relevant changes in stratigraphy, dating the land/sea level changes, was marked in the same way as the radiocarbon ages, using in these cases the chronology of the pollen zoning given on the abscissa. The points dated with pollen analysis have a *P* added to the dot in the diagrams. The arrows next to the dots give the direction of the land/sea level changes as determined by the stratigraphical changes at each site. If a peat bed is overlying marine sediments it means that the sea level dropped below the level of the basal peat before this peat was formed. If, on the other hand, a peat bed is underlying marine sediments, the sea level rose above the level of the topmost peat after it had been formed. Peat beds lying between marine sediments thus date a temporary relative regression of sea level. If a date from organic deposits is not close to a stratigraphical change, reflecting a regression or a transgression, no arrow was drawn in the diagram next to the dot giving the age and altitude of the deposit. As will be seen, the results obtained about the land/sea level changes from the diagrams are not entirely consistent, and a curve for the changes can not directly be drawn on the basis of the arrows from the different sites. The reason for this is that the stratigraphical record at the various sites is not everywhere without gaps.

Thus, when a peat bed is overlying marine sediments it was not necessarily formed immediately after the time when the sea, which deposited the marine sediments, receded from this altitude. A considerable time may have elapsed before the peat began to form on the exposed shore. Similarly, when marine sediments are overlying peats, the rising sea level may have eroded parts of the peat before it deposited marine sediments on top of it. Erosion is less likely to occur when clay is deposited on top of peat in an estuary than when beach sands and gravels are deposited on peat in a more exposed part of the coast. In the interpretation of the stratigraphical changes it must also be taken into account that peat can be overlaid with marine sediments without it being caused by a rise in sea level (Godwin, 1943). In the two above-mentioned cases, when peat is either underlain or overlain by marine deposits, dates obtained for the peats may give too young a date for the regression and too old a date for the transgression. Thus, an arrow pointing downwards in the diagram gives a minimum age and an arrow pointing upwards a maximum age, assuming that the dates in themselves are correct. A possible hiatus in the profile is more easily detectable with the help of a pollen diagram, whereas the dating is perhaps more accurate with radiocarbon determinations. The dated stratigraphical changes between organic and minerogenic deposits in the profiles makes it possible to date a relative rise or fall of sea level; but in the determination of the altitude to which a transgression reached, the highest altitude at which the sediments formed

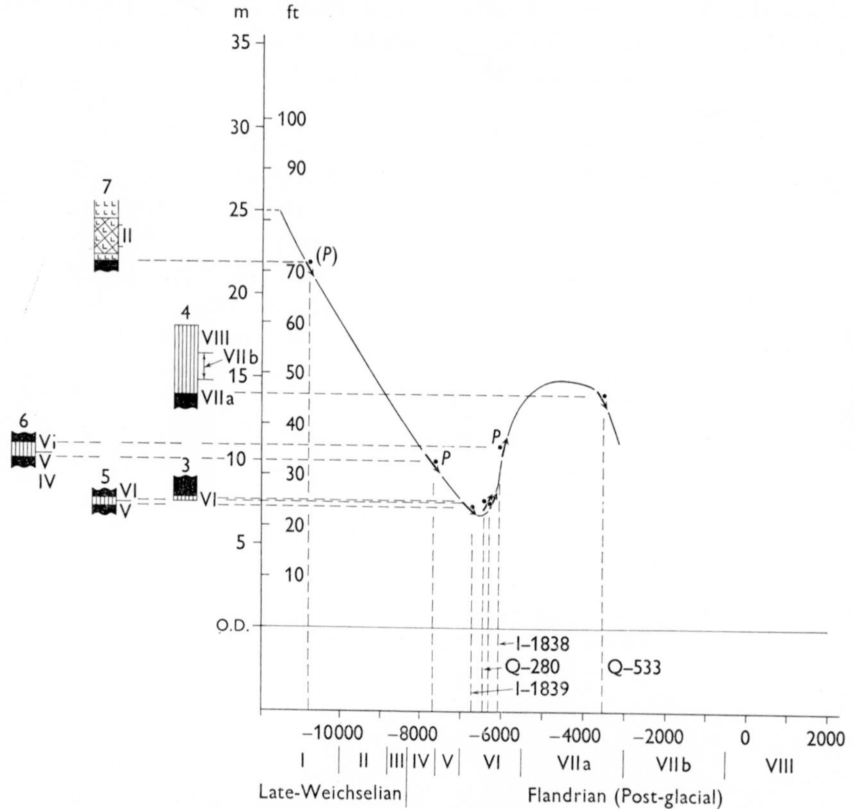


Figure 2. Curve of land/sea level changes in central Scotland (symbols in figure 3). Numbers above stratigraphical columns are site numbers.

during the transgression were deposited has to be taken into account.

During a regression, when peat was formed, it is often not possible to determine how far below the site the sea level receded. In the diagrams in figures 2, 3 and 4 all altitudes were referred to O.D., but it is likely that at many sites peat formation only took place above high water level. In some instances intertidal peats may have formed at any level between mid- and high-tide (Churchill, 1965; Stephens & Synge, 1966). As it is likely, however, that at least some of the stratigraphical changes represented in the diagrams are related to the altitude of the high water level, and not to O.D., the curves showing the land/sea level changes perhaps in some parts give too great values for the altitude at which the mean sea level stood at a certain time in relation to O.D.

The results from sites studied in central Scotland presented in figure 2 are clearer than those from other parts of Scotland, presented in figures 3 and 4. This is probably because the sites in figure 2 are from a relatively small area, in which there were no great differences in the rate of uplift between the sites. The Late-Weichselian regression is only determined on the basis of site 7, Garscadden Mains, where the regression took place in Zone I, from an altitude of at least 25 m, as shown by the marine clay at the site. Three sites, Airth Colliery (3), Kippen (5) and West Flanders Moss (6), determine the regression during which the submerged peats were formed, and one site, Flanders Moss (4), gives a date for the end of the Flandrian transgression. On the basis of all the above-mentioned sites the curve in figure 2 for the land/sea level changes from Late-Weichselian time to *c.* 3000 B.C. could be drawn. One of the samples, Birm-2, with an age of 1622 B.C., from Linwood Moss near Paisley, not used in figure 2 but mentioned earlier, may give a date for the determination of the later Flandrian regression. In spite of the long time gaps between the determinations from the various sites, the curve gives the broad outlines of the relative sea level changes. The shape of the curve presented in figure 2 is in full agreement with the radiocarbon ages from two sites, Eastfield of Dunbarney (1) and Broombarns (2), in which peat beds below Carse Clay were dated to 6461 B.C. and 6394 B.C. respectively. These sites were not included in figure 2 because their altitudes have not been recorded. The two earlier mentioned sites, Heathershot, Bridge of Allan, and Littleward, were not included because of the dates from the peats below Carse Clays being so obviously too young as compared with other dates. In addition, the altitudes of these two sites were not recorded.

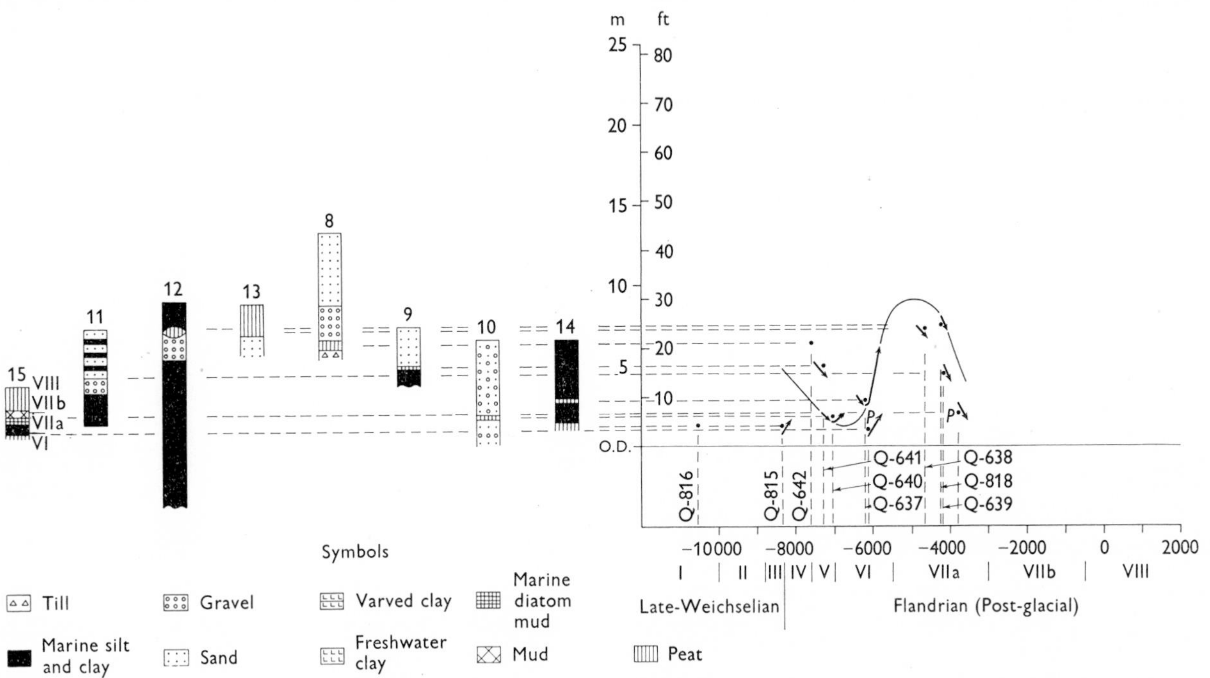


Figure 3. Curve of land/sea level changes in the marginal parts of Scotland and symbols used in figures. Numbers above stratigraphical columns are site numbers.

In figure 3 the results from sites studied in the marginal parts of Scotland are presented. There are no sites which date the Late-Weichselian land/sea level changes; the dates are for the buried peats or for the Flandrian transgression. The trend of the changes is clear and the same as in the previous diagram, but the determinations give a less consistent picture. There is one site in which the results do not agree with those from the other sites. The ages of 8350 B.C. and 10,340 B.C., and its altitude just above O.D., for the peat at Redkirk Point (14), show that the sea level had not reached this altitude during the Late-Weichselian time whereas, as far as can be judged, it was well above O.D. in the areas in which the other sites studied are situated. The difference between Redkirk Point and these other sites is, however, understandable if its position in south-western Scotland, near Cumberland (figure 1), is taken into account. It lies in an area which is outside the area in which Late-Weichselian beaches are found above the present sea level, as seen already from the map drawn by Wright (1937, p. 369) of the distribution of the beaches in the British Isles. The sites at Girvan (9 and 10), in which a regression can be demonstrated to have taken place before the Flandrian transgression, are in Ayrshire and lie already more central in Scotland than Redkirk Point. Thus, the different results concerning the Late-Weichselian and early Flandrian land/sea level changes seen in figure 3 are probably only caused by differences in the geographical position of the sites studied. At Irvine (8) the peat bed probably overlies an old land surface (Jardine, 1964) and does not date the regression mentioned above.

The results presented in figure 3, dating the end of the Flandrian transgression, as well as the regression before the early Flandrian submerged peats were formed, show some variations in age, but do not conflict with the main outlines of the land/sea level changes. The thin submerged peat beds between marine clays, silts, sands or gravels do not necessarily represent the whole period during which the sea level stood relatively low before the Flandrian transgression. All dates together, however, give a fairly accurate dating for the submerged peats. Similarly, the date for the end of the Flandrian transgression varies in the sites used. The lens of peat at Gatehouse of Fleet (12) and the wood fragments at Newton Stewart (11) are connected with the sediments formed during the end of the Flandrian transgression. Compared with the ages obtained from these two sites the age from the basal peat, resting on beach sands, at Lochar Moss (13), seems too old. If it were correct it would be necessary to assume that the conditions at Lochar Moss were different from those in other areas, as pointed out by Jardine (1964). In addition to the sites from south-western Scotland, Loch Mòr, Soay (15), was included in figure 3. The pollen analytical dating, if rightly interpreted, of the Flandrian transgression agrees with the dates obtained from the other sites, even if it is unlikely that the transgression here reached the same altitude as in south-western Scotland.

As seen from figure 3, the tentative curve constructed for the land/sea level changes between

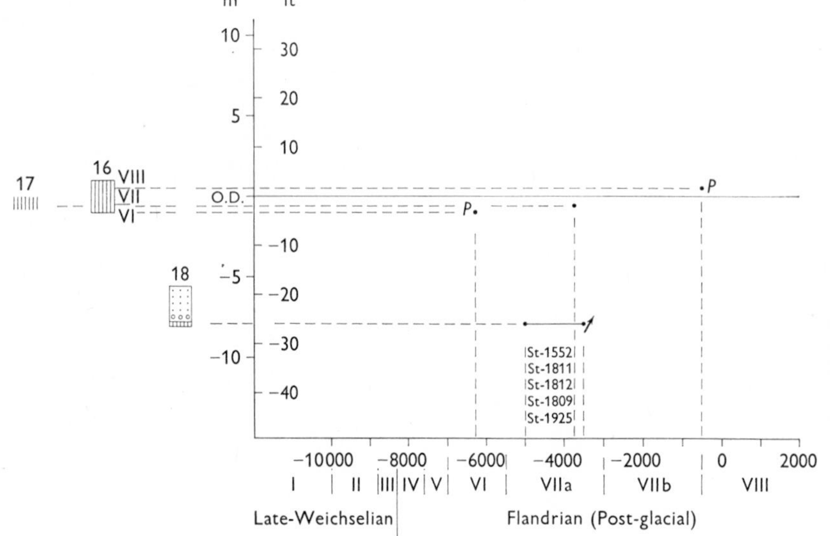


Figure 4. Sites studied in Outer Hebrides and Shetland in relation to O.D. (symbols in figure 3). Numbers above stratigraphical columns are site numbers.

*c.* 8000 B.C. and *c.* 4000 B.C. has essentially the same shape as that in figure 2, but throughout the diagram it is at a lower altitude in relation to O.D. In both diagrams the sites available are still too few to demonstrate possible smaller fluctuations within the Flandrian transgression; fluctuations which would alter the smooth curve now drawn in the diagrams. Further, there are no sites investigated in detail which show the land/sea level changes after *c.* 3000 B.C. Some fluctuations are likely to have occurred even at that time, as indicated by the description of thin peat beds interbedded in marine clays in the Forth area (Sissons, 1965).

In figure 4 the results from the sites studied on the islands of the Outer Hebrides and Shetland are given. No curve for the land/sea level changes can be drawn on the basis of these results, but they clearly show that the changes differ from those in the mainland of Scotland. The inter-tidal peats below O.D. at Borve (17) and Isle of Calvay (16) in the Outer Hebrides already show that the sea level here rose to its present position late in Flandrian time, and the peat bed at Symbister (18), Shetland, shows that it was here still at least 8–9 m below O.D. about 3500 B.C., at the same time as the Flandrian transgression ended in the above-mentioned areas of the mainland. The low relative position of sea level in Shetland may indicate a peripheral submergence related to the isostatic recovery of Scotland (Hoppe, 1965). Further study of the submerged peats on the islands of Scotland will elucidate the question of the crustal movements in these parts.

With the help of the diagrams in figures 2, 3 and 4 it has been shown that the Late-Weichselian and Flandrian land/sea level changes varied in different parts of Scotland as a result of crustal warping taking place in addition to eustatic changes of the sea level. The warping can be further demonstrated by a shoreline diagram in which some of the sites described above are used. In figure 5 those sites are presented which lie in a line from central Scotland, north of Glasgow, to the south-west along the coast of Ayrshire (see figure 1). The altitudes in relation to O.D. are given on the ordinate and the distance in kilometres between the sites on the abscissa. For each site the stratigraphy is given as well as the dates obtained from the radiocarbon determinations. In addition to the seven sites from Scotland, Roddans Port, Co. Down, Northern Ireland, was included because it lies on the same line as the sites from Scotland and because of the detailed studies of its stratigraphy supported by several radiocarbon

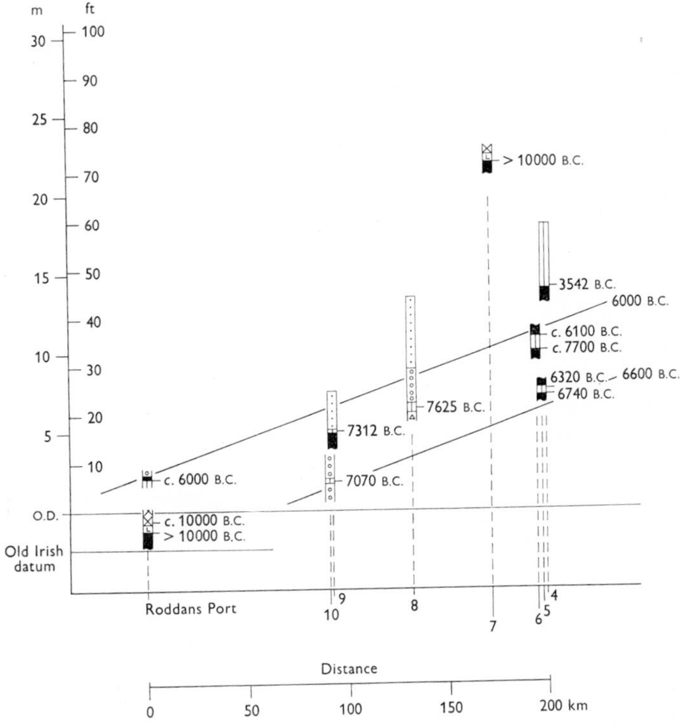

Figure 5. Diagram of the warped zone of the submerged peats in south-western Scotland (symbols in figure 3). Numbers below stratigraphical columns are site numbers.

determinations (Godwin & Willis, 1964, 1965; Morrison & Stephens, 1965). Here marine Late-Weichselian clays from Zone I underlie organic muds from Zone II. The Flandrian transgression in Zone VI is also recorded at Roddans Port by lagoon clay and beach material overlying peat. As the altitudes of the section at Roddans Port were given in relation to the Old Irish Datum its difference with O.D. in Scotland had to be taken into account in figure 5 in order to make a comparison between the sites possible. From figure 5 it can be seen that, in a distance of *c.* 200 km, the buried peat beds formed before the Flandrian transgression lie in a warped zone. Its upper part is at an altitude of *c.* 11 m in central Scotland but only at *c.* 2 m at Roddans Port in Northern Ireland. The peats which were covered by sediments of the Flandrian transgression were formed before 6000 B.C. The sea level was at its lowest position *c.* 6600 B.C., as shown in figures 2 and 3. The highest position of the Flandrian transgression cannot be determined only on the basis of the sites shown in figure 5, but the tilted position of the zone in which the submerged peats occur gives an indication of its position some metres above this zone. At Flanders Moss (4) the surface of the Carse Clay at an altitude of *c.* 14 m is shown as well as the date for the basal peat, also shown in figure 2. The zone in which the submerged or buried peat beds occur, here found to slope down towards the south-west in south-western Scotland, was already found by Geikie (1894) to slope down from the Forth valley also towards Perth and Aberdeen.

In addition to the Flandrian organic and marine sediments there are two sites in figure 5 in which Late-Weichselian clays from Zone I, formed before 10,000 B.C., underlie Zone II freshwater sediments. At Garscadden Mains (7) the clay reaches an altitude of 25 m O.D. (in the profile studied and shown in figure 5 *c.* 22 m), whereas the surface of the clay in the profile from Roddans Port is below O.D. These two sites have already suggested that the position of the shoreline, corresponding to the time when the Late-Weichselian clays were formed, is more steeply inclined at present than the shoreline of the Flandrian transgression. The stratigraphical record presented in figure 5 is not, however, complete enough to permit drawing the position of any Late-Weichselian shorelines.

From the curves of the land/sea level changes in figures 2 and 3 and from the diagram in figure 5 the twofold division of the marine sediments into Late-Weichselian and Flandrian can clearly be seen as well as the chronological position of the peat beds which became submerged during the Flandrian transgression. The Late-Weichselian culmination of sea level was in Zone I, before 10,000 B.C., whereas the Flandrian transgression reached its maximum some time between 6000 B.C. and 4000 B.C. after the early Flandrian regression, during which the peat beds were formed between 8000 B.C. and 6000 B.C. The distribution above sea level of marine sediments from the Late-Weichselian transgression is much more restricted, even if they reach higher altitudes, than for the sediments from the Flandrian transgression, as already recognized in the early stratigraphical work in Scotland (see Wright, 1937). This means that the position of the highest Late-Weichselian shoreline now crosses that of the maximum altitude of the Flandrian transgression.

Further stratigraphical work connected with radiocarbon determinations should help in drawing isobases for certain periods in Scotland, as done by Churchill (1965) for the position of the sea level 6,500 years ago in the south of Britain and Holland. In the following section the evidence of the raised beaches will be viewed against the stratigraphical results discussed above, in order to see to what extent they give additional information about the land/sea level changes.

## RAISED BEACHES

In the central parts of Scotland the sea level was high, in relation to the present sea level, at the time of the Late-Weichselian withdrawal of ice. Then followed a regression, during which those peats were formed which later became buried under the Flandrian transgression. These land/sea level changes, already early recognized in the stratigraphy (Jamieson, 1865) and discussed in the previous section (see figures 2 and 3), led to the separation of the raised beaches into two main groups, the higher Late-glacial group and the Post-glacial group, which has a wider distribution than the former (Wright, 1937). The Late-glacial beaches, which are Late-Weichselian in age, consist of terraces in drift or of outwash deltas formed immediately outside the retreating ice margin. The high beaches are absent from the inner parts of many of the Highland sea-lochs, which shows that these were still occupied by glaciers at the time of the formation of the beaches (Geikie, 1894; Wright, 1937). The two main moraines formed during the Late-Weichselian retreat of the ice are those of the Perth Readvance and Loch Lomond Readvance, of which the former was formed before Zone II, some time in Zone I, and the latter in Zone III, on the basis of the distribution of pollen-analytically studied sites with Late-Weichselian sediments (Donner, 1957). The position of the Perth Readvance moraine is still inadequately known, but the readvance is believed to have extended south-westwards along the coast of Ayrshire and in places to the coast of Solway, and in north-east Scotland the limit probably corresponds to the Dinnet readvance (Synge, 1956), whereas in the north of Scotland its position is uncertain, as shown by the map compiled by Sissons (1965). The dating of the Loch Lomond Readvance was done in the area in which it was studied by Simpson (1933), and the correlation of moraines in other parts of Scotland with the Loch Lomond Readvance moraine is still tentative. It is, however, very likely, especially when taking into account the occurrence of Late-Weichselian sediments at Loch Droma in Ross-shire with a radiocarbon age of 12,814 ± 155 B.P. (Q-457, Godwin & Willis, 1961), that the ice of the Loch Lomond Readvance covered a more restricted area than previously assumed and that it does not correspond to the line drawn by Charlesworth (1955; also used by Donner, 1957, 1959) for his Moraine Glaciation, as shown by the above-mentioned map by Sissons (1965). On the basis of morphological studies it has been concluded that during the formation of the Loch Lomond Readvance moraine the sea level had dropped from its highest Late-Weichselian position and was clearly lower than it had been during the older Perth Readvance. The altitude of the sea level during the Loch Lomond Readvance was determined to be less than 15 m O.D. at Loch Lomond and less than 12 m O.D. near Lake of Menteith and at Loch Etive north of Oban (Simpson, 1933; Wright, 1937; Sissons, 1965, 1966).

If the above-mentioned pollen-analytical dating of the Loch Lomond Readvance moraine near Loch Lomond and Lake of Menteith is correct, it means that the position of the sea level in Zone III, on morphological grounds, was situated at 12–15 m O.D. These conclusions agree remarkably well with those based on the stratigraphical results from central Scotland (see also Donner, 1959). In figure 2 the curve for the sea level changes is at 12–13 m O.D. in Zone III. As the chronological position of the Loch Lomond Readvance seems established, on the basis of the material available, it was included in figure 7 in which the curves of the land/sea level changes are compared with similar curves from Norway.

In Scotland it is difficult to connect Late-Weichselian raised beaches in order to show former positions of sea level and construct isobases for them, unless a particular shoreline can be connected with a dated moraine, as done with the Loch Lomond Readvance. But as long as its position in other parts of Scotland is not well established, and the altitude of the shoreline corresponding in age with this moraine not known in many different areas, isobases cannot be drawn even for this stage. The highest altitude of the Late-Weichselian submergence is shown by raised beaches at *c.* 30–35 m O.D. in central Scotland. This led to the separation of the Late-glacial 100-foot beach, a name referring to its altitude in some parts of Scotland but not taken to indicate that it has the same altitude everywhere. On the contrary, Wright (1937) pointed out that it can be presumed that it is more warped than the younger Post-glacial beach (= Flandrian) and that this is the reason why its distribution above the present sea level is more limited than that of the Post-glacial beach. Another, less clear, Late-glacial beach at a lower altitude was described as the 50-foot beach, but, as pointed out by Wright (1937), it may not be the same shoreline everywhere and merely be one of many beaches formed during the regression of the sea from its highest position. The classification of the Late-glacial raised beaches into the 100-foot and 50-foot beaches was used in the Geological Survey One Inch Maps of Scotland and on the basis of the few available data of the altitude of the 100-foot beach isobases were drawn for this beach (Movius, 1942; see also Lacaille, 1954), showing that it is highest above sea level around the central Highlands. Some measurements of its altitude in various parts of Scotland (Donner, 1959, 1963) gave a somewhat more detailed picture of the isobases, but their altitude and shape were essentially the same as those previously presented. Later the low-lying beaches in Sutherland were connected with the above-mentioned beaches further south (King & Wheeler, 1963).

Because of the difficulties in the determination of the position of former shorelines in the field on the basis of raised beaches and without direct stratigraphical evidence, the interpretation of particularly the Late-Weichselian shorelines in Scotland have varied. In contrast to the probably oversimplified picture with one clear main Late-Weichselian shoreline, the 100-foot beach, and lower less well-developed shorelines of the regression from this level, some detailed investigations in the Forth area (Sissons, 1962*a*, 1962*b*, 1963, 1965, 1966), south-eastern Scotland (Sissons, Smith & Cullingford, 1966), south-western Scotland (Stephens & Synge, 1966; Synge & Stephens, 1966; Synge, 1967) and eastern Fife (Cullingford & Smith, 1966) have led these authors to conclude that there are a number of more steeply warped Late-Weichselian shorelines and that the 100-foot beach is metachronous and thus consists of formations of different age, being younger in the central parts of Scotland than in the marginal parts. As the 100-foot beach does not occur at the altitude its name indicates, and as it is likely to represent a metachronous marine limit, it has been suggested that the use of this term, as well as that of the 50-foot beach, be abandoned (Sissons, 1962*a*). The separation of a number of Late-Weichselian steeply warped shorelines in the Forth valley was doubted on the grounds that the stratigraphical evidence, as well as the altitudes of the main shorelines, does not support such a view (Earp, Francis & Read, 1962). Over a longer distance, as between central Scotland and Northern Ireland, a warping of the Late-Weichselian position of the sea level is, however, indicated by the occurrence of the Zone I marine clays, as seen in figure 5.

From the recent studies of Late-Weichselian shorelines it can be seen that the number of shorelines separated and the conclusions about their warping vary. Some of the differences may be caused by differences in the interpretation of raised beaches: whether any indication of a former position of sea level is taken to represent a beach formation, or whether only clear terraces are interpreted as such. On the basis of the recent studies, mentioned above, dealing with the measurements of the altitudes of Late-Weichselian raised beaches in Scotland, it can be concluded that the formations interpreted as raised beaches occur at various altitudes and that they were formed during the regression of the sea level at the time of the withdrawal of the ice. It is not clear from the present material whether any particular shorelines, for instance representing a transgression, can be singled out among all these raised beaches or whether these were formed at any time in suitable places and at suitable altitudes, depending on exposure to waves, topography and available material, during a quick relative lowering of the sea level. There is no shoreline for which isobases could be drawn, but the distribution of the Late-Weichselian shorelines, taken as a group, can be determined with the help of the uppermost altitude at which raised beaches are found (Synge & Stephens, 1966), and eventually, when some of the main retreat stages of ice are known in detail, they can be correlated with the Late-Weichselian land/sea level changes, as was already done with the Loch Lomond Readvance moraine in a restricted area.

In contrast to the Late-glacial group of beaches, which are Late-Weichselian in age, more is known about the Post-glacial or Flandrian group of beaches. Already the altitudes of the peat beds buried under the minerogenic marine sediments of the Flandrian transgression indicate, as shown in figure 5 for south-western Scotland and earlier observed for eastern Scotland (Geikie, 1894), how great the warping has been after their formation and where the centre of uplift was situated. The uppermost limit of the transgression is marked by an often well-developed terrace representing the shoreline known under the name of the 25-foot beach, the distribution of which was given in a map by Wright (1937). As seen from the stratigraphical results it was transgressive even in the central parts of Scotland, from where its terrace can be followed towards the south-west along the Irish coast to Dublin, and it is morphologically similar, for instance, to the terrace of the Flandrian *Tapes* transgression in northern Jutland in Denmark. It is in the nature of a shoreline representing a transgression that it is metachronous in that way, that it is younger in its marginal parts of distribution than in the central parts, where the isostatic uplift has been greater. This difference in age can be observed along the Irish coast (Morrison & Stephens, 1965), but as seen from figures 2 and 3 no differences in age between the culmination of the Flandrian transgression in the central and marginal parts of Scotland could be observed on the basis of the stratigraphical material, on which the curves were drawn.

The age of the transgression which formed the beach, and during which the Carse Clays were deposited, is determined to 6000–4000 B.C. Whereas the age and approximate altitude of the highest shoreline from this time is known in those areas in which it has been stratigraphically dated, the tracing of it elsewhere has been based on morphological studies of the raised beaches. This means, as in the study of the older raised beaches, that it can be confused with other raised beaches at about the same altitude, especially when comparisons between widely separated areas are made. The shoreline described as the 25-foot beach was recorded to reach altitudes of about 15 m O.D. in central Scotland (see figures 2 and 5) and below this another terrace called the 15-foot beach was sometimes recorded. Measurements of raised beaches thought to represent the 25-foot beach at first gave no clear indication of a warping of the shoreline in the central and northern parts of Scotland, only in the southern parts (Donner, 1959, 1963). Later more numerous measurements, in the investigations in the Forth area and south-western and south-eastern Scotland, mentioned in connection with the Late-Weichselian shorelines, and in Sutherland (King & Wheeler, 1963), eastern Scotland (Smith, 1966), Ross (Kirk, Rice & Synge, 1966) and western Scotland from Firth of Lorne to Loch Broom (McCann, 1966), have made it possible to trace the shoreline in different parts of Scotland and to construct isobases for its altitude, as on the maps published by Sissons (1965) and Synge & Stephens (1966). Even if these maps must be taken as tentative they show how the shoreline reaches an altitude of *c.* 15 m O.D. in central Scotland and declines in all directions, being less steeply inclined towards the south-west than the west and east, as already indicated by the distribution of the raised beaches of this shoreline (Wright, 1937). In the case of this shoreline the measurements of its altitude and careful morphological studies have made it possible to trace it in various parts of Scotland and, as seen from the discussion above, it is so far the only shoreline surface in Scotland for which isobases can be drawn. The determination of their altitude is fully in agreement with the stratigraphical results discussed earlier.

To avoid confusion with other shorelines the term 25-foot beach has been abandoned in the most recent investigations, in which it instead has been called the Main Postglacial Shoreline (Sissons, Smith & Cullingford, 1966; McCann, 1966) or the highest Postglacial Shoreline (Synge & Stephens, 1966). As several shorelines, and not only one, have been separated below the Main Postglacial Shoreline, the term 15-foot beach has also been abandoned. In the determination of these shorelines, formed after the culmination of the Flandrian transgression, there is the same difficulty as in the determination of the younger Late-Weichselian shorelines. As they were formed during a regression of the sea level it is difficult to date any particular position of the receding shoreline, unless the emergence of land was interrupted by transgressions. As seen from figures 2 and 3 no such transgressions have yet been recorded after the main Flandrian transgression in Scotland on the basis of the stratigraphical results used here. The land/sea level changes are, however, not stratigraphically dated after 3000 B.C., and young Flandrian transgressions, such as are recorded further south in Britain, may also have occurred in Scotland.

In addition to the evidence obtained from the distribution and altitude of the raised beaches, discussed above, some evidence about the climatic changes during the formation of the beaches is obtained from the shells found in the marine sediments. There is no detailed stratigraphical division of these sediments on the basis of the shells, but some of the Late-Weichselian marine silts and clays contain an arctic or subarctic fauna, whereas the shells in the beaches of the Flandrian transgression and in the Carse Clays represent a fauna similar to that found today in the same waters, or sometimes indicate a slightly higher temperature of the sea water than at present (see Sissons, 1965). The difference recorded in the marine fauna between the Late-Weichselian and Flandrian sediments is another example of the two episodes of submergence in Scotland after the withdrawal of the Weichselian glaciers.

## GENERAL CONCLUSIONS AND COMPARISONS WITH SURROUNDING AREAS

As seen from the dating of the land/sea level changes and from the determination of the amplitude of these changes, the following main results have been obtained. The high Late-Weichselian position of sea level was followed by a regression interrupted by the Flandrian transgression. The Late-Weichselian submergence was greatest in central Scotland, where also the Flandrian transgression reached its greatest altitude, as seen from figures 2 and 5. Further, the peats buried under the marine sediments of the Flandrian transgression show that the shoreline of this transgression has been warped in south-western Scotland (figure 5), as in eastern Scotland, recorded earlier (Geikie, 1894). On the basis of morphological studies the position of the sea level has been determined for the time during which the Loch Lomond Readvance moraine was formed, which has pollen-analytically been dated to the Late-Weichselian Zone III. The altitude of the sea level determined on the basis of morphology was found to correspond to the stratigraphically determined altitude at the time of Zone III in central Scotland (figure 2). With the help of the available measurements of the raised beaches representing the shoreline of the Main Postglacial Shoreline of the Flandrian transgression, isobases can be constructed for this shoreline, as shown in figure 6, in which the isobases for southern Britain and Holland for 6500 B.P. (Churchill, 1965) were also included to show the down-warping of these areas in relation to the isostatic uplift of Scotland as well as Fennoscandia.

The evidence from Scotland gives curves of land/sea level changes broadly similar to the curves from those parts of other formerly glaciated areas, where the isostatic uplift of the land was not too rapid to stop the eustatically rising sea level from overtaking it during the Flandrian transgression.

To elucidate this similarity two examples were chosen from Norway, from the marginal parts of the Fennoscandian area of isostatic uplift, to the east and

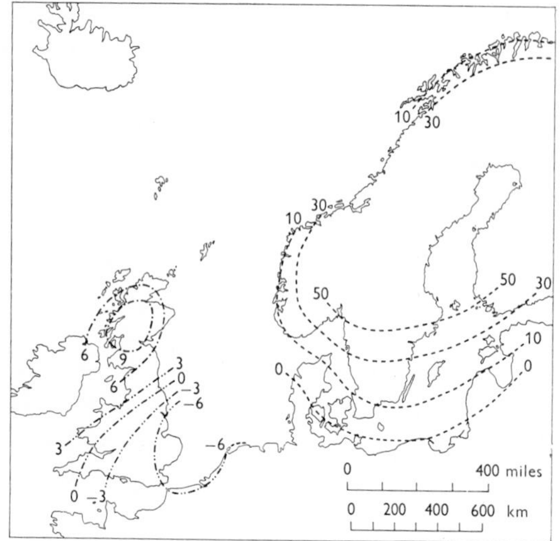


Figure 6. Map of Scotland in relation to surrounding areas, showing isobases in metres for: 1 (-·-·-), uppermost Flandrian shoreline, from 8000–6000 B.P., in Scotland (after Sissons, 1965); 2 (----), uppermost *Tapes–Litorina* shoreline, from 7000–6000 B.P., in Norway, Denmark and the Baltic coasts (based on numerous sources); 3 (·······), shoreline for 6500 B.P. in south Britain and Holland (after Churchill, 1965).

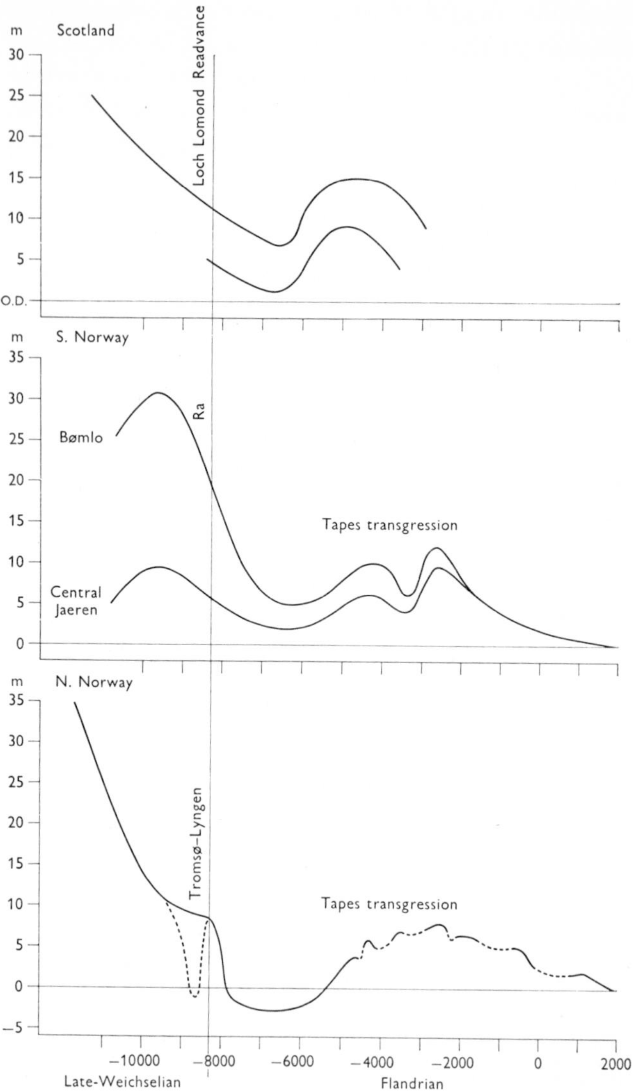


Figure 7. Land/sea level changes in Scotland compared with south Norway (after Faegri as given by Hafsten, 1960) and north Norway (after Marthinussen, 1962). The main moraines in each area are also included in the figure.

north-east of Scotland. In figure 7 all land/sea level changes are drawn to the same scale and uppermost are the two curves from Scotland as determined in figures 2 and 3, with their relationship to the Late-Weichselian Loch Lomond Readvance. Underneath these curves there are two curves from south Norway. Both curves were determined on the basis of pollen-analytical datings of the isolation of lake basins at different altitudes. The upper curve, from an area with greater isostatic uplift than the other is from Bømlo (Faegri, 1944) and the lower curve from central Jaeren (Faegri, 1940), both here drawn as presented by Hafsten (1960). Both curves show that the Flandrian transgression, here called the *Tapes* transgression, was interrupted by a temporary regression. The position of the Ra moraine is also included in the diagram to show its relationship to the Loch Lomond Readvance moraine, even if it has not in this part of Norway been directly connected with the curves of the land/sea level changes, as in the Oslofjord area. The lowermost curve in figure 7 is from Ramså in north Norway and is a tentative curve based on raised beaches and radiocarbon determinations of driftwood, peats and shells found on or in the beaches (Marthinussen, 1962). The curve shows several minor oscillations within the *Tapes* transgression. The Late Weichselian regression may have been interrupted by a temporary regression (in figure 7 drawn with a broken line). The main moraine in the area, which was here formed during the Tromsø-Lyngen substage, was dated to the end of the Late-Weichselian, thus corresponding to the formation of the Ra moraine further south in Norway (Marthinussen, 1962). According to radiocarbon dates from the area the Tromsø-Lyngen substage, however, probably represents both the Late-Weichselian Older Dryas and Younger Dryas periods (Andersen, 1965).

The curves presented in figure 7 show similarities in their broad outlines but differences in details. The main outlines show that there was a general rapid Late-Weichselian emergence of the down-warped land and that later, when the rate of isostatic uplift had slowed down, a transgression of the sea level in Flandrian time, the *Tapes* transgression in Norway (see also figure 6). The differences in detail between the curves in figure 7 may partly be due to inaccuracies in dating and differences in the interpretation of the data, but even excluding these differences it is likely that the land/sea level changes varied somewhat in different areas with isostatic uplift. Each area should, therefore, be treated separately and generalizations cannot yet safely be made in which results from areas widely apart, with different crustal movements, can be combined into one curve. In order to demonstrate differences in the amount of uplift or down-warping it is, however, useful to present a number of curves for land/sea level changes in the same diagram, as was, for instance, done by Andersen (1960) for the areas bordering the North Sea, or by Lundqvist (1965) for different parts of Sweden. In the Baltic Sea the land/sea level changes, however, were also affected by the local changes in the Baltic at those times when it formed an independent lake, and direct comparisons with areas bordering the Atlantic Ocean cannot be made.

It was seen that there are great similarities between Scotland and parts of Norway (figure 7), and some similarities can also be found between Scotland and the coasts of Iceland. After the rapid Late-Weichselian regression down to below the present sea level it gradually rose to its present position (Einarsson, 1961). In north-eastern Iceland, however, there was a Flandrian transgression, the *Nucella* transgression (Kjartansson, Thorarinsson & Einarsson, 1964), of the same age as the *Tapes* transgression in Norway and the Flandrian transgression forming the Main Postglacial Shoreline in Scotland.

The evidence of land/sea level changes obtained from Scotland shows how the changes in central Scotland, the centre of the glaciation of northern Britain, were similar in nature and amplitude to those in the coastal areas of Norway, which form the marginal parts of the greater Fennoscandian glaciation, and also to some extent to those in separately glaciated Iceland. Further research in Scotland will bring out the similarities and differences with surrounding areas in more detail, and allow a better reconstruction of the Late Weichselian and Flandrian land/sea level changes throughout Scotland.

## SUMMARY

The relatively small isostatic recovery of the earth's crust in Scotland after the Weichselian glaciation was overtaken by the Flandrian eustatic rise of the ocean level. The available results from pollen analyses and radiocarbon dating show that the Flandrian transgression reached its maximum in Scotland between 6000 B.C. and 4000 B.C. after a low position of sea level between 8000 B.C. and 6000 B.C., during which peats had been formed, now often lying underneath marine deposits, such as the Carse Clay. The sea level reached its highest position in the Late Weichselian Zone I, before 10,000 B.C. The amount of the relative sea level changes has varied at different sites as a result of crustal warping, as seen from the curves constructed to show these changes, and isobases can be constructed for the shoreline of the Flandrian transgression on the basis of measurements of raised beaches. A comparison with Flandrian land/sea level changes in southern Britain, Holland and the coastal areas of Norway is also possible.

## REFERENCES

Andersen, B. G. (1960). Sørlandet i sen- og postglacial tid (Summary: The Late- and Postglacial history of Southern Norway between Fevik and Åna-Sira). *Norg. Geol. Unders.* **210**, 1–142.

Andersen, B. G. (1965). Glacial chronology of Western Troms, North Norway. *Geol. Soc. Am. Special Paper* **84**, 35–54.

Charlesworth, J. K. (1955). The Late-glacial history of the Highlands and Islands of Scotland. *Trans. R. Soc. Edinb.* **62**, 769–928.

Churchill, D. M. (1965). The displacement of deposits formed at sea-level, 6,500 years ago in Southern Britain. *Quaternaria* **7**, 239–49.

Coffey, G. & Praeger, R. L. (1904). The Antrim raised beach: a contribution to the Neolithic history of the North of Ireland. *Proc. R. Ir. Acad.* **25** C, 143–200.

Cullingford, R. A. & Smith, D. E. (1966). Late-glacial shorelines in Eastern Fife. *Trans. Inst. Br. Geogr.* **39**, 31–51.

Donner, J. J. (1957). The geology and vegetation of late-glacial retreat stages in Scotland. *Trans. R. Soc. Edinb.* **63**, 221–64.

Donner, J. J. (1959). The Late- and Post-glacial raised beaches in Scotland. *Suomal. Tiedeakat. Toim.* A III, **53**, 1–25.

Donner, J. J. (1962). On the Post-glacial history of the Grampian Highlands of Scotland. *Commentat. biol.* **24**, no. 6, 1–29.

Donner, J. J. (1963). The Late- and Post-glacial raised beaches in Scotland II. *Suomal. Tideakat. Toim.* A III, **68**, 1–13.

Durno, S. E. (1956). Pollen analysis of peat deposits in Scotland. *Scott. Geogr. Mag.* **72**, 177–87.

Durno, S. E. (1957). Certain aspects of vegetational history in north-east Scotland. *Scott. Geogr. Mag.* **73**, 176–84.

Durno, S. E. & McVean, D. N. (1959). Forest history of the Beinn Eighe Nature Reserve. *New Phytol.* **58**, 228–36.

Earp, J. R., Francis, E. H. & Read, W. A. (1962). Discussion on J. B. Sissons' paper on Late-glacial shorelines. *Trans. Edinb. geol. Soc.* **19**, 216–20.

Einarsson, T. (1961). Pollenanalytische Untersuchungen zur spät- und postglazialen Klimageschichte Islands. *Sonderveröff. geol. Inst. Köln* **6**, 1–52.

Engstrand, L. G. (1967). Stockholm natural radiocarbon measurements VII. *Radiocarbon* **9**, 387–438.

Erdtman, G. (1923). Iakttagelser från en mikropaleontologisk undersökning av nordskotska, Hebridiska, Orkadiska och Shetländska torvmarker. *Geol. För. Stockh. Förh.* **45**, 538–45.

Erdtman, G. (1924). Studies in the micropalaeontology of Post-glacial deposits in Northern Scotland and the Scotch Isles, with especial reference to the history of the woodlands. *J. Linn. Soc.* (Bot.) **46**, 449–504.

Erdtman, G. (1928). Studies in the postarctic history of the forests of north-western Europe. *Geol. För. Stockh. Förh.* **50**, 123–92.

Faegri, K. (1940). Quartärgeologische Untersuchungen im westlichen Norwegen. II. Zur spät-quartären Geschichte Jaerens. *Bergens Mus. Årb.* 1939–40, **7**, 1–201.

Faegri, K. (1944). Studies on the Pleistocene of Western Norway. III. Bømlo. *Bergens Mus. Årb.* 1943, **8**, 1–100.

Fraser, G. K. & Godwin, H. (1955). Two Scottish pollen diagrams: Carnwath Moss, Lanarkshire, and Strichen Moss, Aberdeenshire. *New Phytol.* **54**, 216–21.

Geikie, J. (1894). *The Great Ice Age and its relation to the antiquity of man.* 3rd ed. London: Stanford.

Godwin, H. (1943). Coastal peat beds of the British Isles and North Sea. *J. Ecol.* **31**, 199–247.

Godwin, H. (1945). Coastal peat-beds of the North Sea region, as indices of land- and sea-level changes. *New Phytol.* **44**, 29–69.

Godwin, H. (1956). *The History of the British Flora.* Cambridge University Press.

Godwin, H. (1960). Radiocarbon dating and Quaternary history in Britain. *Proc. R. Soc.* B **153**, 287–320.

Godwin, H., Suggate, R. P. & Willis, E. H. (1958). Radiocarbon dating of the eustatic rise in ocean-level. *Nature* **181**, 1518–19.

Godwin, H. & Switsur, V. R. (1966). Cambridge University natural radiocarbon measurements VIII. *Radiocarbon* **8**, 390–400.

Godwin, H., Walker, D. & Willis, E. H. (1957). Radiocarbon dating and Post-glacial vegetational history: Scaleby Moss. *Proc. R. Soc.* B **147**, 352–66.

Godwin, H. & Willis, E. H. (1959*a*). Radiocarbon dating of the Late-glacial period in Britain. *Proc. R. Soc.* B **150**, 199–215.

Godwin, H. & Willis, E. H. (1959*b*). Cambridge University natural radiocarbon measurements I. *Am. J. Sci. Radiocarbon Suppl.* **1**, 63–75.

Godwin, H. & Willis, E. H. (1960). Cambridge University natural radiocarbon measurements II. *Am. J. Sci. Radiocarbon Suppl.* **2**, 62–72.

Godwin, H. & Willis, E. H. (1961). Cambridge University natural radiocarbon measurements III. *Radiocarbon* **3**, 60–76.

Godwin, H. & Willis, E. H. (1962). Cambridge University natural radiocarbon measurements V. *Radiocarbon* **4**, 57–70.

Godwin, H. & Willis, E. H. (1964). Cambridge University natural radiocarbon measurements VI. *Radiocarbon* **6**, 116–37.

Godwin, H. & Willis, E. H. (1965). Radiocarbon dates from Roddans Port, Northern Ireland (Appendix IV in Morrison & Stephens, 1965). *Phil. Trans. R. Soc.* B **249**, 249–53.

Godwin, H., Willis, E. H. & Switsur, V. R. (1965). Cambridge University natural radiocarbon measurements VII. *Radiocarbon* **7**, 205–12.

Hafsten, U. (1960). Pollen-analytic investigations in South Norway. *Norg. geol. Unders.* **208**, 434–62.

Harrison, J. W. H. (1948). The passing of the Ice Age and its effect upon the plant and animal life of the Scottish Western Isles. *New Nat.* (1948), 83–90.

Harrison, J. W. H. & Blackburn, K. B. (1946). The occurrence of a nut of *Trapa natans* L. in the Outer Hebrides, with some account of the peat bogs adjoining the loch in which the discovery was made. *New Phytol.* **45**, 124–31.

Hoppe, G. (1965). Submarine peat in the Shetland Islands. *Geogr. Annlr* **47** A, 195–203.

Jamieson, T. F. (1865). On the history of the last geological changes in Scotland. *Q. Jl geol. Soc. Lond.* **21**, 161–203.

Jardine, W. G. (1962). Post-glacial sediments at Girvan, Ayrshire. *Trans. geol. Soc. Glasg.* **24**, 262–78.

Jardine, W. G. (1963). Pleistocene sediments at Girvan, Ayrshire. *Trans. geol. Soc. Glasg.* **25**, 4–16.

Jardine, W. G. (1964). Post-glacial sea-levels in south-west Scotland. *Scott. geogr. Mag.* **80**, 5–11.

Jelgersma, S. (1966). Sea-level changes during the last 10,000 years. In *World Climate from 8000 to 0* B.C. (ed. J. S. Sawyer) 54–71. Roy. Meteorological Society, London.

King, C. A. M. & Wheeler, P. T. (1963). The raised beaches of the north coast of Sutherland, Scotland. *Geol. Mag.* **100**, 299–320.

Kirk, W., Rice, R. J. & Synge, F. M. (1966). Deglaciation and vertical displacement of shorelines in Wester and Easter Ross. *Trans. Inst. Br. Geogr.* **39**, 65–78.

Kjartansson, G., Thorarinsson, S. & Einarsson, T. (1964). C[14]-aldursákvaðanir á sýnishornum varðandi íslenzka kvarterjarðfraeði (Summary: C[14] datings of Quaternary deposits in Iceland). *Náttúrufraeðingurinn* **34**, 97–145.

Lacaille, A. D. (1954). *The Stone Age in Scotland.* Wellcome Historical Medical Museum, New Series 6, Oxford.

Lundqvist, J. (1965). The Quaternary of Sweden. In *The Quaternary*, vol. 1 (ed. K. Rankama). London: Wiley.

Marthinussen, M. (1962). C[14]-datings referring to shore lines, transgressions, and glacial substages in Northern Norway. *Norg. Geol. Unders.* **215**, 37–67.

McCann, S. B. (1966). The main Post-glacial raised shoreline of Western Scotland from the Firth of Lorne to Loch Broom. *Trans. Inst. Br. Geogr.* **39**, 87–99.

Mitchell, G. F. (1952). Late-glacial deposits at Garscadden Mains, near Glasgow. *New Phytol.* **50**, 277–86.

Morrison, M. E. S. & Stephens, N. (1965). A submerged Late-Quaternary deposit at Roddans Port on the north-east coast of Ireland. *Phil. Trans. R. Soc.* B **249**, 221–55.

Movius, H. L. Jr. (1942). *The Irish Stone Age.* Cambridge University Press.

Newey, W. W. (1966). Pollen analysis of sub-carse peats of the Forth Valley. *Trans. Inst. Br. Geogr.* **39**, 53–9.

Olsson, I. U., Stenberg, A. & Göksu, Y. (1967). Uppsala natural radiocarbon measurements VII. *Radiocarbon* **9**, 454–70.

Praeger, R. L. (1888). The estuarine clays at the new Alexandra Dock, Belfast. *Proc. Belfast Nat. Field Club*, Appendix 1886–7, 29–51.

Praeger, R. L. (1892). Estuarine clays of the north-east of Ireland. *Proc. R. Ir. Acad.* **2**, 212–89.

Praeger, R. L. (1897). Report upon the raised beaches of the north-east of Ireland, with special reference to their fauna. *Proc. R. Ir. Acad.* **4**, 30–54.

Ritchie, W. (1966). The Post-glacial rise in sea-level and coastal changes in the Uists. *Trans. Inst. Br. Geogr.* **39**, 79–86.

Shotton, F. W., Blundell, D. J. & Williams, R. E. G. (1967). Birmingham University radiocarbon dates I. *Radiocarbon* **9**, 35–7.

Simpson, J. B. (1933). The Late-glacial readvance moraines of the Highland border west of the River Tay. *Trans. R. Soc. Edinb.* **57**, 633–46.

Sissons, J. B. (1962*a*). A re-interpretation of the literature on Late-glacial shorelines in Scotland with particular reference to the Forth area. *Trans. Edinb. geol. Soc.* **19**, 83–99.

Sissons, J. B. (1962*b*). Reply (to Earp, Francis & Read, 1962). *Trans. Edinb. geol. Soc.* **19**, 221–4.

Sissons, J. B. (1963). Scottish raised shoreline heights with particular reference to the Forth Valley. *Geog. Annlr* **45**, 180–5.

Sissons, J. B. (1965). Quaternary. In *The Geology of Scotland* (ed. G. Y. Craig). Edinburgh: Oliver & Boyd.

Sissons, J. B. (1966). Relative sea-level changes between 10,300 and 8300 B.P. in part of the Carse of Stirling. *Trans. Inst. Br. Geogr.* **39**, 19–29.

Sissons, J. B., Smith, D. E. & Cullingford, R. A. (1966). Late-glacial and Post-glacial shorelines in south-east Scotland. *Trans. Inst. Br. Geogr.* **39**, 9–18.

Smith, A. G. (1965). Problems of inertia and threshold related to Post-glacial habitat changes. *Proc. R. Soc.* B **161**, 331–42.

Smith, J. S. (1966). Morainic limits and their relationship to raised shorelines in the east Scotland Highlands. *Trans. Inst. Br. Geogr.* **39**, 61–4.

Stephens, N. & Synge, F. M. (1966). Pleistocene shorelines. In *Essays in Geomorphology* (ed. G. H. Dury). London: Heinemann.

Synge, F. M. (1956). The glaciation of north-east Scotland. *Scott. geogr. Mag.* **72**, 129–43.

Synge, F. M. (1967). The relationship of the raised strandlines and main end-moraines on the Isle of Mull, and in the district of Lorn, Scotland. *Proc. Geol. Ass.* **77**, 315–28.

Synge, F. M. & Stephens, N. (1966). Late- and Post-glacial shorelines, and ice limits in Argyll and north-east Ulster. *Trans. Inst. Br. Geogr.* **39**, 101–25.

Turner, J. (1965). A contribution to the history of forest clearance. *Proc. R. Soc.* B **161**, 343–53.

Walker, D. (1966). The Late Quaternary history of the Cumberland Lowland. *Phil. Trans. R. Soc.* B **251**, 1–210.

West, R. G. (1963). Problems of the British Quaternary. *Proc. Geol. Ass.* **74**, 147–86.

Wright, W. B. (1937). *The Quaternary Ice Age*, 2nd ed. London: Macmillan.

# VEGETATION HISTORY IN THE NORTH-WEST OF ENGLAND: A REGIONAL SYNTHESIS

by Winifred Pennington (Mrs T. G. Tutin)

*Freshwater Biological Association, Windermere*

Partly by geographical endowment, and partly by historical accident, the north-west of England, here interpreted as the counties of Cumberland, Westmorland and Lancashire north of the sands, has been investigated by pollen-analysts more fully than any area of comparable size in Britain. It has been the scene of four separate research projects, all begun in Professor Godwin's Sub-department of Quaternary Research during the last thirty years, and within the 2671 square miles of the region are included the sites of more than sixty pollen diagrams. It therefore provides a unique opportunity to consider what can be learned from such a detailed study of a single region. In preparing this synthesis, I have drawn, with permission, on the published work of the authors of three of these projects; the fourth is my own.

It was inevitable that a region so ecologically attractive as the country which lies between the Morecambe and Solway sands, between the Pennines and the Irish Sea, should arouse the curiosity of the historical ecologist. The great raised bogs of the Morecambe Bay estuaries had been described, at an early stage in British ecology, by Munn Rankin in Tansley's *Types of British Vegetation* (1911) and illustrated by a picture of a corduroy road stratified into the Foulshaw peat; Kendal had recorded 'interglacial' peat from sections in the St Bees cliffs (1881); a Neolithic settlement site had been described from Ehenside Tarn, with a Forest Bed and a Leaf Bed (Darbyshire, 1874); and in the innermost Lake District valleys lay the moraine hummocks which seemed to mark a renewal of glaciation in the high corries (Marr, 1916). The development of systematic investigation of vegetation history by Professor Godwin presented to us all an opportunity to apply this new technique to investigate these and many other problems of the recent history of north-west England. Dr Smith and Professor Oldfield were concerned with Lowland Lonsdale and the valleys running south to Morecambe Bay, in the area where the Silurian slates of the southern Lake District give way to the Carboniferous limestone of South Westmorland, and worked mainly on the raised bogs built up on the marine clay of the mid-Post-glacial marine transgression, but also on smaller basins in which the deposits extend back into the Late-glacial. Dr Walker has published a major treatise on the Late Quaternary history of the Cumberland Lowland, extending his researches from the Vale of Eden, across the drift-covered Trias of the north Cumberland plain to the coastal strip of West Cumberland. He produced the first zoned and radiocarbon-dated profile in Britain, from Scaleby Moss in north Cumberland, and investigated specific problems in former lakes in Kentmere and Langdale Combe. My own work from this laboratory has tried to link vegetation history with historical limnology in the Lake District proper.

The special interest of north-west England is that it includes in a single rather isolated area a great variety of altitude, parent rock and drift deposits; most of this great variety of habitats has now been investigated, and the sites are shown in figure 1. The Lake District mountains nourished an independent ice-cap in the Weichselian glaciation, so that they and the surrounding lowlands show a rather clear history of retreat and recrudescence of glaciation during the Late-glacial period. In the Post-glacial, the district must have been relatively isolated, by distance and mountain barriers, from the sources of plant immigration. In the relation between climatic changes during the later Post-glacial period and vegetation history, north-west England seems to occupy an intermediate position between Ireland and south-eastern England; that is, though in particular habitats it seems possible to detect the effects of climatic changes since the mid Post-glacial climatic optimum, there is, after 3000 B.C., no perceptible overall climatic effect which could serve as the basis of a zone-boundary. In description of the vegetation history, accordingly, no reference has been made to any zone-boundary after 3000 B.C. This is, of course, somewhat at variance with the standpoint adopted by Walker (1966), but since one of his own conclu-

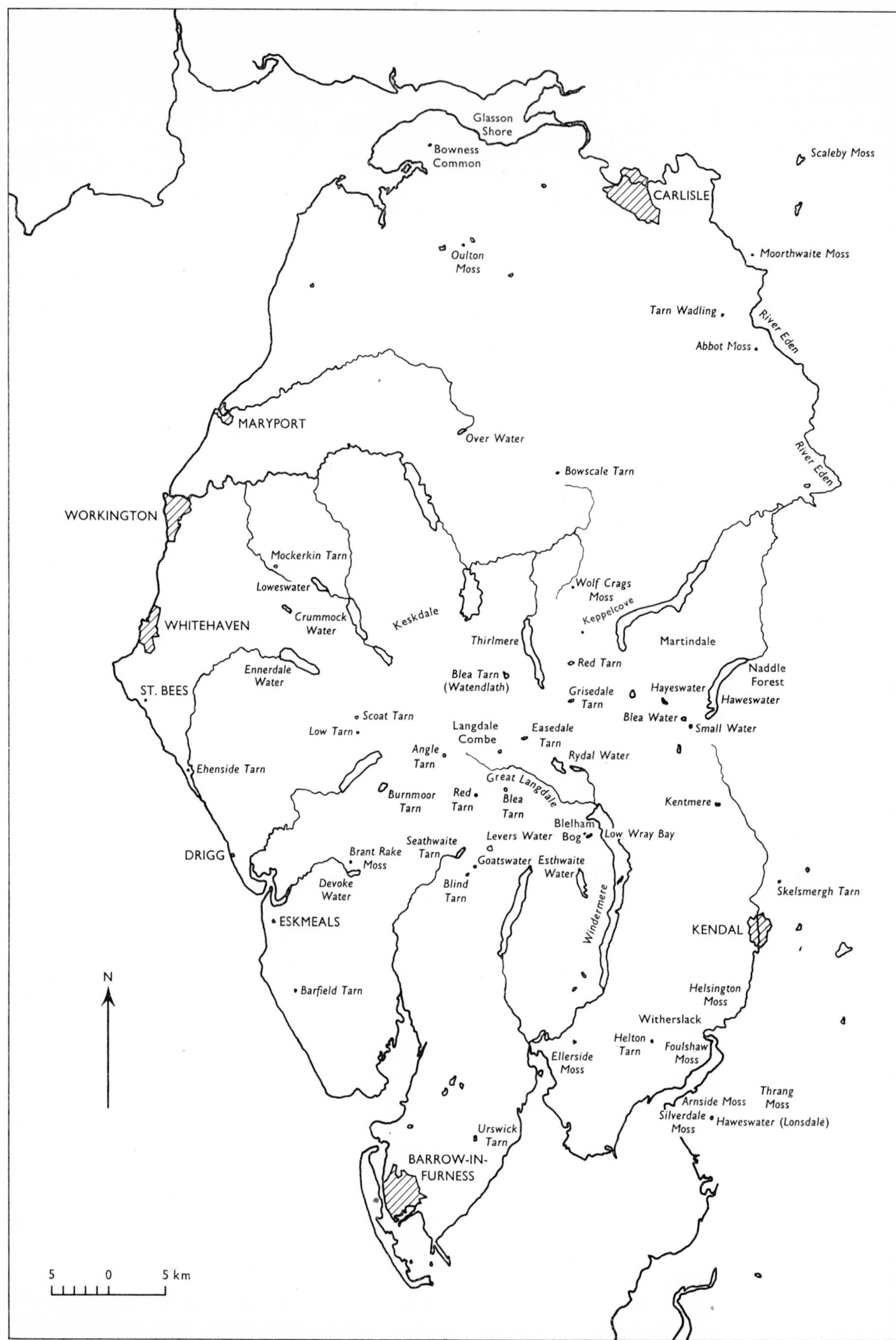


Figure 1. Map of north-west England showing the sites mentioned in the text.

sions was that during the second half of the Post-glacial period 'the dominant determinant of vegetation has been human activity, with edaphic and climatic factors playing but a secondary role' it has seemed reasonable to describe vegetation changes after 3000 B.C. with reference only to dated horizons. A final tentative scheme to set these later changes into a framework based on human history is given (figure 17).

## THE LATE-GLACIAL PERIOD

At the height of the Weichsel glaciation, the Lake District ice was at least 2500 ft (760 m) in thickness; it covered all but the highest mountain tops and extended outwards on to the surrounding plains, where its outward movement was deflected by the Scottish ice on the north and west. During the initial stages of retreat, temporary glacier-lakes of melt-water formed round the western, northern and north-eastern margins of the Lake District ice, and of these some traces remain in the form of beaches and lake deposits of glacial clay, often laminated, but with no organic content. There is no evidence as to how long this phase of the retreat lasted. In both the Lake District and the surrounding lowlands there is a virtual absence of frontal moraine. In the existing large lakes of the district, a great thickness of coarsely laminated clay overlies a stony blue-grey clay which seems to represent the ground-moraine of the ice. This suggests a period of considerable length during which these drift-dammed rock basins were in existence as lakes, receiving glacial melt-water, in conditions which were too severe for any vegetation to flourish. Analysis of these coarse lower varves has shown (Mackereth, 1966*a*) that their calcium and sodium content is almost equivalent to that of the parent rock, of the Borrowdale Volcanic Series, so their material has been scarcely weathered or leached; it must have been buried in the lake sediments after a short period of fluvio-glacial redistribution of fresh rock flour.

Though the stages of the Weichsel glaciation are still controversial, an attempt has been made, in figure 16, to place dated horizons from North Britain, including north-west England, into the later part of the sequence for the 'Last Glacial in the Netherlands' recently given by Van der Hammen *et al.* (1967). It is assumed that during the period called by them the 'Upper Pleniglacial', which in Holland lasted from about 26,000 to 15,000 years ago, Full-glacial conditions prevailed in north-west England, with active glaciation, centred in the Lake District mountains, extending on to the surrounding plains; there are no data as yet for the position of the margin of the Scottish ice at this time. The earliest date for north-west England is from the organic deposit at Blelham Bog, a kettlehole in gravels which had been laid down 'marginal to ice stagnating in the main (Windermere) basin' (Hollingworth, 1951). This organic mud, dated to 12,380 B.C. ± 230 (Cambridge, Q-758) contains the remains of a pioneer vegetation of lichens, plus the remains of algae, Cladocera and other zooplankters: its pollen spectrum (figure 3) indicates a tundra type of vegetation dominated by grasses, sedges and *Rumex*, but containing willows, juniper and *Betula nana*, and therefore approaching the 'shrub-tundra' of Van der Hammen *et al.* (1967). At this date, these authors suppose that 'polar desert' conditions still prevailed in Holland. In Denmark, some kettleholes are known to have remained as blocks of dead ice for another two thousand years (Hartz, 1912). The open kettlehole at Blelham Bog, in which sapropel mud was accumulating nearly 14,500 years ago, suggests that the retreat stages began earlier, and that mean July temperatures were appreciably higher, in north-west England than on neighbouring parts of the continental mainland.

As yet, there are no radiocarbon dates available for any stage in the retreat of the Scottish ice from the Irish Sea basin. In this review, the problem must be faced of the reality or otherwise of the supposed 'Scottish Readvance'. The Geological Survey Memoir (Whitehaven & Workington, 1931) described two boulder clays, a Lower and an Upper, on the north Cumberland plain, and ascribed them to the main glaciation and to a later re-advance of the margin of the Scottish ice-sheet; the supposed 'interglacial' organic layer, described from the cliff section at St Bees, was then attributed to the interstadial between the main glaciation and the Scottish Re-advance. In his work on the Cumbrian lowland, Walker (1966) accepted this geological interpretation, and claimed to have found evidence of the interstadial vegetation in two profiles from basins outside the limits of the Upper Boulder Clay: he was unable to find the original 'interglacial' organic layer at St Bees. By 1960, Mitchell and Synge could not believe that the Scottish ice margin had ever advanced in this way across the Solway estuary; they attributed the main fluctuation of the Scottish ice in the Irish Sea to a readvance to the moraines much further north, at Armoy on the Irish coast and the

Firth of Clyde in Scotland (Mitchell, 1960). The present interpretation by field officers of the Geological Survey, in the Vale of Eden and the Carlisle plain, is that all the boulder clay there can be interpreted as the drift of a glaciation and subsequent deglaciation, and that there is no evidence for the existence of a later glacial episode—i.e. the so-called Upper Boulder Clay is not the result of a separate and distinct glaciation (R. S. Arthurton, pers. comm.).

This recent interpretation of the drift of the Cumberland lowland is easier to reconcile with the data set out in figure 16 (p. 76) than was the previous conception of two glacial stages separated by a 'Cumbrian Oscillation' during which trees grew in north-west England. Walker had pointed out the difficulties of reconciling this supposed sequence with the continental succession.

In recent work on the Lake District, I have defined the beginning of the Late-glacial as the time when micro-fossils indicative of a local pioneer vegetation first appear in the lake sediments. This horizon has now been shown at several sites to coincide with soil changes—the first accumulation of organic matter, a rise in iodine and total halide content interpreted as the result of increased westerly precipitation, and the beginning of leaching of bases, such as sodium—which all agree with the change from a 'polar desert' under predominantly anticyclonic conditions to the beginning of accumulation and maturation of a tundra soil profile with a climatic regime dominated by incoming depressions from the west (Pennington, 1969*b*). From this level, which may or may not be synchronous at the different sites, to the well-known and widely recognized and dated rise in temperature at the beginning of the Alleröd interstadial, at *c.* 10,000 B.C., comprises Zone I of the Late-glacial; within it, there is as yet in Britain no dated evidence for a synchronous and general climatic amelioration of Bölling type, but there are, in north-west England, fluctuations in environment and vegetation; these will be reviewed in the next section.

The threefold division of Late-glacial deposits was early recognized in north-west England, when Dr C. H. Mortimer first obtained cores of the littoral sediments of Windermere, and the organic sediments of Zone II appeared, containing macroscopic remains of *Betula pubescens*, between two layers of varved clay. The correlation of this succession with that at Alleröd (Pennington, 1943, 1947) was confirmed by Godwin (1960) who obtained the date of 9920 B.C. ± 120 for organic mud at the base of the layer containing birch remains, and completed a full pollen analysis of the deposits. Later it was shown that after the Alleröd interstadial, during which both *Betula pubescens* and *B. pendula* flourished in the southern Lake District, there was sufficient recrudescence of glaciation in high corries to produce varved clays in those lakes receiving drainage from them, and that no Late-glacial deposits are present behind the fresh moraine in the high valleys which was described by Marr (1916) and assessed in detail by Manley (1959) (Franks & Pennington, 1961; Walker, 1965; Pennington, in the press, *b*).

## Late-glacial vegetation: Pollen in Late-glacial deposits—a consideration

It has been very general practice, in the division or zoning of Late-glacial deposits, to give considerable weight to the curve representing the Arboreal Pollen/Non-arboreal Pollen ratio. Periods of higher A.P./N.A.P. ratio have been interpreted as periods of milder or ameliorating climate, and periods of reduced percentage of tree pollen have been taken to indicate increased severity of climate. In this review of published descriptions of Late-glacial vegetation in north-west England, it is rather important to consider the supposed origin of the 'tree pollen'. The genera concerned are *Pinus* and *Betula*; *Populus* pollen has not so far been recorded, though a catkin-scale in the Zone II deposits of Windermere testifies to its presence.

*Pinus* pollen has been recorded in amounts of up to 20 % of the total pollen, but there are no firm records of macroscopic remains of *Pinus* in the Late-glacial in north-west England, in spite of intensive search; the record of the supposed tracheid from Zone II in Windermere has been withdrawn, because it might be juniper. This strongly suggests that the *Pinus* pollen is not of local origin, and fluctuations in its percentage frequency (whether of total or of arboreal pollen) do not therefore reflect changes in the local proportion of pine trees, but rather, changes in the relative amounts of pollen supplied by the local vegetation and by distant transport. If the *Pinus* pollen is supplied by distant transport, and the number of grains arriving per year per square centimetre of growing deposit remains constant, then the percentage of pine grains will rise in periods when the supply of pollen from *local* vegetation is reduced (Jessen, 1949). This can be seen in many diagrams for Zone III, including that from Skelsmergh Tarn (Walker, 1955)—an area where there is no suggestion that the severe conditions of this post-Alleröd cold period could have favoured the pine. In my view, it is therefore unsound to interpret a rise in the percentage A.P. in Late-glacial deposits in north-west England as indicating climatic amelioration in the Late-glacial, if *Pinus* is a significant contributor to the A.P.

The second tree genus, *Betula*, presents a further problem. It is still a matter of controversy as to whether or not it is possible to distinguish quantitatively between pollen of *B. nana* and the tree birches. All workers are agreed on the recognition of 'typical' grains of *B. nana*, as described for instance by Terasmäe (1951) and by Walker (1955). It is common experience, however, to find great difficulty in assigning many grains either to *B. nana* or to the tree birches with confidence. Examination of type material of *B. nana* from a variety of Scottish localities has shown that the *range* of pollen morphology in this species is certainly wider than that given by Terasmäe, and it seems very probable that there is considerable overlap between the range of pore-depth and diameter found in *B. nana* and in the tree birches. (The method of effecting quantitative separation, on size-statistics, described by Birks (1968) is possible only in samples where no treatment with HF is required.) Because of this overlap it is potentially unsound to regard as necessarily tree birch pollen that fraction of a Late-glacial *Betula* sample which in pore-depth falls outside Terasmäe's diagnosis of *B. nana*. In many Late-glacial samples from north-west England, *all* the *Betula* pollen falls within the *possible* range of British *B. nana*. It is therefore inaccurate to regard a curve for percentage *Betula* pollen as indicative of the relative importance of trees in the vegetation, even if grains of the obvious *B. nana* type as described by Terasmäe have been subtracted from the total, since many of the remaining grains could well belong to *B. nana* and not tree species. A rise in such a *Betula* curve may therefore indicate an expansion of *B. nana* only.

The significance of absolute pollen diagrams must also be considered. In order to count the numbers of each type of grain incorporated into a growing deposit per square cm per year, either a deposit which has accumulated at a constant rate, or an extremely closely spaced series of radiocarbon dates, is required. As yet it has not proved possible to use this technique in Britain, but the results of Davis & Deevey (1964) from Rogers Lake, Connecticut, have produced two important general ideas. One is that, in times like the early part of the Late-glacial, when absolute frequencies of tree pollen were very low, changes in the percentages of pollen of the various trees, 'previously interpreted as evidence of climatic oscillations, may represent statistical artefacts rather than significant vegetational changes' (Davis & Deevey, 1964). The other striking point to emerge from comparison of percentage and absolute diagrams was that at the transition from a predominantly herbaceous Late-glacial vegetation to one dominated by trees, the herbaceous pollen, e.g. that of grasses and sedges, continued to appear in constant absolute frequencies in the absolute diagram, whereas in the percentage diagram the herbaceous types were all suppressed. This apparent suppression of herbaceous vegetation on the arrival of trees in the locality is now seen as entirely a product of the percentage method, due to the enormous increase in the total yearly production of pollen contributed by the incoming trees.

It therefore seems clear that in the interpretation of Late-glacial pollen diagrams, it will henceforward be more profitable to attempt to analyse the vegetation represented into communities, characteristic of various habitats and tolerances, in the manner exemplified by Godwin's analysis of Late-glacial vegetation at Loch Droma (Kirk & Godwin, 1963) rather than to use tree pollen curves. In the higher parts of north-west England it becomes essential to do this. The incompleteness of the pollen record prevents any analysis of the vegetation into communities by phytosociological methods. Walker (1966) divided the herbaceous pollen of Late-glacial levels at sites in the Cumberland lowland into separate pollen sums of 'dry land herbs' and 'damp land herbs' respectively. Two major difficulties limit the value of this method—in these shallow basins a complicating factor is introduced by the marginal hydrosere, and some taxa of major importance, such as grasses and divisions of the Compositae, must include members of both types of community.

### Late-glacial vegetation: Pre-Alleröd vegetation

The opening of the Alleröd interstadial has been dated by radiocarbon to *c.* 10,000 B.C. at many sites in Denmark, Holland and Germany, and the existing dates for British Alleröd material are in agreement with this (Godwin & Willis, 1959). It seems amply proven that at about this date there was a very general rise in temperature, sufficient to bring about a considerable northward extension of birch woodland, and to put an end to solifluction and cryoturbation in the lowlands of north-west Europe. Before 10,000 B.C. there is no comparably documented horizon of synchronous climatic change, but there are many radiocarbon dates from deposits of a type suggesting a temporary climatic amelioration, within the eleventh millennium B.C. These deposits, generally described as of 'Bölling' type (Iversen, 1954; Firbas, Muller & Munnich, 1955) range in age from 10,700 B.C. in south-west Norway (Chanda, 1965) to 10,450 B.C. in Holland and Denmark (Van der Hammen *et al.* 1967).

In north-west England there is as yet no dated evidence for environmental changes between the early sapropel mud at Blelham Bog—12,380 B.C.—and the base of the Alleröd mud in Windermere—9920 B.C. Many borings in lake basins reach barren clay only a little way below the organic Alleröd deposits (see figure 6, Devoke Water), and it seems that conditions favourable to the preservation of records of the pioneer vegetation of the pre-Alleröd period must have occurred comparatively rarely. In the large lakes, the deposits of the pre-Alleröd period are typically a silty clay in which pollen is either absent or very sparse; where more informative deposits have been found in smaller basins (e.g. Blea Tarn, figure 5), it has only been after extensive probing with a sampler, since these earliest deposits are often restricted to quite a small area of the deepest part of the original basin. This conveys a warning as to the inadvisability of drawing any conclusions from the *absence* of pre-Alleröd—and indeed Alleröd—deposits, unless repeated probes have been made in a basin, because older deposits may always be found on subsequent sampling.

Because of this, I am inclined to view with caution Walker's hypothesis that there is a significant age difference between the lowest deposits in basins on the surface of the 'Scottish Readvance' boulder clay, compared with basins outside the limits of this Readvance. It may be so, but two sites in each position is not a very large sample, and older deposits may well be found in future in North Cumberland or South-west Scotland.

As illustrations of the pre-Alleröd vegetation history of the region, pollen diagrams from Moorthwaite Moss, Blelham Bog and Blea Tarn are shown in figures 2, 3 and 5 (p. 51). At Moorthwaite Moss, and at Abbot Moss, Walker's view is that there is recorded a 'Cumbrian oscillation', covering Zones C1, C2 and C3 of his Cumbrian zonation, when a significant increase in tree pollen (*Betula* and *Pinus*) preceded the deposition of solifluction material, barren of pollen, which he equates with the climatic severity accompanying the Scottish Readvance. This was followed by a period (Cumbrian Zone 4) of predominantly herbaceous vegetation, grasses, sedges, *Rumex* and *Artemisia*, with some *Betula* and *Pinus*, and the first appearance of juniper, which Walker equates with the final stage of Zone I. The fact that these successive periods, or Cumbrian Zones, can be recognized in the diagrams from both sites, suggested to him that they represent comparable and synchronous phenomena.

The only macroscopic record in support of the suggested Cumbrian birch-pine interstadial is, however, a record of *Betula pubescens* fruits and catkin-scales as 'rare' in deposits of this age at Abbot Moss. No macroscopic remains of pine are recorded. If the pollen criteria of Cumbrian Zones C1, C2 and C3 at Moorthwaite Moss (figure 2) and Abbot Moss are objectively regarded, it is possible to formulate an alternative explanation. *Pinus* pollen has been shown to be over-represented in many deposits of periods where local pollen production was sparse. The *Betula* total may be almost wholly of *B. nana* pollen. Furthermore, the considerable quantities of 'Coryloid' pollen, and the almost constant presence of *Alnus* pollen, indicate the presence of secondary pollen. *Corylus* and *Alnus* pollen grains are common in mineral soils of the district at present. The basal deposits at Abbot and Moorthwaite Mosses, included in Cumbrian Zones C1, C2 and C3, and described as sands, sandy clay-muds and silty clay-muds, could equally well be interpreted as formed entirely during the retreat stages of a glaciation, with a very sparse local vegetation, receiving pine pollen from wind transport, and including much secondary pollen from the lowland soils over which the ice had passed. In my view, there is no convincing argument for a 'Cumbrian interstadial' with tree birches and pine present in Cumberland, as suggested by Walker. The whole basal succession at Abbot and Moorthwaite Mosses could represent the early stages of Zone I of the Lateglacial, complicated by secondary pollen. There does not seem to be any evidence of changing plant communities suggestive of environmental changes in Zone I.

In deposits from the small basin of Haweswater in Lowland Lonsdale, Oldfield (1960) recorded a possible pre-Alleröd climatic fluctuation. Since this is based on only four samples, in which the proportion of secondary pollen was probably high because *Corylus*, *Alnus* and *Picea* were present, it does not provide any clear picture of a changing vegetation; moreover it is based entirely on the *Betula* and *Pinus* curves.

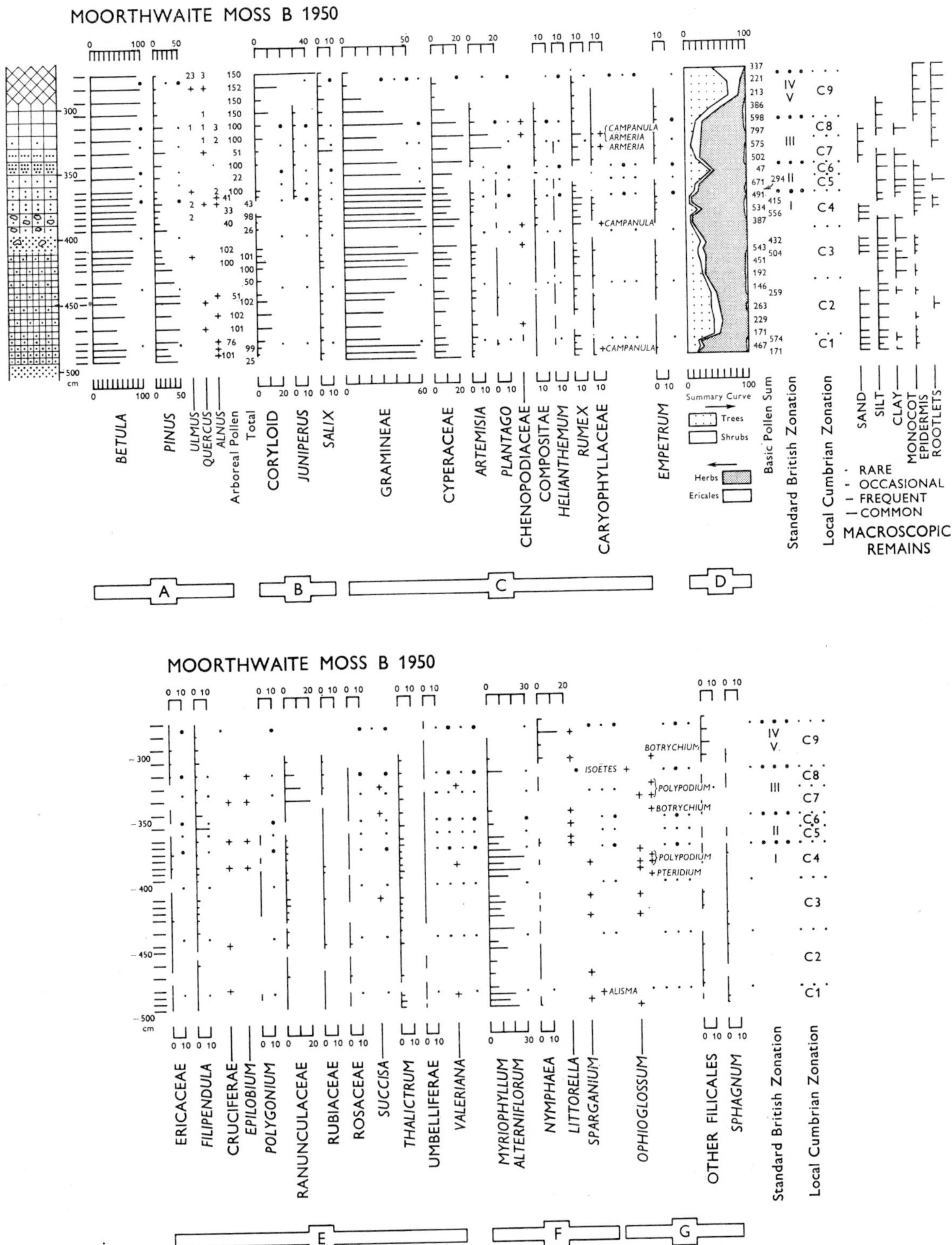


Figure 2. Moorthwaite Moss. (Reproduced from Walker (1966) figure 14.) Pollen diagram through the Late-glacial and early Post-glacial deposits at boring 10. Section A: individual trees. Section B: individual shrubs. Section C: individual land herbs. Section D: summary curves for land flora (*A*+*B*+*C*+*Ericales*). Section E: individual taxa of uncertain ecology. Section F: individual aquatic herbs. Section G: individual Pteridophyta and Bryophyta. The pollen frequencies in Section A are shown as percentages of the sum of Section A pollen (=arboreal pollen total) of the appropriate sample. The pollen and spore frequencies in the other sections are shown as percentages of the basic pollen sum (*A*+*B*+*C*+Ericales) of the appropriate sample.

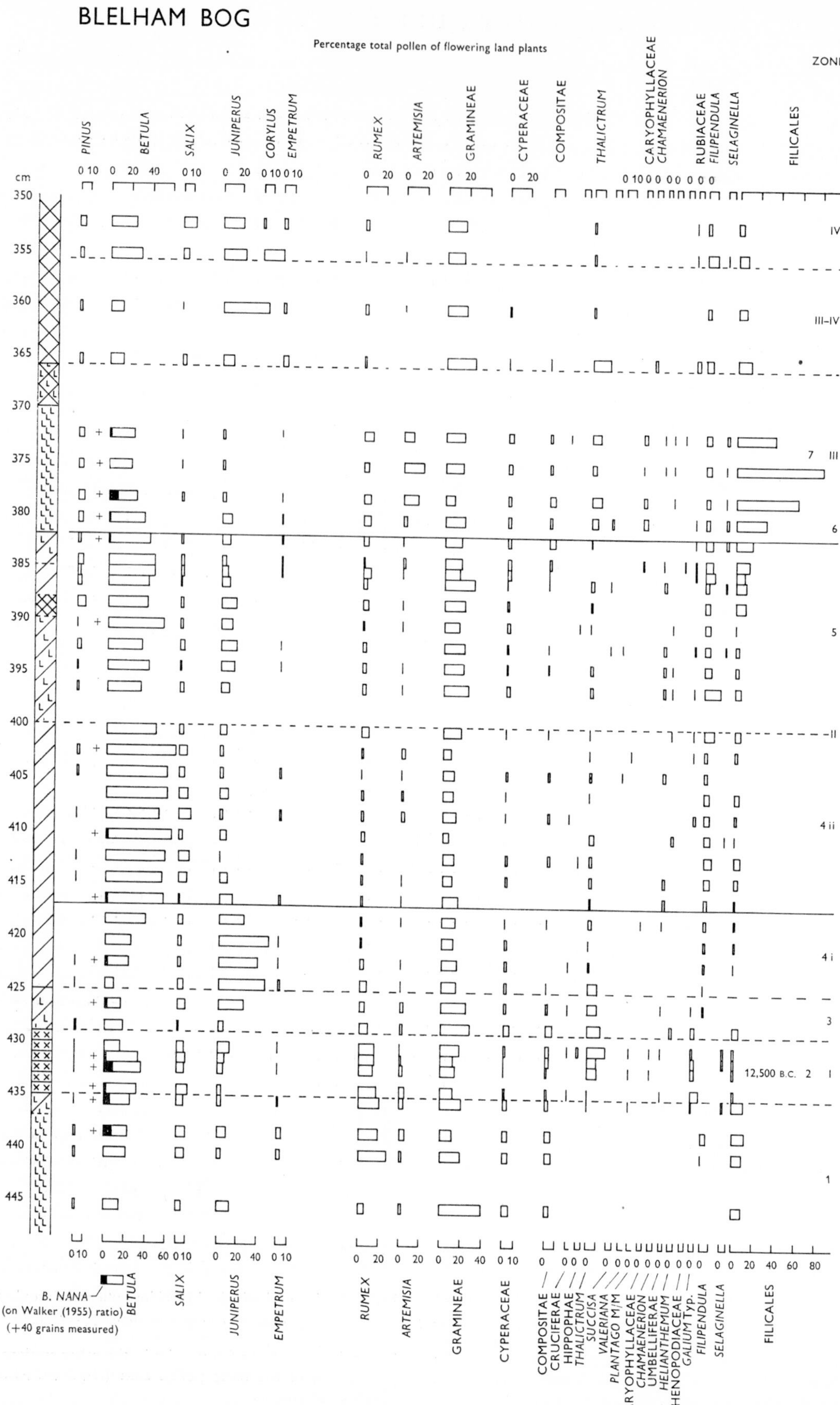


Figure 3. Late-glacial pollen diagram from the Blelham Bog kettle-hole, showing the pre-Alleröd organic layer which has been dated to *c.* 12,500 B.C. (The *Betula* percentage is divided into *B. nana* type and the remainder, on the criteria of Terasmäe (1951) and Walker (1955), at those levels where at least 40 grains were measured.)

Figure 3 illustrates the Late-glacial profile at Blelham Bog from which the date of 12,380 B.C. ± 230 (Cambridge, Q-758) has been published. This site, which is being fully described elsewhere (Pennington, 1969*b*) was discovered by Professor Oldfield in 1961. In the basal silty clay below the dated sapropel mud, the pollen spectra record a vegetation typical of Zone I, dominated by grasses, sedges and *Rumex* (*acetosa*/*acetosella* type) and containing also willows, some juniper, *Betula*, *Artemisia* and *Hippophaë*. Within the sapropel, there is an increase in *Betula* pollen which includes many grains which fall within the possible morphological range of British type *B. nana*, and an enrichment of the grass-sedge vegetation by herbs such as *Thalictrum*, *Epilobium*, *Helianthemum* and *Rubiaceae* (*Galium* type). *Artemisia* decreases within the organic layer. This layer seems to represent an early environmental fluctuation within the lowest part of Zone I, during which this small kettle-hole pool contained many algae and a rich zooplankton. Much of the organic layer is made up of lichen remains—crumbs of interwoven fungal hyphae and unicellular algae—representing an important component of the pioneer vegetation. Above this organic mud, a return to a more silty mud indicates a climatic recession sufficient to increase movements of mineral soils, and in this silty mud the total *Betula* percentage is reduced. This early oscillation within Zone I is distinguished as vegetation phases 1, 2 and 3 (figure 3).

The next radiocarbon date in north-west England is that for the base of the Alleröd mud in Windermere, 9920 B.C. ± 120 (Godwin, 1960). This horizon (figure 4) corresponds with the major rise of the *Betula* pollen curve and the first occurrence in quantity of macroscopic remains of *B. pubescens* (Pennington, 1947; figure 14). At both Blelham Bog and Low Wray Bay, Windermere, sites *c.* 1·2 km apart, there is a major expansion of juniper before the *Betula* phase. In both, the rise in organic content of the mud begins in the juniper phase and is continued into the *Betula* phase. This suggests that in the Lake District there was an uninterrupted succession on stable soils, through a juniper phase, which is here called vegetation phase 4i, to birch woodland called vegetation phase 4ii. The date of the succession to birch woodland at the Windermere site corresponds with the accepted date for the beginning of the Alleröd interstadial. The expansion of juniper must therefore pre-date this, but both the increase in juniper pollen *and* the change

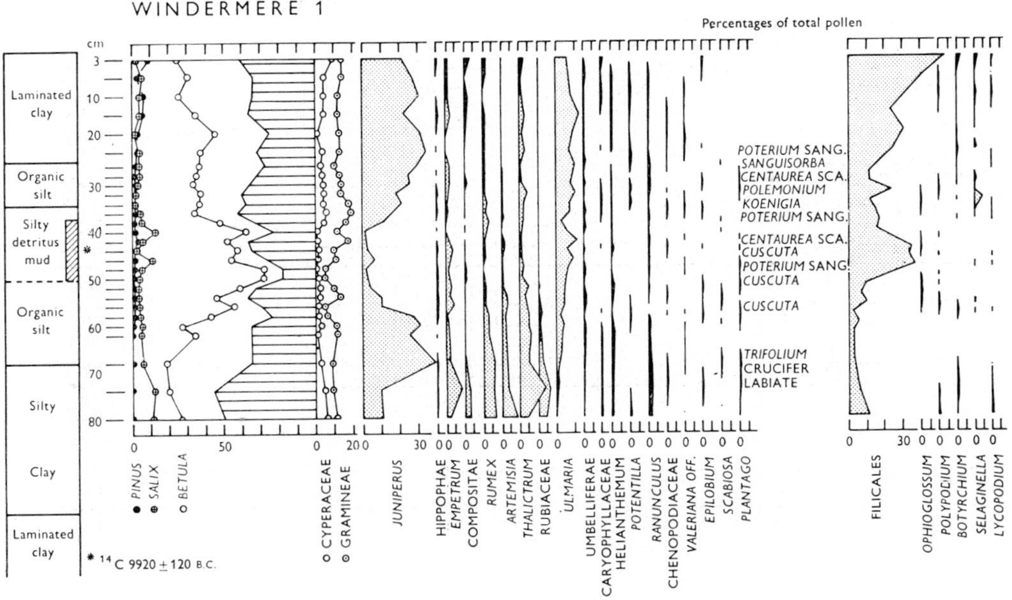


Figure 4. Late-glacial pollen diagram from Low Wray Bay, Windermere. (Reproduced from Godwin (1960), figure 1.) Pollen diagram through a core in the bed of Windermere. The organic silts between the laminated clays were formed during the Alleröd oscillation, as is confirmed by the radiocarbon date. Birch, pine, juniper and willow were abundant throughout these Late-glacial layers but herbs and shrubs typical of very open vegetation are various and abundant. In the left-hand block, the shaded areas represent total frequency of herb pollen, the unshaded area that of tree and shrub pollen. Analyses by Miss R. Andrew.

This profile and date show that a phase dominated by juniper (=4i) occurred *before* 9920 B.C. and is therefore pre-Alleröd. The macroscopic remains of tree birches at this site are found in the mud between 35 and 50 cm (Pennington, 1947, figure 14). It is suggested that much of the *Betula* pollen below 50 cm is probably that of *B. nana*.

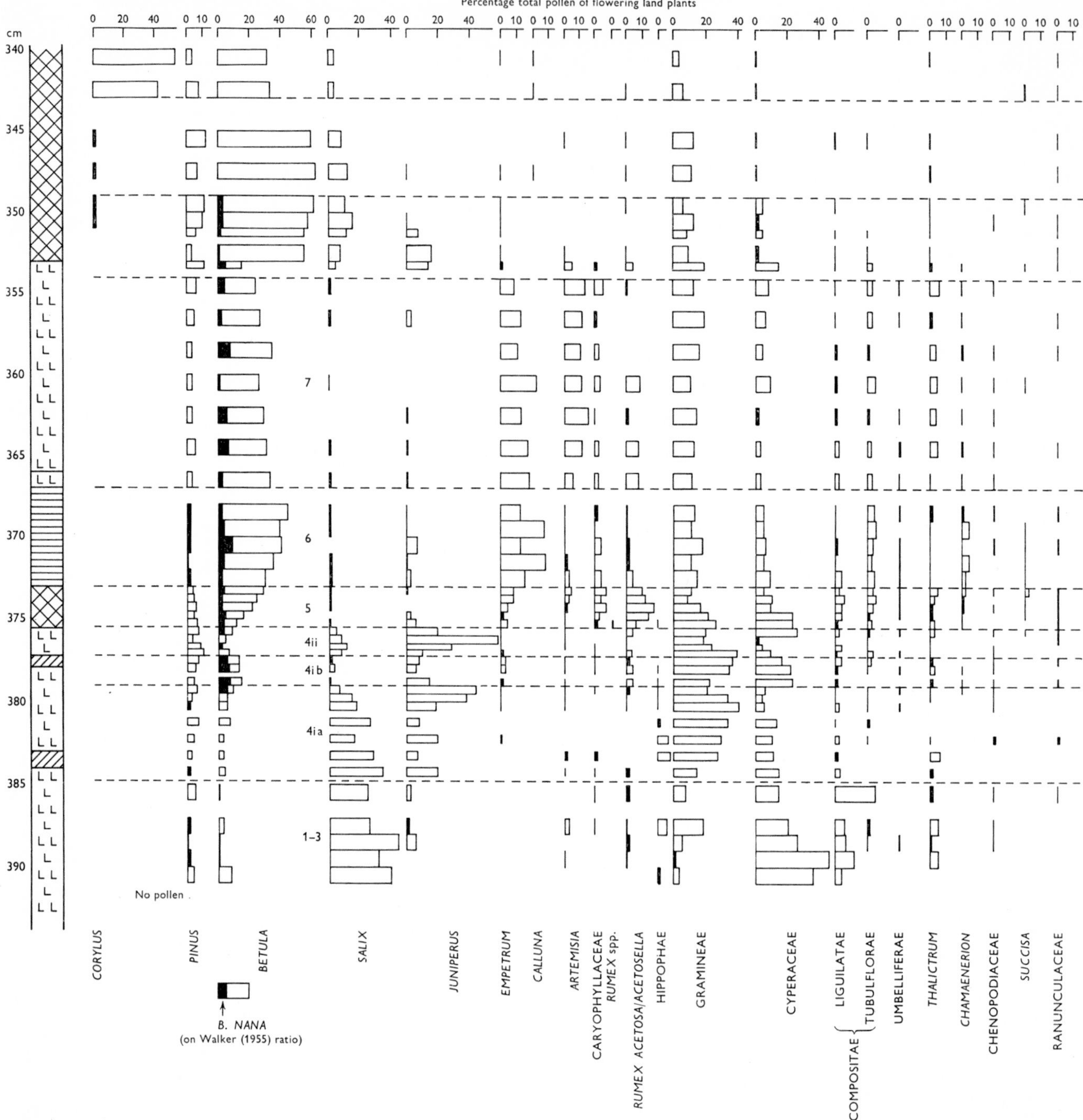


Figure 5*a*. Late-glacial section of the core from Blea Tarn (1964) which was analysed for pollen and for chemical elements. 5*a*. *Pollen diagram*. For description of the vegetation phases, see text pp. 52–7. Phase 4*ib* is tentatively identified as a period of snow-tolerant vegetation marking the Bölling–Alleröd recession, but since it is doubtful whether tree birches were present locally, the Zone boundary I/II is not drawn. Division of the *Betula* percentage as in figure 3·5 *b*. *Chemical analyses*. For comment on these curves, see text pp. 52–4.

to a more organic mud which accompanies it indicate a rise in average or mean temperatures (with fewer days on which freeze–thaw cycles disturbed the soil) *before* the opening of the Alleröd, if Iversen's (1960) interpretation of the role of juniper in a vegetation transitional from shrub-tundra to woodland is accepted.

The most probable date before 9920 B.C. for a pre-Alleröd rise in temperature, sufficient to stimulate stunted juniper to profuse flowering, is during the

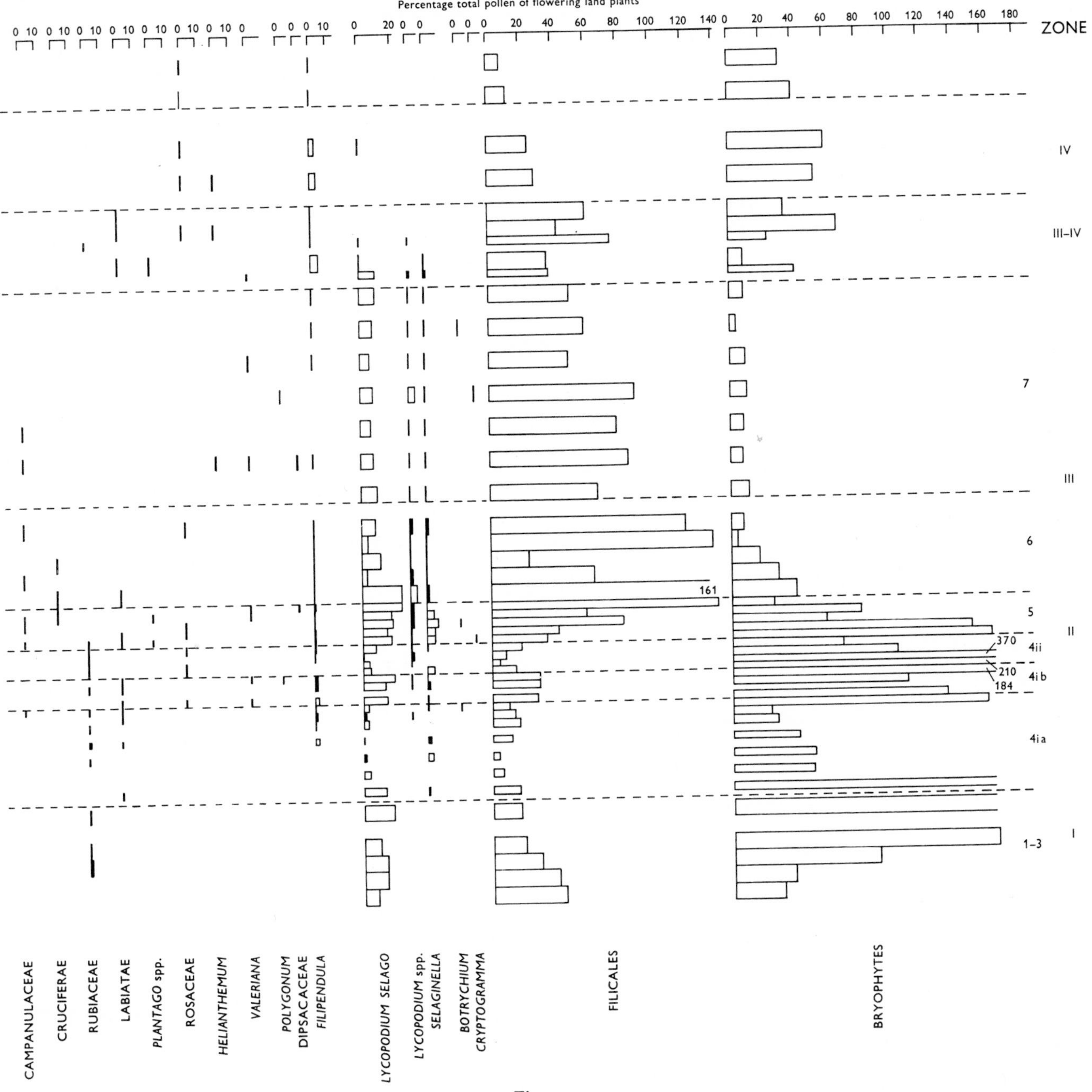


Figure 5*a*.

eleventh millennium B.C., i.e. Bölling *sensu lato*, but as yet there is no confirmation of this from radiocarbon dating. If this correlation is correct, the absence of any evidence for a climatic recession after the Bölling amelioration, in such pollen diagrams as figures 3 and 4, can be explained by supposing that in the southern Lake District no climatic threshold critical for juniper was crossed during the Bölling–Alleröd interval, which did not exceed 200 years and could have been as short as 50 years (Kirk & Godwin, 1963). Figure 5, the pollen diagram from Blea Tarn, Langdale, altitude 612 ft (187 m), among the mountains of the central Lake District, *does* show evidence in the vegetation of a climatic recession which followed this first juniper phase, with a return to a more chianophilous vegetation with *Betula nana* and *Lycopodium selago*, but no evidence in the stratigraphy for any renewed frost-disturbance of the soils. It would seem quite possible that in the oceanic conditions of western Britain, temperatures may not

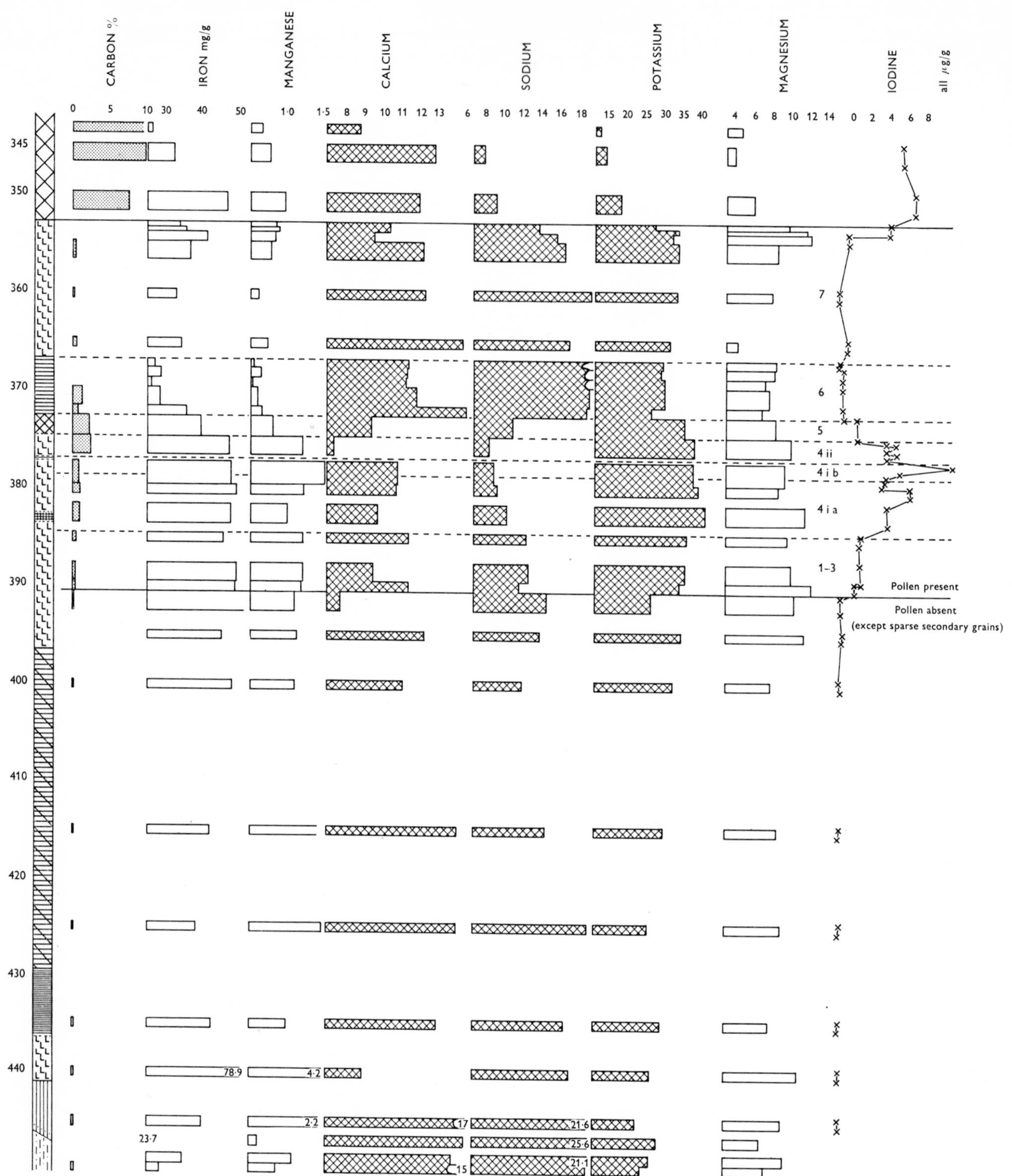
CARBON %
IRON mg/g
MANGANESE
CALCIUM
SODIUM
POTASSIUM
MAGNESIUM
IODINE
all μg/g
Pollen present
Pollen absent
(except sparse secondary grains)
7
6
5
4 ii
4 i b
4 i a
1–3
78·9
4·2
2·2
17
21·6
23·7
25·6
21·1
15

Figure 5*b*.

have fallen very low during the Bölling–Alleröd interval (*Dryas* 2) but that snowfall may have increased in the mountains, with an adverse effect on flowering of juniper.

At all Walker's lowland Cumberland sites, the small percentages of *Juniperus* pollen, in comparison with the Lake District sites, is noteworthy. Walker interprets the vegetation succession in his diagrams, such as Moorthwaite Moss (figure 2) as indicating the absence of competition between woody species, and visualizes the small quantities of juniper, willow and *Rumex* pollen, which persisted into Zone II, as supplied by plants growing in abundant open spaces within the Alleröd birchwoods. The available diagrams from Lowland Lonsdale, on the other hand, resemble those from the Lake District in suggesting a succession from an initial phase of flowering of juniper, the immedate response to the first major rise in temperature, to an expansion of tree birches at the opening of the Alleröd. The decline in juniper from its maximum could be either a real decrease due to shading by tree birches, or a statistical artefact of the percentage method due to abundant pollen production by the birches.

The continuity of soil development through the course of Zone I and into Zone II is shown in figures 5 and 6 from the lake sediments of four tarns in the Lake District, at altitudes from 612 ft (187 m) to 1200 ft (366 m). These profiles were analysed simultaneously for pollen and for the chemical elements carbon, iodine, iron, manganese, calcium, sodium, potassium and magnesium; the sediment column was interpreted as a series of soils washed in from the drainage basins (Mackereth, 1966*a*, *b*). Though clear evidence of strongly erosive processes can be seen in the sodium curves at the *end* of the Alleröd oscillation, there is no evidence for any period of severe soil erosion in the pre-Alleröd section. It is therefore possible to visualize the vegetation history of north-west England from pre-Alleröd into Alleröd times as a succession, in response to temperature changes, on steadily maturing soil profiles.

## Late-glacial vegetation: The Alleröd Interstadial

The Alleröd interstadial began with the well-documented climatic amelioration of *c.* 10,000 B.C., and the criterion adopted for defining this horizon in the lowlands of north-west England is the expansion of tree birches; from the evidence of fruits and catkin-scales the main expansion was of *Betula pubescens*, but the more thermophilous *B. pendula* was present in the southern Lake District (Walker, 1966; Pennington, 1947; Franks & Pennington, 1961). The increase in organic content of the lake sediments is interpreted as another effect of the rise in temperature, namely the virtual cessation of solifluction movements of the soil, and accumulation of soil humus. Macrophytic vegetation was abundant in the lakes, as were benthic and epiphytic diatoms. Manley (1959) has supposed that the presence of tree birches at Windermere would suggest a mean July temperature appreciably exceeding 10 °C in the lowlands. Walker (1966) has discussed the aspects of the climatic change which would be most likely to end solifluction in the lowlands; Pennington (in the press, *b*) discusses differences in this respect found in Zone II deposits at different altitudes in the Lake District.

The pollen diagrams from the higher altitudes (figures 5 and 6) show lower percentages of total *Betula* pollen than in the lowlands (figures 3, 4): the more northerly sites Moorthwaite Moss (figure 2) and Abbot Moss show lower *Betula* percentages than the sites in the southern Lake District and Lowland Lonsdale. These differences are tabulated in table 1 (page 56). In view of the necessary caution in interpreting percentage figures, no firm conclusion can be reached, but this list does suggest that there was a strong altitudinal control of the spread of tree birches during the Alleröd in north-west England, and possibly the latitude is near the northward margin of the general distribution of birch woodland. At Seathwaite Tarn, the highest site where undisturbed Alleröd sediments were found, there is no evidence at all for a local expansion of tree birches. Since it seems probable that at sites at intermediate altitudes (Blea Tarn, Devoke Water and Burnmoor Tarn) the local expansion of tree birches was delayed after the period of expansion in the lowlands, no attempt has been made to draw the lower boundary of Zone II on these pollen diagrams, but the expansion of juniper (vegetation phase 41) is interpreted as the response to the first major rise in temperature.

Using the curves for carbon and sodium shown in figures 5 and 6 as indicating respectively the accumulation of soil humus and the relative erosion rate of mineral soils (Mackereth, 1966*a*), the latter part of Zone I and the whole of Zone II are seen as periods of soil stability, during which a continuous vegetation cover became established. The onset of renewed instability of soils which reached its maximum development in Zone III can be seen to coincide, in the later part of the Alleröd, with vegetation changes.

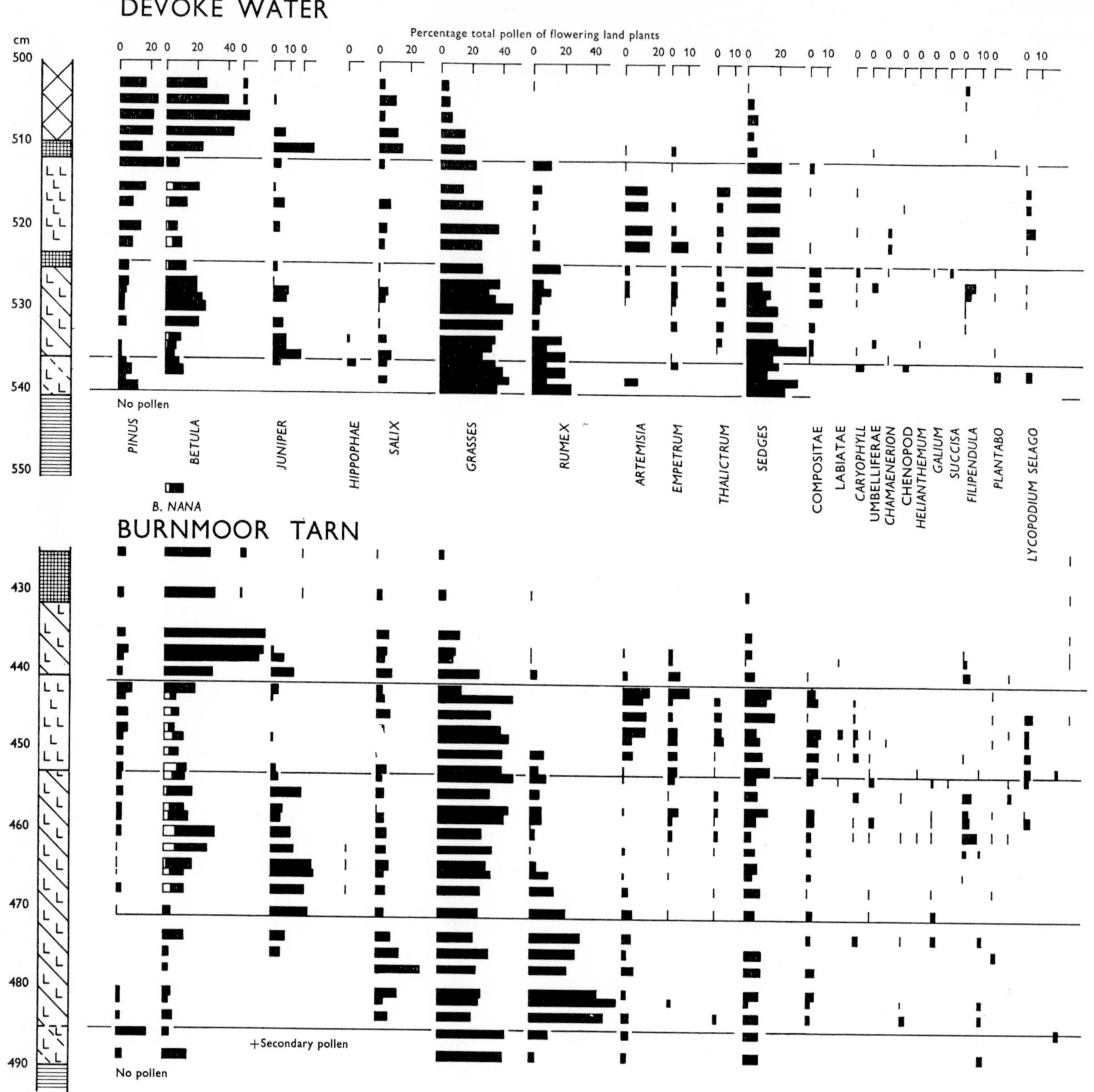


Figure 6. Late-glacial pollen diagrams and selected chemical analyses from Devoke Water (766 ft), Burnmoor Tarn (840 ft), and Seathwaite Tarn (1210 ft). The division of the *Betula* curve is here based on the range of *B. nana* pollen morphology found in British type slides, and not on measurement of grains. Since the expansion of tree birches at Devoke Water and Burnmoor could have been later than at the lowland sites (figures 3, 5), the Zone boundary I/II is not drawn. Seathwaite Tarn must have been above the altitude limit for tree birches in the Alleröd interstadial. Please note that the name beneath the Devoke Water diagram also apply of course to the Burnmoor Tarn and Seathwaite Tarn diagrams.

A fall in the iron and manganese curves during the middle and later part of the Alleröd is interpreted as showing the presence of humic acids and reducing conditions in Late-glacial soils, leading in many areas to podsolization and the spread of *Empetrum* heaths during the transition from Alleröd conditions to those of the following Post-Alleröd cold period (Zone III). This is well seen in the Blea Tarn diagram (figure 5).

In Walker's pollen diagrams from Lowland Cumberland, the vegetational instability characteristic of Zone III becomes apparent in the later part of Zone II, and Walker distinguished this as a separate Cumbrian Zone, C6. In the Lake District

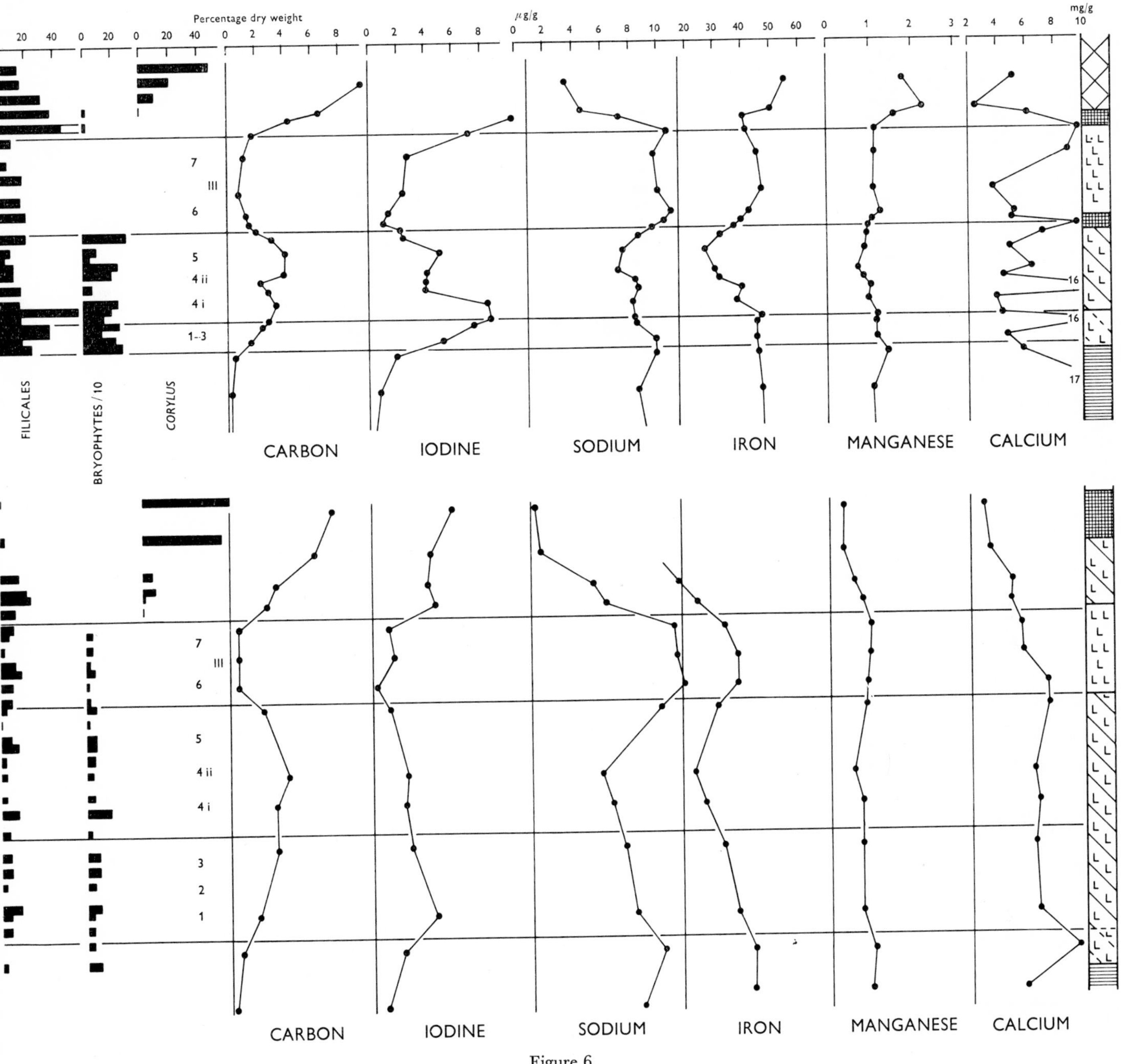


Figure 6.

diagrams it is labelled as vegetation phase 5, and characterized by the reappearance of many of the plants of open habitats such as *Plantago*, *Lycopodium selago* and *Rumex*; *Koenigia* and *Selaginella* appear at this horizon in the Windermere diagram, while tree birch pollen is still abundant. The environmental change involved in the recession from Alleröd to Post-Alleröd conditions must have included other factors than the fall in temperature; edaphic changes must have been considerable. Walker supposed his diagrams from the Cumberland lowland to suggest that precipitation or the precipitation-evaporation ratio increased at the end of the Alleröd period. The very consistent decline in the iodine curves in the Lake District profiles suggests that there precipitation must have decreased during the latter part of the Alleröd oscillation (cf. figures 5 and 6).

On all the evidence, it seems most likely that the

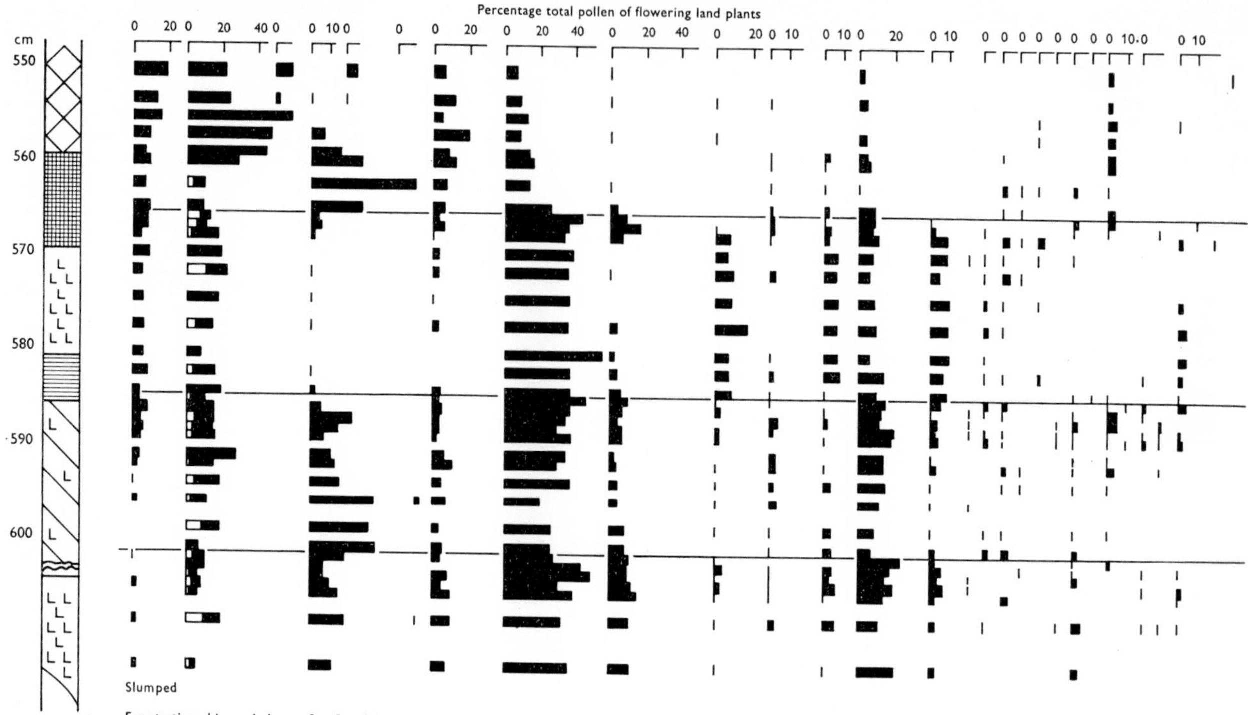


Figure 6.

highest temperatures of the Alleröd period occurred during the earlier part of the birch phase, in Walker's Cumbrian Zone C5. During the later part of the

TABLE 1. *Percentage of the total pollen of land plants contributed by* Betula *in the birchwood phase of Zone II (Walker's Cumbrian Zone C5)*

| Site | Altitude (m) | Total pollen (%) |
|---|---|---|
| Haweswater, Lonsdale | 9 | 40–50 |
| Blelham Bog | 40 | 60 |
| Windermere, Low Wray Bay | 38 | 40 |
| Moorthwaite & Abbot Mosses | 120 | <30 |
| Blea Tarn, Devoke Water and Burnmoor Tarn | 183–243 | 20–30 |
| Seathwaite Tarn | 366 | 10–20 |

Alleröd, the effects of falling temperatures would of course first become apparent at the higher altitudes, and in basins such as Blea Tarn and Windermere, which include land of very different altitudes, this would explain the mixture in the pollen spectra in the second part of Zone II of tree birch pollen with that of plants of open habitats and of those which tolerate low temperatures. During the period of maximum temperatures all ice and permanent snow must have disappeared from the Lake District mountains, because the organic Alleröd muds of the Lakeland Haweswater show that even the great Mardale corries, where the Post-Alleröd ice was slowest to melt, must have been free from ice during the mild Alleröd period.

## LATE-GLACIAL VEGETATION: THE POST-ALLERÖD GLACIATION

The radiocarbon dates from Scaleby Moss show that the breakdown of established communities was apparent there before 8878 B.C. ± 185 (Godwin, Walker & Willis, 1957); at this site this was the date of the beginning of the change to the conditions of Zone III. At Blea Tarn there is an interesting deposit of about 50 narrow laminations in a silty mud at this transition; the laminations are not graded varves, but could represent a series of solifluction silts formed in a period of severe winters. The pollen in this laminated deposit (figure 5) indicates a vegetation of *Empetrum* heath locally, with *Betula* pollen, either from distant transport or secondary in the moving Alleröd soils. Manley (1962) suggested that between 50 and 100 years of cold summers and heavy winter snowfall could have been sufficient to re-establish glaciers in the high corries of the Lake District after the Alleröd oscillation, and suggested that persistent flooding, associated with the disturbed chilly climate, prevailed.

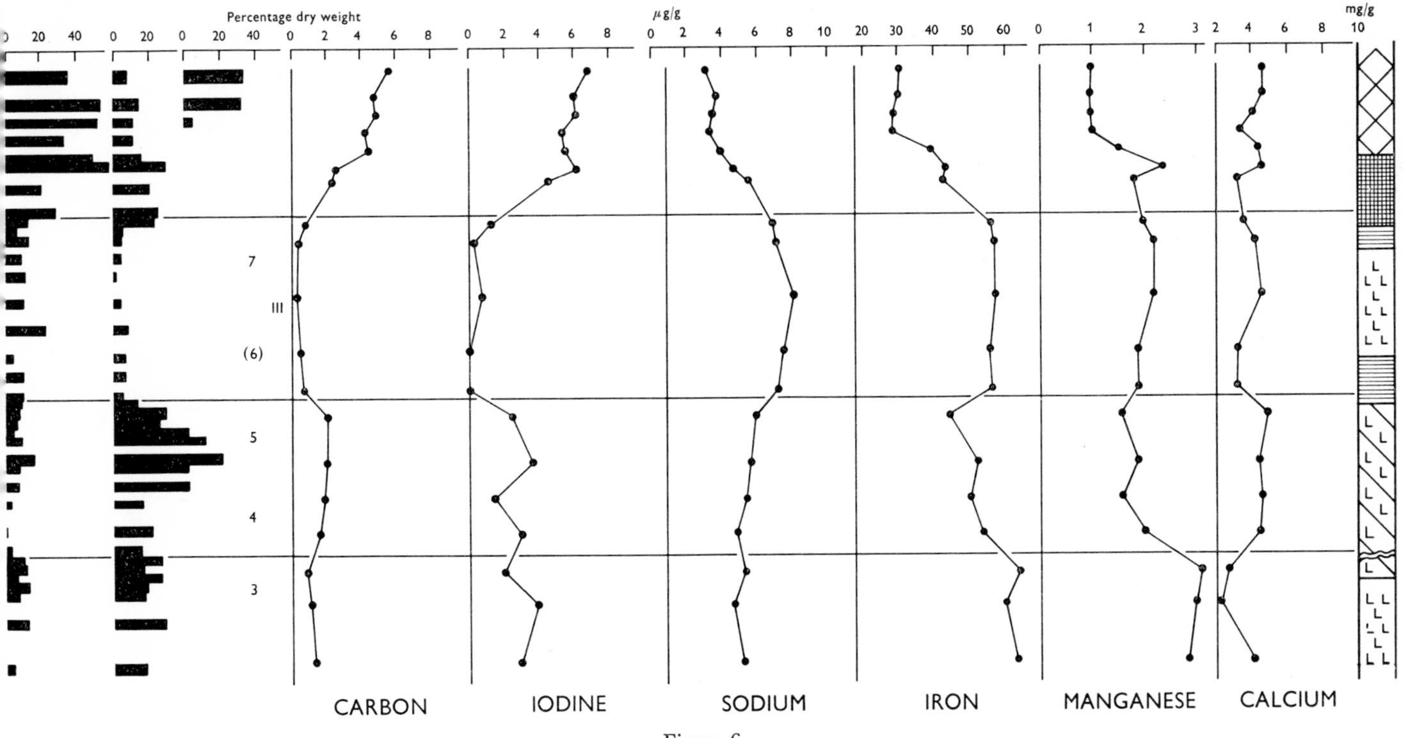


Figure 6.

The chemical analysis from Blea Tarn shows, however, that the input of material into the tarn soon included mineral matter so high in calcium and sodium that it must be regarded as unweathered drift from deeper layers in the soils than those which had been leached during Alleröd times. This is evidence that during the Post-Alleröd period, there was more severe periglacial erosion in the Blea Tarn basin than had occurred since the Full-glacial; that is, since long before any vegetation had been present locally.

In surveying the records of vegetation of that large part of north-west England which was not directly affected by ice during the five centuries between 8800 and 8300 B.C., the striking feature in Cumberland and the Lake District is the universal dominance of *Artemisia* in the pollen spectra from all types of basin and at all altitudes. Such a universal dominance cannot be the result of differential pollen preservation (cf. Maher, 1963) but must record a real change in the composition of the vegetation. Walker (1966) interprets the lowland evidence as showing that all soils were sufficiently disturbed by frost to break up the vegetation cover and expose bare soil to colonization by *Artemisia*. At sites behind the fresh moraine attributed to corrie glaciers of this period, no Late-glacial deposits are present, but at some (e.g. Langdale Combe, Walker (1965)) the basal sediment contains *Artemisia* pollen, showing that there the Post-Alleröd ice melted during the *Artemisia* phase; at other sites (e.g. Small Water (Pennington & Lee, unpublished), the earliest polleniferous deposits date from the beginning of the Post-glacial, indicating longer survival of the local ice (cf. figures 7 and 8).

Walker's division of Zone III into two Cumbrian Zones, C7 and C8, seems to agree with a possible division of the Lake District diagrams into an earlier part with much *Empetrum*, and a later part with an absolute minimum of tree pollen (with consequently increased *Pinus* percentages) and maximum occurrence of *Artemisia*. The vegetation contrast between the Alleröd and Post-Alleröd periods—Zones II and III—which is so marked in the Cumberland lowland and mid-Westmorland at Kentmere and Skelsmergh Tarn, as well as in the glaciated mountains of the Lake District, is less obvious in the sheltered lowlands round Morecambe Bay, at Helton Tarn and Witherslack (Smith, 1958) and at Lonsdale Haweswater (Oldfield, 1960). There is also much less evidence for severe frost disturbance of the soils at the sites round Morecambe Bay. It seemed possible to Oldfield that at sheltered sites like Lonsdale Haweswater, tree and shrub communities were able to persist throughout the Post-Alleröd cold period.

Manley (1959) discusses the fall in temperature required to produce permanent snowfields in equilibrium with the Lake District corrie glaciers. He suggests a June–September mean of about 6·3 °C at Windermere, a January mean there of −7·5 °C, and a mean annual temperature just below freezing point. Walker (1966) considers that this is in reasonably close agreement with the evidence from plant distribution, particularly the presence of the Arctic plants *Salix herbacea* and *Koenigia islandica* near sea level at Oulton and at Scaleby Moss.

## THE EARLY POST-GLACIAL PERIOD, UNTIL 3000 B.C. (ZONES IV TO VIIA)

The juniper maximum, interpreted as a response to rising temperatures and forming a transition stage in the development to closed forest, falls between 8300 B.C. and 7800 B.C. at Scaleby Moss (Godwin, Walker & Willis, 1957). After the development of birch forest which, proceeding at rather different rates at the various sites, succeeded the juniper phase, closed forest became established over the whole region up to an altitude of at least 2500 ft (760 m). This deduction is based on the assumption that a non-arboreal pollen component of less than 10 % of the total tree pollen indicates continuous forest (Faegri & Iversen, 1964), but it seems likely that some herbaceous plants persisted even though their pollen ceased to appear in the pollen sum (Davis & Deevey, 1964). Shade-intolerant plants would, of course, be restricted to refuges on mountain-tops, cliffs and the coast. The pollen diagram from Red Tarn, Langdale (Pennington, 1964) shows that at this altitude, 1700 ft (518 m), members of the Late-glacial herbaceous flora, including *Helianthemum*, persisted into Zone VIa, which is interpreted as evidence that solifluction remained sufficiently active at this altitude to keep the forest open for another thousand years. After this, however, herbaceous pollen disappeared at all sites up to 2500 ft (760 m), except for less than 5 % of grass pollen, until the end of Zone VIIa.

In the early Post-glacial Zones IV, V and VI, the changes in forest composition, in response first to the increasing warmth and immigration of the warmth-demanding trees and then to increasing moisture, show sufficient agreement with other parts of England to suggest that the spread of *Corylus*, *Ulmus* and *Quercus* took place rapidly after the immigration of these trees into eastern England. The special interest of the north-west in this period seems to lie in the behaviour of *Pinus*, for the *Pinus* curve is the feature of the pollen diagrams which differs most from site to site within the region. Comparison of diagrams shows so much variation in the proportion of tree pollen contributed by *Pinus* that little confidence can be placed in the concept of a regional pollen rain. For this reason, no attempt has been made to correlate Walker's zone subdivisions (the Cumbrian Zones) with north-west England outside lowland Cumberland. The only possible conclusion seems to be that whereas the spread of *Betula*, *Corylus*, *Ulmus* and *Quercus* was climatically controlled, *Pinus* must have been edaphically controlled in north-west England all through the early Post-glacial period. The precise situations in which pine trees were growing are still not clear, and the situation is confused by the over-representation of *Pinus* pollen in lake deposits in water of more than *c.* 20 m depth. Two well-established facts emerge. First, the comparative abundance of *Pinus* at some sites (figures 8 and 14) negates any idea that the low values for pine at such sites as Scaleby are due to any limiting climatic factor. Secondly, where pine is present in quantity, its maximum comes in the later part of Zone VI (VIc of the British zonation), much later than the pine stage in south and east England and well after the expansion of *Ulmus* and *Quercus* (Oldfield, 1965). The tree for which north-west England does appear to have been at the climatic limit is *Tilia*, which does not appear anywhere in the region until the end of Zone VI, and then shows a clear limitation to the limestone areas round Morecambe Bay where it still grows.

During the hazel maximum of Zone VI, *Corylus* pollen reaches the generally rather high percentages of total tree pollen which are characteristic of western Britain. The variation in actual percentage reached, from site to site, including that between marginal and central sites in the same lake (Devoke Water, Pennington, 1964), suggests that some aspect of the transport of *Corylus* pollen, on the lines suggested by Tauber's (1967) recent experimental results, may be involved, rather than a real variation in the number of hazel bushes. Further investigation on these lines, when more data are available, may answer the question as to whether *Corylus* is most likely to have been growing as an undershrub, with most of its pollen transported through the trunk space, or as a forest dominant. The persistence of *Betula* as a dominant tree in the pollen rain through most of the Boreal, in contrast to its behaviour in south and east England, is another rather general feature of the Highland zone of Britain.

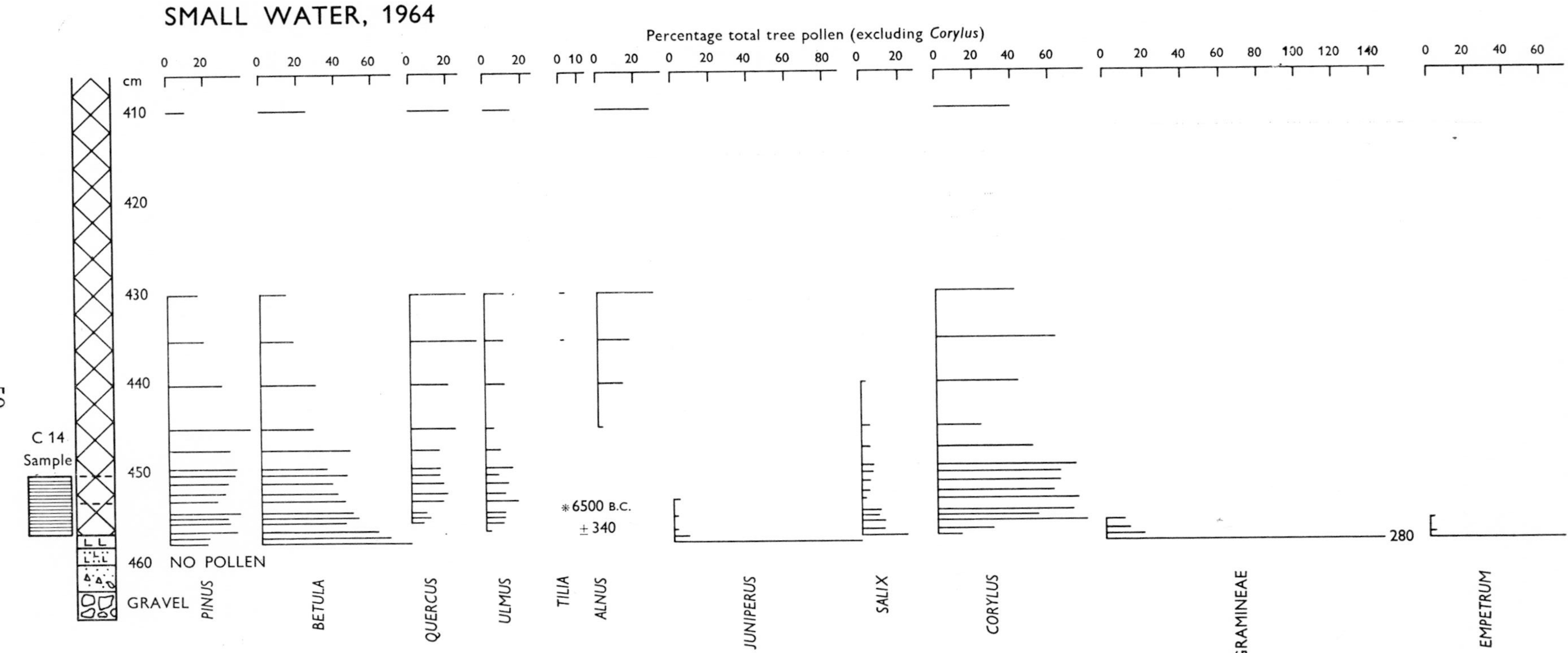


Figure 7. The basal deposits from the high corrie Small Water (1460 ft) where the only Late-glacial deposits are the non-polleniferous gravel and silty clay of the post-Alleröd ice. The pollen diagram shows selected taxa in relation to the radiocarbon date for the basal 6 cm of organic mud, 6500 B.C. ± 350 (U.S. Geol. Survey, W 1460). At this site, inorganic clay was accumulating during the juniper phase at the Zone III/IV boundary.

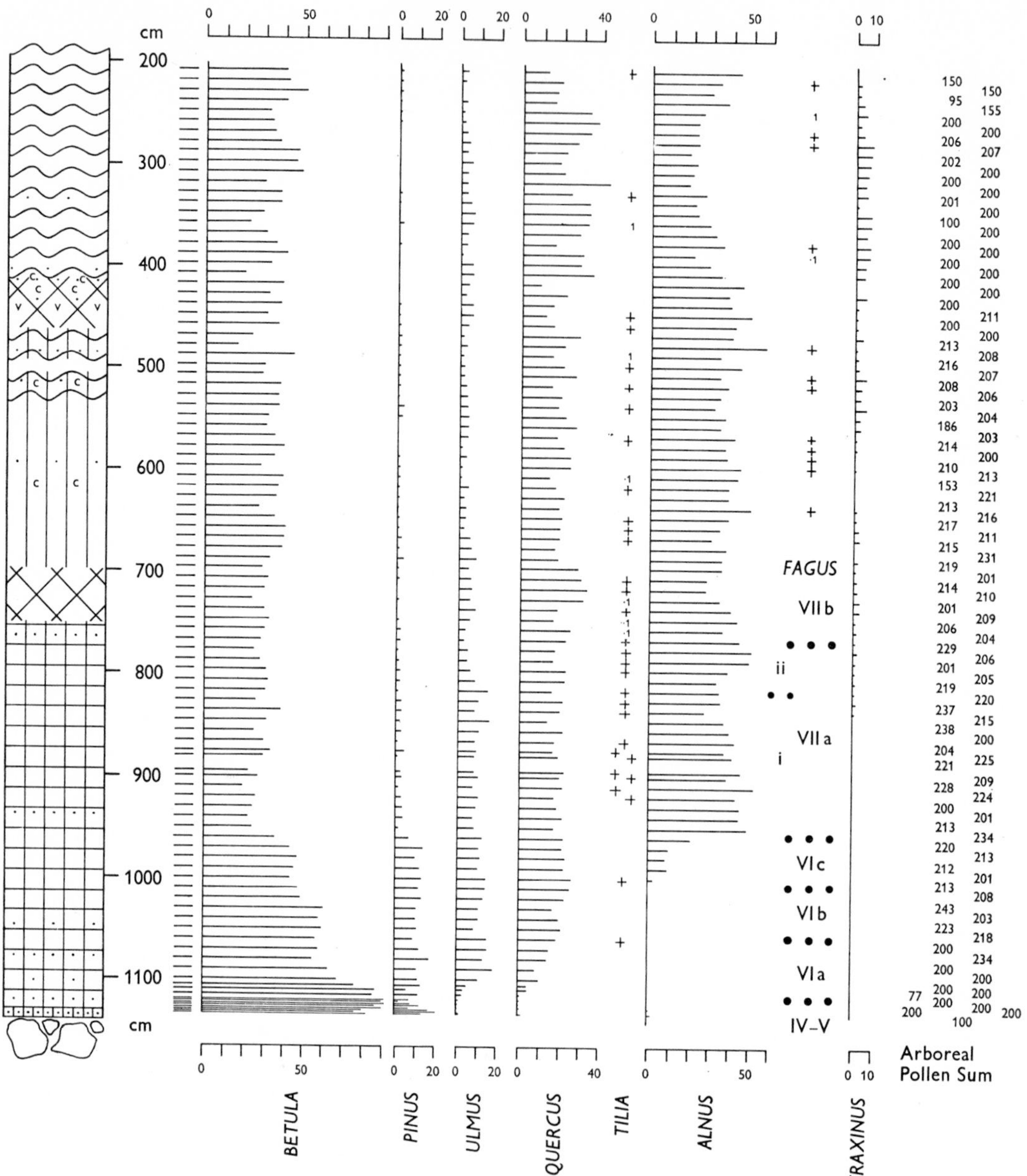


Figure 8. Pollen diagram from lake deposits of Langdale Combe (reproduced from Walker (1965), figure 3). Except where otherwise shown, all pollen frequencies are expressed as percentages of total arboreal pollen excluding *Corylus*. This site lies in a hollow, formerly a tarn, in the surface of kettle-moraine attributed to the post-Alleröd glaciation; organic deposition began here, as shown by the *Artemisia* phase followed by a juniper phase, before the end of Zone III, i.e. much earlier than at Small Water (figure 7).

Iversen's (1960) interpretation of the role of *Corylus* in preventing regeneration of *Betula* would imply that, in many parts of western Britain, edaphic or altitudinal limitation of *Corylus* allowed continued regeneration of the birch forests, yet in western Britain *Corylus* values at many sites are particularly high.

The boundaries between Zones IV, V and VI, as defined at Scaleby, are customarily interpreted as stages in the increasing warmth of the early Post-glacial period, during which competitive interaction between different tree species produced synchronous changes in forest composition. In other parts of the country, during Zone VIc, there are indications of changing precipitation in the lowered water levels recorded in the stratigraphy at such sites as Hockham Mere (Godwin, 1956). Walker (1966) found no evidence in lowland Cumberland for a dry phase in the

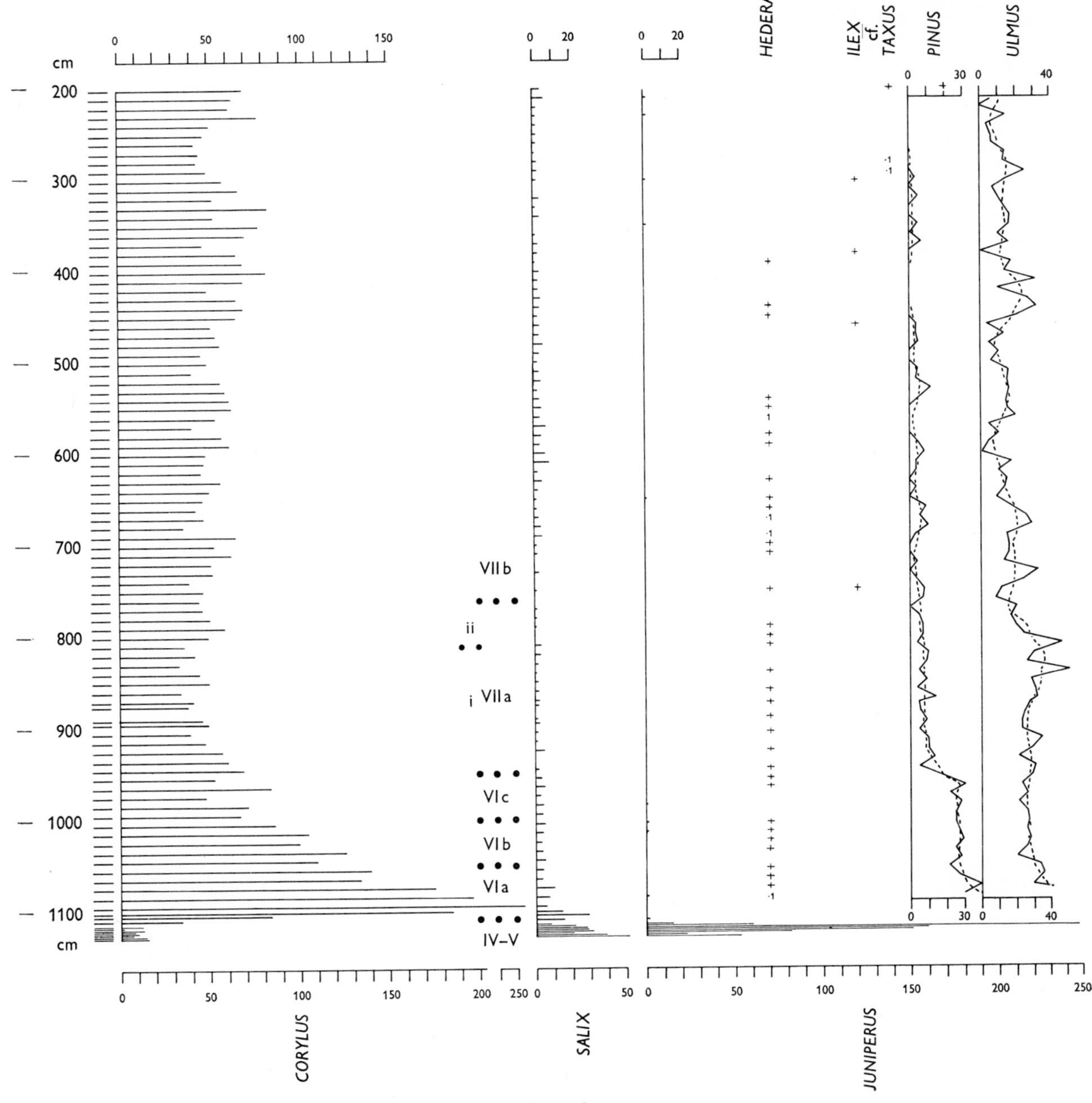


Figure 8.

Boreal, and there is no consistent evidence for a synchronous Late-boreal dry phase in Lowland Lonsdale.

Minerogenic layers possibly attributable to lowered water levels occur within Zone VIc at some Lake District sites, but not at all. Figure 9 shows evidence from this horizon at four sites, three lakes and one topogenous valley mire where changes in the mire surface are recorded. The iodine content of the lake sediments is shown, and tentatively suggested as an index of precipitation (Mackereth, 1966*b*). At all three lake sites there is a consistent recession of the iodine curve in Zone VIc (defined by the *Alnus* curve), and all the evidence is consistent in suggesting the following succession. First, in Zone VIb, there was a wet period during which high iodine values agree with a wet mire surface carrying willow-birch carr and *Sphagnum*. Secondly in Zone VIc, a drier phase of the mire surface with no willow agrees with the recession of the iodine curve, and the consistent increase in *Pinus* and *Calluna* pollen suggests a spread of these plants on dry mires and possibly on to soils partially podsolized during the preceding wet period. Thirdly, on the evidence of the iodine curve, precipi-

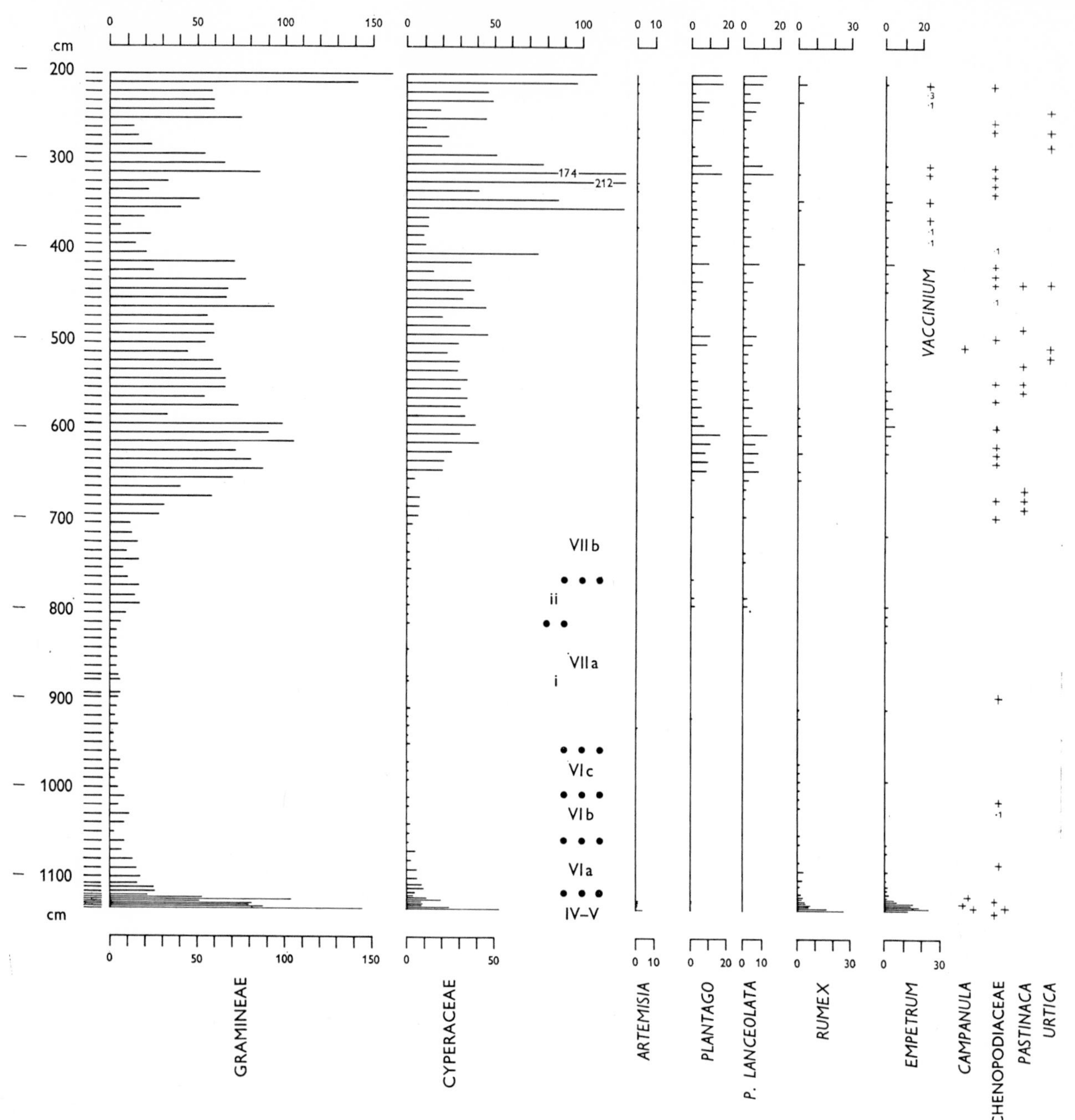


Figure 8.

tation increased once more, and flooded mire surfaces would provide habitats for alder carr, which at such sites replaced the willow-birch carr of Zone VI b; the alder expansion defines the beginning of Zone VII a.

At these upland sites, the poorer soils and incipient podsols occupied by pine and not flushed by ground water at this time presumably passed by retrogressive stages into blanket peat at some later date (Iversen, 1964, and figure 15 which shows the succession at Burnmoor Tarn). In Lowland Lonsdale, Oldfield (1965) has supposed that the pine forest of Zone VI c was replaced by deciduous woodland during the Atlantic period, Zone VII a.

During the Atlantic period in the Lake District, the alder must have predominated in valley mires, which would include the flat floors of the glacial troughs, at all altitudes up to at least 1200 ft (370 m), on the evidence of alder wood in valley peats. In lowland Cumberland, Walker's interpretation of the

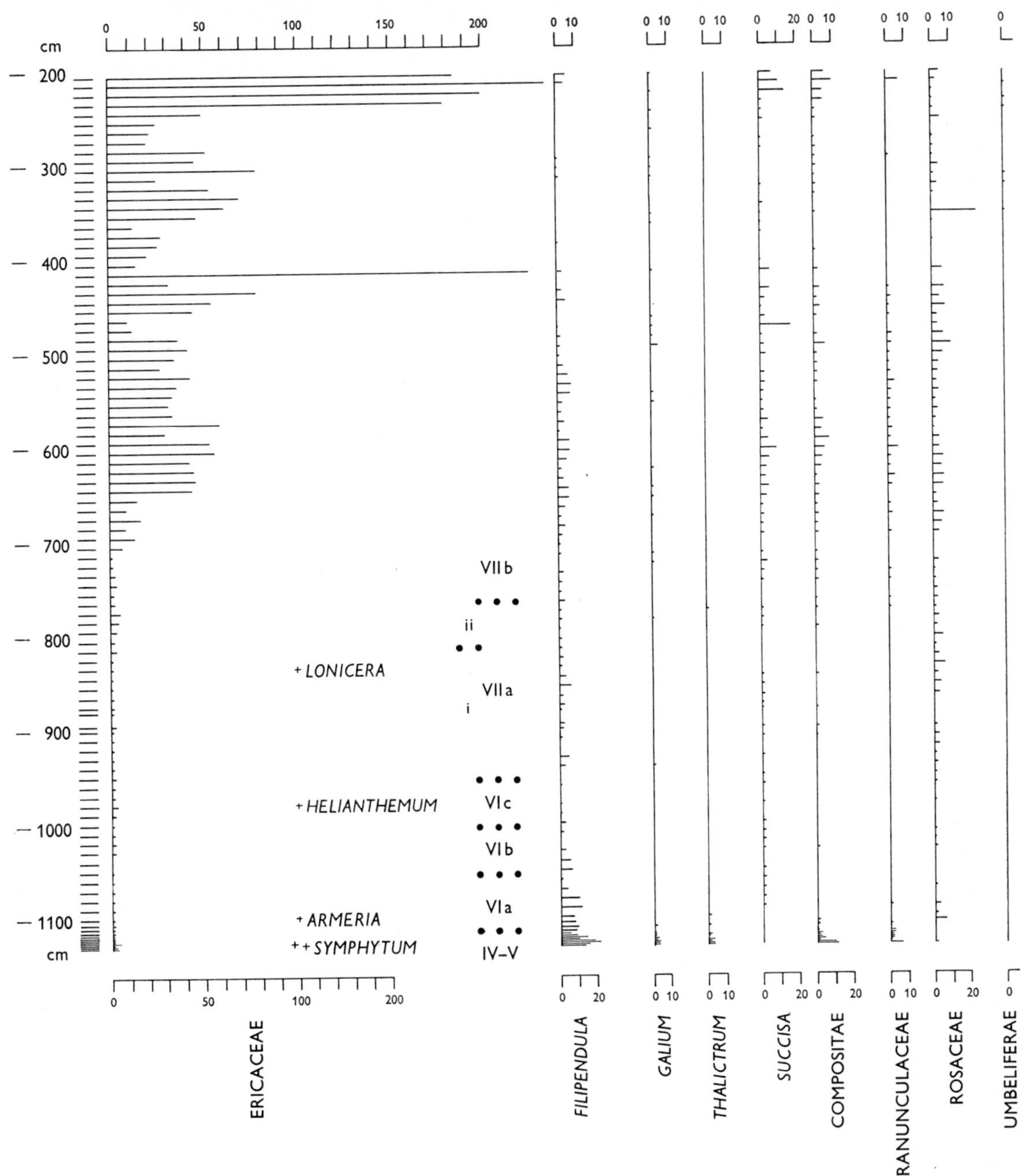


Figure 8.

change in forest composition at the Boreal–Atlantic transition is that alder must have replaced birch in hollows and valleys which became wetter as the water-tables rose. There must have been considerable general extension of alder as a streamside tree and as hillside alder woods on flushed soils.

Throughout the Boreal and Atlantic periods in north-west England, *Quercus* and *Ulmus* are assumed to have occupied most of the well-drained lowland soils, and those deeper soils of the uplands which, on slopes, remained as free-draining brown earths. *Ulmus glabra*, the only possible source of the elm pollen, is more base-demanding than either species of *Quercus*; *Ulmus* contributed a higher percentage of the pollen rain at the higher sites in the Lake District than in either the valley deposits or in lowland Cumberland, suggesting that steep and eroding slopes maintained a higher base-status in soils than the more subdued topography of the drift-covered lowlands. Nevertheless, Mackereth (1966*a*) has

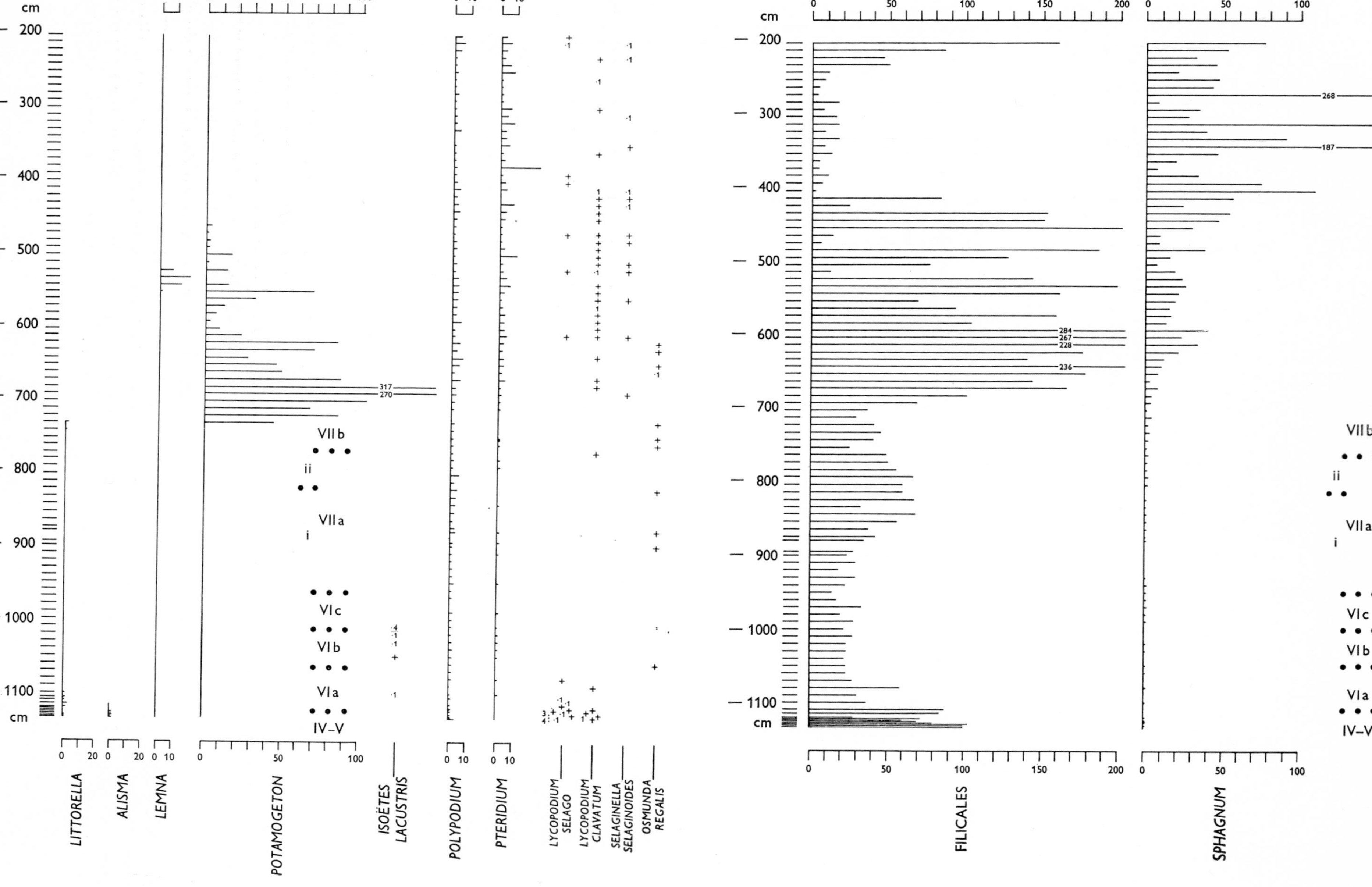
LITTORELLA
ALISMA
LEMNA
POTAMOGETON
ISOËTES LACUSTRIS
POLYPODIUM
PTERIDIUM
LYCOPODIUM SELAGO
LYCOPODIUM CLAVATUM
SELAGINELLA SELAGINOIDES
OSMUNDA REGALIS
FILICALES
SPHAGNUM
cm
VII b
ii
VII a
i
VI c
VI b
VI a
IV–V

Figure 8.

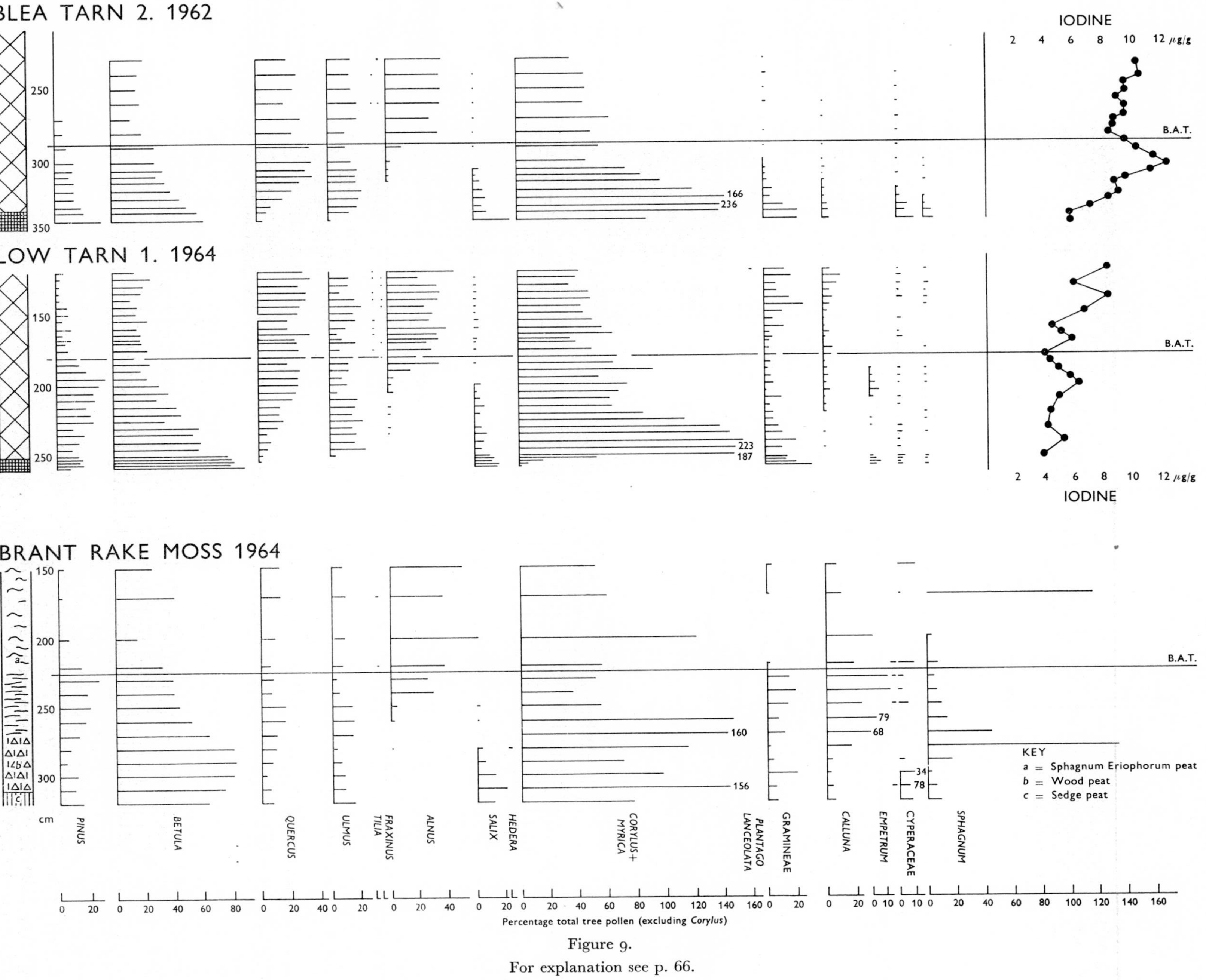


Figure 9.
For explanation see p. 66.

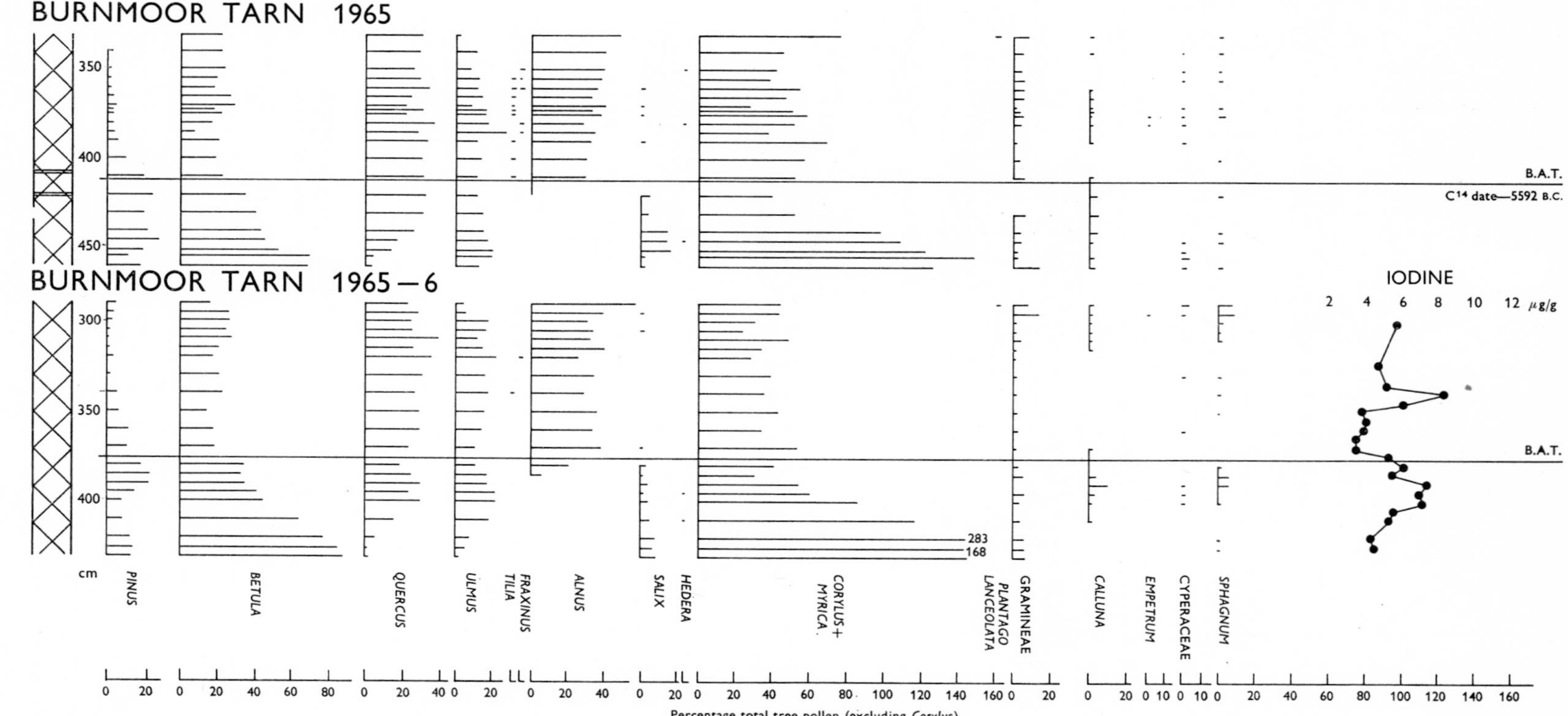


Figure 9. The Boreal–Atlantic transition in the south-west Lake District. Pollen diagrams at four sites compared; three are lake sediments with the iodine curve as a tentative index to precipitation rate, and the fourth, Brant Rake Moss, is a topogenous mire. The pine phase in Zone VIc corresponds with a recession in the iodine curve and with humified peat (carrying a *Pinus–Calluna* phase) in the mire profile. High iodine values before and after the recession correspond with *Sphagnum* peat in the mire profile. ($^{14}$C date, Zone VI c, 5610 B.C. ± 160 (Y–2361) at Burnmoor Tarn.)

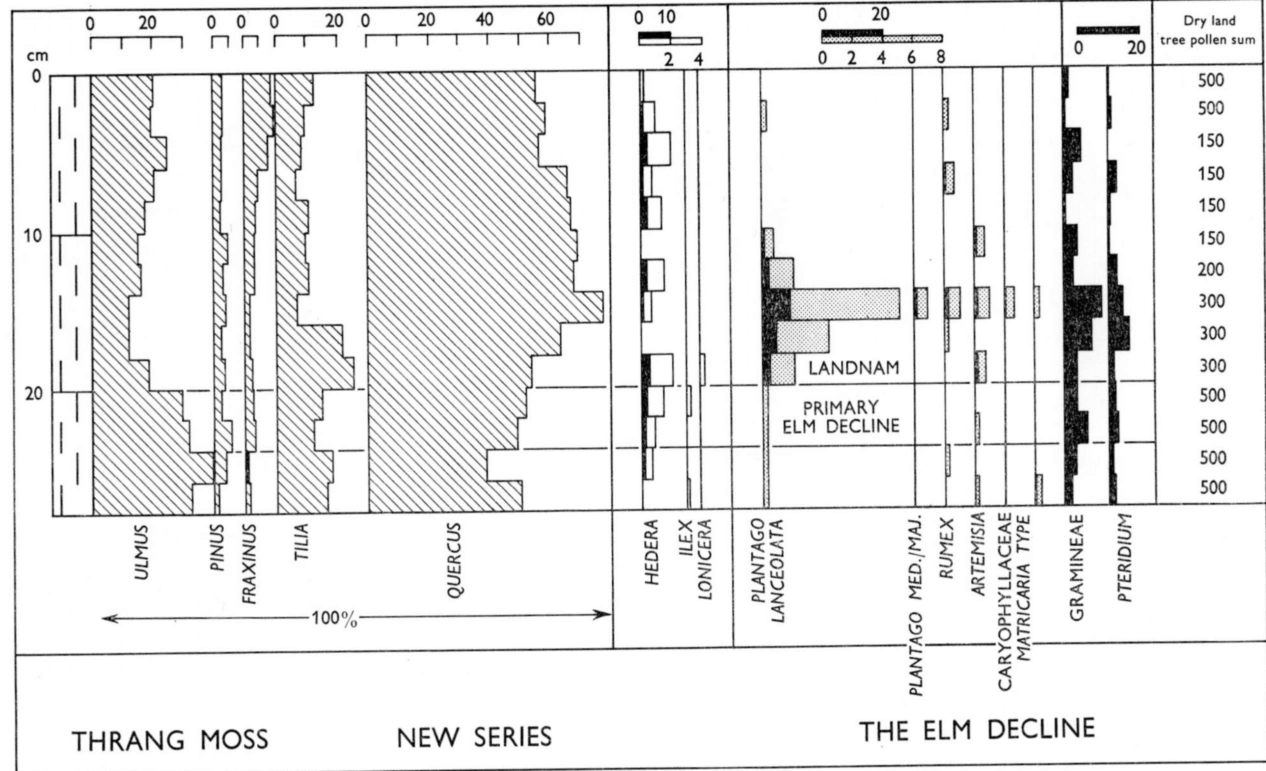


Figure 10. Thrang Moss. New pollen analyses from the Elm Decline. Reproduced from Oldfield (1963), figure 4.

shown on chemical evidence that Zone VIIa was a period of rapid leaching, and that, by the end of this period, many of the soils of north-west England, except on the limestone, must have been approaching the limit for regeneration of elm. Mackereth correlates the high rate of leaching during the Atlantic period with a low erosion rate, and concludes that a stable and continuous forest cover must have occupied the drainage basins of all the Lake District lakes, so that erosion was minimal during this period.

*Fraxinus* pollen appears only rarely in pollen diagrams of this period. The continuity of the forest canopy must have restricted the development of this light-demanding tree, which, however, must have found some habitat in the area, probably in natural clearings resulting from the death of large trees, from which it was able to spread into the secondary forest after forest exploitation by early Neolithic man.

## THE LATER POST-GLACIAL PERIOD, SINCE 3000 B.C.

The first traces of human interference with primary vegetation appear on the West Cumberland coast, as they do in Ireland, some centuries before 3000 B.C. but this date represents the mean about which most of the radiocarbon dates for the zone boundary VIIa, b, the Elm Decline, cluster. Smith (1958) first found that pollen of *Plantago lanceolata* occurred at this horizon in his South Westmorland sites. Later, Oldfield (1963) reviewed the characteristics of this horizon in all the Morecambe Bay bogs, and figure 10 shows how, at Thrang Moss, he was able to distinguish a Primary Elm Decline, affecting only the percentage of tree pollen contributed by the elm, from a later phase where *P. lanceolata* and an expansion of grass pollen indicated real clearance of the forest, an episode which he compares with the Danish Neolithic Landnam (Iversen, 1941). Walker (1966) found evidence in north and west Cumberland for a comparable succession: a first stage, in which the pollen of elm was reduced, with a very small increase in grasses and rare plants of *P. lanceolata*, and at some sites a reduction in oak pollen, followed by a second stage in which the pollen contribution of trees and shrubs of dry land fell markedly, and the increase in pollen of grasses, *P. lanceolata* and *Rumex* was sufficiently great to suggest clearings of significant size in the forests. These horizons also appear in the sediments of Barfield Tarn in south-west Cumberland, and figure 11 shows these two successive episodes in relation to the radiocarbon date of

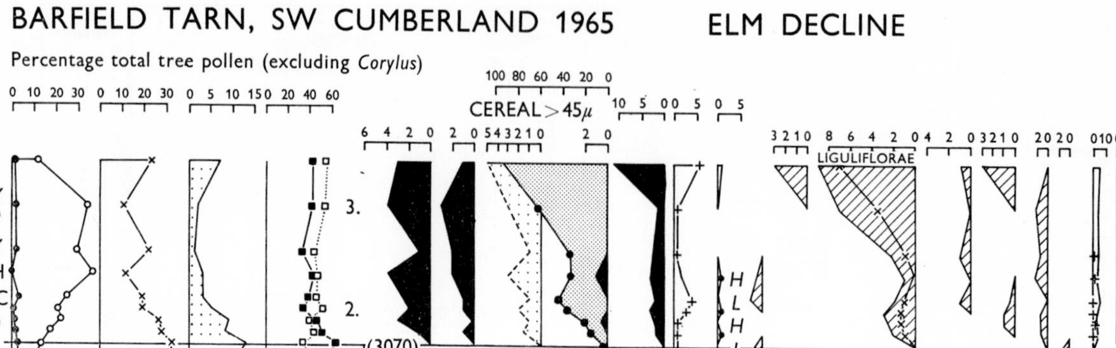


Figure 11. Pollen diagram from Barfield Tarn, south-west Cumberland, at the Zone boundary VII a/b, showing the stratigraphic change and the position of the radiocarbon-dated sample. The approximate dates for the upper and lower boundaries of the dated sample were calculated by assuming a constant rate of sedimentation between the B.A.T. and the middle of the sample. For a description of episodes 1, 2 and 3, see text p. 68–9.

3390 B.C. ± 120 (Copenhagen K 1057). The probable age limits of the section of the core used to obtain the date are shown; these were calculated from the sedimentation rate during the period of deposition of sediment of constant composition since the Boreal–Atlantic transition. This date entirely confirms Walker's supposition that the activity giving rise to the first phase began, on the Cumberland coastal plain, during the fourth millennium B.C.

At all the West Cumberland sites, and at many round Morecambe Bay, the initial episode (1) is followed by the steepest fall in elm pollen (2), which is accompanied by evidence for significant forest clearance at most of the lowland sites, but not at all the upland sites (Pennington, 1965). This suggests an initial settlement on the coasts by seaborne immigrants, followed by a much more widespread, but still not uniform, settlement, resulting in larger clearings and a diminution in the regional pollen rain of elm. Parts of the region, notably in the western uplands, experienced only a change in forest composition, for there is no evidence there of significant clearances. It is not possible on the vegetation evidence to decide whether these two episodes resulted from significantly different farming practices, but both imply a wide-ranging nomadic economy, exploiting elms, and in places oaks, over considerable distances.

At Barfield Tarn the second episode was associated with the presence of pollen grains which I have tentatively identified as of cereal type, on the criterion of a diameter greater than 45μ in fresh glycerine jelly preparations (further work on cereals in north-west England using silicone mounts and phase-contrast microscopy is needed). The presence with these grains of pollen of taxa which include many weeds—composites, Caryophyllaceae and Chenopodiaceae—lends support to the suggestion of cultivation, but most striking is the lithological change accompanying the second episode at Barfield Tarn, from an organic

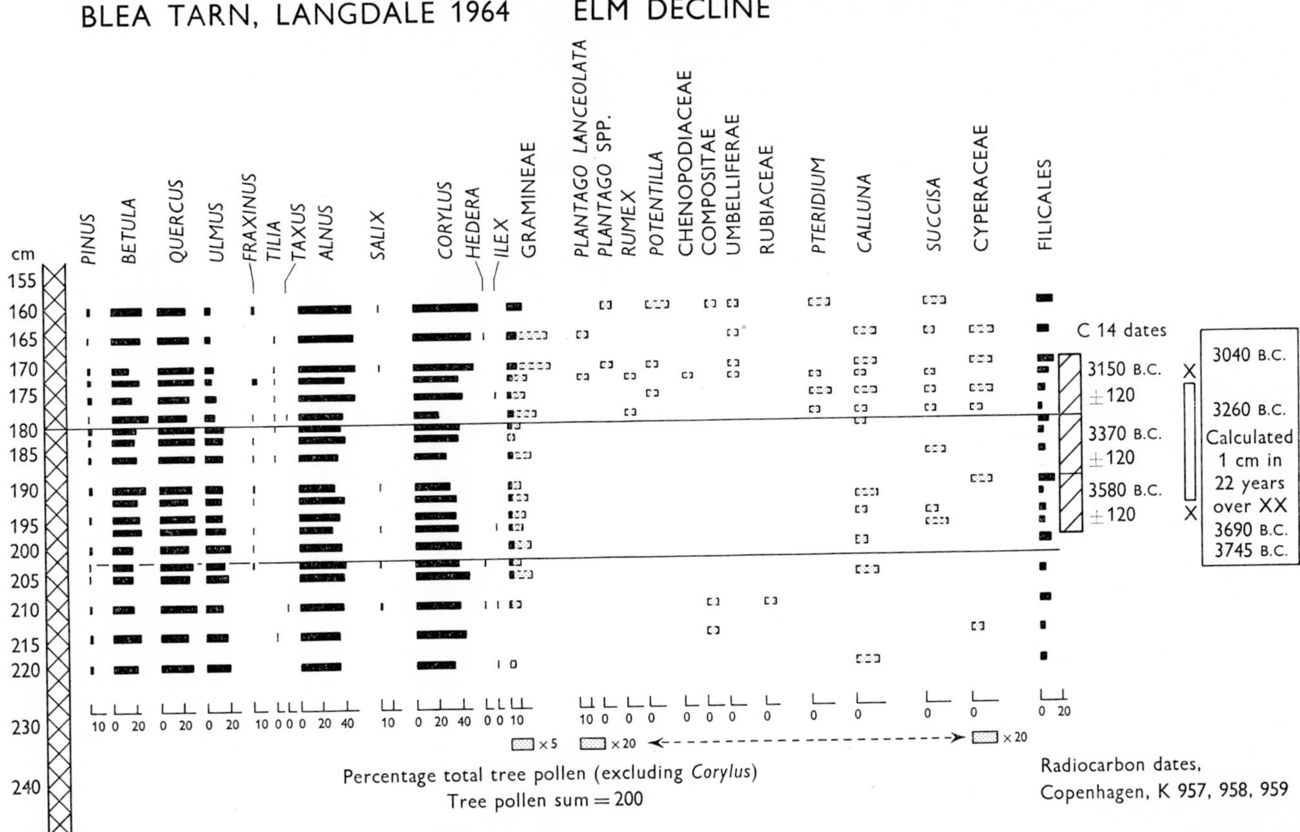


Figure 12. Pollen diagram from Blea Tarn (1962) at the Zone boundary VII a/b, showing the radiocarbon dates transferred from the duplicate core (1964). From the three dates (K957, 958 and 959), the duration of the clearance episode has been calculated as lying between the alternatives of 1 cm in 22 years, which represents the constant accumulation rate over the section covered by the three radiocarbon dates, and 1 cm in 29·5 years, which represents the overall rate of sedimentation between the dated horizon and the surface. On either scale, the clearance episode corresponds in date with the period supposed on archaeological grounds to have been that of greatest activity of the Great Langdale axe-factory.

lake mud to a pink clay which appears to be re-deposited boulder-clay from the rather steep slopes which surround this kettle-hole tarn. The local settlement just before 3000 B.C. must have been sufficiently dense to break up the natural vegetation cover and allow severe soil erosion, so cultivation seems very probable. The steep fall in elm and oak pollen at this episode (figure 11) would therefore indicate clearance for cultivation. At other sites it would seem that the clearances were made for stock-keeping purposes, and Walker suggests that the evidence at his sites is more in favour of the trampling effects of cattle than of a continuous grassy sward grazed by sheep. At no site in north-west England is there any evidence for the use of fire in these centuries around 3000 B.C.

A somewhat later period of very vigorous clearance and cultivation at Barfield Tarn (figure 11, Episode 3) which led to permanent and almost complete deforestation, can be tentatively correlated with a horizon in the Ehenside Tarn profile at which Walker (1966) found the lowest cereal pollen, and to which (on the basis of the Scaleby Moss radiocarbon dates) he assigned a date of about 2200 B.C. This horizon marks the beginning of vigorous and permanent forest clearance in the neighbourhood of Ehenside Tarn. This site, like Barfield Tarn a kettlehole on the West Cumberland coastal plain, was drained in 1869 and archaeological evidence for Neolithic settlement was revealed (Darbyshire, 1874; Piggott, 1954). Walker's work on the vegetation succession at Ehenside is summarized in figure 14, reproduced from Walker (1966); he interprets the evidence as indicating a continuous local occupation of the same type until after 2000 B.C., when a change towards a period of more active clearance and intensive cultivation took place. The four radiocarbon dates which have been obtained, all from wood, from archaeological material at the Ehenside site, span the period from 3010 B.C. ± 300 to 1580 B.C. ± 150. Walker asso-

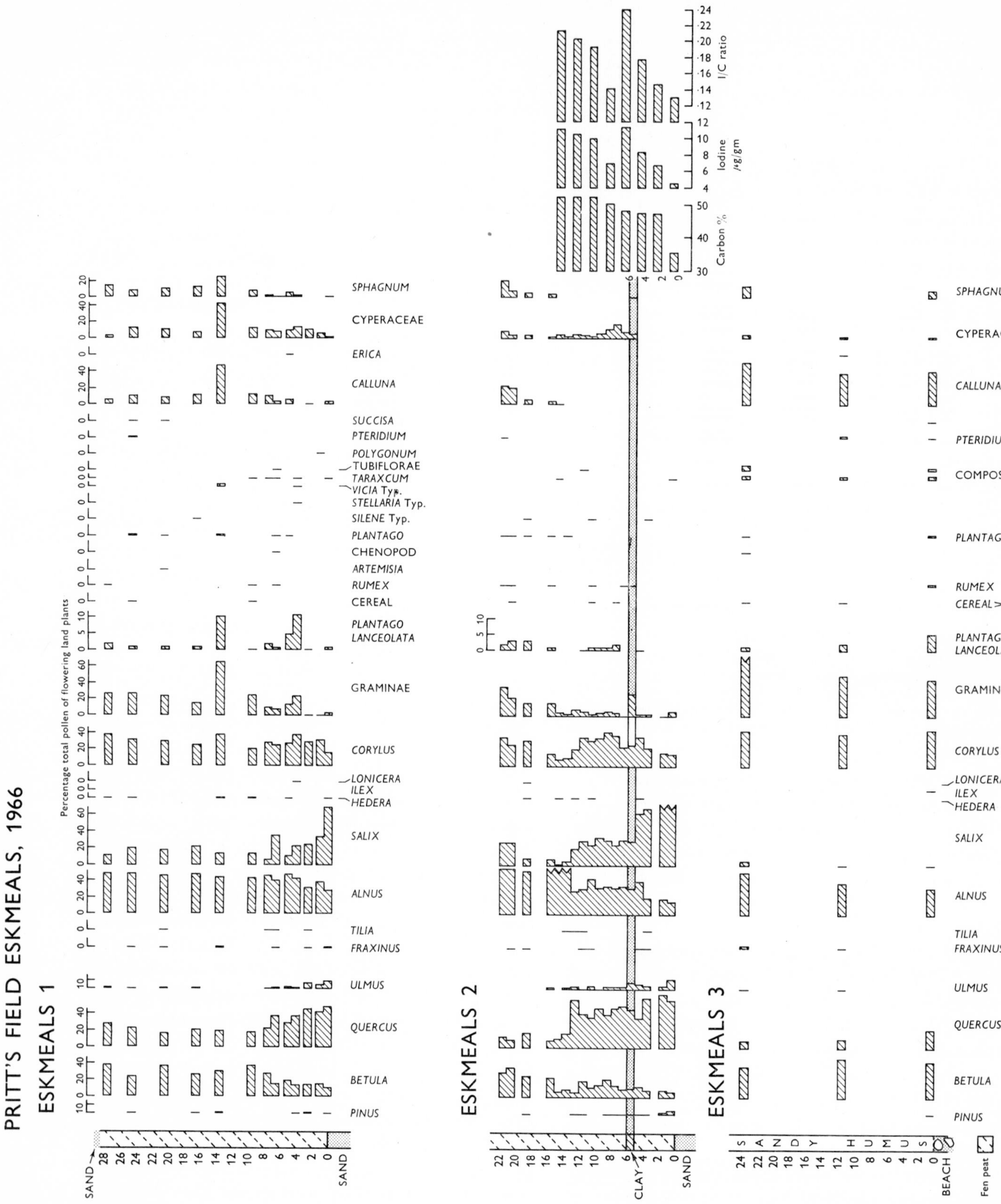

# PRITT'S FIELD ESKMEALS, 1966

## ESKMEALS 1

## ESKMEALS 2

## ESKMEALS 3

Figure 13. (See opposite.)

ciates the whole of the long vegetation episode with an economy associated with the use of Cumbrian stone axes, manufactured at the axe-factories in the Lake District. Such axes, both polished and unpolished, were found at Ehenside (Fell, 1964). Other more recently discovered sites on the Cumberland coastal plain have yielded a wide range of flint artefacts and a few sherds as surface finds; many of these are likely to date from within the long span of Neolithic occupation of the area.

Recent archaeological opinion is now in favour of a date as early as 3000 B.C. for the manufacture of axes from igneous rock (Evens *et al.* 1962), so it is possible that the exploitation of the tuff outcrop at the head of Great Langdale began soon after this date. Figure 12, a detailed pollen analysis of the appropriate level at Blea Tarn near the main Langdale axe factory sites, shows the radiocarbon dates for these two episodes: the limiting dates have been calculated from the accumulation rate indicated by three successive radiocarbon samples. Episode 1, the fall in elm pollen attributed to the collection of leafy branches for fodder, is dated at 3300 B.C. to 3200 B.C. and Episode 2, involving sufficient expansion of grasses and *P. lanceolata* to indicate forest clearance, covers the period from about 3000 B.C. to a little before 2000 B.C., with its peak of intensity at about 2500 B.C. Evens *et al.* (1962) suggested *c.* 2500 B.C. to 2000 B.C. as the probable dates of the main export phase of the axe factories. On the evidence of the pollen diagram, human activity in the Langdale area was maintaining clearings in the forest from *c.* 3000 B.C. until some centuries after 2500 B.C. Stratigraphic evidence at Blea Tarn and Angle Tarn (Pennington, 1969, in the press, *a*) indicates that increased soil erosion, consistent with the effects of forest clearance accompanied the beginning of Episode 2. Walker's pollen diagram from Langdale Combe (figure 8) is in agreement with the results from Blea and Angle Tarns, so the history of this period round the head of Great Langdale is now well established. The percentages of grass and herb pollen are higher at the higher altitudes, suggesting a less dense forest. Troels-Smith (pers. comm.) has suggested that in this type of country, the earliest Neolithic occupants may have moved along the lightly forested ridges, as Mesolithic people are known to have done in the Pennines, and have begun their attack on the forest from its upper margins. Figure 13, pollen diagrams from fen peat, stratified into the sands on a raised beach on the coast near Eskmeals, from which a roughed-out Great Langdale axe was turned up during ditch digging, proves the relationship between the occurrence of these axes and vegetation Episodes 1, 2 and 3 (cf. Barfield Tarn, figure 11) but it is not possible to relate the axe to any one episode.

In what appears to be the later stages of the Neolithic period, at the three sites round the head of Great Langdale above about 1300 ft (400 m), the very great and permanent expansion of grass and sedge pollen, accompanied by such grassland plants as *P. lanceolata* and *Rumex acetosella*, and by *Calluna*, indicates the main deforestation horizon, after which in this area grassland, and communities dominated by heather, must have exceeded forest at this altitude. Walker (1965) found wood charcoal stratified into the muds at Langdale Combe at this horizon (figure 8, at 618 cm), and at Angle Tarn this horizon coincides with a change to a more acid type of inwashed humus. The charcoal evidence suggests that here the forest above 1300 ft (400 m) was cleared by man, using fire, during the latter part of the Neolithic period. This seems to have accelerated soil changes due to leaching at this altitude and permanently altered the habitat from forest to open moorland. After this, changes in relative proportions of tree pollen at these high sites are probably not very significant, because the absolute numbers of grains of tree pollen per year would be very much reduced.

Over great areas of gently sloping upland country above 1700 ft (518 m) remains of former forest can be seen as branches and trunks, usually of birch, now buried beneath 1 to 2 m of blanket peat. Below the wood layer is preserved the acid mor humus of the forest soils, e.g. Red Tarn Moss (Pennington, 1965). Wood from the base of the peat at this site has now been dated to 1940 B.C. ± 90 years (N.P.L. 122). It is

---

Figure 13. Pollen diagrams from Pritt's Field, Eskmeals, where a rough-out Cumbrian axe was found while a ditch was being excavated. Sites *1*, *2* and *3* lie on a transect of the peaty hollow in the surface of the raised beach. Site *3*, the rough-out position, is on the margin of the hollow, and there the organic band (in which the axe was found) consists of sandy humus, in which the pollen content throughout indicates partly cleared ground. Site *2* is at the lowest point on this transect, and contains a clay band suggesting a flooding horizon at the level of Episode 2, the Elm Decline, and the suggestion of increased precipitation is borne out by the increase in iodine/carbon ratio in the organic band, which is here a fen peat. Site *1* is at a slightly higher level, where the organic band is also fen peat but shows no flooding horizon; at this site there is a record of three clearance episodes which corresponds closely with the record at Barfield Tarn (figure 11). The rough-out axe therefore corresponds in date to the period covered by Episodes 1, 2 and 3, but it is not possible to assign it to any one episode.

not yet certain how far this replacement of forest by bog, which was primarily the result of soil degeneration consequent on leaching (Pearsall, 1950; Iversen, 1964), was accelerated by anthropogenic factors or by climatic shifts.

From the horizon which he assumes to date from about 1700 B.C. (cf. figure 14) onwards, Walker's pollen diagrams from lowland Cumberland suggest little change in land use until a few centuries after the birth of Christ. There is no dated evidence of land use in Bronze Age times in the Morecambe Bay pollen diagrams. It seems likely that more detailed investigation, by closely spaced samples and radiocarbon dating, might reveal in north-west England records of 'small temporary clearances' of Bronze Age type, like those described by Turner (1965 and this volume) from northern and western Britain. There is a suggestion of such episodes in the pollen diagrams from Thirlmere (figure 15*b*) and Rydal Water.

The most striking vegetation change yet discovered in the area which can be attributed to Bronze Age times is the permanent change in the upland oak forest, which seems limited to those areas where burial cairns abound; some excavated cairns have yielded urns of Bronze Age type (Collingwood, 1933). This episode has been described from the uplands of south-west Cumberland between 700 ft (215 m) and 900 ft (277 m) (Pennington, 1964, Devoke Water and Seathwaite Tarn) and has since been found at Burnmoor Tarn (figure 15*a*) and Haweswater. At Seathwaite Tarn it has been dated to 1080 B.C. ± 140 (N.P.L. 124) and figure 15*a* shows its position in relation to the later dated horizon at Burnmoor Tarn; it appears to have been contemporaneous at these two sites. Wherever it occurs, this episode shows a decrease in oak forest (the former existence of which round Burnmoor Tarn is proved by remains of oak leaves in the lake sediment) and an expansion of grassland shown by parallel curves for the pollen of grasses and of grassland herbs. No cereal or weed pollen was found; this episode seems to record a pastoral land use in the uplands, in which much of the replacement of trees by grassland could have resulted from the grazing of many animals in the woodlands, with consequent failure of tree regeneration. A significant fall in organic content of the Devoke Water sediment is interpreted as evidence that increased erosion of mineral soils accompanied forest destruction. The decline in oak as a percentage of tree pollen is accompanied by a rise in *Betula*, but this is interpreted as a purely anthropogenic effect with no regional significance.

The Burnmoor Tarn diagram (figure 15*a*) includes one of the few vegetation changes in north-west England which could be attributed to the climatic deterioration at the opening of the Sub-atlantic period (recorded in humification changes in the Morecambe Bay raised bogs, Oldfield, 1963). Between the interposed dates of *c.* 1200 B.C. and *c.* 500 B.C., there is a fall in *Alnus* pollen, and after about 500 B.C. a rise in grass and Coryloid pollen, much of the latter being certainly *Myrica*. This is interpreted as replacement of upland alder woods by *Molinia–Myrica* swamps during the centuries of deteriorating climate at the opening of the Sub-atlantic. The main recurrence surface in the Lonsdale raised bogs, with which the vanished corduroy road was associated by Smith (1959) probably dates from this same period.

The next occupation level in the Lake District uplands, provisionally termed 'Brigantian' (Pennington, 1965) has now been dated to A.D. 390 ± 130 at Burnmoor, and A.D. 200 ± 130 to A.D. 580 ± 190 at Devoke Water (N.P.L. 116 to 120). The traces of this agricultural episode, the earliest at which cereal pollen has been found in the uplands, appear in the sediments of ten upland tarns round the western and northern margins of the Lake District, and it was clearly responsible for the permanent deforestation of these parts. At Devoke Water this horizon has been correlated pollen-analytically with a humus layer at the base of a field-bank forming part of a nearby farming settlement, already provisionally dated to the Romano–British centuries on archaeological grounds. Smith (1959) found strong evidence for clearance and cereal cultivation associated with a retardation layer in the Morecambe Bay mosses which was dated to A.D. 436 ± 100. These dates suggest a period of drier climate in late Romano–British times, which not only saw an extension of farming in the lowlands, but allowed the cultivation of cereals at a higher level in the Lake District uplands than before or since. On the Cumberland lowland, Walker's calculated time-scale for Ehenside Tarn indicated a phase of vigorous agriculture and renewed forest clearance at the end of the Roman period, resulting in the almost complete deforestation of that area by about A.D. 800. Shortly before this time, a new type of farming practice in the lowlands had been shown by the appearance of *Linum usitatissimum* and the *Cannabis–Humulus* pollen type at Ehenside.

In the areas of 'Brigantian' settlement in the Lake District, profound soil changes took place, and quantities of highly organic soil, rich in *Calluna* pollen

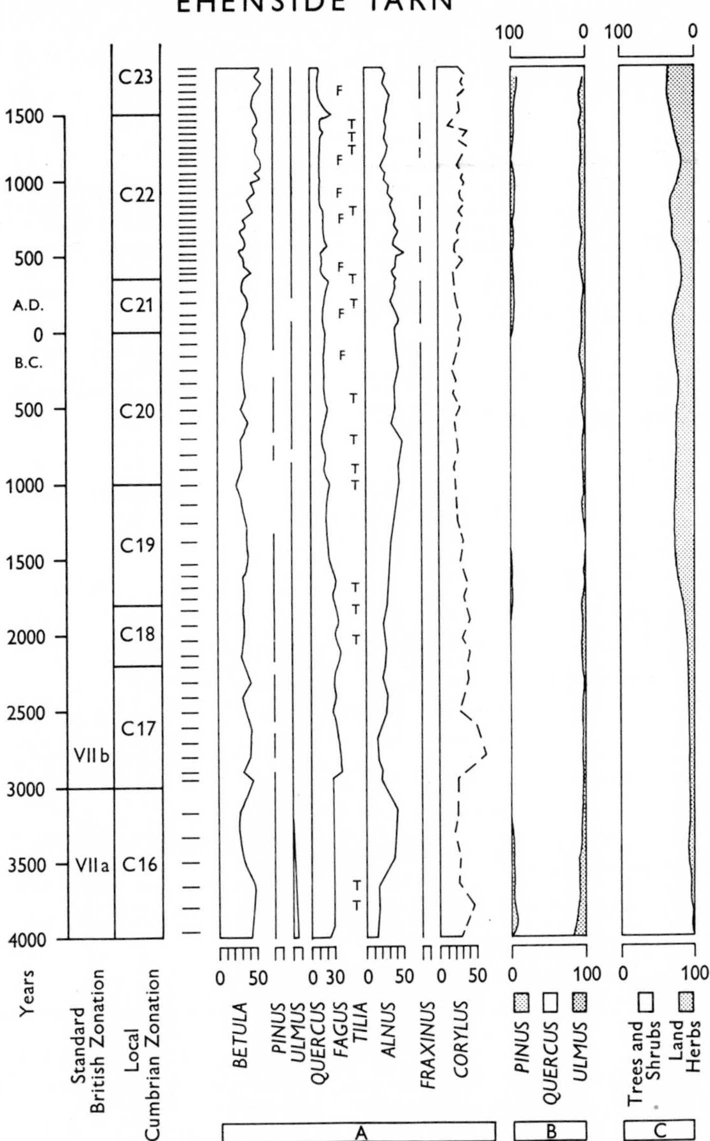


Figure 14. Ehenside Tarn. Summary pollen diagram. (Reproduced from Walker (1966), figure 44.) The time-scale is derived from radiocarbon dates from Scaleby Moss together with considerations of relative accumulation rates from time to time. Sample positions are spread through each zone as indicated by original diagram zonation. In section A, frequencies of all pollen types are shown as percentages of total arboreal pollen, omitting *Corylus*. In Section B, frequencies of *Pinus*, *Quercus* and *Ulmus* are shown as percentages of their combined totals. In Section C the frequencies of tree and shrub pollen are contrasted with those of dry-land herb pollen as percentages of their combined totals.

and composed of highly acid humus (Atherton *et al.* 1967) were washed into the tarns. The radiocarbon dates reinforce the interpretation of this as the result of a change in land use rather than the result of increased run-off in a period of wetter climate. The pollen diagrams suggest that pastoral Bronze Age land use led to deforestation and partial podsolization in the uplands, and then later settlement brought about severe erosion of the soils, truncating soil profiles and depositing a highly organic layer in the tarn sediments. After this episode, continuous leaching and paludification in this area of high rainfall must have transformed the deforested upland soils into their present acid and infertile state.

Later settlements and vegetation history around Morecambe Bay have been described by Oldfield & Statham (1963). Pearsall (1961) pointed out that the place-name evidence suggests that Anglian settlement in north-west England was confined to soils suitable for cultivation. In the Lake District and the upland country of Furness and South Westmorland, the preponderance of place-names of Scandinavian

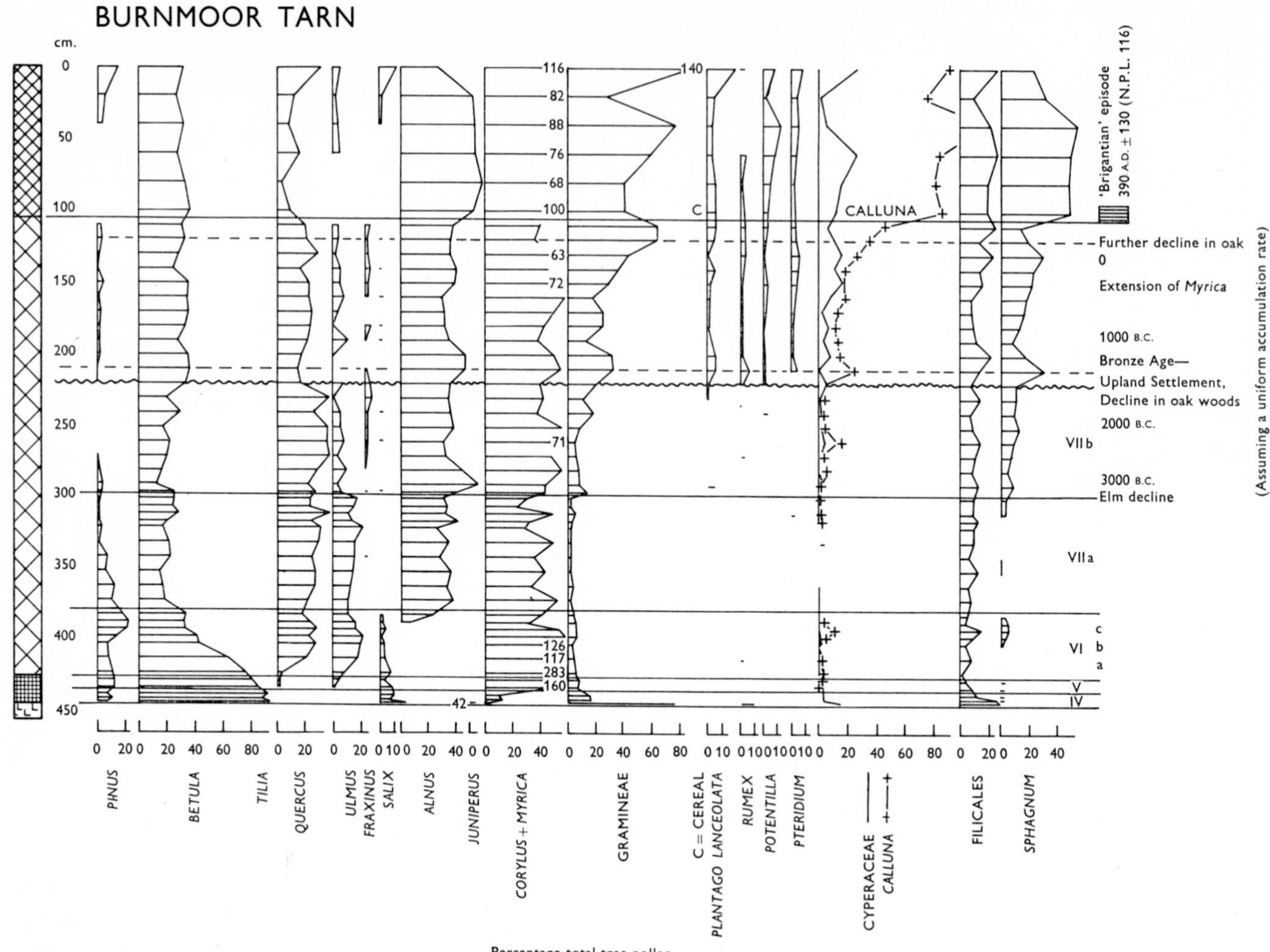


Figure 15*a*. Post-glacial vegetation history in the Lake District, illustrated by pollen diagrams from (*a*) Burnmoor Tarn in the south-west, and (*b*) Thirlmere in the centre.

The wavy line on each diagram indicates the 'revertence' of *Betula*, which was at one time thought to indicate the opening of the Sub-atlantic period. This horizon is now seen to be non-synchronous, by comparison of pollen diagrams, and is interpreted as an anthropogenic effect, resulting from destruction of upland oak woods by the clearance episode which precedes the horizon of sustained expansion of *Betula*. The time-scale on the Burnmoor diagram is calculated from interpolation between the Elm Decline and the radiocarbon date of A.D. 390 ± 130 (N.P.L. 116) for the 'Brigantian' episode.

origin, and the frequency of the suffix 'thwaite', meaning a clearing, indicates a dense settlement of this area in Viking times. Though there are few archaeological remains to substantiate this, it would seem that many of the existing farms in the central and southern Lake District originated in clearances made by Norse immigrants in surviving secondary forest which had never been cleared by Bronze Age or Romano-British activity. The main expansion of grasslands shown in pollen diagrams from the larger lakes and lowland tarns of the Lake District (Pennington, 1965, Blelham Tarn, and figure 15*b*, Thirlmere) is attributed to this episode, though as yet no radiocarbon dates are available. Oldfield (1963) recognized a corresponding episode in Lonsdale.

From the Norse colonization dates the widespread sheep-farming in the Lake District which finally ended any possible regeneration of forest in the uplands, except in isolated patches, as at Blea Tarn (Pennington, 1964) and the few surviving fragments of natural woodland like the ash-oakwoods of Naddle Forest, the high-level oakwoods of Keskadale, and the Martindale alderwoods. From historical sources it is known that Cistercian settlement in the twelfth century encouraged the spread of sheep-farming, and the present poverty of the flora of the Lake District, among mountain floras, is probably a result of continuous overgrazing by sheep during the last thousand years.

*Fagus* and *Carpinus* do not appear in pollen dia-

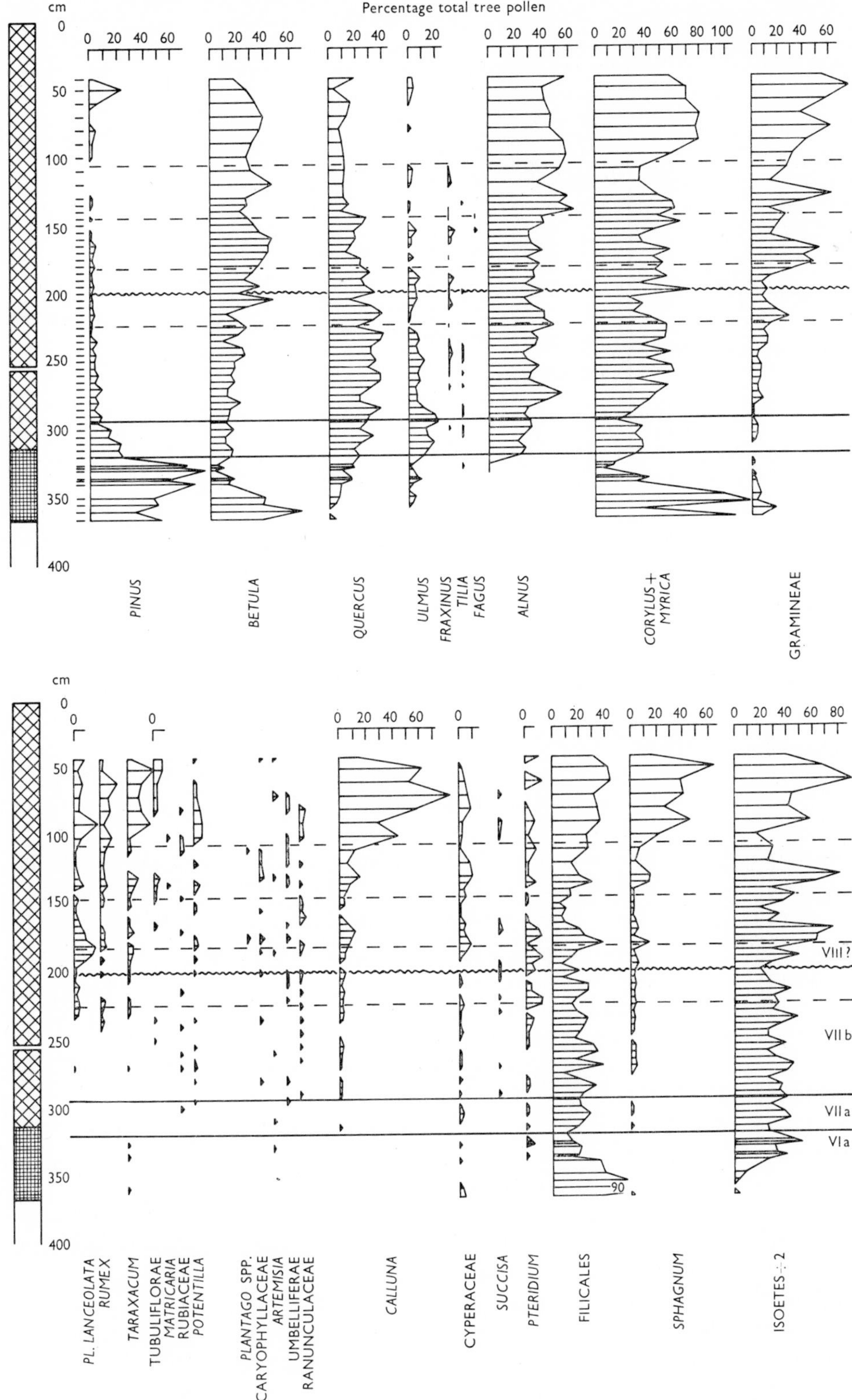
THIRLMERE—old north basin
cm
Percentage total tree pollen
PINUS
BETULA
QUERCUS
ULMUS
FRAXINUS
TILIA
FAGUS
ALNUS
CORYLUS + MYRICA
GRAMINEAE
PL. LANCEOLATA
RUMEX
TARAXACUM
TUBULIFLORAE
MATRICARIA
RUBIACEAE
POTENTILLA
PLANTAGO SPP.
CARYOPHYLLACEAE
ARTEMISIA
UMBELLIFERAE
RANUNCULACEAE
CALLUNA
CYPERACEAE
SUCCISA
PTERIDIUM
FILICALES
SPHAGNUM
ISOETES ÷ 2
VIII?
VII b
VII a
VI a

Figure 15*b*.

| B.C. | MEAN JULY TEMP. 5° 10° 15° 20° C | NW EUROPEAN ZONES, WITH RADIOCARBON DATES FOR NW ENGLAND | SOUTHERN LAKE DISTRICT VEGETATION Blelham Bog Windermere Esthwaite Water | Vegetation Phase | LAKE DISTRICT MOUNTAINS Blea Tarn | | Seathwaite Tarn | | Corries Langdale Combe Small Water | Cumbrian Zones | CUMBERLAND LOWLAND (MOORTHWAITE MOSS) |
|---|---|---|---|---|---|---|---|---|---|---|---|
| 8,000 | | | | | | | | | | | |
| | | III (Dryas 3) | *Artemisia* *Salix herbacea* | 7. | 7. | *Artemisia* | 7. | *Artemisia* | Glacial | C8 | *Artemisia* |
| | | | *Empetrum* | 6. | 6. | *Empetrum* | 6. | *Empetrum* | Deposits | C7 | Breakdown of |
| 9,000 | | * Scaleby Moss | Koenigia Plants of open soils Reduction in tree birches | 5. | 5. | Plants of open soils—inc. *Lycop. selago* | 5. | Plants of open soils | on ROCK | C6 | plant communities *Betula pubescens* *Betula nana* |
| | | II Alleröd | *Betula pendula* | | | ? Tree birches | | | | | |
| 10,000 | | * Windermere | *Betula pubescens* | 4 ii | (4 ii) | Juniper | | Juniper | | C5 | *Betula pubescens* |
| | | Ic (Dryas 2) | | | 4 ib | *Lycop. selago* *B. nana* | | | | C4 | *Betula pubescens* Gramineae |
| | Holland | Ib Bölling | Juniper | 4 i | 4 ia | Juniper | 4 i | Juniper | | | |
| | Norway | | | | | | | | | | |
| 11,000 | | | | | | | | | | | |
| | (Late-glacial / Full-glacial Holland) | Ia (Dryas 1) | Pioneer Vegetation | 3. | 1–3 | Pioneer Vegetation inc. *Lycop. selago* | 1–3 | Pioneer Vegetation | | ? | |
| 12,000 | | | | | | | | (cores do not penetrate further) | | | |
| | | * Blelham Bog (Kettlehole open) | Pioneer vegetation, + *B. nana* + herbs | 2. | | | | | | | |
| 13,000 | | | Pioneer Vegetation | 1. | | | | | | C 1–3 ? | |
| B.C. | | ? | FULL GLACIAL with varved clays forming in the lakes | | | | | | | | |

Figure 16. The Late-glacial period in north-west England. Correlation table.

grams from north-west England until a very recent horizon which corresponds with eighteenth-century planting; *Pinus* reappears in quantity at the same level. The absence of native *Fagus* and *Carpinus*, and the scarcity of *Tilia* except in South Westmorland, has always made it impossible to draw the British zone boundaries on northern pollen diagrams after the end of Zone VIIa. Figure 15 illustrates the unsatisfactory result of attempting to define the climatic deterioration at the opening of the Sub-atlantic period by the revertence of *Betula*; it is now recognized that this is non-synchronous, and almost certainly anthropogenically conditioned, within the region. The only climatic index on which the opening of the Sub-atlantic period could be based would be the overall change in humification of the lowland raised bogs. The upland blanket peat, potentially susceptible to the same climatic threshold, may provide evidence for a comparable horizon when it has been more thoroughly investigated. In theory, it seems likely that the retrogressive processes by which the forest above *c.* 1500 ft (457 m) was replaced by blanket peat went on continuously after *c.* 2000 B.C., and only further radiocarbon dating will show whether there is, in fact, a synchronous horizon at which these processes were accelerated.

## SUMMARY

The subject is reviewed in three sections, dealing respectively with the Late-glacial, the early Post-glacial period, and the later Post-glacial after 3000 B.C. Contrasts in Late-glacial history between the Cumberland Lowland (Walker, 1966), the valleys and mountains of the Lake District (Pennington, 1947, 1964, and in the press) and the lowlands round Morecambe Bay (Smith, 1958; Oldfield, 1960, 1965) are discussed, and the conclusion reached is that the area lay near to the altitudinal and latitudinal limit of continuous woodland. Comparison of many pollen diagrams shows that in the Boreal forests of north-west England there was great variation from site to site in the relative importance of the pine, which seems to have been edaphically controlled. In the

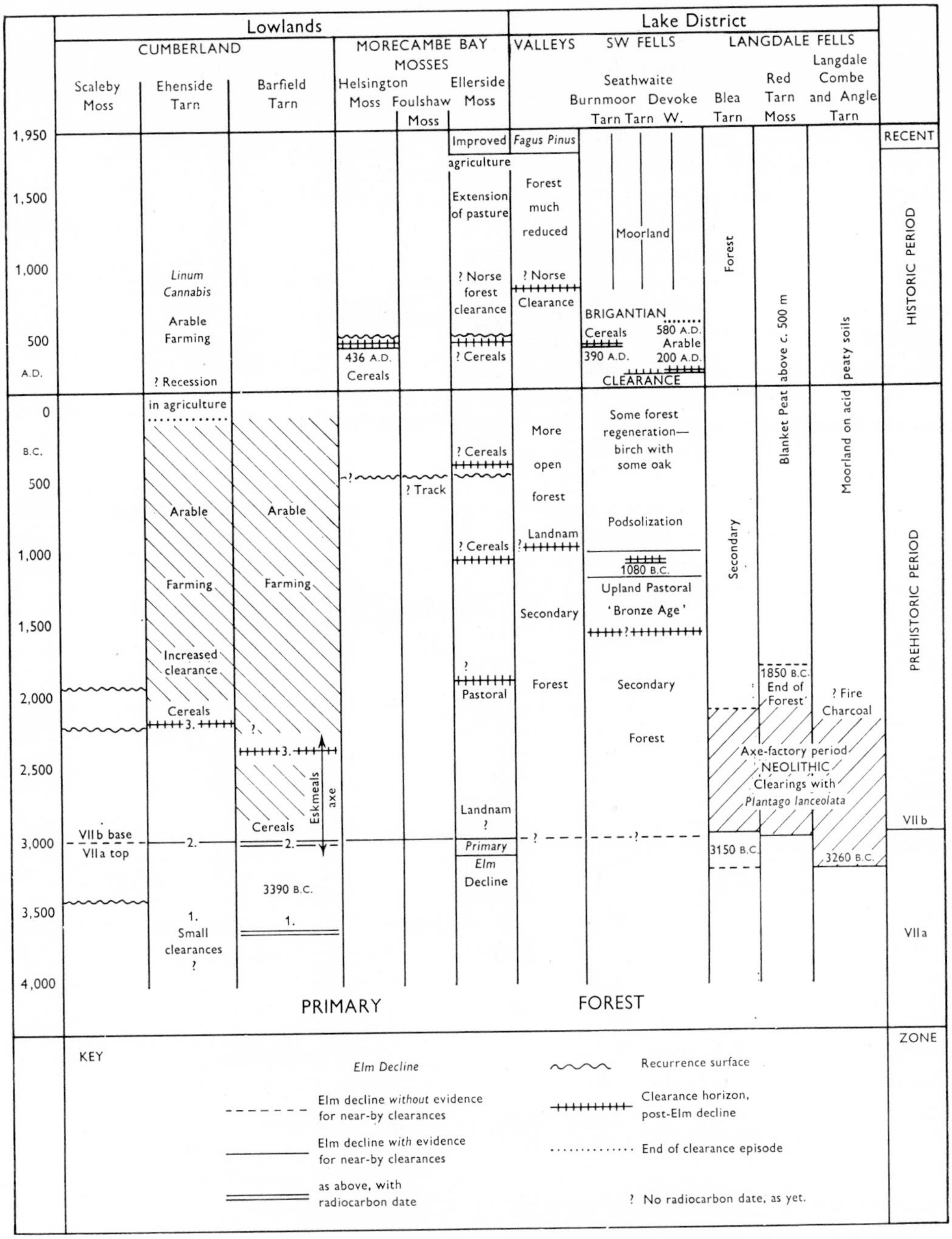

Figure 17. Vegetation history in north-west England after 4000 B.C. A suggested division of the period after 3000 B.C., taking the building of Hadrian's Wall as the opening of the historic period in north-west England.

mid Post-glacial forests, *Tilia* reached its northward limit within the area. After the Atlantic–Sub-boreal transition at 3000 B.C. there is no synchronous and climatically controlled change in vegetation history on which to base a zone boundary; the most important factor controlling vegetation history is shown to have been man. An alternative zonation scheme for this second half of the Post-glacial, for the highlands and for the lowlands, based on human history, is proposed.

In preparing this synthesis I have drawn extensively on the published work of Dr D. Walker, Professor F. Oldfield, and Dr A. G. Smith, and am grateful to them all for their co-operation and permission to reproduce figures. For much of the hitherto unpublished data included, I am indebted to Mrs J. M. Horwood, Mrs C. E. Lee and Mrs J. P. Lishman, who have worked as my research assistants. I acknowledge with gratitude the help of Professor Gordon

Manley and Miss Clare Fell, who read and made suggestions on the sections dealing with, respectively, the Late-glacial period and the later Post-glacial period. I should like also to record my continuing indebtedness to the late Professor W. H. Pearsall, F.R.S., who first stimulated me to think about the historical ecology of north-west England.

## REFERENCES


Atherton, N. M., Cranwell, P. A., Floyd, A. J. & Haworth, R. D. (1967). Humic Acid. I. E.S.R. spectra of humic acids. *Tetrahedron* **23**, 1653–67.

Bartley, D. D. (1962). The stratigraphy and pollen analysis of lake deposits near Tadcaster, Yorkshire. *New Phytol.* **61**, 277–87.

Birks, H. J. B. (1968). The identification of *Betula nana* pollen. *New Phytol.* **67**, 309–14.

Chanda, S. (1965). The history of vegetation of Bröndmyra. *Arb. Univ. Bergen. 1965. Mat.-Naturv. Serie No. 1*, 1–17.

Collingwood, R. G. (1933). An introduction to the pre-history of Cumberland, Westmorland and Lancashire north of the sands. *Trans. Cumb. West. Ant. Arch. Soc.* (N.S.) **33**, 163–200.

Darbyshire, R. D. (1874). Notes on discoveries in Ehenside Tarn, Cumberland. *Archaeologia* **44**, 273– .

Davis, M. B. and Deevey, E. S. (1964). Pollen accumulation rates: estimates from Late-glacial sediment of Rogers Lake. *Science* **145**, 1293–5.

Eastwood, T., Dixon, E. E. L., Hollingworth, S. E. & Smith, B. (1931). The geology of the Whitehaven and Workington district. *Mem. Geol. Survey U.K.* Sheet 28.

Evens, E. D., Grinsell, L. V., Piggott, S. & Wallis, F. S. (1962). Fourth report of the south-western group of museums and art galleries on the petrological identification of stone axes. *Proc. Prehist. Soc.* **28**, 209–66.

Faegri, K. & Iversen, J. (1964). *Text-book of Pollen Analysis.* Oxford: Blackwell.

Fell, C. I. (1964). The Cumbrian type of polished stone axe and its distribution in Britain. *Proc. Prehist. Soc.* **30**, 39–55.

Firbas, F., Muller, H. & Munnich, K. O. (1955). Das wahrscheinliche Alter der späteiszeitlicher Bölling-klimaschwankung. *Die Naturwissenschaften* **42**, 509.

Franks, J. W. & Pennington, W. (1961). The Late-glacial and Post-glacial deposits of the Esthwaite basin, north Lancashire. *New Phytol.* **60**, 27–42.

Godwin, H. (1956). *The History of the British Flora.* Cambridge.

Godwin, H. (1960). Radiocarbon dating and Quaternary history in Britain. *Proc. Roy. Soc.* B **153**, 287–320.

Godwin, H., Walker, D. & Willis, E. H. (1957). Radiocarbon dating and Post-glacial vegetational history: Scaleby Moss. *Proc. Roy. Soc.* B **147**, 352–66.

Godwin, H. & Willis, E. H. (1959). Radiocarbon dating of the Late-glacial Period in Britain. *Proc. Roy. Soc.* B **150**, 199–215.

Hartz, N. (1912). Alleröd–Muld: Alleröd–Gyttjens Landfacies. *Medd. dansk geol. Foren.* no. 4.

Hollingworth, S. E. (1951). The influence of glaciation in the Lake District. *J. Inst. Wat. Engrs.* **5**, 486–96.

Iversen, J. (1941). Landnam i Danmarks Stenalder. *Danm. geol. Unders.* **66**, 1–68.

Iversen, J. (1954). The Late-glacial flora of Denmark and its relation to climate and soil. *Danm. geol. Unders.* **88**, 87–119.

Iversen, J. (1960). Problems of the early Post-glacial forest development in Denmark. *Danm. geol. Unders.* **4**, 1–32.

Iversen, J. (1964). Retrogressive vegetational succession in the Post-glacial. *J. Ecol.* **52** (Suppl.), 59–70.

Jessen, K. (1949). Studies in late Quaternary deposits and flora-history of Ireland. *Proc. R. Irish Acad.* **52** B, 85–290.

Kendal, J. D. (1881). Interglacial deposits of West Cumberland and North Lancashire. *Quart. J. Geol. Soc. Lond.* **37**, 35–8.

Kirk, W. & Godwin, H. (1963). A Late-glacial site at Loch Droma, Ross and Cromarty. *Trans. Roy. Soc. Edinb.* **65**, 225–49.

Mackereth, F. J. H. (1966*a*). Some chemical observations on Post-glacial lake sediments. *Phil. Trans. R. Soc.* B **250**, 165–213.

Mackereth, F. J. H. (1966*b*). In *Ann. Rep. F.B.A.* **34**, 26–7.

Maher, Louis J. (1963). Pollen analyses of surface materials from the southern San Juan Mountains, Colorado. *Geol. Soc. Amer. Bull.* **74**, 1485–1504.

Manley, G. (1959). The Late-glacial climate of north-west England. *Liverpl Manchr Geol. J.* **2**, 188–215.

Manley, G. (1962). The Late-glacial climate of the Lake District. *Weather* **17**, no. 2, 60–4.

Marr, J. E. (1916). *Geology of the Lake District.* Cambridge.

Mitchell, G. F. (1960). The Pleistocene history of the Irish Sea. *Adv. Sci.*, **17**, 313–25.

Oldfield, F. (1960). Studies in the Post-glacial history of British vegetation: Lowland Lonsdale. *New Phytol.* **59**, 192–217.

Oldfield, F. (1963). Pollen-analysis and man's role in the ecological history of the south-east Lake District. *Geogr Annlr.* **45**, 1, 23–40.

Oldfield, F. (1965). Problems of mid-Post-glacial pollen zonation in part of north-west England. *J. Ecol.* **53**, 247–60.

Oldfield, F. & Statham, D. C. (1963). Pollen-analytical data from Urswick Tarn and Ellerside Moss, North Lancashire. *New Phytol.* **62**, 53–66.

Pearsall, W. H. (1950). *Mountains and Moorlands.* London: Collins.

Pearsall, W. H. (1961). Place-names as clues in the pursuit of ecological history. *Namn. och Bygd.* **49**, 1–4, 72–89.

Pennington, W. (1943). Lake sediments: the bottom deposits of the north basin of Windermere, with special reference to the diatom succession. *New Phytol.* **42**, 1–27.

Pennington, W. (1947). Lake sediments: pollen diagrams from the bottom deposits of the north basin of Windermere. *Phil. Trans. R. Soc.* B **233**, 137–75.

Pennington, W. (1964). Pollen analyses from the deposits of six upland tarns in the Lake District. *Phil. Trans. R. Soc.* B **248**, 205–44.

Pennington, W. (1965). The interpretation of some Post-glacial vegetation diversities at different Lake District sites. *Proc. Roy. Soc.* B **161**, 310–23.

Pennington, W. (1969*a*). The usefulness of pollen analysis in interpretation of stratigraphic horizons, both Late-glacial and Post-glacial. *Proc. Int. Symp. Paleolimnology*, Tihany. (In the press.)

Pennington, W. (1969*b*). The Late-glacial period in the English Lake District: 1. (In the press.)

Pennington, W. & Lee, C. E. (Unpublished). The late-glacial period in the English Lake District: 2.

Piggott, S. (1954). *The Neolithic cultures of the British Isles.* Cambridge.

Smith, A. G. (1958). Two lacustrine deposits in the south of the English Lake District. *New Phytol.* **57**, 363–86.

Smith, A. G. (1959). The mires of south-western Westmorland: stratigraphy and pollen analysis. *New Phytol.* **58**, 105–27.

Tansley, A. G. (1911). *Types of British Vegetation.* Cambridge.

Tauber, H. (1967). Investigations of the mode of pollen transfer in forested areas. *Rev. Palaeobot. and Palynol.* **3**, 277–86.

Terasmäe, J. (1951). On the pollen morphology of *Betula nana. Svensk Bot. Tidsk.* **45**, 358–61.

Turner, J. (1965). A contribution to the history of forest clearance. *Proc. Roy. Soc.* B **161**, 343–54.

Van der Hammen, T. (1957). The stratigraphy of the Late-glacial. *Geologie en Mijnbouw* **19**, 250–4.

Van der Hammen, T., Maarleveld, G. C., Vogel, J. C. & Zagwijn, W. H. (1967). Stratigraphy, climatic succession and radiocarbon dating of the last glacial in the Netherlands. *Geologie en Mijnbouw* **46**, no. 3, 79–95.

Walker, D. (1955). Studies in the post-glacial history of British vegetation, XIV. Skelsmergh Tarn and Kentmere, Westmorland. *New Phytol.* **54**, 222–54.

Walker, D. (1965). The Post-glacial period in the Langdale Fells, English Lake District. *New Phytol.* **64**, 488–510.

Walker, D. (1966). The Late Quaternary history of the Cumberland Lowland. *Phil. Trans. R. Soc.* B **251**, 1–210.

# THE INFLUENCE OF MESOLITHIC AND NEOLITHIC MAN ON BRITISH VEGETATION: A DISCUSSION

by A. G. Smith

*Botany Department and Palaeoecology Laboratory, Queen's University, Belfast*

## INTRODUCTION

In *The History of the British Flora*, published in 1956, Godwin remarks of Palaeolithic and Mesolithic man: 'the forest or the tundra dominated him, dictated the nature of his food, his tools, his clothing, his dwellings and directly or indirectly controlled almost every aspect of his existence. He, on the other hand, made small local clearings by stone axe and by fire and provided scattered and impermanent habitats for a few species capable of responding quickly to the local openness of vegetation' (p. 332). The concept, in the words of Iversen (1949), that 'primitive Mesolithic man was entirely dependent on nature; he could not interfere much with the vegetation' would have met with general agreement among British palaeoecologists until quite recently. This view must, however, now be re-examined. Voices are beginning to be raised against the idea that Mesolithic man was dominated by his environment; suggestions are being made that different farming techniques may have spread into Europe at different times, and it is seen that even the use of the terms 'Mesolithic' and 'Neolithic' cloud the real issues.

The argument that man's activities had no marked effect on vegetation in pre-Neolithic times derives to some extent from the premise that, in addition to the small human population, at the economic level of food gathering man did not need to clear forests and was in any event ill-equipped to do so. Thus there has been a tendency to overlook the possibility that the deliberate use of fire, or even accidental burning, may have had quite widespread effects. Again, while attention has been paid to the effectiveness of the polished stone axe for forest clearance, little has been heard of the potentialities of the *tranchet* axe. Radley & Mellars (1964) point out, however, that as opposed to the highland sites, the lowland Maglemosean sites are characterized by transversely sharpened flint axes 'presumably associated with woodworking and probably with felling trees'.

Dimbleby (1962) has distinguished quite marked effects of Mesolithic people on vegetation in his pollen diagrams from mineral soils. It must be remembered, however, that because of the questionable time of origin of the pollen in these deposits, the interpretation of the pollen diagrams is even more difficult than usual. At Iping Common, Sussex (Keef, Wymer & Dimbleby, 1965), Dimbleby considers that, possibly, local replacement of hazel scrub by heath may have been associated with the activities of Mesolithic man. At Oakhanger, Hants (Rankine, Rankine & Dimbleby, 1960), he speaks of increasing destruction of the forest cover during successive Mesolithic occupations in the Atlantic period. Dimbleby suggests here that increasing soil acidity might have resulted in a loss of structure of the sandy soil. This process may have been aggravated by the effects of fire, probably leading finally to a loss of soil stability. After this an increased proportion of heaths is registered in the flora. Dimbleby regards the whole process as one of degradation, due in large measure to human activity. In a discussion of his work at Addington, Kent (Dimbleby, 1963), where little effect can be seen of the Mesolithic occupation, Dimbleby points out that such degradation might not occur on more fertile soils. Clearly, as he says, the effect of man on his environment depends on the intensity and duration of the practices which might lead to modification; in addition, the ability of the forest climax to withstand human influences may be determined by the soil fertility.

It is further pointed out by Dimbleby (1962) that, if Mesolithic man did use fire to drive game, his effect on vegetation might have been out of all proportion to his numbers. Very little is known of the extent of burning in prehistoric times but some idea is given by an old account (De la Pryme, 1701) of trees found underground at Hatfield Chase, Lincolnshire. In quoting this account it is not intended to imply that the destruction of the forests referred to can necessarily be attributed primarily to human activities. The account may be slightly inaccurate but nevertheless contains observations impossible to make at the present day. De la Pryme says:

In the soil of all, or most of the said 180,000 acres of land, of which 90,000 were drained, even in the bottom of the river Ouse, and in the bottom of the adventitious soil of all Marshland, and round about the skirts of the Lincolnshire Woulds...are found vast multitudes of the roots and trunks of trees of all sizes, great and small, that this island either formerly did, or at present does produce; as firs, oaks, birch, beech, yew, thorn, willow, ash, etc., the roots of all or most of which stand in the soil in their natural positions, as thick as ever they could grow, and the trunks of most of them lie by their proper roots...a 3d. part of all being pitch trees, or firs...It is very observable, and manifestly evident, that many of those trees of all sorts have been burnt, but especially the pitch or fir trees, some quite through, and some all on a side; some have been found chopped and squared, some bored through, other half split with large wooden wedges and stones in them and broken axe heads, somewhat like sacrificing axes in shape, and all this at such places and such depths, as never could be opened since the destruction of the forest, till the time of the drainage.

To this account we may add the impression gained by McVean (1964), from superficial observations, that burning in ancient times has probably been widespread in the Scottish Highlands. Another instance of the signs of widespread burning can be found in the Sperrin Mountains in the north of Ireland. There, charcoal fragments are encountered at the base of the blanket peat over a fairly wide area and up to an altitude of at least 520 m. (1700 ft.) (Smith & Pilcher, unpublished observations). Charcoal can also be found in a similar position on the Antrim plateau. It would clearly be of interest to have further knowledge of the age and distribution of these charcoal layers, whether or not man was responsible for starting the fires they represent.

It is possible that during the Mesolithic period sophisticated systems of land use may have developed that involved deliberate interference with the forest cover. The transition between the Mesolithic and Neolithic periods is usually regarded in the British Isles as rather sharp, and connected with the immigration of new peoples bringing with them new techniques and new biological materials. But the possibility exists of the indigenous development of specialized food-collecting, more or less sedentary communities (cf. Braidwood, 1960, IB, 2–3) or communities that discovered means whereby they could manipulate the productivity of their environment to their own advantage. Archaeological evidence for such an ability will, of course, be difficult to find since the tools and techniques required might not differ from those previously in use. Nevertheless, we should examine the biological evidence in the light of these thoughts.

## THE INFLUENCE OF MESOLITHIC MAN

### The early Post-glacial: the abundance of hazel

It has long been known that Mesolithic man collected the nuts of hazel (Godwin, 1956, p. 198) but whether by accident or design he encouraged its spreading is another question. The high percentages of hazel pollen found in the Boreal period are generally taken to represent considerable areas of hazel scrub. While certain difficulties have been seen in explaining such a development, it is generally thought to be a climatic effect or due to the migration rate of hazel outstripping that of the other thermophilous trees (cf. Godwin, 1956; Iversen, 1960). Many Mesolithic sites have been shown to belong to the period of the 'Boreal hazel maximum' in north-west Europe (cf. Jessen, 1935; Clark, 1954). But in the British Isles, only at the sites in the Vale of Pickering, East Yorkshire (Clark, 1954; Walker & Godwin, 1954) can we observe the relationship between the archaeological material and the full course of the pollen curves in the early Post-glacial period. Here, both at Star Carr and Flixton (Moore, 1950), the cultural remains lie just at the point where the hazel curve begins to rise (the transition between Zones IV and V). At the Flixton I site, the occupation level comes just at the point where the birch curve is declining. While there is evidence of the utilization of birch at Star Carr (the felled tree and numerous birch-bark rolls), at the supposed occupation level the birch curve shows no variation; slightly later, however, it has a temporary decline. At the Flixton site II (Moore, 1954), a layer (B) of sand and clay was encountered containing hazel nuts, tree bark and shale, immediately overlying a charcoal layer. No artifacts were found in this layer but, nearby, Moore records charcoal together with Mesolithic flints. Layer B falls in the Flixton site II, pollen diagram (Walker & Godwin, 1954) at the transition between Zones IV and V where the birch curve begins to fall and the hazel curve begins to rise. It seems entirely plausible that this sequence represents an attack on the forest, using fire, destructive enough to cause a certain amount of soil erosion. Such an attack on the forest could, of course, have been quite local and, indeed, the Flixton II pollen diagram does differ from the long diagram (DB 1) taken from a deeper part of the deposits.

The hazel nuts in layer B, and the hoards of hazel nuts in the Seamer excavations, show that hazel must have been relatively abundant in the area when the sites came to be occupied even though the hazel pollen curves run at that time only at *c.* 20 per cent. It is clear that the birch forest was then closed and the shade-tolerant hazel may have naturally replaced the light-requiring birch. But the possibility is raised that human activities hastened this process.

In this country there are no other fully investigated sites comparable in date to those in the Vale of Pickering. At the Danish site Aamosen, however, charcoal dust is present at the beginning of the rise of the *Corylus* curve (Jørgensen, 1963; Verup Complexet, Verup 5, Profil 33, layer 6); here, Jørgensen suggests that a forest fire occurred which caused a temporary decline in the *Pinus* curve and a temporary increase of the *Betula* curve. There are only a few indications of human activity in the region at this time. At another continental site, Hohen Viecheln in Mecklenburg (Schmitz, 1961), the expansion of hazel occurs at the same level as the most dense concentration of Mesolithic artifacts, with the presence of charcoal, in the organic deposits (Hohen Viecheln HV 7, 241–55 cm). The author maintains (pp. 20–1) that this rapid expansion of hazel is only to be explained in terms of a rapid climatic alteration. In these cases we see that, at the expansion of hazel, man was present and there is evidence for fire, either deliberately used by man or naturally occurring. Of course, there are vast areas in which no such association can be shown. This is no reason, however, for discarding the possibility that the expansion of hazel may be connected in some way with human activity. As with the Vale of Pickering sites, however, if there was a human effect, it seems likely to have been to increase the rate of change of an autogenic process.

Whether or not man's activities had any connection with the initial expansion of hazel, a quite different question is whether the apparent prevalence of hazel in the Boreal period in general can be attributed to human activity. Rawitscher (1945), from his experience of the effect of fires in Brazil, says that if fire had been extensively used by Mesolithic man in Europe the existence of a 'prevalent hazel vegetation would not be unexpected'. He quotes the persistence of hazel through recurrent prairie fires in Wisconsin by virtue of its strong root system (Chavannes, 1941). There is no doubt that the European *Corylus avellana* is fire resistant, springing up again readily from burnt stumps, but Rawitscher's suggestion has gone almost without comment in the literature. Presumably most workers find it difficult to envisage a small Mesolithic population seriously affecting the forest vegetation, even if fire was used, when the vast areas that would have been involved are taken into consideration. Attention may be drawn, however, to the abundance of charcoal in the Boreal layers of the Ageröd bog complex in Scania (Nilsson, 1967) which has been extensively excavated. Charcoal occurs both in the late Boreal culture layer and in the earlier Boreal deposits. It is not impossible that charcoal would be found in Boreal deposits elsewhere if similar large-scale excavated sections could be examined.

At White Gill, in the Cleveland Hills, Dimbleby (1961; 1962, pp. 20 and 111 ff.) has found an increase of hazel pollen from 35 to 60 per cent of the total tree pollen between two samples, the first taken from the mineral soil on the surface of which were found flints of a microlithic industry associated with charcoal of *Quercus*, *Alnus*, *Betula* and *Corylus*, and the second from a black humus resting on the mineral soil. Taking this evidence together with that from a later (Bronze Age) site, Dimbleby concludes that 'the great preponderance of hazel in spectra from prehistoric ages is not a climatically induced condition but a direct result of man's impact on the forest'. To be certain that this was really the case at White Gill, and similar Pennine sites, however, it would have to be demonstrated that there was no long time gap between the human occupation and the initiation of humus accumulation. Leaving aside that proviso, however, if Dimbleby's argument is extended to include the Boreal period then the prevailing climatic conditions can hardly be left out of account. Even if the Boreal hazel maximum does represent a fire climax, it could be connected in some way with the radical changes of forest constitution that were in progress, and the dry conditions. There is no clear connection, however, between the highest values for hazel and indications of local dryness (cf. Oldfield, 1965).

### The Boreal–Atlantic transition

It has been found in many cases that where the influence of Neolithic man on vegetation can be distinguished in pollen diagrams, the *Alnus* curve exhibits a marked rise (cf. Mitchell, 1956; Smith & Willis, 1962; Oldfield, 1963; Pennington, 1965). There may be more than one explanation for this behaviour of the *Alnus* curve, as will be seen from later discussion. For the moment, however, it is of interest to look at the first rise of the *Alnus* curve at the Boreal–Atlantic transition, a radical vegetational development which has traditionally been thought of as due to a climatic change. Across the British Isles there are marked differences in the behaviour of the pollen curves in the Boreal and Atlantic periods, particularly in the case of *Pinus* (cf. Smith, 1964; Oldfield, 1965). Of most interest for our present discussion, however, is the finding by Simmons (1964) of a 'forest recession' in his Blacklane diagram from Dartmoor. This recession takes place in what is apparently the Boreal period and Simmons considers that an anthropogenic cause cannot be ruled out; he suggests that since the recession comes before the *Ulmus* decline any such activity would probably be due to Mesolithic man. In other deposits from the same Dartmoor area, charcoal fragments are present in the Boreal deposits, often at a sharp decline of the *Corylus* and a rise of the *Alnus* curves. Simmons suggests that, with increasing wetness, *Alnus* may have been competing with *Corylus* and that where burning had taken place it affected the hazel allowing alder to immigrate faster. There is little doubt that Simmons is correct in assuming a forest recession at the end of Zone VI at Blacklane: *Pteridium* increases and both *Artemisia* and *Plantago lanceolata* are present; the first traces of *Fraxinus* are seen, together with pollen of *Prunus/Sorbus* type (possibly the light-requiring *Sorbus aucuparia*) and at this point the curve for *Alnus* begins to rise. The real question is whether Zone VI, here, is really the Boreal period in a broad chronological sense. *Fraxinus* pollen is not normally present in quantity until the Sub-boreal period. The deposit attributed to Zone VI is an amorphous, or wood, peat in which the values for *Quercus* rise to 80 per cent. It could be that this is a forest peat or raw-humus deposit that supported an oak forest until it was affected by human activity at some date later than the Mesolithic. In his diagram from Taw Marsh, Simmons has recognized such a possibility; here he has tentatively assigned the oak-hazel period to Zone VII even though alder is scarcely represented. There is obviously a good case, however, for considering that, while a climatic effect cannot be entirely ruled out (cf. Frenzel, 1966), fire or human activity, or both, had some effect on the vegetation at a point that looks like the Boreal–Atlantic transition, whether chronologically it is so or not. A series of radiocarbon dates for such diagrams would undoubtedly be of much interest.

The question is now obviously raised as to whether any effect of human activity can be seen at the Boreal-Atlantic transition where this is known to occur at a chronologically normal time. Such a situation exists at Shippea Hill in the East Anglian Fenland (Clark, Godwin & Clifford, 1935; Godwin & Clifford, 1938; Clark & Godwin, 1962). Here the Mesolithic layer falls exactly at the Boreal–Atlantic transition which is bracketed by two radiocarbon dates: Q-587 (Zone VIc), 5650 ± 150 B.C. and Q-586 (Zone VIIa), 4735 ± 150 B.C. The very high values rapidly attained by *Alnus* are considered by Godwin, in view of the presence of macroscopic remains, to indicate the presence of fen woods on the valley floor. It may be that the lime and ash, whose pollen curves rise at the same time, were also growing in similar damp situations. The fall of the pine curve at this point could then be a relative effect due to the local establishment of an alder fen wood. If it were due to the destruction of pine by man, an increase of non-tree pollen might be expected; this is not the case, however, and Godwin states that no clearance phase is indicated by the pollen curves at the Mesolithic level. Nevertheless, two processes may have been going on: the invasion of wetter situations by alder, and perhaps other tree species, together with the ousting of pine from such localities which could have been emphasized by human activities. It is striking that the *Corylus* curve declines during the Mesolithic layer at Shippea Hill; if we were to argue that hazel was affected by human activity, we should have to consider it to have been physically cleared, or to have been un-resistant to burning in this particular case. Physical clearance seems unlikely and in most other cases in the British Isles it seems that hazel has been favoured by human activity. Thus, in the present instance, we have to conclude that the relative decline of hazel was part of an autogenic process unless the dry conditions then prevailing in that region, as exemplified by the low lake levels at Hockam Mere (Godwin & Tallantire, 1951) for instance, somehow prevented the regeneration of burned-off hazel.

A series of changes similar to those at Shippea Hill occur in the pollen diagram from the Swedish site Baremosse II (Nilsson, 1967) where a Mesolithic layer falls at the end of the Boreal period. In this case an estimate has been made of the abundance of charcoal particles and there is a close parallel between the decline of pine and hazel with the appearance of charcoal in the culture layer and its subsequent slow disappearance. In view of this behaviour of the *Corylus* curve it is the more interesting to note that its first steep rise comes exactly at the point, earlier in the Boreal period, where charcoal particles first became really abundant. This provides another example of the association of fire with the expansion of hazel which can be added to those quoted earlier.

Another instance of the decline of pine in the East Anglian Fenland can be seen at Old Decoy (Godwin, 1940) where a sand layer occurs at its final demise, alder having expanded slightly earlier. This sand layer was thought first to be a local effect of dryness and exposure of the soil surface, but later (Godwin, 1956, p. 26) to be a 'Mesolithic occupation level'.

Whether or not the decline of the *Pinus* curve, and indeed the *Corylus* curve, can in some instances be attributed, in part, to human activity, it is clear that Mesolithic communities did utilize these species. Two instances may be cited. Pieces of hazel and pine wood showing possible signs of working were found by Mitchell (1955) at Toome Bay (Northern Ireland) in the late Boreal layer containing early Larnian flint implements. At the top of this layer (Toome Bay, pit 'i') the hazel curve falls sharply; slightly later the *Alnus* curve rises steeply. In a rather more elaborate diagram from the site, however (Toome Bay, point 3, where the occupation layer was not present) the fall of the hazel curve coincides exactly with the rise of the *Alnus* curve. As with the Fenland sites that have just been discussed, the probability is that *Alnus* was invading locally the marshy valley floor. The sharpness of the changes at the Toome Bay site are suggestive, however, of an unconformity in the sedimentary sequence. At a nearby site in the Bann Valley (Newferry, Co. Antrim; Smith, 1964 and unpublished) charred and uncharred pine wood was found in a Mesolithic layer coming exactly at the Boreal–Atlantic transition; charcoal of hazel and other trees such as oak, willow and birch was also found in the late Boreal peat. At this site, the hazel and birch curves fall and the pine curve rises just before the rise of the alder curve. As at Mitchell's Toome Bay site there is a depositional change at this point, though the likelihood of an erosional gap is less. It is evident that complex physiographic changes were in progress. The depositional change proceeds from a *Phragmites* mud through muddy diatomite to pure diatomite and it has been supposed that this sequence is a consequence of changes of water level, or a change from standing water to seasonal flooding (Mitchell, 1955; Jessen, 1936, 1949). It is clear that such a change was rather gradual, however, and it remains striking that the *Alnus* curve rises exactly at the level of an occupation layer. It is hardly conceivable, however, that man so affected the tree cover and soil conditions as to alter the whole ecology and hydrography of the region. While interpretation is complicated by the change of sedimentation, the indications are that the forest remained closed.

At the sites mentioned there is evidence of both human occupation and fire. Evidence of fire alone, beginning at an apparently early date, may also be quoted. In their investigation of forest history in the neighbourhood of the Beinn Eighe nature reserve (West Ross-shire), Durno & McVean (1959) found three layers of burnt pine stumps and charcoal, and a probably recent charcoal layer. McVean (1964) takes this as evidence of the destruction of at least four successive forests in Boreal, Atlantic and recent times. Whether the earlier fires were started by man is imponderable, though Steven & Carlisle (1959) consider that Mesolithic people could have had little effect on the forests even if they used fire to round up game. They point out that limited burning might stimulate natural regeneration of pine forests. Durno & McVean, however, consider that at Beinn Eighe there has been a gradually diminishing forest cover since Boreal times due to successive forest fires. Two Irish examples may also be given. At Ballynagilly, Co. Tyrone (Pilcher & Smith, unpublished), considerable quantities of charcoal have been found in a reedswamp deposit, some hundreds of metres away from the nearest dry land, just at the point where the *Alnus* curve is rising. No Mesolithic remains are known from the immediate vicinity though the area was occupied in Neolithic and later times. At Sluggan Bog, Co. Antrim (cf. Erdtman, 1928), pine stumps at the base of the *Sphagnum* peat can be seen to have been burnt and charcoal is present in the peat at the base of the stumps. The stumps can be referred (on the basis of a pollen diagram prepared as a student exercise) to the earliest part of the Atlantic period.

The number of sites in the British Isles where Mesolithic artifacts have been found stratified into pollen bearing deposits is rather small. We see, however, from those mentioned above, that up to the beginning of the Atlantic period there is often a remarkable coincidence of vegetational change with indicators of the presence of man and the occurrence of fire. Such correlations do not provide proof that man's activities caused the vegetational changes. We should face the possibility, however, that changes of vegetation that were under way as a result of differential migration rates, soil development, and climatic change, as has generally been supposed, may have been in some instances accelerated by human activity.

### The Atlantic period

The Atlantic period, unlike the periods that preceded it, was presumably one of relative vegetational stability. During this period, except at the very beginning, there are few well-marked changes in the pollen diagrams until the decline of the *Ulmus* curve that has been used to mark its end. In the Atlantic period, it should be much easier, therefore, to distinguish effects of man on vegetation where suitably stratified archaeological material is encountered.

Unfortunately the ideal conditions for such a study have rarely been found. At Dozemare Pool on Dartmoor, where a microlithic industry has been discovered (Wainwright, 1960; Conolly, Godwin & Megaw, 1950), charcoal and ash layers are associated with sharp temporary fluctuations of the pollen curves in Zone VII. No stratigraphical connection has been established, however, between the artifacts and the charcoal layers in the peat. The pollen diagram from Dozemare Pool is rather an old one and the authors consider it insufficiently detailed to warrant analysis of the pollen curves in vegetational terms. It seems likely, however, that some marked, if local, change of the forest vegetation occurred; there is a birch maximum at the level of the charcoal layers with a temporary decline of the alder and willow curves. Elm does not seem to have been affected—the curve falling steadily—and it appears that the features referred to are to be assigned to the Atlantic period.

The site at Stump Cross, near Grassington, Yorkshire investigated by Walker (1956) provides the only example of an early Atlantic occupation where the full course of the pollen curves can be observed. At this site several flint flakes were found, from the pollen spectra of the mud adhering to them, to belong to the first half of the Atlantic period and perhaps going as far back as the Boreal–Atlantic transition. It is interesting to see that, at the level of a flake found associated with charcoal close to the point of pollen sampling, the curves for *Corylus* and Ericaceae rise sharply, and there is a grain of *Artemisia* pollen. At this point the *Pinus* curve has fallen to a low level and the *Alnus* curve is still rising. Walker regards the increase of heath pollen as probably indicating the beginning of blanket bog development in the area but there seems equal reason to suppose the woodland vegetation was being locally affected by human activity. The rise in the hazel curve is temporary, as it is in the Neolithic *landnam* phases in some Irish pollen diagrams (Smith, 1958*a*) but *Plantago lanceolata* is apparently absent. As is well known *Plantago lanceolata* pollen has been found frequently to be associated with Neolithic clearance phenomena since Iversen's pioneering work in Denmark (Iversen, 1941) but in that country it has not been found in Atlantic deposits. This is not the case in the British Isles, however, as we shall see shortly. For the moment it is of interest, in view of what has been said of the Stump Cross site, to note that in the diagram from the nearby Malham Tarn (Pigott & Pigott, 1963) *Plantago lanceolata* pollen is present not only in the Late-glacial but also in the Atlantic period. Moreover, as far as the diagrams can be compared, this *Plantago* pollen comes at approximately the same stage as the possible signs of human activities at Stump Cross. *Plantago* is first present, together with *Rumex* and *Urtica*, in small quantities just at the fall of the *Pinus* curve and the rise of the *Alnus* curve that mark the Boreal–Atlantic transition. A few centimetres above this transition it appears in greater quantity. There is a *Corylus* maximum at the transition which persists until the higher *Plantago* values a few centimetres above where there are also peaks in the curves for Gramineae and Cyperaceae. Furthermore, it is also in this earliest part of the Atlantic period that *Fraxinus* pollen forms a continuous curve, a feature which is not usually observed until the Sub-boreal period, and *Taxus* pollen is found for the first time in the diagram. The authors of this work do not see any evidence of forest destruction until the late Neolithic period (late in Zone VIIb); the features of the diagram just described appear to the present writer, however, most strongly to suggest that the forest was locally opened by human activity, and probably that regeneration took place, in the earliest part of Atlantic time. If the Boreal–Atlantic boundary has any chronological significance (an open point, however, *vide* Smith, 1964; Oldfield, 1965) this activity is that of a Mesolithic people. Indeed, the authors record the presence of a Mesolithic site with Sauveterrian microliths on a mound at the shore of the present fen, and thin seams of charcoal (carbonized twigs of *Calluna*) apparently in the late Boreal and early Atlantic deposits. No stratigraphic connection between the site and the charcoal has been established, however. We see again, then, some suggestion that human activity affected vegetational development close to the Boreal–Atlantic transition, and a strong indication of the opening of the forests, allowing *Fraxinus* and possibly *Taxus* to expand or flower more freely in the earliest part of the Atlantic period.

### The secondary *Corylus* maximum at the Boreal–Atlantic transition

A secondary *Corylus* maximum, coming close to the Boreal–Atlantic transition as at Malham Tarn, can be observed in several diagrams from the British Isles. It can be seen, for instance, in the pollen diagram from Elstead, Surrey (Seagrief & Godwin, 1960) where it occurs at the *Pinus* decline late in Zone VI at a level where sand and clay layers occur in the deposit. It can be seen in diagrams from Northumberland: Muckle Moss (Pearson, 1960), Bradford Kaims (Bartley, 1966); in the Lake District at Thrang Moss (Oldfield, 1960); and in several Scottish diagrams: L. Creagh, near Aberfeldy, Perthshire (Donner, 1962) Netherly Moss, Kincardineshire and other diagrams from the Grampian Highlands (Durno, 1956). In Ireland a secondary *Corylus* maximum can be seen, for instance, in such diagrams as (34) Lisnolan, Co. Mayo and (36) Roundstone, Co. Galway (Jessen, 1949). The secondary *Corylus* maximum at these sites falls either at the end of Zone VI or at the Zone VI–VII boundary, though, in view of the slight discrepancies between the criteria for demarcation of the zone boundary used by the different authors, no real significance can be attached to the apparent differences. What can be said is that the *Corylus* maximum generally falls just before the point where the *Alnus* curve steeply rises to high values (though at Muckle Moss it is at the same time as the *Alnus* rise) or during the slow rise of the *Alnus* curve in the Irish diagrams. Thus, the secondary *Corylus* maximum usually comes at a point where the *Pinus*, or *Betula*, curve falls. Now the actual vegetational changes at the Boreal–Atlantic transition have always been difficult to understand (cf. Conway, 1948; Smith, 1964; Oldfield, 1965) though the increase of *Alnus* has usually been regarded as a response to increasing wetness. The finding of a widespread, but not universal, *Corylus* maximum at this point complicates the issue still further. We have seen reason at Stump Cross and Malham Tarn to regard the *Corylus* maximum during the *Alnus* rise as suggestive of the opening of the forest by human activity: what of the other sites? At none of them is there any indication of human activity from the non-tree pollen; indeed, this is a period almost devoid of non-tree pollen. In some of the Scottish diagrams, however, there is a rise in the curve for heath pollen, a feature which could be explained by assuming it indicated the local initiation of acid bog conditions. Whatever the reason for the resurgence of hazel at these sites, it does appear to represent a retrogressive succession. If we assume that the hazel was earlier reduced in quantity, or that its flowering was suppressed, by the growth of shade tolerant trees (cf. Iversen, 1960) then the increase of hazel pollen at the Boreal–Atlantic transition relative to that of other trees demands an explanation in terms of the reduction of high forest cover. Furthermore, since the *Corylus* maximum is not a constant feature and since its position in relation to the rise of the *Alnus* curve is somewhat variable, we might suspect that it does not reflect an autogenic development. The secondary *Corylus* maximum comes at roughly the same stage as the secondary birch maximum described by Welten (1944) from Switzerland which is discussed by Conway (1948). The explanation given for this pattern of events, which includes a rise of the hazel curve at the same time as the birch maximum, is that it is an expression of a time lag between a climatic change and the establishment of a new vegetation-climate equilibrium. A similar explanation may hold for this country; if pine (or birch) was growing on damp valley floors and basins and then became excluded either directly or indirectly by a rising water table, then hazel along the margins may have temporarily contributed relatively more pollen to the total before alder invaded the wetter situations. The possibility of such a sequence receives a little support from the finding at Woodgrange, Co. Down (Singh, 1964; Singh & Smith, 1966) of pine cones together with alder wood just before the rise of the *Alnus* curve. It is also noteworthy that hazel nuts are so commonly encountered in Boreal deposits that hazel shrubs may well have been growing quite close by. Detailed investigation of the mire stratigraphy at a site showing a secondary hazel maximum would clearly be of assistance in interpretation. The actual recolonization of areas previously occupied by pine as envisaged by Oldfield (1965) is another possibility. Where the secondary *Corylus* maximum comes together with indicators of open conditions after the rise of the alder curve, however, and particularly where there is the supporting, though circumstantial, evidence of the presence of man, there seems no reason to change the opinion previously expressed that we are looking at the effect of human activity.

### THE PENNINE PEATS

Dimbleby (1962) has put forward a strong, though not incontrovertible, argument that the origin of heaths in the areas he has studied is due in large measure to the removal of forest cover, particularly in the Bronze Age. It is not fitting in this essay to enter into a discussion of this idea, but another point raised by Dimbleby does call for consideration. Dimbleby says 'it may be more than a coincidence that the spread of bog, heath and other open communities at the expense of woodland took place in the Pennines at a time contemporary with the microlithic industries'. The problems of dating the Pennine microlithic industries are, of course, very great, as has been emphasized by Davies (1963). Except at Stump Cross (Walker, 1956) we have always to bear in mind that since the artifacts are found at the surface of the mineral soil below the peat, there may well have been an interval without deposition before the beginning of peat accumulation. Thus, there is a possibility that pollen will have washed into the soil after the occupation. While blanket peat formation appears often to have begun close to the Boreal–Atlantic transition, Tallis (1964) has emphasized that, in the southern Pennines, peat formation has probably started at intervals throughout a long period in topographically different areas. Thus, while agreeing that there is evidence of the opening up of high altitude woodlands in the Pennines before the Neolithic, there is as yet nothing to show a causal connection between the initiation of heath and bog and the activities of Mesolithic man. This is not to deny that a causal connection may exist, and there is a clear need for the investigation of the age of the Pennine and other upland peats using the radiocarbon dating method.

### THE SIGNIFICANCE OF *PLANTAGO LANCEOLATA* POLLEN IN PRE-NEOLITHIC DEPOSITS

In many pollen diagrams from the British Isles, *Plantago lanceolata* pollen appears in small quantity in the Atlantic and earlier periods. It is difficult to know when this pollen is connected with prehistoric agriculture, as seems fairly clearly to be the case from the elm decline onwards. This difficulty is greatest in the later part of the Atlantic period; in earlier times we can be reasonably certain that agriculture, including stock-raising, was not practised. One might point, for instance, to the occurrence of *Plantago lanceolata* in the Boreal period in Scotland (Donner, 1962, p. 23) a time surely before any possibility of prehistoric agriculture but not too early for forest clearance to have been carried out. Boreal *Plantago lanceolata* has been found at coastal sites in south-west Norway by Hafsten (1965, p. 53) who considers that since this pollen occurs together with that of Chenopodiaceae, *Rumex* and *Artemisia*, *Plantago lanceolata* was originally a constituent of the strand flora. A similar explanation of early occurrences of *Plantago lanceolata* pollen in Norway has been advanced by Egede Larssen (1949). Sagar and Harper (1964) remark that *Plantago lanceolata* is 'present on sand-dunes, spray-washed and unstable cliffs and cliff tops'. Indeed, occasional plants of *Plantago lanceolata* can be found in the highest salt marsh in Co. Down. As is the case with other weeds and ruderals, it may have persisted in such localities from the Late-glacial period through the closed forests of the early Post-glacial. Van Zeist (1964) has offered the suggestion that Atlantic *Plantago lanceolata* in Brittany might be regarded as a Late-glacial relict, though, alternatively, it might be a result of farming activities. *Plantago lanceolata* has been found in north-west Germany by Müller (1947) to be referable to the Mesolithic period and while Tauber (1965) has taken this as evidence of farming in the Atlantic period, Müller (p. 81) regarded the question as to whether it was introduced by man as unanswerable. On the other hand, Florin (1961, p. 390), dealing with the Swedish site Mossbymossen, writes of quite clear signs of cultivation at a point in time which, so far as we can decide at present, falls in Zone VII at the end of Atlantic time.

Roux & Leroi-Gourhan (1965) have pointed recently to several French pollen diagrams in which possible signs of forest clearance have been found in the Atlantic period. These signs of forest clearance are attributed by the various authors (with varying degrees of confidence) to the activities of Neolithic peoples. Roux and Leroi-Gourhan consider that clearances were made by a stock-raising Proto-Neolithic people and that this form of subsistence was disseminated from the Middle East before either pottery or (arable) agriculture. Without entering into a detailed discussion of these French diagrams, it may be noted that a good deal of the evidence for pre-elm decline forest clearance depends on the presence of *Plantago lanceolata* pollen or a maximum of the hazel curve, that the elm decline may be rather slight (e.g. Elhai, 1960) and that the major post-glacial rise of the *Alnus* curve may not occur in some French diagrams until apparently rather late (Van Zeist, 1964, *v.* esp. pp. 161–2; Oldfield, 1964) approaching very close to the elm decline in one case (Saint Michel de Brasparts I; Van Zeist, 1963, 1964) which is radiocarbon dated to 3450 ± 60 B.C. (GRN-1983). It is possible, therefore, that where radiocarbon dates have not been obtained the chronology may not be certain. Nevertheless, it is clear that *Plantago lanceolata* pollen does occur in French deposits before *c.* 3400 B.C. and in the British Isles before the major forest clearance activities usually attributed to Neolithic peoples.

Whether forest clearance in the Atlantic period in the British Isles can in any instance be regarded as implying the practice of any form of agriculture is a question that must now be examined. We have already noted the presence of *Plantago lanceolata* pollen in Boreal deposits, presumably too ancient for any form of agriculture, in areas so distant from each other as Scotland and Cornwall. The presence of *Plantago lanceolata* pollen at this early date in the Dartmoor area (Simmons, 1964) has been amply confirmed by re-analysis of the deposits at Hawks Tor (Conolly, Godwin & Megaw, 1950) where *Plantago lanceolata* forms a continuous curve from the end of the Boreal period onwards (Smith & Crowder, unpublished). We may also mention the occurrence of *Plantago lanceolata* pollen noted by Churchill & Wymer (1965) immediately above the Mesolithic layer at Westward Ho!, Devon, which has been radiocarbon dated to 6585 ± 130 B.P. (Q-672). The *Plantago lanceolata* pollen early in the Atlantic period at Malham Tarn may also be recalled. As with the French diagrams mentioned above, there are difficulties of interpretation (particularly with the Dartmoor diagrams, for instance) and too few radiocarbon dates are available for us to be sure that the Boreal–Atlantic transition has any chronological significance (cf. Smith, 1964; Oldfield, 1965). It would be difficult to imagine, however, that the occurrences of *Plantago lanceolata* pollen in the cases here quoted have anything to do with agricultural practices. That the species entered clearings at quite an early date, however, is entirely plausible. While we have seen reason to believe that clearings were made by pre-Neolithic man, it is possible that clearings were created by indigenous herbivores (cf. Smith, 1958*c*). The example is brought to mind of the soil disturbances caused by herds of bison in the Minnesota area mentioned by McAndrews (1966) which provided habitats for various chenopodiaceous species. Another possibility is the deliberate clearance of forest by man to provide pasture for native grazing animals as an adjunct to hunting, perhaps even decoy hunting. Such areas could have remained open for long periods. Equally there may have been complex interactions between the effect of animals and the use of fire by man as is the case with the present-day vegetation in parts of Africa (cf. Boughey, 1963).

Turning to the later Atlantic period, brief mention may be made of recent work in the Lower Bann Valley, Northern Ireland. Here, at Newferry (Movius, 1936; Jessen, 1936; Smith, 1964 and unpublished) and Ballyscullion Bog (Jessen, 1949; Smith & Crowder, unpublished), indications of forest clearance have been found before the elm decline that marks the end of the Atlantic period. At Newferry, these indications take the form of increased percentages of grass pollen and the presence of *Plantago lanceolata*. At Ballyscullion, the grass pollen curve rises temporarily and *Plantago lanceolata* pollen is consistently present for a short period, together with pollen such as that of *Polygonum*, *Urtica* and Chenopodiaceae. In these two cases there appears to have been no certain destruction of any particular tree species and the clearance was apparently rather general. At Ballyscullion, *Taxus* in particular appears to have been favoured; it may have taken part in the regeneration, or have flowered more freely with the opening of the canopy (cf. Hafsten, 1965, p. 46). The possibility is raised at this site that the forest was opened up by more or less indiscriminate lopping or pollarding followed by a regeneration from the mutilated plants so rapid as to exclude the full development of scrub. The Bann Valley is an area in which it is conceivable that a more or less sedentary way of life evolved, based on a regular pattern of seasonal hunting and fishing. Certainly the river bank at Newferry was repeatedly visited over some hundreds, if not thousands, of years, as is shown by the ash layers containing flint and stone implements stratified in the diatomite deposits. The indigenous development of some form of land use involving deliberate forest clearance after a long period of occupation appears to be a strong possibility. Whether such a system could have involved domesticated or semi-domesticated animals is an open question. But the possible slow emergence of the domesticated cattle from the native aurox demonstrated for Denmark by Degerbøl (1963) and the similar possibility suggested by Jewell (1963) for England serves to emphasize that grazing animals may have been under human control even at this stage.

### The influence of Mesolithic man on vegetation: general summary

There is a mounting body of relevant information which, despite the difficulties of interpretation, force us to open our minds to the possibility that before the introduction of agriculture man was not so much dominated by closed forests as has hitherto been supposed. His activities may have had considerable effects on the physiognomy, floristic constitution and change in the vegetation. In Boreal and Pre-boreal times human occupation levels frequently coincide with marked vegetational changes. At the moment is is virtually impossible to decide whether man was responsible for any *general* vegetational change, but the possibility appears rather strong that human activities did have an effect in particular cases. Perhaps they were to hasten changes that were under way for other reasons; man may have tipped the balance where tension existed in the vegetation-environment relationship. There is fairly clear evidence of human influence on vegetation from early in the Atlantic period onwards. The actual date of these activities requires further elucidation, as does the scale of the effects. The impression is given from the magnitude of the changes in the pollen curves, however, that man's activities affected the vegetation well outside the immediate vicinity of his living-places, and the possibility cannot be rejected that the Atlantic vegetation of the British Isles was at times composed in part of secondary communities.

## THE INFLUENCE OF NEOLITHIC MAN

### The elm decline

The effect of Neolithic man on vegetation has been a lively topic since the publication by Iversen of his *Landnam i Danmarks Stenalder* in 1941. Evidence of forest clearance has now been found throughout Europe, beginning close to the Atlantic–Sub-boreal transition. One of the main features of pollen diagrams at this time is, of course, the decline of the elm curve. This has been so much discussed in recent years (cf. Iversen, 1949, 1960; Troels-Smith, 1953, 1955, 1960; Godwin, 1956, 1959, 1965; Mitchell, 1956, 1965*b*; Smith, 1961*b*, 1965; Tauber, 1965; Frenzel, 1966, and many other authors) that only a few comments will be made here.

The idea of a marked, and maintained, climatic change at the end of the Atlantic period receives much less support than hitherto, though Nilsson (cf. Nilsson, 1960, for instance) has always held that the decline of elm in Sweden was due to climatic change. Frenzel (1966) has made a case for a short term cold spell from about 3400 to 3000 B.C. in the Northern Hemisphere. The pollen analytic records for the British Isles are not inconsistent with this idea (cf. Smith, 1961*b*, 1965) but the problem can hardly be resolved until a new type of measurement of climatic change can be obtained. One point to be borne in mind in the consideration of Frenzel's argument is that, while high altitude vegetation is clearly in a sensitive condition to react to a change of climatic conditions, it is quite conceivable that vegetational changes restricted to high altitudes could have been caused by prehistoric man. The author maintains his opinion (Smith, 1961*b*), which is not unsupported by Frenzel's conclusion, that at the Atlantic–Sub-boreal transition we are dealing with a complex of effects and that, in different areas, different factors or combinations of factors may have been critical for the vegetation. The demonstration of a brief Late-atlantic clearance in County Antrim (Newferry and Ballyscullion) allows of the conjecture that the *landnam* phase might be one of a series of such clearances and that during the *landnam* phase the vegetational changes are intensified because of a synergistic climatic effect. This speculation clearly needs to be tested by the construction of very detailed pollen diagrams in selected areas.

The long-term effect of the early clearances, as Iversen (1949) implies, probably depended to a large extent on the soil conditions. In infertile areas (e.g. Cannons Lough, Co. Londonderry, Smith, 1961*a*) the first marked clearance appears to have had lasting effects on the forest cover, though it is possible that earlier small scale clearances took place which did not register clearly in the diagram. In most of the pollen diagrams from Great Britain, once the elm curve has fallen at the point where pollen of *Plantago* and other open habitat plants appears in some frequency, it does not recover. We may not be able to interpret the full significance of this in vegetational terms, but it does imply that a permanent change occurred. Exceptions can be found, however, in the lowlands of the English Lake District (Smith, 1958*c*, 1959, 1961*b*; Oldfield & Statham, 1963; Oldfield, 1963; Walker, 1965, 1966) and in some areas in Scotland (Durno, 1965).

The idea of selective utilization of elm as cattle fodder by Neolithic peoples, propounded by Troels-Smith (1960) is beginning to receive some support as an explanation of the elm decline in some areas of the British Isles (Pennington, 1964, 1965; Walker, 1966). There is no direct evidence for such utilization in the British Isles, but it would serve as an explanation of a particular feature of the *landnam* phases that have been worked out in detail in Ireland (Fallahogy: Smith, 1958*a*, Smith & Willis, 1962; Ballyscullion:

Smith & Crowder, unpublished). At the beginning of stage 1 (the clearance stage of Iversen) the elm curve declines only gently; at the beginning of stage 2 (the farming stage) the elm curve declines steeply. If at this point of time elm leaves were collected there would be no need to assume in every case that elm was growing in relatively pure stands that were deliberately sought and cleared, as is described as a possibility by Morrison (1959) and considered virtually a certainty by Mitchell (1956, 1965*a*, *b*).

### Oak and alder

It is curious that so far little evidence has been found of the clearance of oak by Neolithic peoples and, while they may have been as selective as is supposed by Mitchell (1956, 1965*a*, *b*), it could be that the mode of construction of the pollen diagrams at present available masks an effect on this species. In several Irish *landnam* phases the oak curve rises rather than falls; this is hardly to be interpreted as an actual increase of oak, but rather that it was less affected by clearance than elm, or not affected at all. But if the total amount of tree pollen, on an absolute basis, became reduced during the *landnam* phase then it could well be that oak, and other trees, were affected. Iversen (1941) shows that in Denmark there was a decrease of pollen frequency of all tree pollen types at the beginning of the *landnam* phase; Oldfield (1963) records indications of a decrease of absolute tree pollen frequency immediately above the 'primary' elm decline. An increase in the proportion of microspores of bog plants during clearance phases noticed in some Irish diagrams (Fallahogy: Smith, 1958*a* and Ballyscullion: Smith & Crowder, unpublished) could have similar implications. It may be that the decline of elm is over-emphasized both in the pollen diagrams and in the literature! This problem can be overcome only when so-called 'absolute' pollen diagrams are available. A similar question arises in relation to the rise of the *Alnus* curve during the *landnam* phases; is this an effect of clearance of trees other than alder or did it enter into forest regeneration (cf. Oldfield, 1963; Smith, 1964; Walker, 1965)?

### Lime, pine and ash

Aside from elm, there are indications, albeit variable, that human activity may have affected both lime and pine in the Neolithic period. The curve for *Tilia* in some cases shows a decline where the first herb pollen indicative of clearance occurs (e.g. at Hatfield Moors, Yorkshire: Smith, 1958*b*). On the whole such a decline appears to be more common in the north (for a discussion of a possible climatic effect see Smith, 1961*b*). In the south of the English Lake District, Oldfield (1963) sees an effect on lime at some sites, followed by a recovery, and Turner (1962) has found evidence of a human effect on lime in Somerset, though perhaps not a selective one, at about 2000 B.C.

The final decline of the pine curve in some Irish diagrams comes exactly at the point where intensive clearance activity is indicated for the second time in the Sub-boreal period (around 2500 B.C. for instance at Fallahogy: Smith & Willis, 1962) and there seems little doubt that it too was reduced or even locally eliminated by man.

It has become increasingly clear in the last few years that some change occurred in the status of *Fraxinus* during the period of the early forest clearances. An increase in the frequency of the occurrence of ash pollen can be seen at the elm decline in many diagrams, for instance at Holcroft Moss, Lancashire and Lindow Moss, Cheshire (Birks, 1965) and at Haweswater, Lancashire (Oldfield, 1960). In cases such as these it seems likely that *Fraxinus* entered into forest regeneration or came to occupy habitats previously held by *Ulmus* and that it was able to maintain a foothold in the woodland from that time onwards. In other diagrams the behaviour of *Fraxinus* is not so clear-cut; sometimes it makes very little contribution to the pollen rain until quite recent times, as in Bartley's diagrams from Radnorshire (Bartley, 1960). Differences in soil conditions do not seem to provide an entirely satisfactory explanation of these discrepancies. In Irish pollen diagrams, *Fraxinus* fairly consistently forms a closed curve some little time after the elm decline, probably around 2500 B.C. (Watts, 1960; McAulay & Watts, 1961; Smith & Willis, 1962). As Mitchell (1956) points out, it may be that the opening up of artificial clearances gave *Fraxinus* the opportunity to expand. Whether, before this and during the early *landnam* type of clearance, ash was confined to few habitats (thin base-rich soils or limestone rock slopes), as Mitchell (1965*a*, *b*) supposes, is more of a problem. From its present-day behaviour we might expect *Fraxinus* to enter immediately into the regeneration cycle on base-rich soils. Perhaps, during the early clearances, it was used for cattle fodder (it is apparently a desirable species: *v.* Troels-Smith, 1960) and it made little headway. The main increase of *Fraxinus* in the Irish diagrams (cf. especially, Fallahogy: Smith, 1958 and Littleton Bog: Mitchell, 1965*a*, *b*) occurs where the *Ulmus* curve begins to decline slowly for the second time; just at this point there are few indications of human activity and a climatic effect cannot be ruled out (cf. Smith, 1964).

## Some further problems

The question as to whether the early clearances of the *landnam* type really represent a shifting type of agriculture demands the examination of many more radiocarbon dated profiles and, in particular, a number of such profiles from sites in close proximity. The radiocarbon dates so far available (for summaries see Godwin, 1960*a*; Nilsson, 1964; Clark, 1965; further discussion is given by Seddon, 1967) do not permit of any firm conclusion on this point. In Ireland, it is apparent that in many cases, soon after the regeneration consequent on the first of the *landnam* type clearances, a second clearance was carried out. This second clearance appears almost always to have had permanent effects on the physiognomy of the vegetation. It is perhaps likely that these second clearances belong to the end of the Neolithic period (cf. Smith & Willis, 1962; McAulay & Watts, 1961; Mitchell, 1965*a*, *b*); the two clearances may represent separate attacks by man on the forest vegetation as a result of economic or other pressures rather than separate brief clearances in a system of shifting agriculture. The actual size of the clearances is also a matter for discussion; it may be, as Turner (1964) points out, that they were on a much larger scale than has generally been imagined.

The possibility of some climatic effect at the Atlantic–Sub-boreal transition will not be pursued further here, but the idea that factors other than human activity affected the vegetation cannot be dismissed. Pennington (1964, 1965) and Mackereth (1965) have pointed out that, in the English Lake District, soil conditions may have been adversely affected by human activity but Walker (1966, p. 196) has seen reason to suggest that, in the Cumberland lowland, a progressive naturally occurring soil depauperation was the underlying cause of the decline of elm, though human activities much overshadowed the effects of this process in populated areas. Soil conditions will undoubtedly have contributed to the final result of any disturbance by man of the balance between the vegetation and the environment. The instance of a permanent vegetational change after the first clearance in the sandy area at Cannons Lough, Co. Londonderry (Smith, 1961*a*) already referred to, again comes to mind. It may be difficult, however, to distinguish effects of edaphic conditions from effects of continuous usage by man. As Godwin (1965) points out, from the time of the earliest Neolithic forest clearances we may expect in some instances that the openings would not have been allowed to revert to woodland. Evidence for continued utilization of clearings has indeed been found by Walker (1966) in the Cumberland lowland, where he deduces a second phase of clearance of shrub growth after the primary clearance. It now appears reasonably clear that in those cases in Ireland where regeneration took place it was hazel above all that sprang up in the clearings (Smith & Willis, 1962). In Somerset, Godwin has found incontrovertible evidence that this was the case in the large quantities of straight thin hazel stems utilized in the construction of Neolithic trackways across the Levels (Godwin, 1960*b*; Dewar & Godwin, 1963). Despite the importance of hazel in this respect it will be one of the main points of interest in future work to attempt to delineate the different types of secondary community that sprang up after clearance and to relate these to underlying factors.

## SUMMARY

There can be little doubt that Neolithic man had profound effects on the vegetation of the British Isles. In some areas, depending perhaps mainly on the soil conditions, the forests were able at first to recover from his depredations. But by the end of the Neolithic period almost everywhere the character of the forest vegetation had been changed. We should beware of placing too much emphasis on the decline of elm at this time, however, as there are difficulties of interpretation which can be resolved only when reliable absolute pollen diagrams become available.

The question as to whether the Neolithic clearances described in the British Isles represent a system of short-term shifting agriculture is seen still to be open. There are some grounds for doubting that this was the case, and evidence is beginning to accumulate that there was considerable variety in the types of land use. The duration of utilization of particular small areas is almost completely unknown. The possibility of clearance not associated with agriculture must be borne in mind; the presence in pollen diagrams of *Plantago lanceolata* pollen alone should no longer be regarded as indicative of anything more than open conditions. The idea that man had no marked effect on the forest vegetation until the Neolithic period must be held in question.

How far naturally-occurring soil degradation and climatic variation were responsible for the vegetational changes during the Neolithic period remains a debatable question. Both soil and climatic conditions, however, and any changes they underwent, presumably much affected the final outcome of the Neolithic clearances: the amount of regeneration, the specific constitution of the secondary woodland and the possibilities of continued utilization of the cleared areas.

Just as in the Neolithic period man's exploitation of his environment seems to have had a variable effect, so we are beginning to see that, even in the Atlantic period, man's activities may have induced the formation of secondary communities of various kinds. It is the scale and duration of the effects that we know least about. This is a problem that can be tackled by looking at the present-day pollen productivity of areas under different types of land use, and by an attempt in favourable circumstances actually to delimit the areas cleared in prehistoric times.

In the earlier part of the Post-glacial period, we have remarked on the often close correlation between Mesolithic occupation layers and marked vegetational changes which have usually been explained as climatic effects or as natural developments. Examples of this association are probably as numerous, and certainly as striking, as for the Neolithic period. Is it possible that a small Mesolithic population could have brought about these radical changes? The problem is one both of ecological feasibility and of credibility. Certainly the use of fire, particularly under conditions drier than those of today, could have had widespread and irreversible effects. And indeed there is commonly evidence of fire both in the Mesolithic settlements and, in some instances, where little indication of the presence of man has been found. But we do not know how widespread these fires were. If the vegetational changes were due to rapid climatic change, and involved the actual death of forest trees, conditions would have been ideal for the propagation of fires over wide areas. Whether such fires occurred, and if so whether they were started naturally or by man, is imponderable. Nevertheless, that man's activities tipped the scales in favour of vegetational change in quite early times where the vegetation was under tension is perhaps more than a possibility. In the Neolithic period the difficulty of knowing in what ways and in what areas the scales were weighted remains a major problem. The characterization of the secondary communities, however, is something that can be tackled.

In preparing this article the author has had the benefit of continuing discussion with Mr J. R. Pilcher.

## REFERENCES

Bartley, D. D. (1960). Rhosgoch Common, Radnorshire: Stratigraphy and pollen analysis. *New Phytol.* **59**, 238–62.

Bartley, D. D. (1966). Pollen analysis of some lake deposits near Bamburgh, Northumberland. *New Phytol.* **65**, 141–56.

Birks, H. J. B. (1965). Pollen analytical investigations at Holcroft Moss, Lancashire, and Lindow Moss, Cheshire. *J. Ecol.* **53**, 299–314.

Boughey, A. S. (1963). Interaction between animals, vegetation, and fire in Southern Rhodesia. *Ohio J. Sci.* **63**, 193–209.

Braidwood, R. J. (1960). Levels in prehistory: a model for the consideration of the evidence. In *Evolution after Darwin* (ed. S. Tax), **2**, pp. 143–51. Chicago.

Chavannes, E. (1941). Written records of forest succession. *Scient. Monthly* July 1941, 76–80.

Churchill, D. M. & Wymer, J. J. (1965). The kitchen midden site at Westward Ho!, Devon, England: ecology, age and relation to changes in land and sea level. *Proc. Prehist. Soc.* **31**, 74–84.

Clark, J. G. D. (1954). *Excavations at Star Carr*, p. 200. Cambridge.

Clark, J. G. D. (1965). Radiocarbon dating and the expansion of farming culture from the Near East over Europe. *Proc. Prehist. Soc.* **31**, 58–73.

Clark, J. G. D. & Godwin, H. (1962). The Neolithic of the Cambridgeshire Fens. *Antiquity* **36**, 10–23.

Clark, J. G. D., Godwin, H. & Clifford, M. H. (1935). Report on recent excavations at Peacock's Farm, Shippea Hill, Cambridgeshire. *Antiq. J.* **15**, 284–319.

Conolly, A. P., Godwin, H. & Megaw, E. M. (1950). Studies in the Post-glacial History of British vegetation. XI. Late-glacial deposits in Cornwall. *Phil. Trans. R. Soc.* B **234**, 397–469.

Conway, V. M. (1948). Von Post's work on climatic rhythms. *New Phytol.* **47**, 220–37.

Davies, J. (1963). A Mesolithic site on Blubberhouses Moor, Wharfedale, West Riding of Yorkshire. *Yorks. Archaeol. J.* **41**, 61–70.

Degerbøl, M. (1963). Prehistoric cattle in Denmark and adjacent areas. *Roy. Anthrop. Inst. Gt Brit. Occ. Paper No. 18*, 69–79.

De la Pryme, A. (1701). Part of a letter concerning trees found underground in Hatfield Chase. *Phil. Trans. R. Soc.* **22** (no. 275), 980–92.

Dewar, H. S. L & Godwin, H. (1963). Archaeological discoveries in the Somerset Levels, England. *Proc. Prehist. Soc.* **29**, 17–49.

Dimbleby, G. W. (1961). The ancient forest of Blackamore. *Antiquity* **35**, 123–8.

Dimbleby, G. W. (1962). *The development of British heathlands and their soils* (*Oxford Forestry Memoirs No. 23*). Oxford.

Dimbleby, G. W. (1963). Pollen analysis of a Mesolithic site at Addington, Kent. *Grana Palynologica* **4**, 140–8.

Donner, J. J. (1962). On the Post-glacial history of the Grampian Highlands. *Soc. Scient. Fennica, Commentat. Biol.* **24**, 6, 1–29.

Durno, S. E. (1956). Pollen analysis of peat deposits in Scotland. *Scot. Geogr. Mag.* **72**, 177–87.

Durno, S. E. (1965). Pollen analytical evidence of 'landnam' from two Scottish sites. *Trans. Bot. Soc. Edinb.* **40**, 13–19.

Durno, S. E. & McVean, D. N. (1959). Forest history of the Beinn Eighe nature reserve. *New Phytol.* **58**, 228–36.

Egede Larssen, K. (1949). Pollenanalytiske undersøkelser i indre Østfold. *Univ. Bergen Årb., Naturvit.* **13**, 1–16.

Elhai, H. (1960). La tourbière de Gathémo (Manche-Normandie). *Pollen et Spores* **2**, 263–74.

Erdtman, G. (1928). Studies in the Post-arctic history of the forests of northwestern Europe. I. Investigations in the British Isles. *Geol. Foren. Stockh. Förh.* **50**, 123–92.

Florin, S. (1961). De äldsta skogarna och det första åkerbruget. *Pub. Inst. Quat. Geol. Univ. Uppsala No. 19*, 329–430.

Frenzel, F. (1966). Climatic change in the Atlantic/Sub-boreal transition on the Northern Hemisphere: botanical evidence. In *World Climate from 8000 to 0 B.C.* (ed. J. S. Sawyer), pp. 89–123. Roy. Meteorological Society, London.

Godwin, H. (1940). Studies of the Post-glacial history of British vegetation. III. Fenland pollen diagrams. *Phil. Trans. R. Soc.* B **570**, 239–85.

Godwin, H. (1956). *The History of the British Flora.* Cambridge.

Godwin, H. (1959). The history of weeds in Britain. *Brit. Ecol. Soc. Symp. No. 1*, 1–10.

Godwin, H. (1960*a*). Radiocarbon dating and Quaternary history in Britain. *Proc. Roy. Soc.* B **153**, 287–320.

Godwin, H. (1960*b*). Prehistoric wooden trackways of the Somerset Levels: their construction, age and relation to climatic change. *Proc. Prehist. Soc.* **26**, 1–36.

Godwin, H. (1965). The beginnings of agriculture in north-west Europe. In *Essays on crop plant evolution* (ed. J. B. Hutchinson), pp. 1–22. Cambridge.

Godwin, H. & Clifford, M. H. (1938). Studies of the Post-glacial history of British vegetation. I. Origin and stratigraphy of Fenland deposits near Woodwalton, Hunts. II. Origin and stratigraphy of deposits in southern Fenland. *Phil. Trans. R. Soc.* B **229**, 323–409.

Godwin, H. & Tallantire, P. A. (1951). Studies in the post-glacial history of British vegetation. XII. Hockham Mere, Norfolk. *J. Ecol.* **39**, 285–307.

Hafsten, U. (1965). The Norwegian *Cladium mariscus* communities and their Post-glacial history. *Årb. Univ. Bergen, Mat. Naturvit. Ser. No. 4*, 1–55.

Iversen, J. (1941). Land occupation in Denmark's Stone Age. (Landnam i Danmarks Stenalder.) *Danm. Geol. Unders.* (II), **66**, 1–68.

Iversen, J. (1949). The influence of prehistoric man on vegetation. *Danm. Geol. Unders.* (IV), **3**, 6, 1–25.

Iversen, J. (1960). Problems of the early Post-glacial forest development in Denmark. *Danm. Geol. Unders.* (IV), **4**, 3, 1–32.

Jessen, K. (1935). The composition of the forests in Northern Europe in Epipalaeolithic time. *Kg. Danske Videnskab. Selskab., Biol. Medd.* **12**, 1–64.

Jessen, K. (1936). Palaeobotanical report on the Stone Age site at Newferry, County Londonderry. Appendix I in Movius, H. L. (1936), A Neolithic site on the River Bann. *Proc. R. Ir. Acad.* **43** C, 17–40.

Jessen, K. (1949). Studies in Late Quaternary deposits and vegetational history of Ireland. *Proc. R. Ir. Acad.* **52** B, 85–290.

Jewell, P. (1963). Cattle from British Archaeological sites. *Roy. Anthrop. Inst. Gt Brit., Occ. Paper No. 18*, 80–101.

Jørgensen, S. (1963). Early Post-glacial in Aamosen. *Danm. Geol. Unders.* (II), **87**, 1–36.

Keef, P. A. M., Wymer, J. J. & Dimbleby, G. W. (1965). A Mesolithic site on Iping Common, Sussex, England. *Proc. Prehist. Soc.* **31**, 85–92.

Mackereth, F. J. H. (1965). Chemical investigations of lake sediments and their interpretation. *Proc. Roy. Soc.* B **161**, 295–309.

McAndrews, J. H. (1966). Post-glacial history of prairies, savanna, and forest in northwestern Minnesota. *Mem. Torrey Bot. Club* **22**, 1–72.

McAulay, I. R. & Watts, W. A. (1961). Dublin radiocarbon dates, I. *Radiocarbon* **3**, 26–38.

McVean, D. N. (1964). History and pattern of Scottish vegetation: prehistory and ecological history, in *The vegetation of Scotland* (ed. J. H. Burnett), pp. 561–7. Edinburgh.

Mitchell, G. F. (1955). A Mesolithic site at Toome Bay, Co. Londonderry. *Ulster J. Archaeol.* **18**, 1–16.

Mitchell, G. F. (1956). Post-boreal pollen diagrams from Irish raised bogs. *Proc. R. Ir. Acad.* **57** B, 185–251.

Mitchell, G. F. (1965*a*). Littleton Bog, Tipperary: and Irish vegetational record. *Geol. Soc. Amer. Special Paper 84*, 1–16.

Mitchell, G. F. (1965*b*). Littleton Bog, Tipperary: an Irish agricultural record. *J. R. Soc. Antiq. Ir.* **95**, 121–32.

Moore, J. W. (1950). Mesolithic sites in the neighbourhood of Flixton, North-East Yorkshire. *Proc. Prehist. Soc.* **16**, 101–8.

Moore, J. W. (1954). Excavations at Flixton, site 2. Appendix, pp. 192–4, in J. G. D. Clark (1954) *Excavations at Star Carr*, Cambridge.

Morrison, M. E. S. (1959). Evidence and interpretation of 'Landnam' in the North-East of Ireland. *Bot. Notiser* **112**, 185–204.

Movius, H. L. (1936). A Neolithic site on the River Bann. *Proc. R. Ir. Acad.* **43** C, 17–40.

Müller, I. (1947). Der pollenanalytische Nachweis der Menschlichen Besiedlung im Federsee und Bodenseegebeit. *Planta* **35**, 70–87.

Nilsson, T. (1960). Ein neues Standardpollendiagramm aus Bjärsjoholmssjön in Schonen. *Lunds Univ. Årsskr.* N.F. **56**, 18, 1–34.

Nilsson, T. (1964). Standardpollendiagramme und $C^{14}$-Datierungen aus dem Ageröds mosse im mittleren Schonen. *Lunds Univ. Årsskr.* N.F. **59**, 7, 1–52.

Nilsson, T. (1967). Pollenanalytische Datierung mesolithischer Siedlungen im Randgebiet des Ageröds mosse im mittleren Schonen. *Acta Univ. Lund. II, No. 16*, 1–80.

Oldfield, F. (1960). Studies in the Post-glacial history of British vegetation: Lowland Lonsdale. *New Phytol.* **59**, 192–217.

Oldfield, F. (1963). Pollen analysis and man's role in the ecological history of the south-east Lake District. *Geogr. Annlr.* **45**, 23–40.

Oldfield, F. (1964). Late-Quaternary deposits at Le Moura, Biarritz, south-west France. *New Phytol.* **63**, 374–409.

Oldfield, F. (1965). Problems of mid-Post-glacial pollen zonation in part of north-west England. *J. Ecol.* **53**, 247–60.

Oldfield, F. & Statham, D. C. (1963). Pollen analytical data from Urswick Tarn and Ellerside Moss, North Lancashire. *New Phytol.* **62**, 53–66.

Pearson, M. C. (1960). Muckle Moss, Northumberland. I. Historical. *J. Ecol.* **48**, 647–66.

Pennington, W. (1964). Pollen analyses from the deposits of six upland tarns in the Lake District. *Phil. Trans. R. Soc.* B **248**, 205–44.

Pennington, W. (1965). The interpretation of some Post-glacial vegetation diversities at different Lake District sites. *Proc. Roy. Soc.* B **161**, 310–23.

Pigott, C. D. & Pigott, M. E. (1963). Late-glacial and Post-glacial deposits at Malham, Yorkshire. *New Phytol.* **62**, 317–34.

Radley, J. & Mellars, P. (1964). A Mesolithic structure at Deepcar, Yorkshire, England, and the affinities of its associated flint industries. *Proc. Prehist. Soc.* **30**, 1–24.

Rankine, W. F., Rankine, W. M. & Dimbleby, G. W. (1960). Further investigations at a Mesolithic site at Oakhanger, Selbourne, Hants. *Proc. Prehist. Soc.* **26**, 246–302.

Rawitscher, F. (1945). The hazel period in the post-glacial development of forests. *Nature, Lond.* **156**, 302–3.

Roux, I. & Leroi-Gourhan, A. (1965). Les défrichments de la période atlantique. *Bull. Soc. préhist. fr. 61*, 309–15.

Sagar, G. R. & Harper, J. L. (1964). Biological Flora of the British Isles. *Plantago major* L., *P. media* L. and *P. lanceolata* L. *J. Ecol.* **52**, 189–221.

Schmitz, H. (1961). Pollenanalytische Untersuchungen in Hohen Viecheln am Schweriner See. *Deutsche Akad. Wissensch. Berlin, Schr. Sekt. Vor-Frugesch.* **10**, 14–38.

Seagrief, S. C. & Godwin, H. (1960). Pollen diagrams from southern England: Elstead, Surrey. *New Phytol.* **59**, 84–91.

Seddon, B. (1967). Prehistoric climate and agriculture: a review of recent palaeoecological investigations, in J. A. Taylor (1967). *Weather and Agriculture*, Oxford.

Simmons, I. G. (1964). Pollen diagrams from Dartmoor. *New Phytol.* **63**, 165–80.

Singh, G. (1964). *Studies of the Late Quaternary vegetational history and sea-level changes in Co. Down, N. Ireland.* Unpublished Ph.D. thesis, Queen's University, Belfast.

Singh, G. & Smith, A. G. (1966). The Post-glacial marine transgression in Northern Ireland—conclusions from estuarine and 'raised beach' deposits: a contrast. *Palaeobotanist* **15**, 230–34.

Smith, A. G. (1958*a*). Pollen analytical investigations of the mire at Fallahogy Td. Co. Derry. *Proc. R. Ir. Acad.* **59** B, 329–43.

Smith, A. G. (1958*b*). Post-glacial deposits in south Yorkshire and north Lincolnshire. *New Phytol.* **57**, 19–49.

Smith, A. G. (1958*c*). Two lacustrine deposits in the south of the English Lake District. *New Phytol.* **57**, 363–86.

Smith, A. G. (1959). The mires of south-western Westmorland: stratigraphy and pollen analysis. *New Phytol.* **58**, 105–27.

Smith, A. G. (1961*a*). Cannons Lough, Kilrea, Co. Derry: stratigraphy and pollen analysis. *Proc. R. Ir. Acad.* **61** B, 369–83.

Smith, A. G. (1961*b*). The Atlantic–Sub-boreal transition. *Proc. Linn. Soc. Lond.* **172**, 38–49.

Smith, A. G. (1964). Problems in the study of the earliest agriculture in Northern Ireland. *Rep. VI Int. Cong. Quat., Warsaw, 1961*, **2**, 461–71.

Smith, A. G. (1965). Problems of inertia and threshold related to post-glacial habitat changes. *Proc. Roy. Soc.* B **161**, 331–42.

Smith, A. G. & Willis, E. H. (1962). Radiocarbon dating of the Fallahogy landnam phase. *Ulster J. Archaeol.* **24–5**, 16–24.

Steven, H. M. & Carlisle, A. (1959). *The native pinewoods of Scotland*, p. 368. Edinburgh.

Tallis, J. H. (1964). The pre-peat vegetation of the southern Pennines. *New Phytol.* **63**, 363–73.

Tauber, H. (1965). Differential pollen dispersion and the interpretation of pollen diagrams. *Danm. Geol. Unders.* (II), **89**, 1–69.

Troels-Smith, J. (1953). Ertebøllekultur–bondekultur. *Aarb. Nord. Oldkynd. Hist.* 1953, 1–62.

Troels-Smith, J. (1955). Pollenanalytischen Untersuchungen zu einigen schweizerischen Pfahlbauproblemen. In *Das Pfahlauproblem* (ed. W. U. Guyan) *Monograph. Ur- und Fru-gesch. Schweiz. II.* Basle.

Troels-Smith, J. (1960). Ivy, Mistletoe and Elm: climatic indicators-fodder plants. *Danm. Geol. Unders.* (IV), **4**, 4, 1–32.

Turner, J. (1962). The *Tilia* decline: an anthropogenic interpretation. *New Phytol.* **61**, 328–41.

Turner, J. (1964). Surface sample analysis from Ayrshire, Scotland. *Pollen et Spores* **6**, 583–92.

Van Zeist, W. (1963). Researches palynologiques en Bretagne occidentale. *Norois* **37**, 5–19.

Van Zeist, W. (1964). A palaeobotanical study of some bogs in western Brittanny (Finistère), France. *Palaeohist.* **10**, 157–80.

Wainwright, G. J. (1960). Three microlithic industries from south-west England and their affinities. *Proc. Prehist. Soc.* **26**, 193–201.

Walker, D. (1956). A site at Stump Cross, near Grassington, Yorkshire, and the age of the Pennine microlithic industry. *Proc. Prehist. Soc.* **22**, 23–8.

Walker, D. (1965). The Post-glacial period in the Langdale Fells, English Lake District. *New Phytol.* **64**, 488–510.

Walker, D. (1966). The Late-Quaternary history of the Cumberland Lowland. *Phil. Trans. R. Soc.* B **251**, 1–210.

Walker, D. & Godwin, H. (1954). Lake stratigraphy, pollen analysis, and vegetational history, in J. G. D. Clark (1964) *Excavations at Starr Carr.* Cambridge.

Watts, W. A. (1960). $C^{14}$-dating and the Neolithic in Ireland. *Antiquity* **34**, 111–16.

Welten, M. (1944). Pollenanalytische, stratigraphische, und geochronologische Untersuchungen aus dem Faulenseemoos bei Spiez. *Veröff. geobot. Inst. Rübel* **21**, 1–201.

# POST-NEOLITHIC DISTURBANCE OF BRITISH VEGETATION

by Judith Turner

*St Aidan's College and Department of Botany, University of Durham*

The idea of our British countryside covered for mile after mile with primaeval forest easily fires the imagination. How different it would have been from today's patchwork of green fields and hedges, broken only by the sprawling urban centres and their network of connecting roads, or even from the desolate stretches of rough pasture and moorland such as those in the Pennines, Lake District, Wales and Scotland.

Little wonder, then, that those interested in the history of vegetation should seek not only to understand the part played in the evolution of our present vegetation by physical factors such as the nature of the underlying rocks, changes in climate and development of soils, but also the overriding determining influence which man himself has had in transforming it. For he has altered our primaeval woodlands from when he first began using the products of the forests in sufficient quantity to modify their nature until the present day when we ourselves must bear the responsibility for so much of the highly artificial nature of our so-called natural environment.

This topic has always been an integral part of studies in the history of our flora and along with other aspects took a major step forward with the application of pollen analysis to Post-glacial deposits in the fourth decade of this century. It seems quite unnecessary to summarise the results obtained by this technique prior to 1956, for Professor Harry Godwin has given such a fascinating account of them in *The History of the British Flora.* To him we owe the first recognition from pollen diagrams of the effects of prehistoric man in clearing the British forests at Hockham Mere on the edge of the Breckland, where shortly after the beginning of pollen Zone VIIb the tree pollen decreases and the pollen of grasses and species of plantain become important constituents of the total assemblage. Godwin (1944) attributed this replacement of trees by grasses and other plants of open habitats to the Neolithic people who are known to have lived on the Breck. But for palynological evidence for clearance in other parts of the country Godwin was at that time only able to whet the appetite with references to Dimbleby's work on the Yorkshire Moors, Conway's on the Pennines, and his own from the Somerset levels. I would like simply to take up the theme at this point and try to show how far his prophetic comment 'The work of British palynologists in this field has hardly begun but the richness of results available is not open to doubt' (1956, p. 333) has subsequently been justified.

The stage was well set for such investigations in the mid-fifties, because the general changes which had been brought about in our flora since the beginning of the Neolithic were well documented. Forest trees and their associated herb communities were known to have decreased in importance although, at least until comparatively recently, scrub communities had expanded. The plants of wet habitats had decreased as more and more areas had been drained and brought under cultivation. On the other hand our flora had become progressively enriched with weeds and ruderals, either by species which had survived the forested period in a restricted number of more open habitats and had subsequently found opportunity to spread, or by the aliens which man had brought into the country with him. Crops, wheat, barley, rye, flax and hemp among them, had begun to play a more important role in the vegetation and even their relative abundance in the various archaeological periods was apparent.

The stage was well set. It had, however, a backcloth somewhat different from that of today, namely the unconscious assumption that the climatic factors, which were thought to have been mainly responsible for the earlier changes in the Post-glacial forests, were still of considerable import. In Godwin's words 'In its present state of development pollen analysis is unable to decide whether the forest clearance in Britain was essentially climatic in cause or anthropogenic' (1956, p. 339). Today few people would question the idea that anthropogenic factors were the more important and that the climatic changes which undoubtedly occurred had a less direct effect in causing forest clearance.

But if the backcloth changes as the story unfolds, the techniques do not. Pollen analysis is still as important as ever, and its potentialities just as far-reaching, although, more and more, radiocarbon dating has been stealing the limelight in recent studies by providing an exact chronology of episodes of deforestation. Furthermore, a series of dates from one profile can give the rate of formation of the deposit and hence the approximate number of years represented in any one pollen sample. Combined with close sampling over really critical horizons, very fine details can thereby be won from deposits. It is possible to slice deposits of a good cheesy consistency into $\frac{1}{2}$ cm or even, with care, $\frac{1}{4}$ cm thick pieces so that the pollen rain of only a few years can sometimes be separated from that of the next few years quite successfully and very small fluctuations in the vegetation revealed.

Since 1956 there has been a large number of regional palynological studies, often backed by radiocarbon dating of critical horizons. These have provided, as it were, the jigsaw pieces which are gradually being fitted together to form a complete picture. Some of these studies have had objectives other than elucidating man's influence on the vegetation of the region and have simply unearthed pieces for the puzzle in passing. Others have been aimed more specifically at providing a completed section of the puzzle and, although the picture is still far from finished, its outlines are slowly becoming clearer.

Before actually considering how much of this jigsaw has been pieced together for the various periods, starting with the Bronze Age, we will look briefly at the way in which Neolithic man appears to have interfered with the natural forest vegetation.

First there was the making of small cleared areas in the forests in which a few crops, such as cereals, were grown and around which domesticated or semi-domesticated animals were allowed to browse (Iversen, 1949). These clearings were abandoned after a few years allowing the forest to regenerate gradually. Such clearings are marked by a temporary decrease in total tree pollen together with substantial increases in the pollen of grasses and plantain and also bracken spores, as well as the occasional grain of other ruderal and crop species in relevant pollen diagrams. After the clearings were abandoned, the pollen spectrum soon returns to something like its former composition.

The second kind of impact, described by Troels-Smith (1960), was the selective use of the leaves and shoots of certain trees, including elm, as fodder for cattle which were kept tethered for at least part of the year. This simply causes an alteration in the composition of the tree pollen, a decrease in elm being the most common, with only a very occasional ruderal pollen grain appearing in the diagram. It is really a modification of the forest rather than a clearance. Another modification which has been recognized is the coppicing of hazel (Dewar & Godwin, 1963), which has the effect of increasing the hazel pollen frequency.

On the European continent these two effects have been closely associated with two archaeologically distinct cultures and whilst this, admittedly, is extremely satisfying, we must be careful not to assume the same to be true in Britain, for, as we shall see, some of our pollen diagrams do not show these two effects so clearly. For example, Godwin's diagram from Hockham Mere (1944) shows that Neolithic man was affecting the forests of the Breck in a manner quite different from either of those described from Denmark. The diagram has continuous high frequencies of pollen types indicating open areas, showing that parts at least of the Breck were permanently clear of forests. Whether this means that it was always the same area that was cleared is quite another question, and one which will be discussed later, but, certainly, a proportion of the land from which the pollen was derived was always free from forest which was not the case in Iversen's *landnam*.

Therefore in considering the Bronze Age vegetation of different parts of the country, we will also briefly consider in each case the effect which Neolithic man had already had there, in order to see if the Bronze Age pattern was any different and, if it was, when and how it became so.

## BRONZE AGE CLEARANCE

We are all familiar with the idea of regional contrasts and that such contrasts may have existed throughout pre-history is not altogether surprising. Certainly with regard to man's effect on the forest vegetation, a regional contrast had developed by the beginning of the Bronze Age; a contrast, as far as one can tell from the inadequate number of jigsaw pieces, between the south-eastern chalklands of England and the whole of the rest of the country.

It is rather unfortunate that the south-east chalkland, long regarded on archaeological grounds as an area densely settled by Neolithic man, should be so

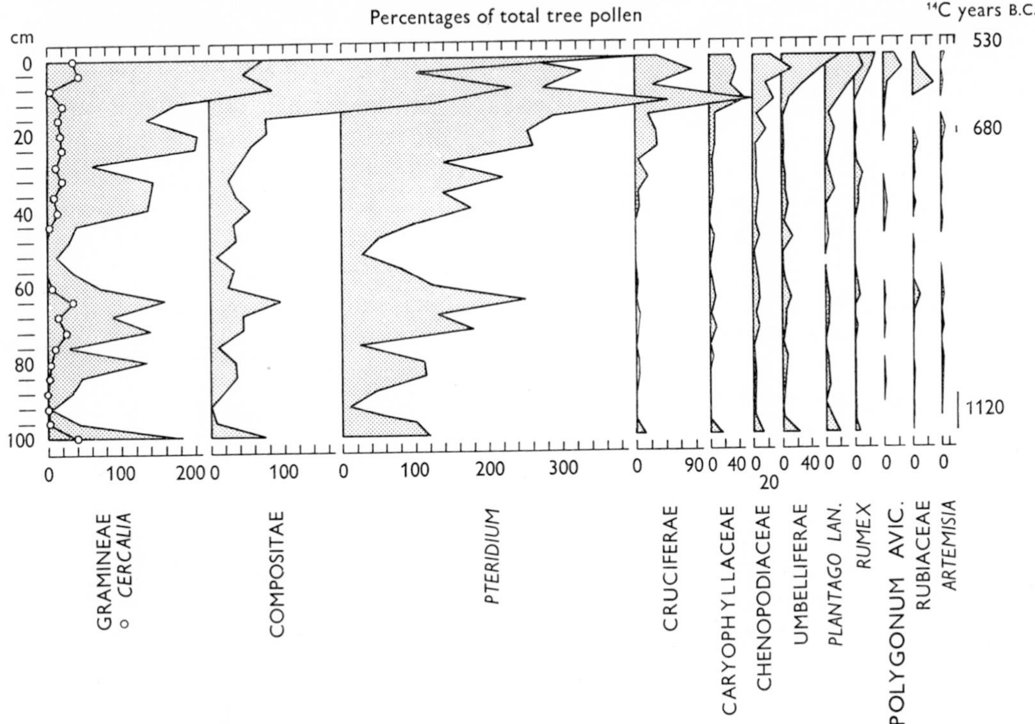


Figure 1. Pollen diagram from Frogholt, Kent: pollen of plants indicative of forest clearance and agriculture (Godwin, 1962, figure 4).

poor in plant preserving deposits. However, organic muds laid down during the Bronze Age have been found in a sewer trench near Wingham in Kent, and also in the valley of the Seabrook stream at Frogholt near Folkestone. The pollen diagrams from these muds (Godwin, 1962) are our best source of information on the Bronze Age vegetation of the region and one only regrets that they do not also cover the preceding Neolithic period.

The muds at Wingham began accumulating about 1700 B.C. and the pollen in them is predominantly herbaceous even at the base. Gramineae and Compositae pollen and *Pteridium aquilinum* spores are common and there is also pollen of a wide variety of ruderals characteristic of mixed farming. There is no notable change in the tree/non-tree pollen ratio throughout the muds, which went on accumulating until A.D. 200.

At Frogholt, the organic muds cover a shorter period, namely 1100–500 B.C. The pollen diagram (figure 1) is much the same as that from Wingham, except that the frequencies for the herbs show more marked fluctuations and, expressed as % T.T.P. (total tree pollen), have average values considerably higher than those at Wingham; Gramineae 100 % instead of 60 %, Compositae 50 % instead of 20 %, and *Pteridium* 100 % instead of 12 %.

So, as Godwin has shown from these two sites, parts at least of the chalklands of south-eastern England had been largely disforested by the Bronze Age. There also seems to have been considerable variation in the actual amount of land being farmed, both from time to time and from one part of the Downs to another; the higher and more fluctuating frequencies of herb pollen at Frogholt than at Wingham suggests this. Kerney, Brown & Chandler (1964) have recorded similar local variation on the Devil's Kneadingtrough near Brook, a steep sided coombe in the chalk escarpment some 13 km from Frogholt. In the upper part of the coombe, the molluscan fauna indicates two periods of forest clearance separated by one during which the woods regenerated: an earlier light or partial clearance and a later more drastic and thorough one. These levels are not radiocarbon dated but sherds found in the section indicate that the lower partial clearance was Bronze Age or earlier. Contrasting with this, just over 300 m away, in the entrance to the coombe, the mollusca indicate only one period of clearance and, although it was a complete clearance, it has been correlated with the lower one from further up that coombe and attributed to the Bronze Age. Kerney envisages the lower part of the coombe at this time being completely cleared by early pastoralists, possibly as a paddock for their herds, whilst the upper part was not so seriously affected.

By mid-way through the second millennium B.C. considerable tracts of the Downs had been cleared, with a good deal of local variation from place to place. How much earlier than 1700 B.C. this had begun one cannot tell from these diagrams. Whether it was simply a continuance of the Neolithic way of life or a distinct advance which took place at the very beginning of the Bronze Age are questions about which there is no direct palynological evidence in the area. Nor is there yet such evidence of the early prehistoric vegetation of the rest of the chalklands of England, although it would seem reasonable, in view of the similarly dense Neolithic and Bronze Age settlement over them, to suppose that they too had been cleared of woodland to a considerable extent by the Bronze Age.

The light sandy soils of the East Anglian Breckland do not appear to have been cleared so thoroughly. Godwin's diagram from Hockham Mere does indeed show the grass and heath land created during the Neolithic persisting throughout the Bronze Age, but if the actual frequencies for grass and heather pollen are any guide to the amount of open land present, then there was far less on the Breck than on the chalk.

But it is now fairly clear that even this degree of clearance was not typical of the whole of Britain. Pollen diagrams from other parts of the country do not show consistently high herb frequencies during the Neolithic or Bronze Ages. Instead, there are either extremely low values for herb frequencies or series of small peaks suggesting a number of temporary clearances after each of which the forest regenerated near each site. Although the rest of the country is much better documented by pollen diagrams than the south-east, no additional regional contrast has emerged so far. Generally, the pollen diagrams cover the earlier Neolithic period, and what they do show are differences between it and the succeeding Bronze Age.

These two periods are shown in reasonable detail in the pollen diagram from Bloak Moss, Ayrshire, in southern Scotland (Turner, 1965) a diagram (figure 2) which has also been radiocarbon dated (Godwin, Willis & Switsur, 1965).

From the level of the elm decline until the radiocarbon dated level of 1875 B.C., the area was well forested and only the slowly rising *Corylus* curve and the occasional grain of *Plantago* can be regarded as evidence for a slight lightening of the tree canopy, probably due to Neolithic people supporting themselves on the produce of the forest without disturbing its basic structure as envisaged by Troels-Smith.

A much less controversial interpretation may be applied to the changes in pollen frequencies between the 1875 B.C. level and the 425 B.C. level, 94 cm higher, where the Gramineae, *Plantago* and *Pteridium* curves all show series of small peaks at 110, 134 and 166 cm respectively, and *Fraxinus* appears for the first time as a small but persistent member of the woodland canopy. These episodes are so similar to those originally described by Iversen that one cannot doubt but that they represent temporary clearances in the woods, a short period of occupation followed by the regeneration of the full forest canopy. Cereal pollen grains have been recorded in very low frequencies but, on the whole, the cleared areas seem to have been dominated by grasses and bracken. The radiocarbon dates indicate that these temporary clearances began some time between 1800 and 1300 B.C. so there can be no doubt that during the Bronze Age there was a distinct intensification of man's effect on the woodlands, an economy based on deliberate forest clearance replacing one in which the woods had only been slightly modified by their Neolithic inhabitants. Similar evidence for the beginning of deliberate forest clearance at this time has been obtained from further north in Scotland at Duartbeg in Sutherland (Moar, 1964) where an initial maximum beginning the continuous curve for *Plantago lanceolata* has been radiocarbon dated to 1740 B.C.

Further south, Walker (1966) has described a very interesting pattern of clearance as the result of detailed Quaternary ecological studies of a number of mosses and tarns in the Cumberland lowlands. These diagrams demonstrate considerable variation in the pattern of woodland modification within the region. Walker proposes that the high forest trees were peeled and felled for their bark and wood, and leaves and young shoots gathered for fodder, whilst domesticated or semi-domesticated animals grazed in the occasional clearing thus created. This way of life seems to have begun as early as 4000 B.C. along the coastal strip and spread gradually through the region during the next thousand years. In places, the occasional cereal pollen grain indicates that these husbandmen were experimenting with the crops, though only on a very small scale, in their predominantly forested environment. This Neolithic economy lasted

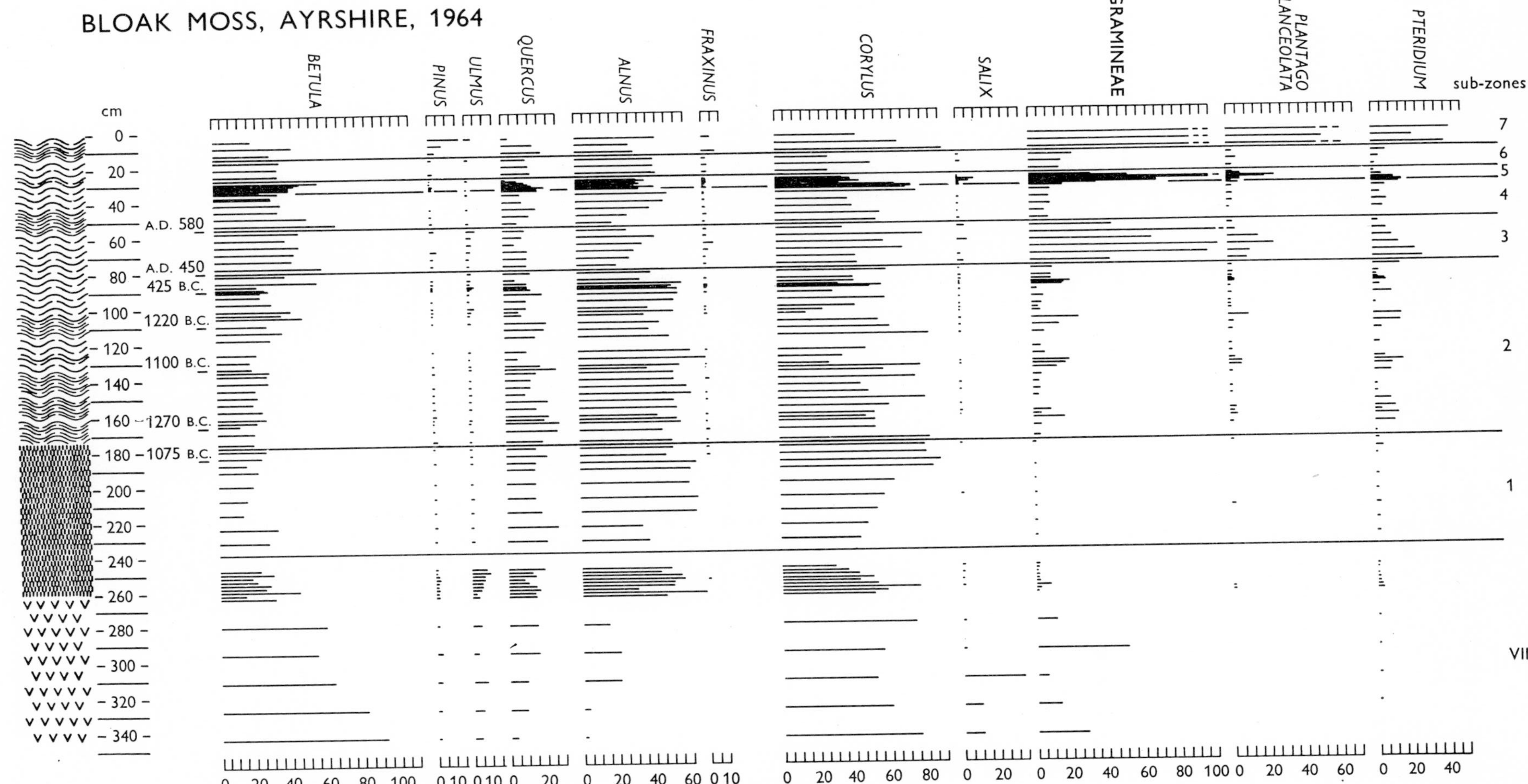


Figure 2. Pollen diagram from Bloak Moss, Ayrshire: pollen of trees and plants indicative of forest clearance (Turner, 1965).

for some time but, at all the sites examined, its indications are followed by unequivocal evidence of definite forest clearance with the widespread cultivation of cereals. The exact time at which this occurs at each site varies, but Walker's summary of his findings speaks for itself: 'The general indications are, therefore, that further clearing activity spread through the Cumberland lowland from about 2000 B.C. This was given new impetus during the period 1750 B.C. to 1400 B.C., when a new economy, more sedentary and utilizing cereals widely, became established.'

Another region adjacent to the Cumberland lowlands which has been studied in great detail is, of course, the Lake District. It differs from the areas previously discussed in being mountainous and Pennington (1965) has shown how the elevation and steep topography there have affected the vegetational history. Unlike its lowland counterpart, once the original upland forest was attacked, the mull soils tended to be lost by leaching and erosion, and the trees had difficulty in regenerating. As a result the forest was generally replaced by grassland, heath and peatland communities. Pennington has described the variation in the pattern of disforestation from one tarn or lake basin to another and has shown how the diversity is clearly associated with prehistoric activity.

Around Devoke Water, there were two episodes of interference with the forest. The first, some time after the elm decline, is marked by a fall in oak pollen accompanied by a small increase in the grass and grassland herb pollen, presumably caused by the development of a grassy ground flora in slightly opened oak woods under grazing pressure. The second, 190 cm higher up the diagram, is marked by further increases in the grass and grassland herb pollen, and indicates a marked extension of true grassland. Pennington attributes the first to the pastoral nomads who are thought to have built the great stone circles of south-west Cumberland about 1800 B.C., and the second to later Bronze Age peoples.

From south of the Lake District there is a number of diagrams which cover the Neolithic and Bronze Ages, including Oldfield's (1960) from Lowland Lonsdale and Oldfield & Statham's (1963) from Lancashire-north-of-the-sands. They have recognized three horizons related to forest clearance in the upper parts of the diagrams, labelled, from the oldest to the youngest, the UF, the UT and the BU horizons. The UF is thought to correspond to the elm decline of other regions and shows an initial forest modification followed very quickly by a little forest clearance. The UT horizon marks the beginning of a period with further and more extensive forest clearance, which, although not radiocarbon dated, almost certainly occurred during the Bronze Age.

That the pattern is similar in the central Pennines is attested by Pigott's diagram from Malham (Pigott & Pigott, 1963) in which there is an increase in the agricultural indicators, half way through Zone VIIb. From the Cheshire plain there is evidence from Lindow and Holcroft mosses (Birks, 1965) for two periods during which the forest was partially cleared and pastures created, both falling between the elm decline and the main recurrence surface which, at Chat Moss not very far away, has been dated to 695 B.C. Although pollen analytically similar, the two clearances are differentiated by higher dry land herb values in the younger than the older one. Birks attributes the first period of clearance to occupation by the Secondary Neolithic folk and the second to the early or middle Bronze Age.

In a pollen diagram from Llanllwch in Carmarthenshire (Thomas, 1965), *Plantago lanceolata* occurs spasmodically from the level of the elm decline upward but, about half way between the elm decline and the radiocarbon dated level of 1200 B.C., it shows a series of pronounced peaks with values of up to 20% T.T.P. At the same level the pollen of other ruderal species increases. Although it would be unwise to date this level by extrapolation, particularly as the stratigraphy is not uniform, it is quite clear that there was a new impetus to the original Neolithic economy some time before 1200 B.C. Simmons (1964) has found a similar rise in cultural indicators half way through Zone VIIb in several of his diagrams from Dartmoor, where he contrasts the sporadic clearings of the Neolithic people with the much increased clearance that took place during the Bronze Age.

For Somerset, Godwin (Dewar & Godwin, 1963) has summarized the information from the numerous pollen diagrams which throw light on the forest history of the surrounding hills. The first traces of interference with the forest occurred, as elsewhere, at about 3000 B.C. with the elm decline and the appearance of occasional grains of *Plantago lanceolata*, *Rumex* species, *Artemisia* and *Pteridium*. In some diagrams the *Corylus* curve also rises strikingly at this point, in others it does so a little higher up, but in either case it seems to reflect a period of definite woodland management during which hazel was being coppiced and some, at least, of the poles so produced were being used in the construction of the trackways which crossed the bogs. A little later there is a decrease in the pollen of *Tilia*, accompanied by a very slight increase in herbaceous pollen, which has been radiocarbon dated to about 2000 to 1900 B.C. It was not until the late Bronze Age, however, that the agricultural indicators really expanded, cereal grains then accompanying substantial amounts of arable weeds. Godwin is inclined to the view that this late expansion represents a sharp cultural break between the middle and late Bronze Age.

To all these examples can be added the clearance of woods in other parts of the country radiocarbon dated in association with the decline in *Tilia* pollen at the VII/VIII pollen-zone boundary (Turner, 1962). In the East Anglian Fens these woods were cleared about 1450 B.C., in Shropshire about 1275 B.C. and just south of the Humber about 1200 B.C.

Many of Mitchell's (1956) diagrams from widely scattered localities in Ireland have a higher or more continuous *Plantago lanceolata* curve above than below his Zone VIII a/b boundary which he places in the early Bronze Age. There is a similar *Plantago* curve in Smith's diagram from Canonns Lough (Smith, 1961) and also in one from Fallahogy (Smith, 1958) where radiocarbon dates for the early Neolithic *landnam* (Smith & Willis, 1961) are consistent with the chronology of Mitchell's zones. On Morrison's (1959) diagram from Parkmore the continuous curve begins only slightly earlier and more recently a similar pattern of clearance has been found in Littleton Bog, Tipperary, where Mitchell (1965) has described how, during the Neolithic and early Bronze Age, small patches of the primary woodland dominated by elm were being cleared and then quickly invaded by secondary stands of elm and hazel. Later on the secondary woods were cleared but on a much more extensive scale than before. By comparison with dated horizons in other diagrams, Mitchell has attributed this increased activity to 1800 B.C. or later.

Taken as a whole, then, these diagrams give a remarkably consistent picture. More and more forest was being cleared everywhere during the Bronze Age, from about 2000 B.C. onward. It is difficult to tell from some of the diagrams whether it was a sudden or a gradual change in each region, though in the areas most thoroughly studied, such as the Cumberland lowlands, the evidence points to it being a fairly gradual one. If one regards the occasional cereal and ruderal pollen grain as the first sign of a rudimentary form of agriculture associated with a nomadic way of life, and the later *landnam* episodes as representing an economy still based on shifting rather than settled agriculture, then one may well infer that this increasing degree of clearance was caused by a slowly increasing population, which only gradually became more skilful in felling trees and clearing scrub, and also more dependent for subsistence upon agriculture and perhaps stock breeding than on fishing, hunting and gathering.

## IRON AGE AND LATER CLEARANCE

If the Bronze Age saw only an extension and intensification of a basically Neolithic way of life, the same is certainly not true of the Iron Age. It may well be that bronze never became cheap enough on the margins of Europe to be widely used in preference to stone for such rough activities as tree felling. Certainly Childe (1950) considers it could only have done so in the late Bronze Age at the earliest. The discovery and utilization of iron, however, was a different matter, not only because iron axes were more effective in clearing forest but also because wood was required to make the charcoal used in smelting the ore. The effect of this demand for timber, that is clearance on a scale hitherto known only from the chalklands, has been well documented in pollen diagrams from a few areas of Britain which are known to have been occupied by people with an iron technology. For example, in west Wales the pollen diagram from Tregaron Bog (figure 3) shows extensive disforestation and the spread of grassland following a series of small temporary clearances typical of the Neolithic and Bronze Age (Turner, 1965). This disforestation has been radiocarbon dated to about 400 B.C. (Godwin & Willis, 1962). Only four miles to the north-east, commanding a fine view over the Teifi valley and Tregaron Bog, is the hill of Pen-y-Bannau capped by a trivallate hill fort.

Godwin's (1960) non-tree pollen diagram from the Vipers A site in the Somerset levels has a similar rise in grass, plantain and other herb pollen at 115 cm, indicating a considerable increase in the amount of grass- and arable-land in the region. This major episode of disforestation has been radiocarbon dated to about 300 B.C. and, as Godwin has pointed out, occurs at the very time the Glastonbury and Meare Lake villages were occupied.

In the Humber region, too, there is similar evidence of the destruction of extensive tracts of woodland (Turner, 1962), at a level radiocarbon dated to about 350 B.C., yet another Iron Age date. Although not radiocarbon dated, the first really substantial

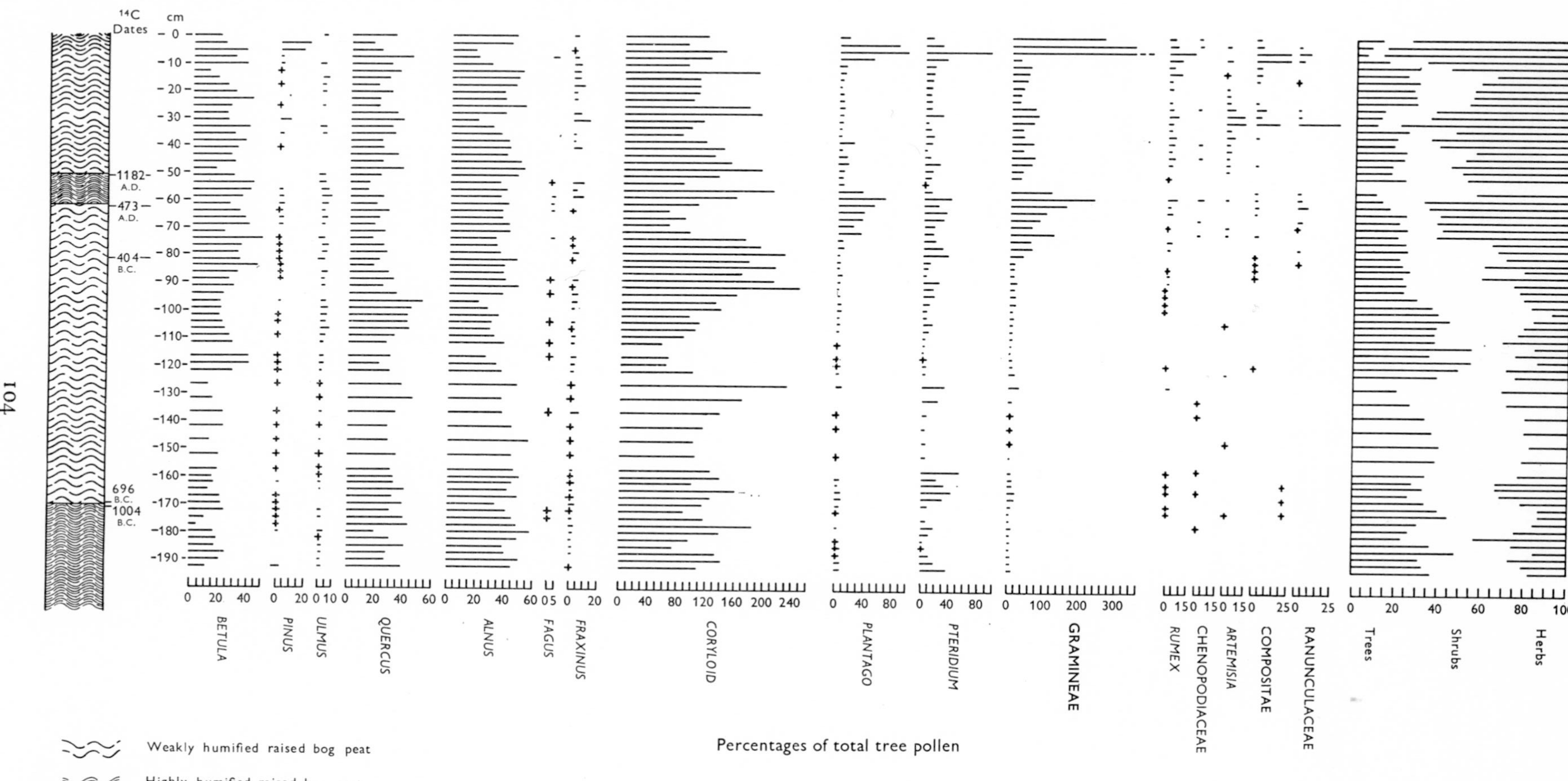


Figure 3. Pollen diagram from Tregaron Bog, Cardiganshire: pollen of trees and plants indicative of forest clearance and agriculture (Turner, 1965, figure 23).

clearance of forest indicated in the Old Buckenham Mere diagram from south Norfolk (Godwin, 1967) almost certainly belongs to the same period. Similarly, at Malham Tarn, where there is archaeological evidence in the form of hut circles and 'corn plots' for Iron Age occupation of the area, the first major clearance occurs early in Zone VIII and has been attributed by Pigott & Pigott (1963) to this period. Simmons (1964), too, has attributed to the Iron Age the increased clearance recorded in his Dartmoor diagrams.

In all these areas archaeological evidence of Iron Age cultures exists side by side with pollen diagrams attesting extensive disforestation at the appropriate time. Elsewhere, regions notable for their paucity of Iron Age relics have yielded pollen diagrams lacking in evidence of clearances. For example in Scotland, in neither the Forth Valley nor Ayrshire is there evidence for wholesale clearance until well into the first millennium A.D. Walker (1966) concluded, for the Cumberland lowlands, that 'the archaeological developments of the Bronze and Iron Age...did not significantly influence the type of agriculture which had been practised in the lowland since middle Neolithic times'. Similarly, Pennington (1964) recorded little change in the Neolithic–Bronze Age economy until well into the first millennium A.D., a stagnation unrelieved in the data of Oldfield (1960), Smith (1959) or Birks (1965) for the rest of the northwest Britain.

There is certainly a regional contrast between the southern part of Britain which was widely cleared of forest and the northern part which was not. This contrast, however, is in the amount of land cleared of forest, and should not necessarily be related with Piggott's (1961) between the Woodbury type of Iron Age economy associated with intensive corn growing and restricted to the southern part of Britain, and that north and west of the Jurassic ridge based more on pastoral farming, possibly with an element of nomadism in it, which he calls the Stanwick type. That difference is concerned more with the different uses to which the cleared land was put than with the actual amount available. Although within the extensively cleared southern part of Britain, there are also differences between places such as that around Tregaron Bog, where the cleared land was used mainly for pasture, and places further east and away from the highland zone, such as Apethorpe in Northamptonshire, where Sparks & Lambert (1961) have found very high frequencies for cereals and arable weeds during Zone VIII.

It is perhaps rather surprising to find that this pattern of cleared land, established during the Iron Age, continues with very few changes right through the Roman occupation of Britain for which period there is as yet comparatively little palynological evidence that new tracts of land were being cleared and brought into cultivation. The only suggestion of this is in a diagram from Flanders Moss in the Forth Valley (Turner, 1965). There, a few samples with high Gramineae and *Plantago* frequencies, representing an extensive though short-lived clearance, lie between two levels over half a metre apart, both of which have been radiocarbon dated to the same period, namely A.D. 200 ± 100 years. But this remains an isolated example. It may be, indeed, that we have not looked in the right places but, with the data now available, there remains little positive evidence for further clearance until much later.

From several sites in Ireland, Mitchell (1956, 1965) has shown further clearing of the secondary stands of forests from A.D. 300 onwards. This increase, shown in the tree pollen curves by a fall in elm, marks the beginning of his Christian period, and is attributed by him to the spread of new agricultural ideas with Christianity into Ireland.

Many diagrams from the north-west of Britain show extensive clearance around A.D. 400. As far as changes in the pollen frequencies are concerned this is virtually indistinguishable from the clearance which took place in the south of Britain during the Iron Age. Only where it has been radiocarbon dated can it be seen that it actually occurred some 700–800 years later. Smith's (1959) diagram from Helsington Moss shows this clearance across the level of an upper retardation layer which has subsequently been radiocarbon dated to about A.D. 436 (Godwin & Willis, 1960) and the diagram from Ellerside Moss (Oldfield & Statham, 1963) also shows extensive clearance at the level of an upper retardation layer. Birks (1965) correlates his major clearance at Holcroft and Lindow Mosses, also associated with an upper retardation layer, with that of Helsington. Several of Pennington's diagrams from the Lake District show an extensive clearance representing the farthest penetration of cereal cultivation into the uplands, up to about 225 m O.D. on the western margin. It corresponds in distribution and intensity with the stone hut circles and farmsteads thought to have been occupied during the Brigantian phase. She tentatively correlates this penetration with the extensive clearance described by Smith and others from further south. From the Cumberland lowlands,

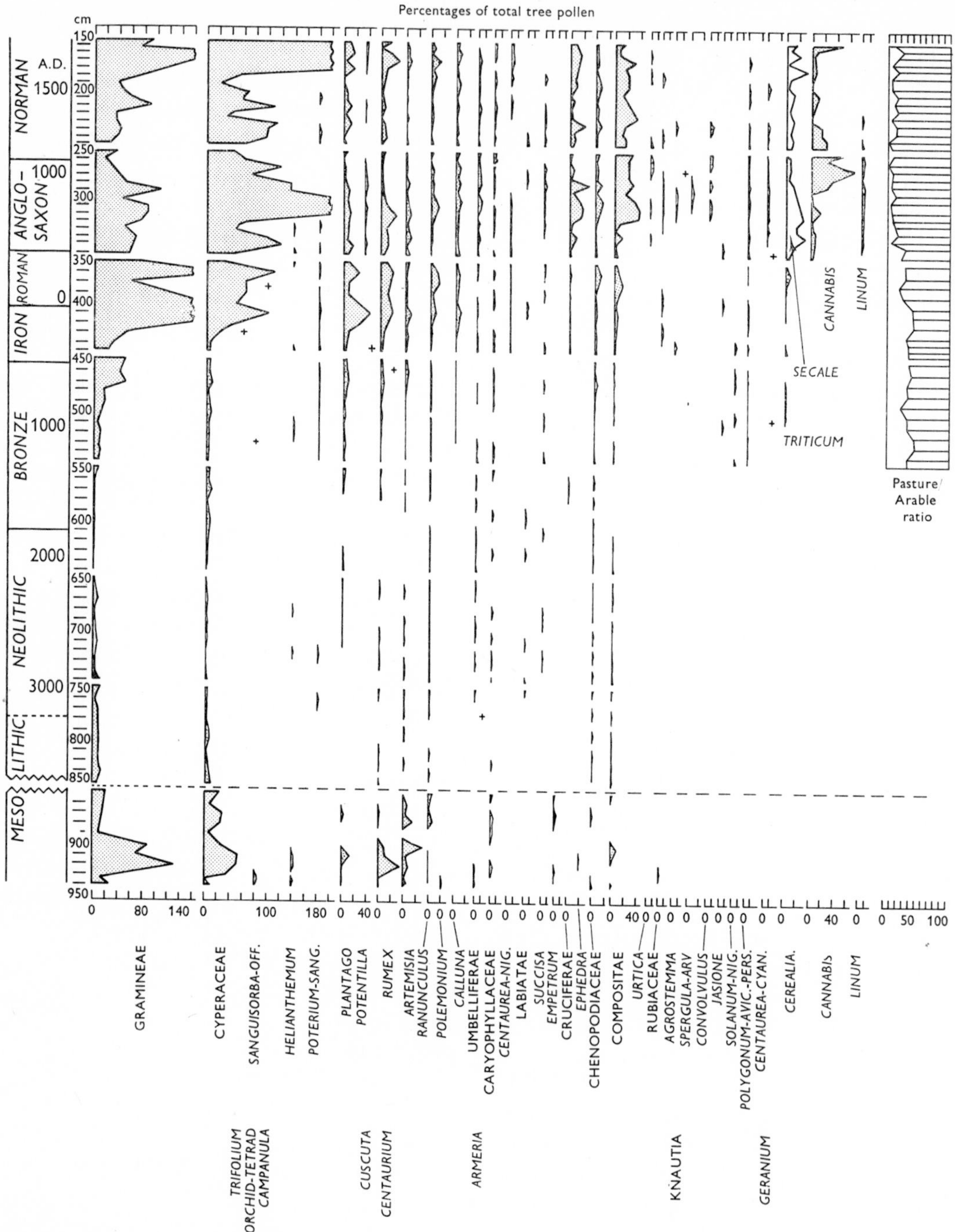


Figure 4. Pollen diagram from Old Buckenham Mere, Norfolk: pollen of dry land herbaceous plants (Godwin, 1967, figure 3).

Walker (1966) also describes a new impetus in forest clearance at about A.D. 400, following after 300 years or so of reduced occupation during the Brigantian Wars, and from further north still the Bloak Moss diagram shows the first extensive clearance, dated by radiocarbon, at about A.D. 450.

There can be very little doubt, therefore, that, at this time, the north-west of Britain was cleared of forest to much the same extent as the south had been during the Iron Age, even though in places the land was allowed to revert to forest again some centuries later.

In East Anglia, the Old Buckenham Mere diagram (figure 4) shows a little more clearance at a level which Godwin (1967) attributes to the beginning of the Anglo-Saxon occupation of the area. But much more interesting is the suite of herb pollen present which changes radically at this level. Hemp and flax pollen appear in the diagram for the first time together with increased Compositae, Cruciferae and Chenopodiaceae frequencies. At the same time, the frequencies of the pastoral weeds, species of *Plantago*, *Rumex* and Ranunculaceae, decrease. This change in emphasis from pastoral to arable farming fits in well with what is known from archaeological sources, the Anglo-Saxon farmers with their multiple ox-teams capable of ploughing the heavier soils following after the peasant farmers of the Iron Age who kept sheep and oxen as well as cultivating cereals. This diagram illustrates very well how it is possible to recognize a distinctive change in the main emphasis of the farming. Perhaps the fact that such changes in the nature of the farming have not been very widely recognized in British diagrams indicates how important the physical geography, particularly the contrast between highland and lowland zones, has been in determining the type of farming practised in this country; greater emphasis on the cultivation of crops in the south east has long contrasted with the more mixed and, further north, predominantly pastoral farming of the highland zone.

From this time onwards the clearance of forests is best considered locally, because it becomes progressively harder to date palynological horizons, either by radiocarbon or by correlation through distinct horizons comparable with other pollen zone boundaries. Dating is more a matter of fitting successive periods of intense and less intense interference with the vegetation with what is known from local historical sources about the density of settlement in a particular area. On the whole this has worked very well, and there are a number of diagrams which cover the whole of the last two thousand years in this way. They show such effects as those of the Scandinavian and Norman invaders and that of the monastic movement. Even recent afforestation shows up. But I do not want to consider these in detail, although it is worth noticing that, although many of the details of when the forests were cleared in different parts of the country have indeed been worked out during the last decade or so, as Godwin predicted in 1956, the pattern which has emerged is still far from complete. There are several obvious gaps upon which it would be well worth concentrating future effort, such as the southern chalklands during the Neolithic, or the north-east of England and much of Scotland where comparatively little work of this kind has yet been done.

But, leaving aside the need for accumulating more data and extending, confirming and modifying the pattern which is now emerging, there is another aspect with which little progress has so far been made, an aspect which may well be equally as productive in the next decade or so. That is the problem of defining the area from which the pollen in a diagram is derived and hence *locating on the ground* the vegetational changes that occurred. For the earlier Boreal and Atlantic forest periods this is admittedly a pretty academic exercise, because of the relatively uniform nature of the vegetation, but when one comes to the Sub-boreal or later, when local variations in the vegetation have resulted from one or more clearances in an otherwise uniform forest, it is less trivial a problem. Two questions immediately come to mind when one looks at a clearance episode on a pollen diagram. How much land was cleared at that time, and where was it?

Similar questions are being asked in a much wider context than that of forest clearance and it is unfortunate that space will not allow consideration of the progress that has already been made in this field. Within the present more limited context it must needs be restricted to but one possible approach to the problem, the use of three-dimensional pollen diagrams.

The idea behind these diagrams is as follows. If the source area of the pollen is limited, and Tauber's studies (1965, 1967) indicate that for at least a sizeable proportion of the pollen it can be, and enough diagrams scattered at critical distances throughout the region are available, diagrams from close to a cleared area should exhibit the results of that clearance whereas those from further away would not. In this way it should be possible to deduce something about the location and size of the area cleared. One would expect the critical distance for spacing such pollen diagrams to be closer for the earlier presumably smaller Neolithic or Bronze Age clearances, and much larger for later more extensive ones.

Stated like this, one is of course assuming an impossibly ideal situation in which one *well* delimited area of forest had been cleared with a number of pollen preserving deposits at convenient distances from it, and the rest of the forest in the region left intact. Nevertheless the approach promises some return as can be shown from its application to the later clearances of the forests of the Ayrshire lowlands, north-west Kilmarnock, in the triangle of land between Stewarton, Beith and Irvine. This area was

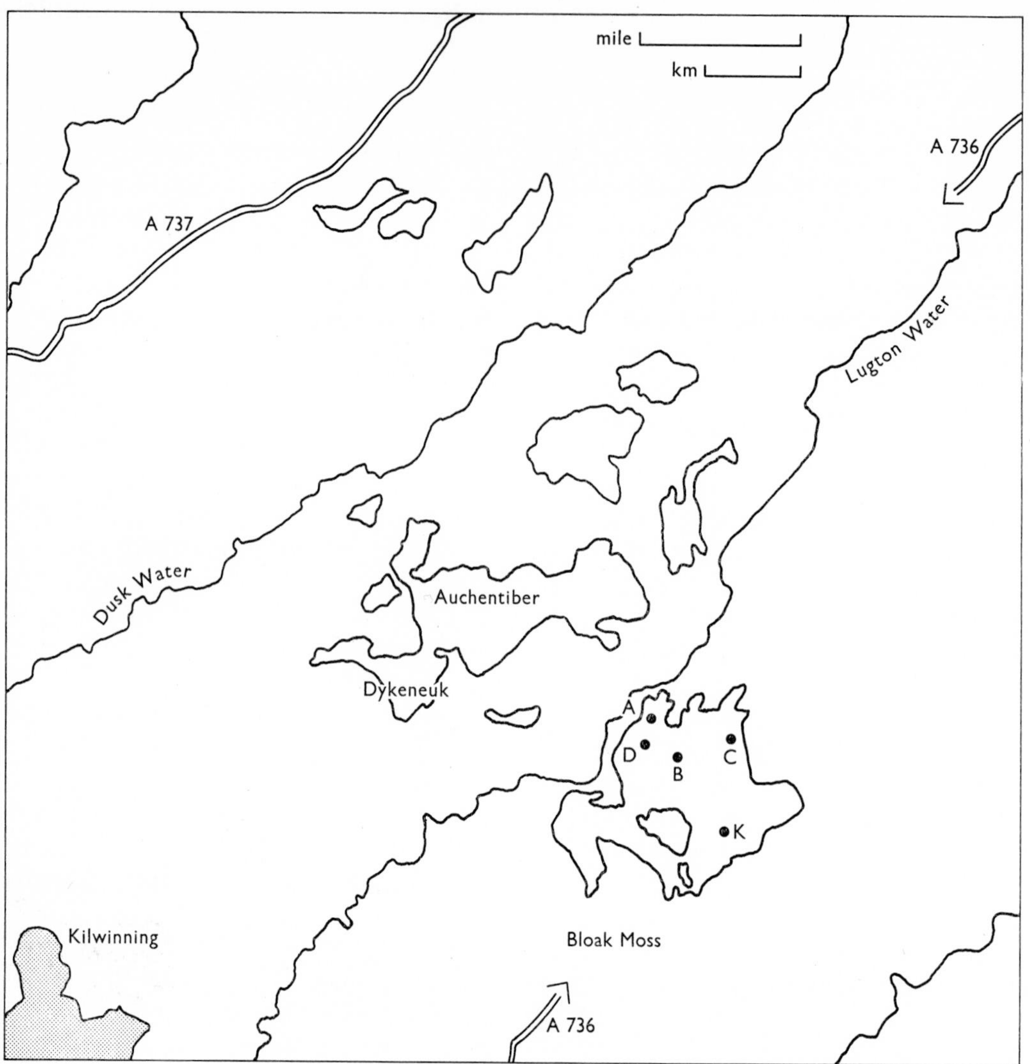


Figure 5. Sketch map of Bloak Moss and neighbouring peat deposits, showing the positions of the ordinary pollen diagrams from Bloak Moss.

chosen for the purpose principally because of its mosaic of raised bogs which approximated, as nearly as could be hoped, to the ideal (figure 5). Outcropping over the area are sediments of Carboniferous age extending in faulted valleys into a higher plateau of lavas of calciferous sandstone age to the north-west. Boulder clay with east–west and north east–south west oriented drumlins covers the older sediments and it is in the depressions in this boulder clay that peat began accumulating during the Atlantic period. Although these peat mosses have been disturbed and are no longer as large as they once were, it is comparatively easy to reconstruct their former size from the distribution of peaty soil in the fields which surround them. In places, however, up to four metres of peat remains relatively undisturbed, although the surfaces are either afforested or are heather-covered and beginning to erode with birch trees spreading in from the edges.

The largest and least disturbed of these mosses is Bloak Moss and a three-dimensional diagram has been constructed for its upper levels. In order to add the third dimension, so that the pollen spectrum could be considered not only in relation to its depth in the deposit, as is customary, but also in relation to its position laterally in the moss, it was first necessary to prepare several ordinary diagrams and then to correlate, from one diagram to another, the levels at which there were regional changes in the pollen frequencies.

Althogether, five ordinary diagrams were prepared from the moss. Their positions are shown in figure 5 by the letters A, B, C, D and K. At A, there was an exposed peat face about 2 m high, and so blocks of peat were cut from it in the field and the samples for pollen analysis and radiocarbon dating sliced from these blocks in the laboratory. By this method contiguous 0·5 cm samples were obtained. The peat below the exposed face was sampled with a Hiller borer. Diagrams B, C, D and K were all constructed from cores collected with a Hiller borer. For diagrams B, D and K the cores were sliced in 1 cm thick samples in the laboratory but, for diagram C, samples at 5 cm intervals were taken from the borer in the field. Normally samples at 8 cm or 16 cm intervals were counted, followed by intercalation of analyses at from 4 to 0·5 cm intervals where important changes in pollen frequencies seemed to be taking place. The possibility that some vegetational changes have been omitted must remain but, so striking is the overall consistency between the five diagrams, that this possibility is extremely remote.

The diagrams show that some time during the Atlantic it became wet enough in this part of Ayrshire for peat to begin forming in the lower lying areas. A birch wood peat began forming at A, D and C fairly early during Zone VII a, although at B it did not begin until the very end of that zone. But, by the beginning of Zone VII b, the formerly separate areas of peat in the eastern and western parts of the present moss were coalescing, and the presence of *Sphagnum* spores, *Calluna* and Cyperaceae pollen throughout the diagrams suggest that it was always a fairly acid peat. The moss appears to have been completely surrounded by forests containing species of *Betula*, *Quercus*, *Alnus* and *Corylus*. There do not seem to have been any breaks in the forest canopy, nor is there any evidence for herbaceous communities, other than those of the moss itself, in the neighbourhood at that time. There was virtually no grassland.

Above the level of the elm decline, the five pollen diagrams are so similar that they were easily divided into seven sub-Zones on the basis of the curves of pollen types indicative of forest clearance, that is the Gramineae, *Plantago* and *Pteridium* curves which rise and fall together throughout the diagram. These seven sub-Zones are shown on figure 2.

Sub-Zone I is the period just after the elm decline in which there is no evidence at all for open spaces in the forested landscape; it lasted from about 3000 B.C. until 1875 B.C. Sub-Zone II, which followed it, began about 1875 B.C. and lasted until A.D. 415. During it, there were a number of small forest clearances, at least four and possibly more, each followed by a period of forest regeneration.

Sub-Zone III, the earliest for which a three dimensional diagram has been constructed, was the time of the first extensive clearance of forest and is characterized by high frequencies of Gramineae and *Plantago lanceolata* pollen and *Pteridium aquilinum* spores (figure 2). Sub-Zone III is bracketed by radiocarbon determinations on diagram A, A.D. 415 (Q-722) marking the beginning and A.D. 495 and A.D. 580 (Q-721) the end, Q 721 *x* between (A.D. 455) providing a confirmatory check. Allowing for the standard deviation of each of these determinations (90 years), the sub-Zone could hardly have covered more than 350 years, and most probably lasted somewhat less than this.

The three-dimensional diagram for this, and other, sub-Zones (figure 6) shows the variation in only the three most significant pollen frequencies; significant in the sense that it was apparent from the ordinary diagrams that these particular frequencies varied quantitatively from one diagram to another.

1. Tree pollen as a percentage of total land pollen. *Corylus* has not been included among the trees, mainly because this has been customary in drawing up British diagrams.

2. Local, as a percentage of total, land pollen. 'Local pollen' includes only those pollen types which can reasonably be supposed to have come from the bog surface itself, namely *Calluna vulgaris*, other Ericaceae, Cyperaceae and *Rhynchospora alba* pollen. *Sphagnum* spores have been omitted because the frequency changes are so erratic as to mask trends shown by the selected local pollen types.

3. Gramineae as a percentage of total tree pollen. In this particular diagram, this is a good measure of the amount of cleared land in the area, because, as has already been mentioned, there is virtually no grass pollen in sub-Zone I, and in sub-Zone II and subsequent sub-Zones its frequency curve closely parallels those of *Plantago lanceolata* and *Pteridium aquilinum*, and with them almost certainly reflects forest clearance and regeneration.

In figures 6 to 9, all the samples from the appropriate sub-Zones at Sites A, B and K are shown. Within a group of samples, depth of sample increases from left to right along the *x*-axis and percentage frequency increases up the *y*-axis. The positions of the sites, although correctly ordered, are arbitrarily distributed across the figure. This three-dimensional presentation has been preferred to the fossil vegetation maps with clock face diagrams or the resolved maps which have been used so effectively for regional comparisons, because it shows the individual fre-

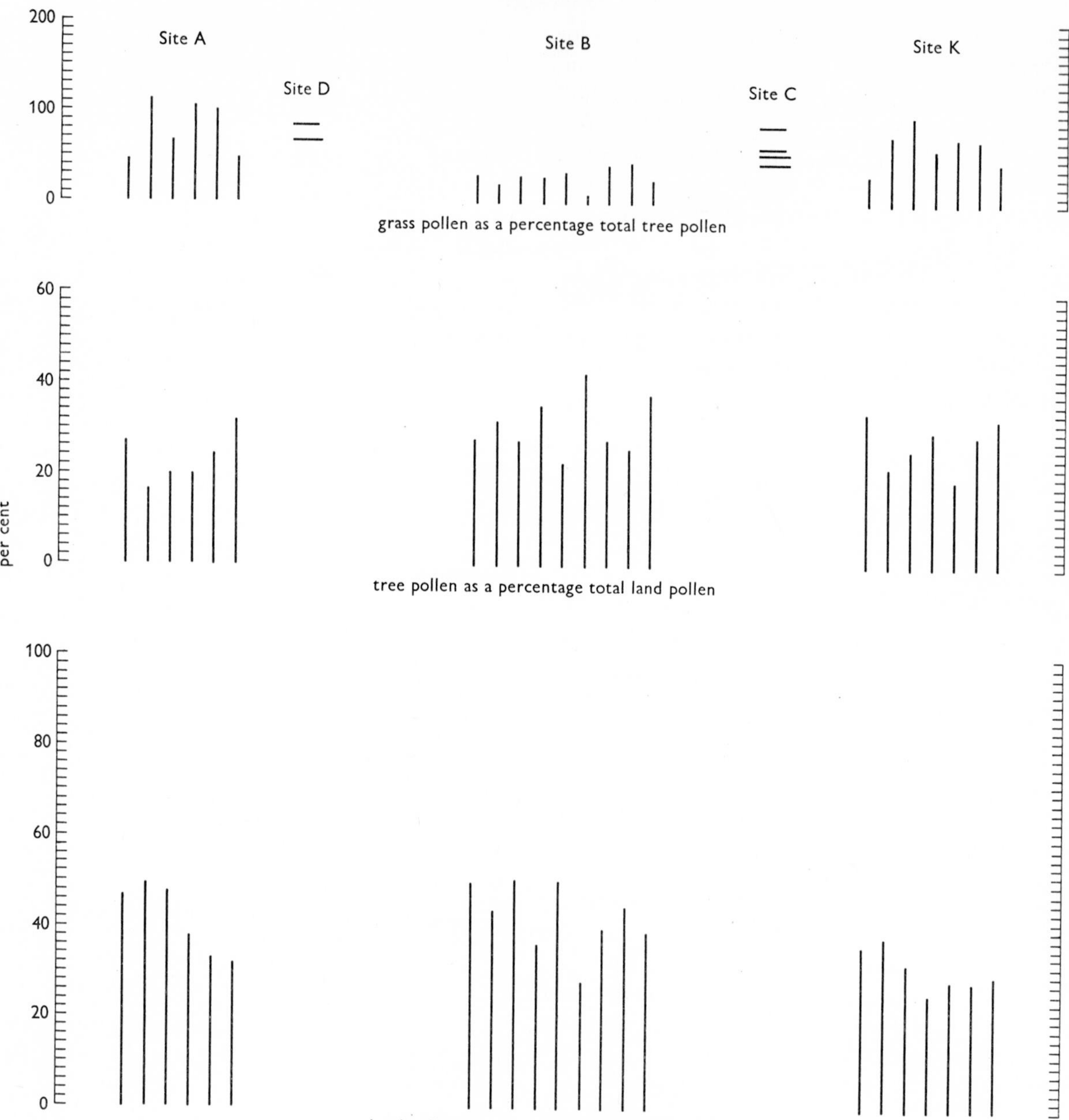


Figure 6. A three-dimensional pollen diagram from sub-Zone III, Bloak Moss, Ayrshire. (See text for explanation.)

quencies upon which the average of a clock face diagram would have been based, and thus enables the reader to judge for himself the significance of the lateral variation in the moss against the variation with depth within a sub-Zone at a given spot. Also, in this particular case, the sites of the ordinary diagrams lie approximately on a straight line, so there is no need to complicate the presentation of results with the extra dimension of area.

What is abundantly clear from the three dimensional diagram of sub-Zone III is that the grass frequency is much lower in the centre of the moss than near either edge, and high frequencies have also been found in the few samples which were counted from this level on diagrams C and D, both of which are from fairly near the edge of the moss. (These frequencies, two from diagram D, and four from diagram C, are shown by bars at the appropriate levels, D near block A and C near block K.) This lower frequency in the centre of the moss is thus well enough substantiated to require an explanation.

Tauber (1965), in re-evaluating the area effectively represented in pollen diagrams in the light of up-to-date information on the transport of airborne particles, has concentrated on the vegetation of closed deciduous forests; forests, in fact, which must have been like those covering the lowlands of Ayrshire until the beginning of sub-Zone II. He has emphasized that the pollen falling on a deposit does so in several different ways: some is washed down with the rain, some, having drifted above the canopy of the forest, settles out only slowly under gravity, whilst yet another fraction travels laterally through the trunk space of the forest before settling on the accumulating deposit. Each of these components consists of pollen derived from different sources, the rained out component being the most regional containing pollen which may even have travelled hundreds of miles; it is thought to comprise up to 20 % of the total pollen. The canopy component varies with the distance of the sampling site from the edge of the forest. For grains of low density sampled 500 m from the edge, 50 % would have come from within 7 km of the deposit, and 50 % from beyond. For denser grains the critical distance is less. But for grains of low density sampled 1000 m from the forest edge 50 % would have come from within 10–11 km of the deposit and 50 % from beyond. The trunk component is the most local, not only is most of it deposited within a few hundred metres of the edge of the forest but most of it would also be derived from within a similar distance of the forest edge, and so would undoubtedly represent much more local vegetation than the canopy component. The relative amounts of canopy and trunk space pollen in any one diagram are therefore critical for determining the area from which the pollen came. Tauber suggests that, for a small bog or lake 100–200 m in diameter, 80 % of the pollen would be from the trunk space of the surrounding forest and only 10 % from above the canopy, whereas, for a lake or bog some kilometres in diameter, only 10 % would be from the trunk space whilst 70 % would come from above the canopy.

Applying these ideas to Bloak Moss, it would seem eminently reasonable to postulate, in general terms at least, that the central part of the moss (B) was receiving a higher proportion of pollen from above the canopy, and so from more regional sources, than the parts near the edge of the moss represented by A, C, D and K. Very little of Tauber's trunk space component would reach B, whereas a reasonable amount could be expected to reach A, C, D and K. Therefore diagram B would give a more regional and the others a more local representation of the pattern of the forest clearance. And it follows from this that, during the fifth and sixth centuries A.D., forest clearance was more intense locally around Bloak Moss, including the areas near both A and K, than regionally in other parts of the county. In reaching this conclusion one rather questionable assumption has been made, namely that the sub-Zone III forests broken by tracts of cleared land may be treated as a similar pollen source to the unbroken deciduous forest that Tauber envisaged. Although he does not discuss pollen dispersal during periods of agricultural activity in very great detail, he is of the opinion that small clearings in a region, provided they are not so close to the bog as to affect the trunk space component strongly, will simply contribute modestly to the canopy component. If, however, they are close enough to affect the trunk space component, they will contribute reasonably large quantities of pollen to it. So there is no reason at the moment to suppose that the relative contribution of the three components at any one site differs significantly from that of pre-agricultural days.

During sub-Zone IV, much of the land cleared during sub-Zone III became forested again. The Gramineae and associated frequencies fall in all the ordinary diagrams but the three-dimensional one (figure 7) shows no significant difference in the grass frequency between the centre and the periphery of the moss. Assuming that one can distinguish between local and regional patterns of clearance, then regeneration of the local woodland has apparently brought the local and regional patterns more into line with each other; hence the similar grass frequencies.

But there are some differences between the central and peripheral parts of the moss during this sub-Zone and also during sub-Zone V. The tree-pollen frequency is higher and the local moss-pollen frequency lower at the edges than in the centre. There is no real reason why the average value of the local moss pollen should vary from one part to another; it seems more likely that there is less tree pollen falling on the centre of the moss. This would certainly be in line with the idea of the trunk space component decreasing towards the centre.

During sub-Zone V, there was a second extensive clearance of woodland but, either because the peat was forming very slowly or because the sub-Zone was comparatively short, this clearance is only represented by a few centimetres on each of the ordinary diagrams. However, contiguous samples were analysed from across the level of the highest grass frequency so that this maximum could be shown

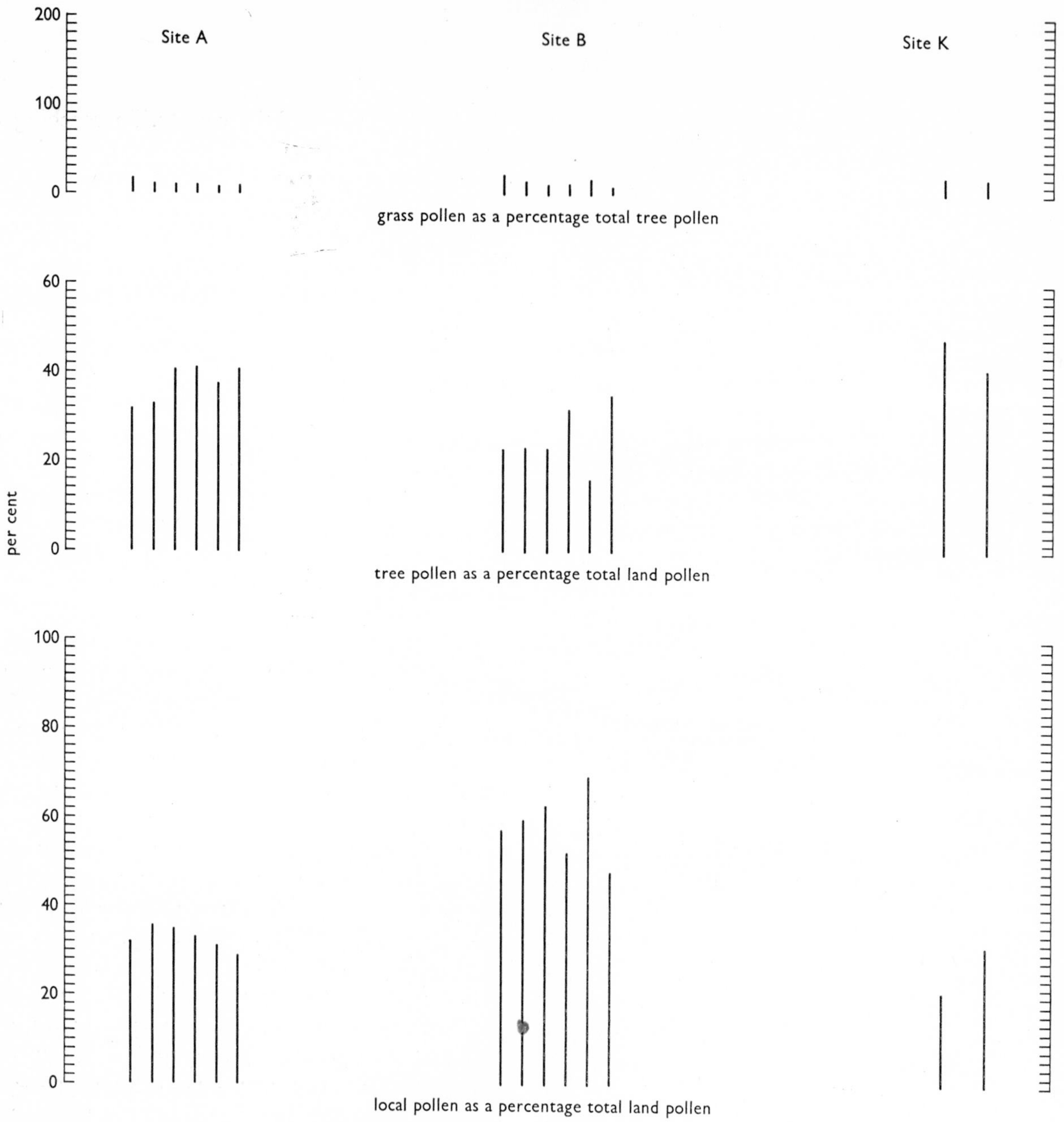

Site A Site B Site K

200
100
0

grass pollen as a percentage total tree pollen

60
40
20
0

per cent

tree pollen as a percentage total land pollen

100
80
60
40
20
0

local pollen as a percentage total land pollen

Figure 7. A three-dimensional pollen diagram from sub-Zone IV, Bloak Moss, Ayrshire. (See text for explanation.)

with greater accuracy on the three-dimensional diagram (figure 8). In this sub-Zone the grass maximum is not lowest in the centre of the moss, but drops from one side of the moss to the other, from 184 % at A to 138 % at B, and finally to 105 % at K. The tree-pollen frequency also differs in the three sites. It is almost uniform throughout the sub-Zone at K, but at A it drops sharply from the beginning to the middle of the sub-Zone and rises equally as sharply towards the end. The same minimum, but less pronounced, occurs at B.

Such grass and tree-pollen frequencies might well have been caused by the clearance of a large tract of forest near A at a time when the forests around K were relatively undisturbed, and when there was also a considerable amount of clearing in the region as a whole. Diagram B again shows the regional clearance pattern, but this time diagrams A and K show a strong contrast in the amount of clearance on each side of the moss. It is perhaps going a little further than the evidence allows at present to suggest that it was the area between Bloak Moss and the Auchentiber–Dykeneuk Moss complex that was cleared (figure 5) but had it been much further away

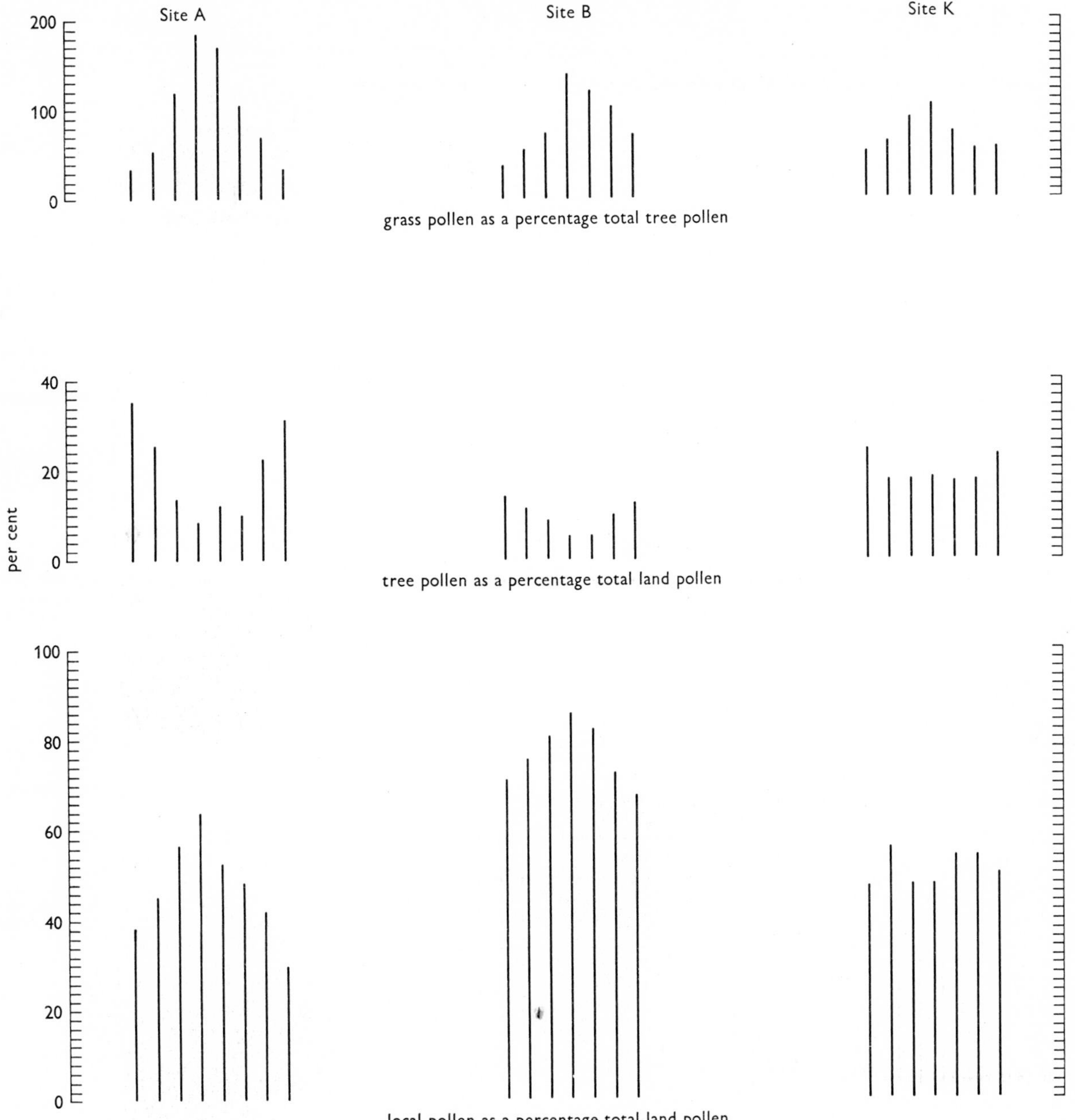


Figure 8. A three-dimensional pollen diagram from sub-Zone V, Bloak Moss, Ayrshire. (See text for explanation.)

it is difficult to see why there should be such a sharp downward gradient in the grass pollen frequency right across Bloak Moss from that side.

During sub-Zone VI this cleared land was abandoned, and the woodland spread to much the same extent as it had done during sub-Zone IV. The Gramineae frequency is about the same as then, and so is that of the tree pollen. The three-dimensional diagram shows much the same: no significant variation in the grass frequency and a tendency for higher local and lower tree pollen in the centre.

Sub-Zone VII is too near the surface of the moss and the peat too disturbed to permit such comparisons as have been made for the earlier periods, but it shows the final attack on the forests which brought about the present landscape.

Whilst the interpretation of these results can only be tentative at this stage and subject to revision when more experimental data on pollen transport are available, the results themselves for sub-Zones III and V are, I think, sufficiently encouraging to allow one to hope that three-dimensional diagrams will

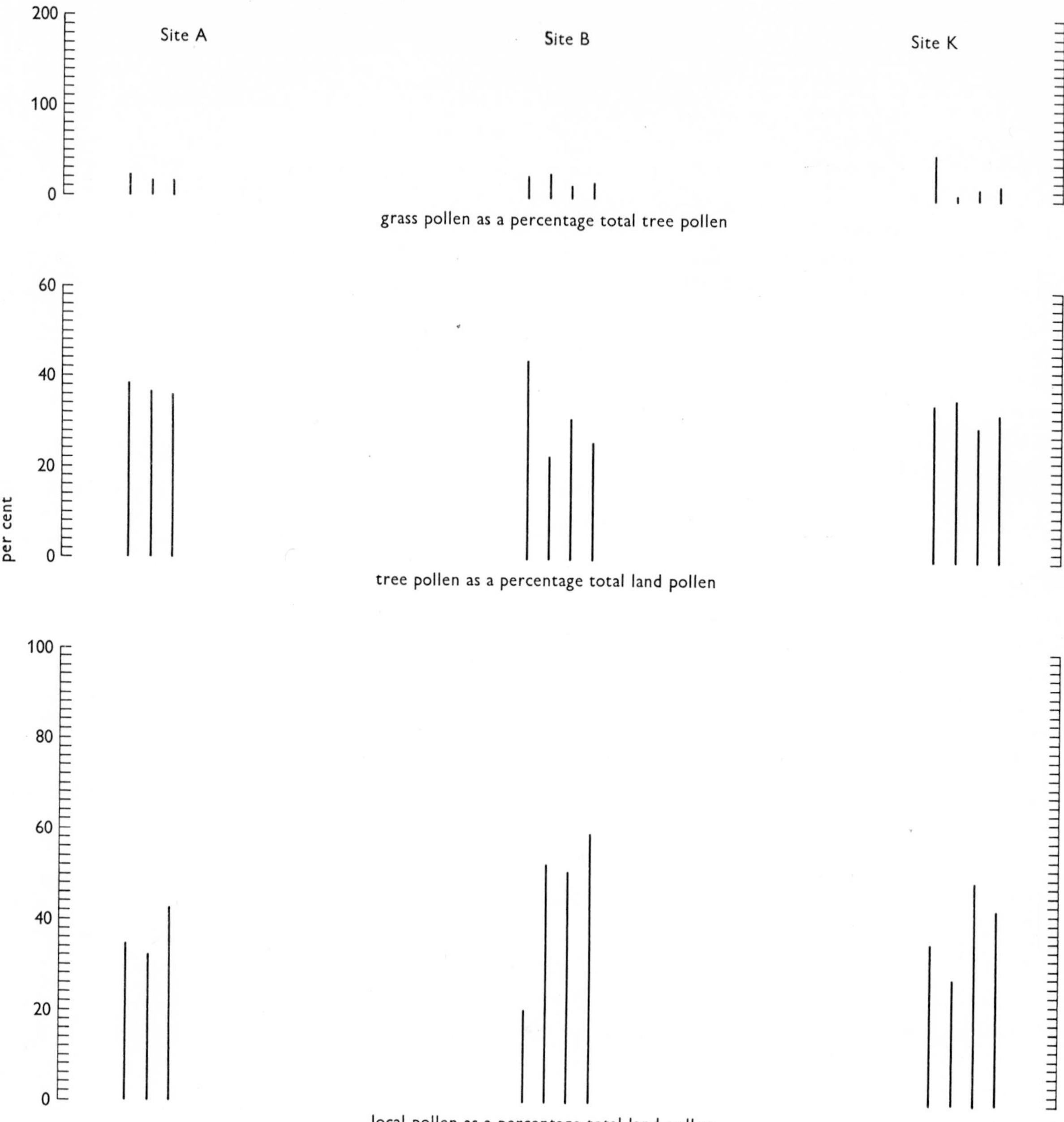


Figure 9. A three-dimensional pollen diagram from sub-Zone VI, Bloak Moss, Ayrshire. (See text for explanation.)

become a means of finding out more about prehistoric forest clearance than has been possible in the past, in particular by locating more closely the areas which were actually affected at different times. On the one hand it seems they allow a distinction to be made between local and regional patterns of clearance, as in sub-Zone III, and on the other, as sub-Zone V illustrates, between two local patterns. Spacing the diagrams at distances of about a quarter to one mile apart seems to have worked reasonably well for determining variations in these fairly extensive clearances, but whether similar distances will prove suitable for the earlier and smaller Bronze Age clearances has yet to be seen. In some ways of course these are the most interesting, lying as they do between two extremes: that of extensive clearance and presumably settled agriculture on the one hand and the minimum of forest modification on the other. The sub-Zone V results, if the interpretation is correct, show the effect of clearing one piece of land near the moss. On the other hand the early Neolithic sub-Zone I vegetation, a minimally disturbed forest, is unlikely to show up differentially on a three-dimensional diagram whatever the distances between the individual sites, since the interference probably affected only the regional pollen rain and that but slightly. But between these two extremes lie the *landnam* peaks of the Bronze Age. What do they mean? Are they the effect of clearing and using, perhaps only for a few years, a single piece of land, as with the sub-Zone V clearance? If so one would expect to be able to locate the particular area cleared from a three-dimensional diagram, either of the scale used at Bloak Moss or smaller. Or do they reflect the cumulative regional effect of many such individual clearances when the region as a whole was occupied by a culture which, several decades or perhaps even centuries, later moved on somewhere else? In this case a three-dimensional diagram would not show any variety, either between local and regional pollen components or between any two local components. And if this were so, then the level at which such a diagram ceases to be uniform and begins to show local and regional differences may correspond with the change from shifting to settled agriculture in the region. But, tempting though it is to speculate, we can only await such diagrams in the hope that they will continue to give interesting results and will eventually justify not only their initial promise but the labour involved in preparing them.

## SUMMARY

During the last fifteen years or so many of the details of the times at which forests of different parts of Britain were cleared have been worked out using pollen analysis and radiocarbon dating. By the beginning of the Bronze Age the Downs had been extensively cleared although the rest of the country remained much as it had been during the Neolithic. Large tracts of forest in southern Britain were cleared during the Iron Age whereas in the north-west clearance on a similar scale was delayed some seven to eight hundred years. There are still parts of the country which need investigating by these methods but closer analysis of the areas represented in pollen diagrams would also be useful, particularly for studying the local variations determined by man's disturbance of vegetation. In this connection a three-dimensional pollen diagram from Bloak Moss, Ayrshire has allowed local and regional patterns of clearance to be distinguished at two levels. During the fourth and fifth centuries A.D. there was much more clearance near the moss than in the region as a whole whereas, during a later period of extensive clearance, it appears that the area to the west of the moss was cleared whilst that to the east was not. The potential of such three-dimensional diagrams is briefly discussed.

I would like to thank most sincerely all my friends and colleagues who have helped in so many ways in making this essay possible. And not having had the opportunity to do so elsewhere, I particularly want to say how grateful I am for Dr Ann Mullinger's practical assistance and for a grant from Newnham College, Cambridge, towards the cost of field work at Bloak Moss.

## REFERENCES

Birks, H. J. B. (1965). Pollen analytical investigations at Holcroft Moss, Lancashire, and Lindow Moss, Cheshire. *J. Ecol.* **53**, 299–314.

Childe, V. G. (1950). *Prehistoric migrations in Europe*, Oslo: Aschehoug, p. 181.

Dewar, H. S. L. & Godwin, H. (1963). Archaeological discoveries in the raised bogs of the Somerset levels, England. *Proc. prehist. Soc.* **29**, 17–49.

Godwin, H. (1944). Age and origin of the 'Breckland' heaths of East Anglia. *Nature, Lond.* **154**, 6.

Godwin, H. (1956). *The History of the British Flora.* Cambridge University Press.

Godwin, H. (1960). Prehistoric wooden trackways of the Somerset levels: their construction, age and relation to climatic change. *Proc. prehist. Soc.* **26**, 1–36.

Godwin, H. (1962). Vegetational history of the Kentish chalk downs as seen at Wingham and Frogholt. Festschrift Franz Firbas. *Veröff. Geobot. Inst. Rübel, Zürich*, **37**, 83–99.

Godwin, H. (1967). The ancient cultivation of Hemp. *Antiquity*, **41**, 42–50.

Godwin, H. & Willis, E. H. (1960). Cambridge University natural radiocarbon measurements, II. *Am. J. Sci. Radiocarbon Suppl.* **2**, 62–72.

Godwin, H. & Willis, E. H. (1962). Cambridge University natural radiocarbon measurements, V. *Radiocarbon*, **4**, 57–70.

Godwin, H., Willis, E. H. & Switsur, V. R. (1965). Cambridge University natural radiocarbon measurements, VII. *Radiocarbon* **7**, 205–12.

Iversen, J. (1949). The influence of prehistoric man on vegetation. *Danm. geol. Unders.* (IV), **3**, 1–25.

Kerney M. P., Brown, E. H. & Chandler, T. J. (1964). The Late-glacial and Post-glacial history of the chalk escarpment near Brook, Kent. *Phil. Trans. R. Soc.* B **248**, 135–204.

Mitchell, G. F. (1956). Post-boreal pollen diagrams from Irish raised bogs. *Proc. R. Ir. Acad.* **57** B, 185–251.

Mitchell, G. F. (1965). Littleton Bog, Tipperary: an Irish agricultural record. *J. Roy. Soc. Antiq. Ireland*, **95**, 121–32.

Moar, N. T. (1964). Ph.D. Thesis. Cambridge University.

Morrison, M. E. S. (1959). Evidence and interpretation of 'landnam' in the north-east of Ireland. *Bot. Notiser*, **112**, 185–203.

Oldfield, F. (1960). Studies in the Post-glacial history of British vegetation. Lowland Lonsdale. *New Phytol.* **59**, 192–217.

Oldfield, F. & Statham, D. C. (1963). Pollen-analytical data from Urswick Tarn and Ellerside Moss, North Lancashire. *New Phytol.* **62**, 53–66.

Pennington, W. (1964). Pollen analyses from the deposits of six upland tarns in the Lake District. *Phil. Trans. R. Soc.* B **248**, 205–44.

Pennington, W. (1965). The interpretation of some Post-glacial vegetation diversities at different Lake District sites. *Proc. Roy. Soc.* B **161**, 310–23.

Piggott, S. (1961). Native economics and the Roman occupation of north Britain, in *Roman and Native in north Britain* (ed. I. R. Richmond). London: Nelson.

Pigott, C. D. & Pigott, M. E. (1963). Late-glacial and Post-glacial deposits at Malham, Yorkshire. *New Phytol.* **62**, 317–34.

Simmons, I. G. (1964). Pollen diagrams from Dartmoor. *New Phytol.* **63**, 165–80.

Smith, A. G. (1958). Pollen analytical investigations of the mire at Fallahogy Td., Co. Derry. *Proc. R. Ir. Acad.* **59** B, 329–43.

Smith, A. G. (1959). The mires of south-western Westmorland: stratigraphy and pollen analysis. *New Phytol.* **58**, 105–27.

Smith, A. G. (1961). Cannon's Lough, Kilrea, Co. Derry: stratigraphy and pollen analysis. *Proc. R. Ir. Acad.* **61** B, 369–83.

Smith, A. G. & Willis, E. H. (1961–2). Radiocarbon dating of the Fallahogy landnam phase. *Ulster J. Archaeol.* **24/25**, 16–24.

Sparks, B. W. & Lambert, C. A. (1961). The Post-glacial deposits at Apethorpe, Northamptonshire. *Proc. malac. Soc. Lond.* **34**, 302–15.

Tauber, H. (1965). Differential pollen dispersion and the interpretation of pollen diagrams. *Danm. geol. Unders.* (II), **89**, 1–69.

Tauber, H. (1967). Investigations of the mode of pollen transfer in forested areas. *Rev. Palaeobotan. Palynol.* **3**, 277–86.

Thomas, K. W. (1965). The stratigraphy and pollen analysis of a raised peat bog at Llanllwch, near Carmarthen. *New Phytol.* **64**, 101–17.

Troels-Smith, J. (1960). Ivy, mistletoe and elm. Climatic indicators—fodder plants. *Danm. geol. Unders.* (IV), **4**, 1–32.

Turner, J. (1962). The *Tilia* decline: an anthropogenic interpretation. *New Phytol.* **61**, 328–41.

Turner, J. (1965). A contribution to the history of forest clearance. *Proc. Roy. Soc.* B **161**, 343–54.

Walker, D. (1966). The late Quaternary history of the Cumberland lowland. *Phil. Trans. R. Soc.* B **251**, 1–210.

# DIRECTION AND RATE IN SOME BRITISH POST-GLACIAL HYDROSERES

by D. Walker

*Research School of Pacific Studies, The Australian National University, Canberra*

## INTRODUCTION

The amount of stratigraphic information now available from British Late- and Post-glacial deposits, chronologically controlled by radiocarbon dating or pollen analysis, prompts an examination of its contribution to the theory of hydroseres. Following Godwin's example, most British investigators of vegetation history have paid particular attention to stratigraphic detail mainly because of its usefulness in selecting the points from which to collect samples for pollen analysis.

The first account in the English language of a hydroseral process is probably that of William King who attributed the origin of Irish bogs to the choking of springs by vegetation and to the infilling of stream beds and lakes resultant from the damming of stream courses (King, 1685; Gorham, 1953). More relevant to the present essay, however, is Gough's (1793) account of the conversion of lakes to drier land:

> The cavity which is, at present, the receptacle of a pool, will, in process of time, be occupied by a stratum of solid matter, which will consist of the remains of its own produce gradually accumulated and preserved by the water which is intimately mixed with them, and which protects them from decay. The substance with which it is constantly filling will acquire a compactness nearly uniform in every part, by the plants of each generation interweaving their fibres with the remains of their predecessors: and by the depositions of the water, which, falling to the bottom, will be lodged in its interstices. All foreign bodies, brought hither by accident, will in time be buried in the increasing soil, where they will remain for ages, without undergoing any changes, besides those, which are produced by the solvent power of water on particular substances. Should the water be most shallow at the sides, and increase in depth as you advance to the middle, which is generally the case, the margin of the pond will be progressively advanced, and its surface contracted in proportion. If any part of it be too cold to favour vegetation, that part will still remain a pool surrounded with a flat, sedgy border. If it be supplied and emptied by two rivulets, the intermediate current will preserve itself a channel through the growing land. Lastly, the solid plain, thus produced, will, in time, be covered with a bed of vegetable earth, whose thickness will determine the difference of high and low water-mark; for the matter between those two limits, being alternately wet and dry, will, at particular periods, be exposed to the action of the air and will, consequently, be decomposed, and changed into mold.

The general theory of the hydrosere as we know it today was formulated in modern terms by Clements (1916) but is perhaps most succinctly stated by Tansley (1939):

> Another type of sere, beginning on submerged soil, is met with on the margins of lakes and slow rivers. These are colonized first by water plants (hydrophytes), and then, when the water is shallow enough, by reeds and bulrushes and other plants of similar life form. Accumulation of the products of decay of aquatic and reedswamp plants eventually raises the submerged soil surface above the water and enables marsh and fen plants to get a footing: by their partial decay and the formation of peat from their remains the soil surface is raised still higher. Shrubs and trees which can tolerate a saturated soil settle down between the fen plants and gradually close up to form a fen or marsh scrub or woodland (now generally known in this country as *carr*), the soil level gradually rising all the time by the deposition upon it of dead leaves and twigs. The process of raising the soil is often substantially accelerated by flood silting along the lower courses of rivers and along lake shores adjoining the mouths of streams which bring down silt, adding mineral constituents to the organic soil, or even replacing it by an almost purely mineral soil. Eventually the surface layers of the soil become dry enough for the growth of mesophytic trees which cannot tolerate waterlogged soil, and this *hydrosere* (Greek ὕδωρ water), as it is called, beginning in water, culminates, like the xerosere, in the establishment of climax forest.

This essay is concerned almost exclusively with autogenic processes beginning in more or less confined fresh water. Successions which are strongly influenced by allochthonous materials (e.g. silt accumulation) are not considered nor are the origins

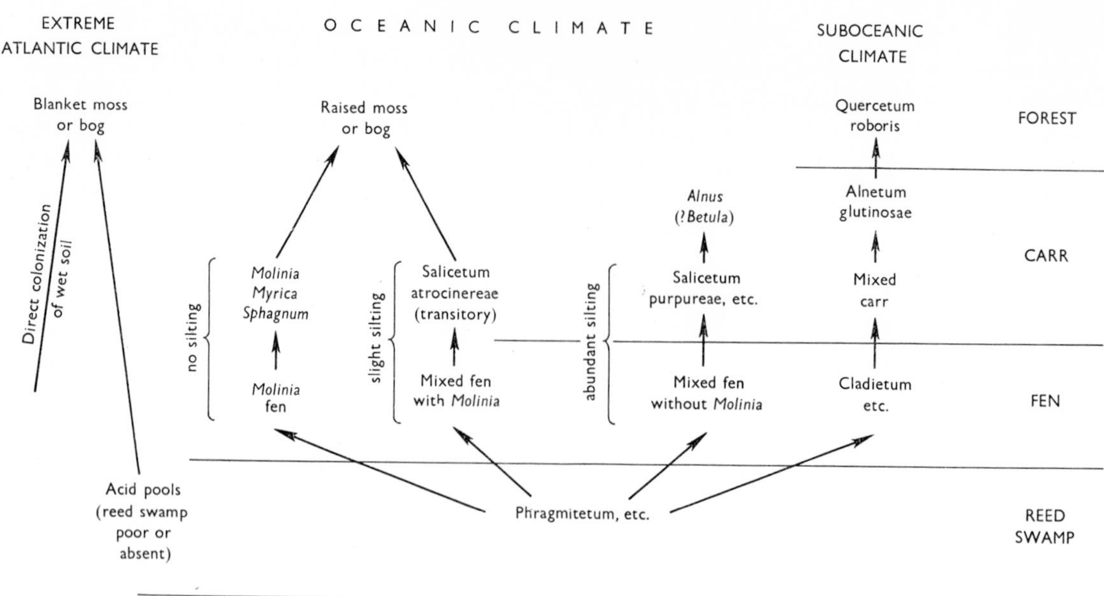


Figure 1. Climatic relations of hydroseres, after Tansley (1939) p. 671.

and development of such strictly ombrogenous mires as blanket bogs.

Tansley's general statement was based primarily on the work of Pearsall (1917, 1918, 1920, 1921) in the English Lake District, Godwin (1931, 1936; Godwin & Bharucha, 1932) at Wicken Fen, Pallis (1911) on the Norfolk Broads, Matthews (1914) on White Moss Lough and White (1932) in Northern Ireland. It led him to formulate a generalized scheme of hydrosere development in relation to climate (figure 1) the important features of which are the sequential relations of reedswamp, fen and carr followed by forest in sub-oceanic climates and bog in oceanic climates in localities free of silting or nearly so. Except in the genesis of blanket bog, *Sphagnum*, as a precursor of bog vegetation, entered late in the fen stage or, later still, in the carr. The ultimate succession from fen carr to climax oak woodland was hypothetical, deriving mainly from the observation of oaks growing on the periphery of fen carr at Calthorpe Broad (Godwin & Turner, 1933).

In a series of papers, Lambert (Lambert, 1946, 1951; Buttery & Lambert, 1965; Buttery, Williams & Lambert, 1965; Lambert & Jennings, 1951) has examined the present distribution and status of vegetation in the Norfolk Broads, particularly those of the Bure Valley. In outline, lake mud is colonized by reedswamp (usually *Typha angustifolia* sometimes preceded by *Schoenoplectus lacustris*) which is followed by an 'early fen' stage dominated by *Phragmites communis*. The course of events then depends on whether *Carex paniculata*, *C. acutiformis* or *Cladium mariscus* is next established. The first leads to swamp carr, the second to semi-swamp carr and the third to more stable fen carr. It is uncertain whether swamp carr and semi-swamp carr ever stabilize into true fen carr. Where the initial substratum is of solid peat or clay (as in old peat cuttings) the succession runs directly to *Phragmites* fen and so to fen carr. That these stages are indeed sequential is borne out by the detailed analysis of the contents of the organic deposits beneath each one. Spence's (1964, 1967) extensive survey of Scottish lowland hydroseres revealed that successional change at most sites was very slow and that, where it did occur, it resulted from the accumulation of mainly allogenic materials. Indeed, in the least turbulent situations, where reedswamp had become established fifty years previously, little if any change had occurred, possibly because of reduced nutrient supply. Therefore, although the distribution of vegetation could be related to environmental factors, such as water depth and water chemistry, evidence for autogenic change was negligible.

At Sunbiggin Tarn in Westmorland, reedswamp (mainly *Carex rostrata*) and fen are related to water depth, access of marginal drainage water from limestone, grazing and gull congregation. Succession is inhibited by a continuously rising water table resulting from the natural blockage of the outflow and bog development is apparently prevented by alkaline drainage water (Holdgate, 1955).

The development of the valley bog at Thornton

Mire, Yorkshire, has been shown, stratigraphically, to have proceeded from an initial phase of aquatic *Equisetum* dominance to a *Phragmites–Juncus* fen as the drainage was impeded by peat accumulation. Into this fen, shrubs occasionally intruded. Increasing depth of peat led to the localization of drainage and isolated the centre of the mire from the marginal base-rich water. *Sphagnum*, *Eriophorum* and *Betula* became established near the middle of the fen, the original species of which were finally replaced by those of *Sphagnum–Calluna* bog (Harley & Yemm, 1942).

The development of rafts of vegetation directly over water is often recorded at the lakeward margins of fens and is evidently one way in which fen can be established without the necessity for the accumulation of reedswamp detritus up to the mean water level. At Wybunbury Moss, Cheshire, such a 'schwingmoor', initiated entirely by aquatic *Sphagnum*, has covered a body of water more than 2 hectares in surface area and exceeding 10 m in depth at least in places. The floating mass of peat is about 5 m thick and bears a variety of vegetation types including woodland (Poore & Walker, 1958–9).

Numerous additional studies of aquatic and mire vegetation (e.g. Godwin, 1941; Lind, 1948–9; Lambert, 1946; Haslam, 1965) are useful in the present context mainly for the light they throw on the importance of direct and indirect human activity on the initiation or deflection of vegetation succession and the maintenance of certain vegetation types in artificial stability.

The likelihood that salt marsh is replaced by reedswamp, under conditions which Tansley (1941) regards as allogenic, first suggested by Chapman (1941), and specifically that communities characterized by *Scirpus maritimus* are replaced by *Phragmites communis* with other species in its wake, is now firmly established from studies of vegetation patterns (e.g. Gillham, 1957) but the actual process of replacement is virtually unrecorded.

Current ecological theory accepts the essence of Tansley's system and somewhat sceptically maintains the view that mire zonation recapitulates hydrosere history. In Sculthorpe's (1967) words, 'Direct evidence for such postulated hydroseres is by no means as abundant as is often thought'. The stratigraphic record, however, might well be expected to resolve some of the uncertainties. In particular, it might help to answer the following questions:

1. To what extent does the commonly observed zonation: submerged macrophytes, floating-leaved macrophytes, reedswamp, fen, fen carr, and its variants, parallel an autogenic succession?

2. To what degree, at which points, and under which conditions, may such a succession be deflected?

3. At what rates do hydroseres progress?

4. To what terminal vegetation types have hydroseres led and how stable have these proved under Post-glacial environmental conditions?

## THE NATURE OF THE DATA

Hydroseres are the only vegetational successions which leave behind them, in the form of partially preserved and often identifiable remains, a substantial record of the plant communities which have contributed to them. Moreover, the physical constitution of the matrix in which these plant remains are incorporated is often a clue to the dominant environmental conditions under which the plants grew. Nevertheless, the interpretation of stratigraphic and associated data is not without its pitfalls and before any attempt is made in this direction it is prudent to consider the limitations of the materials and the main sources of geographical variation which might be expected grossly to affect the course of a hydrosere.

### Geographic setting

The topography of the basin, channel or plain in which the hydrosere is initiated, the size of the catchment from which it draws and the design of its drainage, the chemical and physical properties of the rocks and the soil which mantles them, as well as the stature and composition of the surrounding vegetation, all combine to set limits on the possibilities for sedimentation and vegetation in the hydroseral locality itself. Inextricably confounded with these determinants are, of course, the effects of climate on rock weathering, soil stability, precipitation: evaporation ratio, the specific composition and luxuriance of the vegetation itself and the hydrological conditions in the parent water body.

In a general survey of this kind all these parameters can only be incorporated in the most sketchy manner. One major simplification has arisen from the restriction of the study to primarily topogenous mires, i.e. to hydroseres which owe their origins, if not their present maintenance, to drainage water rather than direct rainfall and are not directly related to riverine situations. For the purposes of this discussion only three main types of site are recognized:

(*a*) small inland basins, the greatest flooded horizontal dimension of which has not exceeded 0·5 km during the period under review;

(*b*) large inland basins, the greatest flooded horizontal dimension of which exceeded 0·5 km and has usually been very much larger during the period under review;

(*c*) large depressions or other areas of impeded drainage in coastal or estuarine regions, usually based on marine sediments but receiving fresh drainage water during the period under review.

The small inland basins include kettles (e.g. Abbot Moss), corrie lakes (e.g. Llyn Dwythwch), other flooded depressions in drift landscapes (e.g. Whattall Moss) some of them in upland valleys (e.g. Blea Tarn), impeded valleys unrelated to glacial deposits (e.g. Cothill) and some of unknown origin (e.g. Elstead). Their maximum depths vary but the gradients of many of their banks exceed 1:10 and are usually very much steeper (e.g. 1:3). They draw their water from small catchments through small streams and drain through single, sometimes intermittent, overflow channels. The flow of water through these basins is relatively diffuse and wave action is minimal.

The large inland basins are in valleys which have been glacially overdeepened (e.g. Windermere) or dammed (e.g. Tregaron), in extensive depressions in drift landscapes (e.g. Scaleby) and probable solution hollows (e.g. Hawes Water). The greatest depths of deposit in these basins are often unplumbed but the gradients of their edges are usually less steep and the extents of more-or-less flat bottoms very much greater than in the small inland basins. The shores of the large basins are often physiographically diverse and some are susceptible to considerable wave action. Their catchments are often relatively large; water enters them by one or many, often well-defined, streams and usually leaves by a single perennially flowing river.

Large coastal mires occupy a variety of topographic settings; some began in lagoons behind spits or raised beaches, others as elevated salt marshes fed with fresh drainage water. Some received little drainage water from the surrounding countryside and, quickly growing above the level of the regional water table, became predominantly ombrogenous. In their topogenous phases they were usually not subject to extreme channelled water movement but wave action along some shores was considerable. Drainage conditions were usually directly determined by geomorphic processes of regional or universal significance, namely isostatic and eustatic changes in the relative levels of land and sea.

This crude classification takes no account of the rock and soil types in the catchments of the basins, important though these are in determining the nutrient status of the lake water. In small basins where water enters somewhat diffusely around the edges, the effects of inborne nutrients are usually limited to a narrow peripheral zone once a marginal buffer of mire vegetation has become established. In large basins, entered by forceful rivers, the mineral content of these, reinforced by that derived from eroding parts of the shore, is clearly important in determining the trophic status of the water body, at least in the epilimnion. In such lakes, however, hydroseres, as distinct from Tutin's (1941) limnoseres, are usually limited to the most sheltered shores and therefore receive a small, perhaps negligible, share of the potential nutrient replenishment.

Geographical location within the British Isles has not been taken into account in the analyses which follow although some consideration is given to this variable in discussion. The main sources of variation related to location are probably rainfall, site topography and soil type.

## Vegetation stages

The data are analysed in terms of 12 vegetation stages, perhaps better called ecological units. The floristic and structural composition of each stage is derived from what seems to be its analogue in contemporary vegetation. Just as present-day vegetation types are rarely completely distinct spatially, stages in the history of a hydrosere rarely succeed one another suddenly. For convenience of description and analysis, however, the following stages are recognized.

1. Biologically unproductive water.
2. Micro-organisms in open water.
3. Totally submerged (or only flowers emerged) macrophytes (e.g. *Myriophyllum* spp., *Littorella uniflora*, *Potamogeton* spp.)
4. Floating-leaved macrophytes, usually with some intervening open water (e.g. *Nymphaea alba*, *Potamogeton natans*).
5. Reedswamp, rooted in the substratum and standing in more-or-less perennial water, with aerial shoots and leaves (e.g. *Phragmites communis*, *Shoenoplectus lacustris*, *Carex rostrata*).
6. Sedge tussock, rooted in the substratum and standing in more-or-less perennial water (e.g. *Carex paniculata*, *C. acutiformis*).
7. Fen, dominated by grasses (e.g. *Phragmites com-*

*munis*, *Molinia caerulea*) or sedges (e.g. *Carex flava*, *C. nigra*) with a variety of acid-intolerant herbs (e.g. *Comarum palustre*, *Filipendula ulmaria*, *Valeriana dioica*), all rooting in organic deposits which are waterlogged for the greater part of the year.

8. Swamp carr formed by trees (e.g. *Salix atrocinerea*, *Alnus glutinosa*) growing on unstable sedge tussocks with some fen herbs, the intervening pools often harbouring thin reedswamp or floating-leaved macrophytes.

9. Fen carr, dominated by trees (e.g. *Alnus glutinosa*, *Frangula alnus*, *Betula pubescens*, *Fraxinus excelsior*) with an undergrowth rich in fen herbs and ferns (e.g. *Thelypteris palustris*), all rooting in a physically stable peat mass.

10. Aquatic Sphagna, floating very closely below or on the water surface (e.g. *S. subsecundum*, *S. cymbifolium*).

11. Bog, usually distinguished by a variety of *Sphagnum* species (notably *S. palustre* and *S. imbricatum*) and acid-tolerant phanerogams (e.g. *Oxycoccus quadripetalus*, *Erica tetralix*, *Myrica gale*, *Eriophorum* spp.) growing in an organic substratum.

12. Marsh, composed of 'fen' species (e.g. *Filipendula ulmaria*, *Eleocharis palustris*) growing on waterlogged mineral soil.

### Stratigraphic interpretation and types of material

The materials which accumulate in lakes and mires are carried there by inflowing water or are produced by organisms living in the water or on deposits earlier accumulated and now above the free water surface. Allochthonous materials are usually, but not necessarily, inorganic and form exclusively sedimentary deposits. The products of free-living lake organisms themselves are also sedimentary although they may accumulate very close to their points of origin. Intermixed with these materials are fragments and products of the organisms growing in other parts of the lake or mire borne to the site of sedimentation by wind or, more commonly, water; the 'detritus' fraction in this context. The establishment of rooted phanerogamic hydrophytes in relatively shallow water leads to the accumulation of deposits in which the relative contributions of sedimentary and sedentary fractions vary limitlessly, the latter becoming progressively more important as the access of free water to the site of accumulation diminishes. Totally sedentary deposits can only be produced where there is no free water on the mire surface capable of redistributing detritus in pools and are therefore commoner in ombrogenous mires. Virtually sedentary deposits, however, are commonly produced by fens and fen carrs.

The contents of sedentary deposits, or the sedentary fraction of mixed deposits, provide unassailable evidence of the nature of the vegetation which formerly grew at the site under investigation. The interpretation of sedimentary materials requires greater caution and needs tempering by considerations of position, grain size, structure and oxidation if the locational origins of the contained detritus are to be determined. A case of special difficulty in the interpretation of sedimentary deposits arises as a result of the erosion, redistribution and reaccumulation of organic deposits. Instances occur where the detrital material is of such distinct colour, texture or content as to be readily discernible amongst the autochthonous matrix but more often, one suspects, they are so similar as to defy separate recognition in the field. Other traps for the unwary stratigrapher include failure to recognize the simultaneous accumulation of a more-or-less sedentary fen peat on a floating mat and of a sedimentary mud in the water below it and failure to distinguish between plant parts truly contemporary with the matrix which contains them and those, particularly roots, which have secondarily penetrated the matrix from above.

Both sedimentary and sedentary deposits are modified after their initial deposition, chemically by humification and physically by compaction. Because it needs an aerated water medium, humification is most characteristic of situations in which the water table is close to the accumulating surface. A rising water table leaves little time for extensive humification and ensures the rapid accumulation of little altered plant remains. Periodically, or permanently, dry surfaces, however, allow the dry oxidative breakdown of dead plants, or of deposits formerly produced by them, destroying the structures entirely and facilitating erosion by both wind and water.

In most lakes and mires dry oxidation is limited and humification virtually ceases as the newly deposited material sinks into the anaerobic zone. There the accumulation of more material above leads to some compaction whilst the activities of benthic organisms and the seasonal renewal of the hypolimnion may produce further physical and chemical changes. The degree of compaction to which any given mixture of deposited material is susceptible, however, depends largely on the bulk and physical properties of the material itself and the physical properties of water with which it is more-or-less fully imbibed and surrounded. Compaction is soon almost totally accomplished and increases very little as more deposit accumulates above. However, compaction is the inevitable result of the removal of water from a deposit, either by a natural fall in water table or by drainage and, together with secondary humification and surface oxidation which usually accompany it, may reduce the thickness of a formerly waterlogged organic deposit to a quarter or less.

In the analysis which follows, sites have been selected which seem to lack most, if not all, of the potentially more misleading stratigraphic phenomena. Authors' own descriptions of deposits have been used to place their materials in the following general classification of deposit types.

a. Sedimentary deposits.

(i) Coarse inorganic (gravel and sand).

(ii) Fine inorganic (silt and clay).

(iii) Nekron mud (gyttja), commonly accumulating in deep water lacking macrophytes.

(iv) Fine detritus mud (nekron matrix containing plant detritus mostly $<2$ mm in greatest dimension), commonly accumulating beneath water in which floating-leaved and submerged macrophytes are growing.

(v) Coarse detritus mud (nekron matrix containing plant detritus mostly $>2$ mm in greatest dimension), commonly accumulating in shallow water along the lakeward edge of reedswamp and in 'drift'-filled bays.

b. Mixed deposits.

(vi) Fen or swamp mud (detritus mud with some sedentary material often introduced secondarily by growth from above), commonly accumulating under reedswamp.

c. Sedentary deposits.

(vii) Fen peat (often including small mud lenses resulting from periods of inundation), accumulating beneath fens.

(viii) Bog peat (occasionally including small mud lenses resulting from impeded surface drainage), accumulating beneath bogs.

## Pollen analytical interpretation

Modern pollen diagrams commonly contain data which derive from the aquatic and mire vegetation of the site from which the samples have been taken and which reflect the course of the hydroseral changes which have taken place at that site. These pollen curves can often be used to amplify the stratigraphic record of change and to afford more certain identification of at least some of the plants involved. Because of the ease with which pollen is transported by wind and water, there is a danger that the pollen curves from a particular sampled point might record the vegetation at another point, upstream or upwind, or the resultant of all the vegetation of the mire which might represent many different stages of the hydrosere. For the purpose of identifying the plants growing at, or very close to, a particular point, pollen analytical data can only be used with confidence where they correspond with the stratigraphy, i.e. where the macroscopic plant remains broadly confirm the probable presence of the plants represented by pollen. Some sites, particularly small inland basins, seem not to suffer very greatly from the transport of pollen, most of it being incorporated close to the plants which produce it. Circumstantial confirmation of this may be found in diagrams from different depths of deposit at the one site in which equivalent vegetation changes have taken place at different times and are chronologically separated in the two diagrams. Whilst, to be most useful, pollen analytical data must correspond with the broad indications of stratigraphy, exact correspondence, species for species or genus for genus, is not to be expected because of factors such as differential production and preservation of pollen grains and identifiable macroscopic parts. Rather should both sources be used judiciously to attain the fullest possible picture, still incomplete, of the total vegetation involved.

## Chronology

Making allowance for the possibility of occasional simultaneous deposition at two vertically separated levels and for the secondary incorporation of eroded and transported material, the principle of superposition provides the framework within which more sophisticated chronological estimates may be fitted. Still the most useful of these is that implied by the zonation of pollen diagrams, based primarily on fluctuations in the curves for the pollen of anemophilous forest trees. Any lack of contemporaneity between the same pollen analytical zone boundaries is likely to be insignificant for present purposes within the area under consideration, and they have therefore been dated from radiocarbon dated pollen diagrams (table 1). A few of the sites used have one or more radiocarbon dates attributed to them directly and these serve as a check on the pollen-analytical chronology or provide dated points additional to the pollen-zone boundaries. Where any anomaly exists between a pollen-analytically determined age and a radiocarbon date, the latter is preferred. Supposed chronological markers which depend entirely on stratigraphy (e.g. recurrence surfaces) have not been used.

TABLE I. *Approximate dates attached to pollen-analytical zones.*

| Pollen-analytical zone | Approximate date of boundary, B.C. | Approximate duration of zone in years |
|---|---|---|
| VIII | | 2,500 |
| | 500 | |
| VIIb | | 2,500 |
| | 3,000 | |
| VIIa | | 2,400 |
| | 5,400 | |
| VI | | 1,600 |
| | 7,000 | |
| V | | 700 |
| | 7,700 | |
| IV | | 600 |
| | 8,300 | |
| III | | 500 |
| | 8,800 | |
| II | | 1,200 |
| | 10,000 | |
| I | | |

## THE COURSE OF SUCCESSION

Only 20 published pollen diagrams proved sufficiently informative and free from obvious allogenic influences to allow the reconstruction of the hydroseral sequences at the points from which they were drawn. In all, 71 vegetation transitions were identified and classified according to the pair of vegetation stages they related (table 2). The figures obtained do not lend themselves to sophisticated mathematical analysis but suggest that the sequence

micro-organisms (2) or submerged macrophytes (3)
→ floating-leaved macrophytes (4)
→ reedswamp (5)
→ fen (7)
→ fen carr (9)
→ bog (11)

is the 'preferred' course in that 46 % of all recorded transitions, in all but 3 of which at least one of the above stages is represented, are accounted for by it. The remaining 54 %, however, indicate the possibilities for variety in succession, particularly from reedswamp onwards. Thus, although 40 % of transitions from reedswamp are to fen, 27 % are to bog and 33 % directly to carr of some kind. Also, stable fen and swamp carr are each recorded directly antecedent to bog (table 2, figure 2). Of all the transitions recorded, 17 % are 'reversed', i.e. in the direction contrary to that indicated by the great majority of the records. Most notable, quantitatively, amongst these are transitions from floating-leaved macrophytes and totally submerged macrophytes to open water with only micro-organisms. Such reversals are usually short-lived and might frequently be due to small changes in water level, temperature or trophic status of the lake water; equally, however, they could represent fragments of cyclical changes autochthonously controlled and overcome only when a later vegetation stage supervenes.

When vegetation successions over complete sites are considered, on the basis of authors' interpretations of stratigraphy and pollen diagrams, many more data become available. For sites where marginal and deep water sequences have undoubtedly differed, each has been separately scored. A total of 159 transitions has been recorded (table 3) but, in spite of the large sample, rigorous mathematical treatment has been inhibited by the problem of weighting the distribution of transitions between a single stage and a variety of stages according to the total number of observations to which the stages contribute. The basic sampling method and the interdependence of the scores do not allow the use of any of the standard techniques of matrix manipulation. For present purposes, however, it is enough to restrict consideration to the most obviously dominating quantities.

The most impressive feature of these data is the variety of transitions which have been recorded and which must reflect the flexibility of the succession. It is impossible to select a 'preferred' sequence, but the most commonly occurring transitions, accounting for 52 % of the total, are those shown in figure 3. In that abstraction from the data, the reed-swamp is emphasized as a critical nodum through which the great majority of hydroseres must pass. Whilst this is true, 30 % of all transitions from predominantly open water situations are to fen, swamp carr, fen carr or aquatic *Sphagnum* stages and do not involve reedswamp. Another remarkable feature is that, although bogs develop equally commonly from aquatic *Sphagnum* communities and fen carr, significant numbers of transitions to bog take place directly from reedswamp, fen and swamp carr. In any particular hydrosere, the point of entry of *Sphagnum* is clearly of critical significance. Salt marsh or freshwater marsh (12) are most commonly replaced by reedswamp or fen presumably depending on the depth of fresh water established, arguably an allochthonous factor.

If the data from small inland lakes are isolated (table 4, figure 4), all of them from kettles, there is a clear tendency for open water communities to be replaced by aquatic *Sphagnum* from either the floating-leaved macrophyte or reedswamp stage. Swamp carr is frequently recorded and bogs, the most common terminal stage, derive from it directly or from aquatic *Sphagnum*. There are several examples of transitions from a less wet to a wetter facies, e.g. from fen to swamp carr or even to aquatic *Sphagnum*, and it may be that these are the results of externally determined increases in water supply to which these almost land-locked basins must be particularly susceptible.

TABLE 2. *Frequencies of transitions between vegetation stages derived from 20 pollen diagrams. See text p. 120 for identification of stages.*

| ANTECEDENT VEGETATION | SUCCEEDING VEGETATION | | | | | | | | | | | | |
|---|---|---|---|---|---|---|---|---|---|---|---|---|---|
| | 1 | 2 | 3 | 4 | 5 | 6 | 7 | 8 | 9 | 10 | 11 | 12 | T |
| 1 | . | . | . | . | . | . | . | . | . | . | . | . | 0 |
| 2 | . | . | 2 | 4 | 1 | . | 3 | . | . | . | 1 | . | 11 |
| 3 | . | 4 | . | 3 | 3 | . | . | . | . | . | . | . | 10 |
| 4 | . | 4 | 1 | . | 8 | . | 1 | . | . | . | 1 | . | 15 |
| 5 | . | . | . | . | . | . | 6 | 3 | 2 | . | 4 | . | 15 |
| 6 | . | . | . | . | . | . | . | . | . | . | . | . | 0 |
| 7 | . | 1 | . | 1 | . | . | . | . | 6 | . | 3 | . | 11 |
| 8 | . | . | . | . | 1 | . | . | . | . | . | 2 | . | 3 |
| 9 | . | . | . | . | . | . | . | . | . | . | 6 | . | 6 |
| 10 | . | . | . | . | . | . | . | . | . | . | . | . | 0 |
| 11 | . | . | . | . | . | . | . | . | . | . | . | . | 0 |
| 12 | . | . | . | . | . | . | . | . | . | . | . | . | 0 |
| T | 0 | 9 | 3 | 8 | 13 | 0 | 10 | 3 | 8 | 0 | 17 | 0 | 71 |

TABLE 3. *Frequencies of transitions between vegetation stages at all sites derived from stratigraphy and pollen diagrams. See text p. 120 for identification of stages.*

| ANTECEDENT VEGETATION | SUCCEEDING VEGETATION | | | | | | | | | | | | |
|---|---|---|---|---|---|---|---|---|---|---|---|---|---|
| | 1 | 2 | 3 | 4 | 5 | 6 | 7 | 8 | 9 | 10 | 11 | 12 | T |
| 1 | . | . | 3 | 2 | . | . | . | . | . | 1 | . | . | 6 |
| 2 | . | . | . | 2 | 2 | . | . | . | . | . | . | . | 4 |
| 3 | 1 | . | . | 4 | 7 | . | . | . | . | 1 | . | . | 13 |
| 4 | 1 | . | 1 | . | 9 | . | 3 | 1 | 3 | 5 | . | . | 23 |
| 5 | . | . | . | 2 | . | 1 | 8 | 6 | 7 | 11 | 4 | . | 39 |
| 6 | . | . | . | . | . | . | . | 1 | . | . | . | . | 1 |
| 7 | . | . | . | . | 2 | . | . | 2 | 8 | 2 | 3 | . | 17 |
| 8 | . | . | . | . | 1 | . | 1 | . | 1 | 2 | 3 | . | 8 |
| 9 | . | . | . | 1 | 2 | . | 1 | . | . | 1 | 10 | . | 15 |
| 10 | 1 | . | . | 1 | 1 | . | . | 2 | . | . | 10 | . | 15 |
| 11 | . | . | . | . | . | . | . | . | 1 | 1 | . | 1 | 3 |
| 12 | . | . | . | 1 | 9 | . | 4 | . | 1 | . | . | . | 15 |
| T | 3 | 0 | 4 | 13 | 33 | 1 | 17 | 12 | 21 | 24 | 30 | 1 | 159 |

In the estuarine and coastal areas, where the initial phase of the hydrosere is usually the flooding of a salt marsh or salt-water lagoon with fresh water, the earliest recorded stage is usually reedswamp but subsequent development is as varied as in almost any other class of site, terminating always in bog (table 5, figure 5).

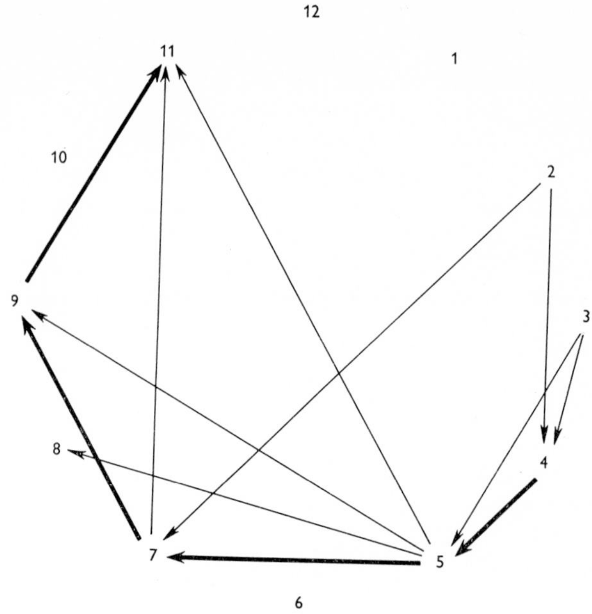


Figure 2. Commoner transitions between vegetation stages drawn from table 2. Thick lines denote dominant courses.

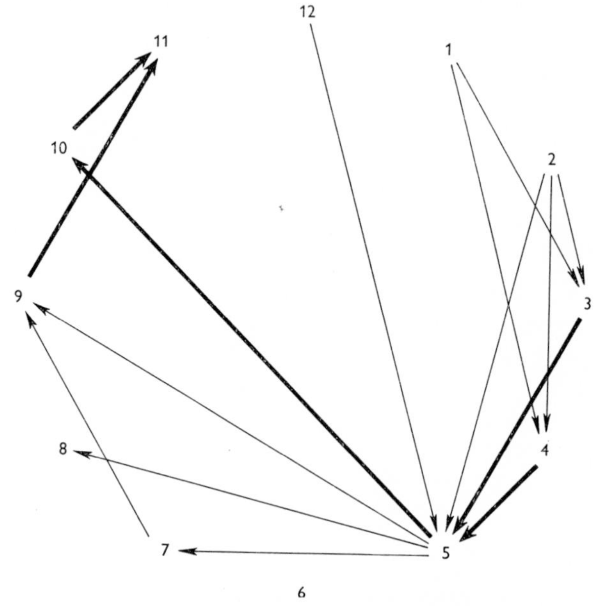


Figure 3. Commoner transitions between vegetation stages at all sites drawn from table 3. Thick lines denote dominant courses.

The large freshwater inland sites expectedly demonstrate a great diversity of succession in which, however, reedswamp is crucial (table 6, figure 6). The entry of *Sphagnum* into predominantly open water is insignificantly represented and terminal bogs are derived almost equally from fen carr, reedswamp and fen. Swamp carr is hardly represented but the frequency with which fen carr is recorded directly following reedswamp suggests the possible mis-identification of swamp carr products as those of fen carr.

TABLE 4. *Frequencies of transitions between vegetation stages in small inland basins derived from stratigraphy and pollen diagrams. See text p. 120 for identification of stages.*

| ANTECEDENT VEGETATION | SUCCEEDING VEGETATION 1 | 2 | 3 | 4 | 5 | 6 | 7 | 8 | 9 | 10 | 11 | 12 | T |
|---|---|---|---|---|---|---|---|---|---|---|---|---|---|
| 1 | . | . | 3 | 2 | . | . | . | . | . | 1 | . | . | 6 |
| 2 | . | . | . | 1 | 1 | . | . | . | . | . | . | . | 2 |
| 3 | 1 | . | . | 2 | 2 | . | . | . | . | 1 | . | . | 6 |
| 4 | 1 | . | . | . | 4 | . | . | 1 | . | 3 | . | . | 9 |
| 5 | . | . | . | . | . | . | 1 | 1 | . | 5 | . | . | 7 |
| 6 | . | . | . | . | . | . | . | . | . | . | . | . | 0 |
| 7 | . | . | . | . | 2 | . | . | 2 | . | 1 | . | . | 5 |
| 8 | . | . | . | . | . | . | . | . | . | 1 | 2 | . | 3 |
| 9 | . | . | . | . | . | . | . | . | . | . | . | . | 0 |
| 10 | 1 | . | . | . | 1 | . | . | 2 | . | . | 3 | . | 7 |
| 11 | . | . | . | . | . | . | . | . | . | . | . | . | 0 |
| 12 | . | . | . | . | . | . | 2 | . | . | . | . | . | 2 |
| T | 3 | 0 | 3 | 5 | 10 | 0 | 3 | 6 | 0 | 12 | 5 | 0 | 47 |

TABLE 5. *Frequencies of transitions between vegetation stages in coastal and estuarine sites derived from stratigraphy and pollen diagrams. See text p. 120 for identification of stages.*

| ANTECEDENT VEGETATION | SUCCEEDING VEGETATION 1 | 2 | 3 | 4 | 5 | 6 | 7 | 8 | 9 | 10 | 11 | 12 | T |
|---|---|---|---|---|---|---|---|---|---|---|---|---|---|
| 1 | . | . | . | . | . | . | . | . | . | . | . | . | 0 |
| 2 | . | . | . | . | . | . | . | . | . | . | . | . | 0 |
| 3 | . | . | . | . | . | . | . | . | . | . | . | . | 0 |
| 4 | . | . | . | . | 2 | . | 2 | . | 1 | 1 | . | . | 6 |
| 5 | . | . | . | 2 | . | . | 3 | 2 | 2 | 4 | . | . | 13 |
| 6 | . | . | . | . | . | . | . | . | . | . | . | . | 0 |
| 7 | . | . | . | . | . | . | . | . | 4 | . | . | . | 4 |
| 8 | . | . | . | . | 1 | . | . | . | . | 1 | . | . | 2 |
| 9 | . | . | . | 1 | 1 | . | . | . | . | . | 5 | . | 7 |
| 10 | . | . | . | 1 | . | . | . | . | . | . | 3 | . | 4 |
| 11 | . | . | . | . | . | . | . | . | . | . | . | . | 0 |
| 12 | . | . | . | 1 | 7 | . | 1 | . | 1 | . | . | . | 10 |
| T | 0 | 0 | 0 | 5 | 11 | 0 | 6 | 2 | 8 | 6 | 8 | 0 | 46 |

Excluding from the analysis the data from the very extensive sites (e.g. Norfolk Broads) and the upland glacial channels (e.g. Thornton Mire), the sequence

totally submerged macrophytes (3)
→ reedswamp (5)
→ fen (7)
→ fen carr (9)
→ bog (11)

accounts for 35 % of the recorded transitions, with the route

reedswamp (5)
→ aquatic *Sphagnum* (10)
→ bog (11)

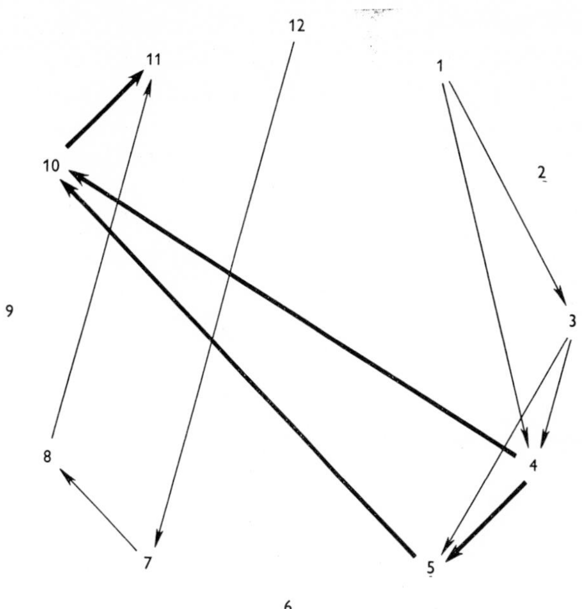


Figure 4. Commoner transitions between vegetation stages in small inland basins drawn from table 4. Thick lines denote dominant courses.

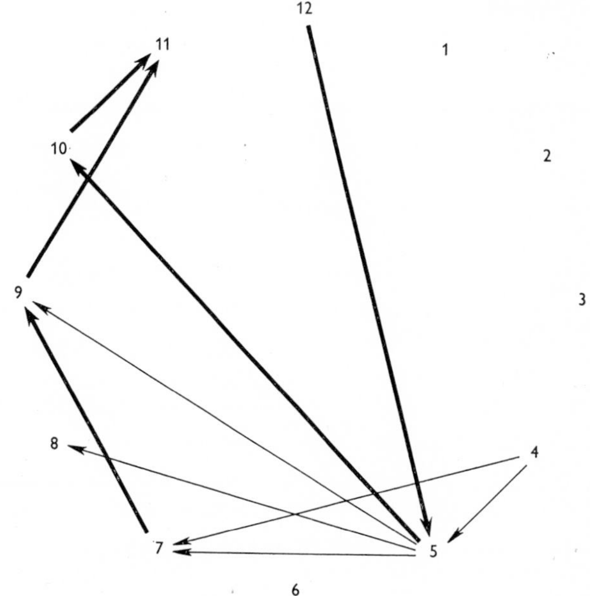


Figure 5. Commoner transitions between vegetation stages in coastal and estuarine sites drawn from table 5. Thick lines denote dominant courses.

playing a small but significant role. In the upland channels the direct transition from floating-leaved macrophyte and reedswamp communities to fen carr (perhaps swamp carr) and the ultimate replacement of all other stages by bog is important.

TABLE 6. *Frequencies of transitions between vegetation stages in large inland basins derived from stratigraphy and pollen diagrams. See text p.* 120 *for identification of stages.*

| ANTECEDENT VEGETATION | SUCCEEDING VEGETATION 1 | 2 | 3 | 4 | 5 | 6 | 7 | 8 | 9 | 10 | 11 | 12 | T |
|---|---|---|---|---|---|---|---|---|---|---|---|---|---|
| 1 | . | . | . | . | . | . | . | . | . | . | . | . | 0 |
| 2 | . | . | . | 1 | 1 | . | . | . | . | . | . | . | 2 |
| 3 | . | . | . | 2 | 5 | . | . | . | . | . | . | . | 7 |
| 4 | . | . | 1 | . | 3 | . | 1 | . | 2 | 1 | . | . | 8 |
| 5 | . | . | . | . | . | 1 | 4 | 3 | 5 | 2 | 4 | . | 19 |
| 6 | . | . | . | . | . | . | . | 1 | . | . | . | . | 1 |
| 7 | . | . | . | . | . | . | . | . | 4 | 1 | 3 | . | 8 |
| 8 | . | . | . | . | . | . | 1 | . | 1 | 1 | . | . | 3 |
| 9 | . | . | . | . | 1 | . | 1 | . | . | 1 | 5 | . | 8 |
| 10 | . | . | . | . | . | . | . | . | . | . | 4 | . | 4 |
| 11 | . | . | . | . | . | . | . | . | 2 | 1 | . | . | 3 |
| 12 | . | . | . | . | 2 | . | 1 | . | . | . | . | . | 3 |
| T | 0 | 0 | 1 | 3 | 12 | 1 | 8 | 4 | 14 | 7 | 16 | 0 | 66 |

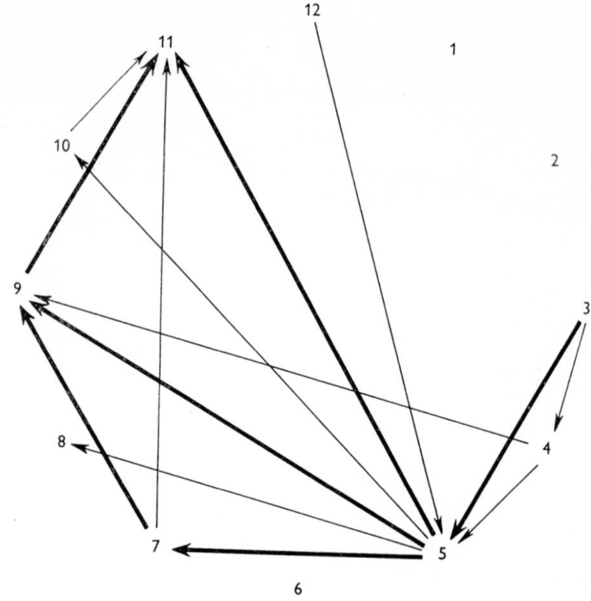


Figure 6. Commoner transitions between vegetation stages in large inland basins drawn from table 6. Thick lines denote dominant courses.

## THE RATE OF SUCCESSION

In a single profile or a section through a deposit the stratigraphic, and even the pollen-analytic, changes which record vegetation shifts are usually fairly sudden. From such data the only way of estimating rates of change is to consider the length of time occupied by the intervening stage. In a few cases, where a number of pollen diagrams are available from different points at the same site, it might be possible to express rates of change as rates of *horizontal* spread of a particular vegetation type. Given the nature of the majority of the data, however, the former method seems preferable. The data, although fragmentary (table 7), expose the probability that totally submerged macrophyte, floating-leaved macrophyte and reedswamp stages are usually short-lived, often less than 500 years. By contrast, unvegetated water on the one hand and fens on the other, and even swamp carr, appear to be able to persist longer. The method of abstracting the data does not allow terminal stages to be included but it is evident that open water and bogs are the most persistent stages of all. Again, although not susceptible to precise estimate, all the indications are that *Sphagnum*, once established, expands very rapidly indeed and quickly imposes its dominating influence on any further development.

These suggestions can be investigated further by examining the periods of time intervening between the end of the open water stages (1, 2 or 3) and the beginning of bog or fen growth at a number of sites (table 8). Naturally, the results vary widely but all suggest that, once a hydrosere is initiated to the extent of the floating-leaved macrophyte stage, conversion to fen may frequently take less than 1000 years and will rarely take more than twice that time. Direct succession to bog, through reedswamp and sometimes swamp carr but *not* through stable fen, can be equally rapid and rarely takes longer than 2500 years. Naturally, conversion to bog through an intervening fen stage, although it can happen in less than 1000 years, is normally much slower, 2500–4000 years representing a 'modal' rate.

## THE RATE OF ACCUMULATION

Although grain size of substratum, the base status of the water in lake or soil, temperature conditions and so on at least partially determine the identity of the species which take part in a hydrosere, the factor of overriding importance leading to the replacement of one group of plants by another is water level or water depth. The overgrowth of open water by swamp plants must await the time when accumulation of bottom deposits has so shallowed the water that bottom rooting phanerogams can live in it. A fen must remain more or less stable vegetationally if it is too wet for forest trees yet sufficiently dry in summer to prevent effective accumulation of peat. Water depth and water movement not only largely determine the plants which will grow at a site but also profoundly affect the kind of deposit which is built at any point. Because plants have fairly well-marked tolerances of water depth both above and below soil level, the rate at which this is changed by accumula-

TABLE 7. *Persistence of vegetation stages at 16 pollen-analytical sites. The sequence 2-4-5 in the column headed 501–1000 implies that stage 4 in the sequence 2-4-5 persisted for between 501 and 1000 years. See text p. 200 for identification of stages.*

| Site | Author | Years in middle stage | | | | | | | | |
|---|---|---|---|---|---|---|---|---|---|---|
| | | < 500 | 501–1000 | 1001–1500 | 1501–2000 | 2001–2500 | 2501–3000 | 3001–3500 | 3501–4000 | > 4000 |
| Cranes Moor (Little Bog) | Seagrief (1960) | — | — | — | 13-7-11 | — | — | — | — | — |
| Cwm Idwal | Godwin (1955*b*) | 4-3-4 | — | — | 5-8-11 | — | — | — | — | — |
| | | 3-4-5 | — | — | — | — | — | — | — | — |
| | | 4-5-8 | — | — | — | — | — | — | — | — |
| Hockham | Godwin & Tallantire (1951) | — | — | — | — | — | — | — | — | 1-4-5 |
| Langdale Combe | Walker (1965) | — | — | — | — | 2-4-11 | — | — | 3-2-4 | — |
| Littleton | Mitchell (1965) | — | — | — | — | — | — | 4-7-11 | — | — |
| Loch Creagh | Donner (1962) | 4-5-7 | 2-4-5 | 5-7-11 | — | — | — | — | — | — |
| Mickleden | Walker (1965) | 2-3-5 | 3-5-8 | — | — | — | 5-8-5 | — | — | — |
| Moorthwaite | Walker (1966) | 4/10-5/10-8 | 3-4-10 | 5/10-8-11 | — | — | — | — | — | — |
| Muckle Moss | Pearson (1960) | 3/4-5-8 | — | — | — | — | — | — | — | — |
| | | 5-8-11 | — | — | — | — | — | — | — | — |
| Nichol's Moss | Smith (1959) | — | — | — | — | 13-5-8 | — | — | — | — |
| Oulton Moss | Walker (1966) | 3-4-10 | — | — | — | — | — | — | — | — |
| Rhosgoch | Bartley (1960) | — | 5-8-7 | 1-5-8 | 8-7-11 | 3-7-11 | — | — | — | — |
| Scaleby | Walker (1966) | 3-5-3 | — | — | — | — | — | — | — | — |
| | | 5-3-5 | — | — | — | — | — | — | — | — |
| | | 3-5-11 | — | — | — | — | — | — | — | — |
| Seamer | Walker & Godwin (1954) | — | 3-4-5 | 4-5-7 | — | — | — | — | — | — |
| Skelsmergh | Walker (1955) | — | — | — | — | — | — | — | — | 4-2-7 |
| Tregaron | Godwin & Mitchell (1938) | — | — | — | — | 4-5-11 | — | — | — | — |

TABLE 8. *Periods intervening between certain vegetation stages at 25 pollen-analytical sites.*

| Site | Author | Years between stages (to nearest half century) | | |
|---|---|---|---|---|
| | | Open water–fen | Open water–bog ('direct') | Open water–bog (through fen) |
| Abbot Moss | Walker (1966) | — | 1500 | — |
| Agher | Mitchell (1951) | ?1000 | — | ?3000 |
| Aros Moss | Nichols (1967) | — | — | 3500 |
| Bettisfield | Hardy (1939) | — | — | 5000 |
| Bowness Common | Walker (1966) | — | 500+ | — |
| Cranes Moor (Little Bog) | Seagrief (1960) | — | — | ?1500 |
| Cwm Idwal | Godwin (1955b) | 1500–2000 | — | — |
| Ellerside | Oldfield & Statham (1963) | — | — | 500 |
| Foulshaw | Smith (1959) | — | 500 | — |
| Helsington | Smith (1959) | — | 1500 | — |
| Hockham | Godwin & Tallantire (1951) | 7000 | — | — |
| Langdale Combe | Walker (1965) | — | ?2500 | — |
| Littleton | Mitchell (1965) | < 500 | — | 4000 |
| Llanllwch | Thomas (1965) | < 500 | — | — |
| Moss Lake, Liverpool | Godwin (1959) | — | 2500 | — |
| Muckle Moss | Pearson (1960) | — | 1000–2500 | — |
| Oulton Moss | Walker (1966) | — | 1500 | — |
| Racks Moss | Nichols (1967) | — | — | 3500 |
| Rhosgoch | Bartley (1960) | < 500 | — | 3000 |
| Scaleby | Walker (1966) | — | < 500 | — |
| Seamer A16 | Walker & Godwin (1954) | 2000 | — | — |
| Seamer DB1 | Walker & Godwin (1954) | 1000 | — | — |
| Skelsmergh | Walker (1955) | 2000 | — | — |
| Tadcaster | Bartley (1962) | 1000 | — | — |
| Treanscrabbagh | Mitchell (1951) | — | 3000 | — |

TABLE 9. *Rates of accumulation of deposits from specified periods at pollen analytical sites. See text p.* 122 *for identification of deposit types; ao = calcareous mud, *interpolated date, †mixed deposit.*

| Site and Diagram | Author | Period in years B.C. unless shown otherwise | Accumulation rates in cm per 1000 years in deposit types *ao* | *a* (ii) | *a* (iii) | *a* (iv) | *a* (v) | *b* (vi) | *c* (vii) | *c* (viii) |
|---|---|---|---|---|---|---|---|---|---|---|
| SMALL INLAND BASINS | | | | | | | | | | |
| Oban 1 a | Donner (1957) | 7800–5400 | — | — | — | 40 | — | — | — | — |
| | | 5400–3000 | — | — | — | 33 | — | — | — | — |
| Oban 1 b | | 7800–5400 | — | — | — | 18 | — | — | — | — |
| | | 5400–3000 | — | — | — | 44 | — | — | — | — |
| Oban 3 | | 7800–5400 | — | — | — | 31 | — | — | — | — |
| | | 5400–3000 | — | — | — | 18 | — | — | — | — |
| Oban 4 | | 7800–5400 | — | — | — | 44 | — | — | — | — |
| | | 5400–3000 | — | — | — | 34 | — | — | — | — |
| Abbot Moss A+B | Walker (1966) | 10000–8800 | — | 20 | — | — | — | — | — | — |
| | | *8600–8300 | — | — | 65 | — | — | — | — | — |
| | | 8300–7000 | — | — | — | 61 | — | — | — | — |
| | | 7000–5400 | — | — | — | 50 | — | — | — | — |
| | | 5400–*4400 | — | — | — | 13 | — | — | — | — |
| | | *4400–*3500 | — | — | — | — | 16 | — | — | — |
| | | *3500–*2500 | — | — | — | — | — | 15 | — | — |
| Ehenside A | Walker (1966) | 7000–5400 | — | — | — | 29 | — | — | — | — |
| | | 5400–3000 | — | — | — | 29 | — | — | — | — |
| | | 3000–*2000 | — | — | — | 55 | — | — | — | — |
| Tadcaster T 14 | Bartley (1962) | 10000–8800 | 21 | — | — | — | — | — | — | — |
| | | *6800–5400 | — | — | — | — | 25 | — | — | — |
| Langdale Combe | Walker (1965) | 8300–7000 | — | — | 19 | — | — | — | — | — |
| | | 7000–5400 | — | — | 100 | — | — | — | — | — |
| | | 5400–3000 | — | — | 185 | — | — | — | — | — |
| Mickleden | Walker (1965) | 8300–7000 | — | — | — | 30 | — | — | — | — |
| | | *6500–5400 | — | — | — | — | 160† | — | — | — |
| | | 5400–3000 | — | — | — | — | 55 | — | — | — |
| Langdale Red Tarn | Pennington (1964) | 7000–5400 | — | — | — | 32 | — | — | — | — |
| | | 5400–*3400 | — | — | — | 63 | — | — | — | — |
| Langdale Blea Tarn | Pennington (1964) | 7800–7000 | — | 32 | — | — | — | — | — | — |
| | | 7000–5400 | — | 25 | — | — | — | — | — | — |
| | | 5400–3000 | — | — | — | 40 | — | — | — | — |
| Skelsmergh A | Walker (1955) | 10000–8800 | 30 | — | — | — | — | — | — | — |
| | | 8800–*8400 | — | 50 | — | — | — | — | — | — |
| | | *8400–7000 | 25 | — | — | — | — | — | — | — |
| | | 7000–*6000 | — | — | — | — | 25 | — | — | — |
| | | *6000–5400 | — | — | — | 29 | — | — | — | — |
| | | 5400–*3500 | — | — | — | — | 36 | — | — | — |
| Urswick Tarn | Oldfield & Statham (1963) | 7000–5400 | 210 | — | — | — | — | — | — | — |
| | | 5400–3000 | — | — | — | 50 | — | — | — | — |
| Cwm Idwal | Godwin (1955*b*) | 8300–7000 | — | 23 | — | — | — | — | — | — |
| | | 7000–*5600 | — | — | — | 59 | — | — | — | — |
| Llyn Dwythwch | Seddon (1962) | 7000–5400 | — | — | — | 56 | — | — | — | — |
| | | 5400–3000 | — | — | — | 56 | — | — | — | — |
| Whattall Moss 7 | Hardy (1939) | 7800–7000 | — | — | — | — | 75 | — | — | — |
| | | 7000–5400 | — | — | — | 59† | — | — | — | — |
| | | 5400–3000 | — | — | 27 | — | — | — | — | — |
| Cothill | Clapham & Clapham (1939) | 8300–7000 | — | — | 73† | — | — | — | — | — |
| | | 7000–5400 | — | — | 163† | — | — | — | — | — |
| Elstead | Seagrief & Godwin (1960) | 8300–7800 | — | — | — | 76 | — | — | — | — |
| | | 7800–7000 | — | — | 20 | — | — | — | — | — |
| Cranes Moor CY | Seagrief (1960) | 7800–7000 | — | — | — | 188 | — | — | — | — |
| Cranes Moor 1 | | 7000–5400 | — | — | — | — | — | — | — | 128 |
| Cranes Moor C 12 | | 7000–*5800 | — | — | — | — | 116 | — | — | — |
| LARGE INLAND BASINS | | | | | | | | | | |
| Loch Creagh | Donner (1962) | 7800–5400 | — | — | — | 44 | — | — | — | — |
| | | 5400–3000 | — | — | — | 37 | — | — | — | — |
| | | 3000–*2300 | — | — | — | 100 | — | — | — | — |
| Loch Mahaik | Donner (1962) | 7800–5400 | — | — | — | 28 | — | — | — | — |
| | | 5400–3000 | — | — | — | 42 | — | — | — | — |
| Bloak Moss | Turner (1965) | 1300–A.D. 500 | — | — | — | — | — | — | — | 60 |
| Flanders Moss | Turner (1965) | 3250–A.D. 220 | — | — | — | — | — | — | — | 68 |

TABLE 9. (*cont.*)

| Site and Diagram | Author | Period in years B.C. unless shown otherwise | Accumulation rates in cm per 1000 years in deposit types *ao* | *a* (ii) | *a* (iii) | *a* (iv) | *a* (v) | *b* (vi) | *c* (vii) | *c* (viii) |
|---|---|---|---|---|---|---|---|---|---|---|
| LARGE INLAND BASINS (*cont.*) | | | | | | | | | | |
| Muckle Moss A | Pearson (1960) | *5000–3000 | — | — | — | — | — | — | — | 55 |
| | | 3000–500 | — | — | — | — | — | — | — | 70 |
| Scaleby B | Walker (1966) | 8800–8300 | — | — | — | 80 | — | — | — | — |
| | | 8300–7800 | — | — | — | — | 50 | — | — | — |
| Scaleby A | | 7700–7000 | — | — | — | — | — | — | — | 93 |
| | | 7000–5400 | — | — | — | — | — | — | — | 78 |
| | | 5400–3000 | — | — | — | — | — | — | — | 69 |
| | | 3000–1936 | — | — | — | — | — | — | — | 65 |
| Devoke Water | Pennington (1964) | 8300–7000 | — | — | — | 8 | — | — | — | — |
| | | 7000–*5800 | — | — | — | 60 | — | — | — | — |
| Seathwaite Tarn | Pennington (1964) | 8300–7000 | — | 19 | — | — | — | — | — | — |
| | | 7000–5400 | — | 34 | — | — | — | — | — | — |
| | | 5400–3000 | — | — | — | 23 | — | — | — | — |
| | | 3000–1080 | — | — | — | 73 | — | — | — | — |
| Kentmere A | Walker (1955) | 8300–7000 | — | — | 15 | — | — | — | — | — |
| | | 7000–5400 | — | — | 22 | — | — | — | — | — |
| Kentmere B | | 7000–5400 | — | — | 34 | — | — | — | — | — |
| | | 5400–3000 | — | — | 50 | — | — | — | — | — |
| Windermere, Low Wray | Pennington (1947) | 7800–7000 | — | — | 40 | — | — | — | — | — |
| | | 7000–5400 | — | — | 110 | — | — | — | — | — |
| Windermere, Ecclerigg | | 7800–7000 | — | — | 50 | — | — | — | — | — |
| | | 7000–5400 | — | — | 90 | — | — | — | — | — |
| Helton Tarn A | Smith (1958*b*) | 3000–500 | — | — | — | — | 30 | — | — | — |
| Helton Tarn B | | 8300–7000 | — | — | — | 42 | — | — | — | — |
| | | 7000–5400 | — | — | — | 37 | — | — | — | — |
| Witherslack Hall | Smith (1958*b*) | 7000–5400 | — | — | — | 25 | — | — | — | — |
| | | 5400–3000 | — | — | — | — | 114 | — | — | — |
| | | 3000–500 | — | — | — | — | 84 | — | — | — |
| Hawes Water | Oldfield (1960) | 8300–7000 | 38 | — | — | — | — | — | — | — |
| | | 7000–5400 | 206 | — | — | — | — | — | — | — |
| | | 5400–3000 | 75 | — | — | — | — | — | — | — |
| Malham Tarn Moss F | Pigott & Pigott (1963) | 7800–7000 | 37 | — | — | — | — | — | — | — |
| | | 5400–3000 | — | — | — | — | — | — | — | 56 |
| | | 3000–500 | — | — | — | — | — | — | — | 96 |
| Seamer A 16 | Walker & Godwin (1954) | 8300–7800 | 90 | — | — | — | — | — | — | — |
| | | 7800–7000 | — | — | — | — | 37 | — | — | — |
| | | 7000–*5700 | — | — | — | — | 73 | — | — | — |
| Flixton DB 1 | Walker & Godwin (1954) | 8300–7800 | 130 | — | — | — | — | — | — | — |
| Chat Moss | Birks (1963–4) | 3000–1120 | — | — | — | — | — | — | — | 93 |
| Bagmere | Birks (1965) | 10000–8800 | — | — | — | 19 | — | — | — | — |
| | | 8800–8300 | — | — | — | 40 | — | — | — | — |
| | | 8300–7000 | — | — | — | 11 | — | — | — | — |
| Nant Ffrancon | Seddon (1962) | 10000–8800 | — | — | — | 54 | — | — | — | — |
| | | 8300–7800 | — | — | — | 180 | — | — | — | — |
| | | 7800–7000 | — | — | — | 88 | — | — | — | — |
| | | 7000–5400 | — | — | — | 62 | — | — | — | — |
| | | 5400–3000 | — | — | — | — | 66 | — | — | — |
| | | 3000–500 | — | — | — | — | 52 | — | — | — |
| Bettisfield | Hardy (1939) | 7000–5400 | — | — | — | 37 | — | — | — | — |
| Whixall | Turner (1964), Hardy (1939) | 1290–760 | — | — | — | — | — | — | — | 42 |
| | | 760–350 | — | — | — | — | — | — | — | 25 |
| | | 350–50 | — | — | — | — | — | — | — | 43 |
| Tregaron W 27 | Godwin & Mitchell (1938) | 7000–5400 | — | — | — | — | — | — | 77 | — |
| Tregaron SE 10 | | 7800–7000 | — | — | — | — | — | — | 100 | — |
| | | 7000–5400 | — | — | — | — | — | — | 81 | — |
| | | 3000–*1000 | — | — | — | — | — | — | — | 54 |
| Tregaron | Turner (1964) | 696–404 | — | — | — | — | — | — | — | 300 |
| | | 404–A.D. 473 | — | — | — | — | — | — | — | 23 |
| | | A.D. 473–A.D. 1182 | — | — | — | — | — | — | — | 18 |
| Rhosgoch RG 14 | Bartley (1960) | 10000–8800 | 8 | — | — | — | — | — | — | — |
| | | 8800–8300 | 26 | — | — | — | — | — | — | — |
| | | 7700–7000 | — | — | — | — | — | — | 28 | — |
| | | 7000–5400 | — | — | — | — | — | — | 19 | — |
| | | 5400–3000 | — | — | — | — | — | — | — | 12 |

TABLE 9. (*cont.*)

| Site and Diagram | Author | Period in years B.C. unless shown otherwise | Accumulation rates in cm per 1000 years in deposit types | | | | | | | |
|---|---|---|---|---|---|---|---|---|---|---|
| | | | *ao* | *a* (ii) | *a* (iii) | *a* (iv) | *a* (v) | *b* (vi) | *c* (vii) | *c* (viii) |
| LARGE INLAND BASINS (*cont.*) | | | | | | | | | | |
| Rhosgoch RG18 | | 7700–7000 | — | — | — | — | — | — | 43 | — |
| Llanllwch | Thomas (1965) | 5400–3000 | — | — | — | — | — | — | — | 17 |
| Hockham DB5 | Godwin & Tallantire (1951) | 8300–7800 | — | — | 210 | — | — | — | — | — |
| | | 7800–7000 | — | — | — | 50 | — | — | — | — |
| | | 7000–5400 | — | — | — | 110 | — | — | — | — |
| | | 5400–3000 | — | — | 40 | — | — | — | — | — |
| Littleton | Mitchell (1965) | 7000–5400 | — | — | — | — | — | — | 100 | — |
| Treanscrabbagh | Mitchell (1951) | 8300–7000 | — | — | — | 31 | — | — | — | — |
| | | 3020–1715 | — | — | — | — | — | — | 70 | — |
| Parkmore | Morrison (1959) | 5400–3000 | — | — | — | — | 27 | — | — | — |
| Ballynakill | Mitchell (1951) | 5400–3000 | — | — | — | — | — | — | — | 32 |
| Agher | Mitchell (1951) | 3000–500 | — | — | — | — | — | — | — | 56 |
| COASTAL AND ESTUARINE SITES | | | | | | | | | | |
| Bowness Common | Walker (1966) | 3000–500 | — | — | — | — | — | — | — | 60 |
| Hatfield | Smith (1958*a*) | 3000–500 | — | — | — | — | — | — | — | 52 |
| Hatfield | Turner (1965) | 265–A.D. 560 | — | — | — | — | — | — | — | 60 |
| Nichol's Moss | Smith (1959) | *4000–3000 | — | — | — | — | 98 | — | — | — |
| | | 3000–500 | — | — | — | — | 32 | — | — | — |
| Helsington | Smith (1959) | *4000–3000 | — | — | — | — | 50 | — | — | — |
| | | 500–A.D. 436 | — | — | — | — | — | — | — | 90 |
| Foulshaw | Smith (1959) | 3000–500 | — | — | — | — | — | — | — | 56 |
| Shapwick DB3 | Clapham & Godwin (1948) | *5000–3000 | — | — | — | — | 60 | — | — | — |
| Meare Heath Trackway | Clapham & Godwin (1948) | 5400–3000 | — | — | — | — | 65 | — | — | — |
| Shapwick Heath Trackway | Clapham & Godwin (1948) and Godwin & Willis (1959) | *5000–3000 | — | — | — | — | 90 | — | — | — |
| | | 3000–1350 | — | — | — | — | — | — | — | 60 |
| Westhay Trackway | Clapham & Godwin (1948) and Godwin & Willis (1959) | *5000–3000 | — | — | — | — | 120 | — | — | — |
| | | 3000–850 | — | — | — | — | — | — | — | 70 |
| Glastonbury Lake Village | Godwin (1955*a*) | 5400–3000 | — | — | — | — | 62 | — | — | — |
| | | 3000–*800 | — | — | — | — | 95 | — | — | — |

tion of different deposit types is of paramount importance in determining the rate of a hydrosere.

Table 9 lists accumulation rates of all the deposit types (p. 122), plus calcareous mud, which can be determined from the stratigraphic descriptions accompanying pollen diagrams from chosen sites. The periods used for the assessment of accumulation rates normally run from the beginning to the end of a pollen-analytical zone or from one zone boundary to a radiocarbon dated horizon. Occasionally, dates have been interpolated at other points in a diagram but each of these rests on considerable, if circumstantial, evidence. Accumulation rates are defined in cm per 1000 years and are naturally applicable only to the section from which the pollen analysed materials were drawn.

As an initial check on the suggestion that compaction is not a significant factor in undisturbed waterlogged deposits, mean depth of sample in the deposit is compared with accumulation rate, for all sites together (figure 7). Whether considered as separate deposit types or collectively, there is no correlation between sample depth and apparent accumulation rate confirming that, beyond the earliest stages of deposition, progressive compaction does not normally occur. At a very few sites (e.g. Abbot Moss, Ehenside Tarn), there is some slight indication to the contrary but whilst this should remind of the possibility of the process under particular circumstances, it cannot reverse the conclusion from the general analysis.

Collectively, the 151 recorded accumulation rates are quite narrowly distributed: 63 % of the records lie between 21 and 60 cm per 1000 years, 78 % between 21 and 80 cm per 1000 years and only 10 % above 100 cm per 1000 years (figure 8). The samples for which these very high rates are recorded come from a variety of sites (in terms of size, altitude, location etc.), deposits and periods. Separately considered, but without regard for age, there are no very striking differences between the distributions of the rates for the different deposit types. Fine detritus mud rates are less well represented in the higher categories than are those of other deposits, e.g. nekron mud and coarse detritus mud each have a third of their recorded rates above 80 cm per 1000 years whilst only a tenth of those for fine detritus mud fall in the same

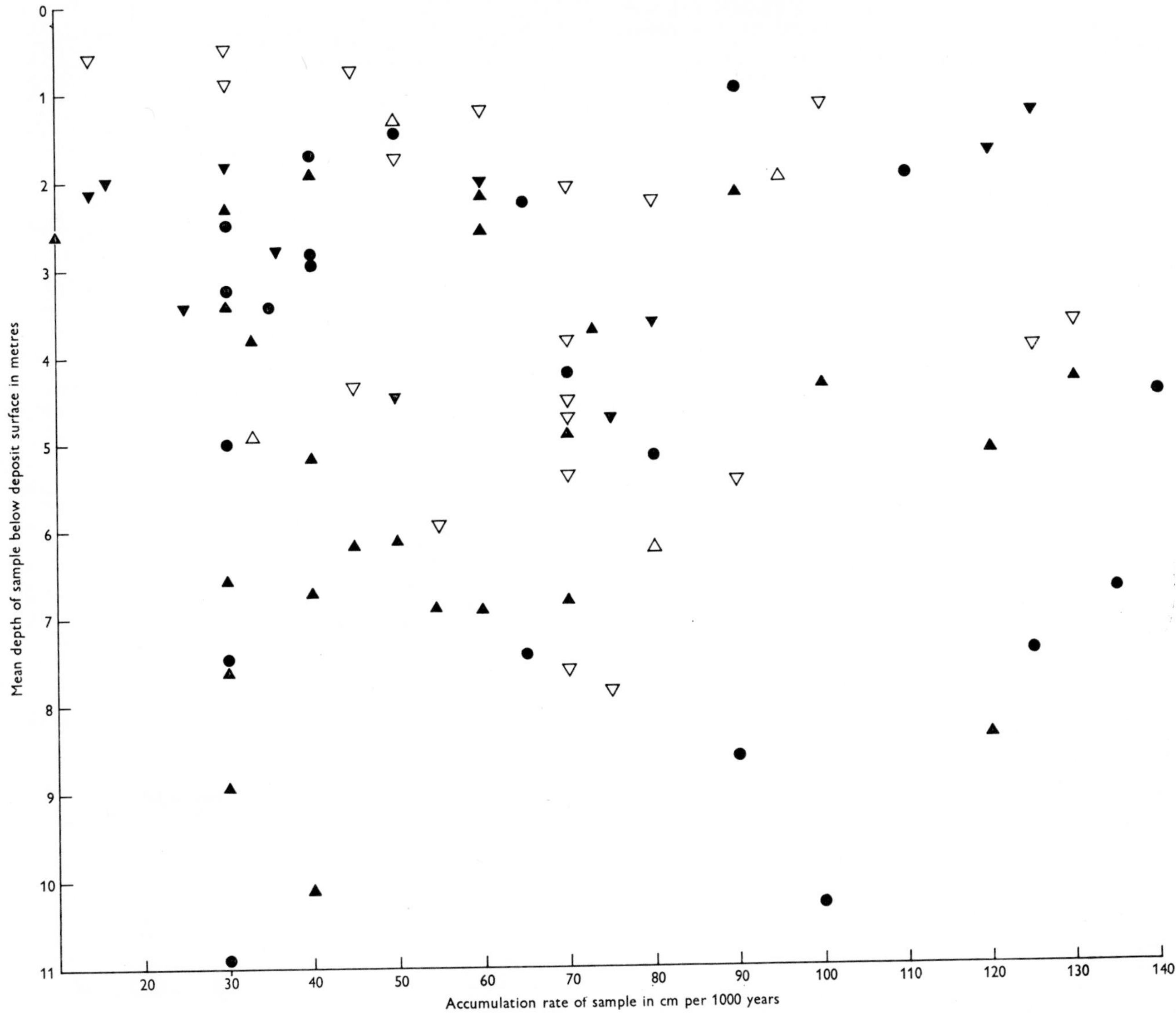


Figure 7. The relationship between mean depth of sample and accumulation rate of sample as a test for compaction of deposits. ● = nekron mud, ▲ = fine detritus mud, ▼ = coarse detritus mud, △ = fen peat, ▽ = bog peat.

category. If the samples of coarse detritus mud from coastal and estuarine sites are excluded, however, a large measure of the difference between the rate distributions of fine detritus mud and coarse detritus mud is lost. The eight rates for fen peat lie between 11 and 100 cm per 1000 years.

When analysed period by period, the data for the separate deposit types are scanty (table 10). In Zone VI the tendency for fine detritus mud to accumulate less rapidly than other materials first becomes apparent. In Zone VIIa the rates for coarse detritus muds of large basins or plains are noticeably greater than those for other materials but it must be remembered that this was the period during which many coastal areas were emerging from the sea and that it is the coarse detritus muds of these coastal sites which contribute the fast rates; they are the products of a particular ecological régime and cannot be used to set a standard for coarse detritus mud in general. There are only 9 rate records for the sedimentary materials and fen peat from Zone VIIb but, such as they are, they suggest that slow accumulation (< 50 cm per 1000 years) of these deposits was very uncommon during that period as it was also for bog peat.

Bog peats accumulate at notoriously varied rates and, moreover, conditions under which secondary modification of the original material, its bulk as well as its chemical composition, takes place (e.g. temporary drying) must occur more frequently than in

TABLE 10. *Mean rates of accumulation of deposits in selected periods. See text* *p.* 122 *for identification of deposit types; r = mean rate in cm per 1000 years, n = number of observations.*

| Period in years B.C. | Deposit type: *ao* | | *a* (ii) | | *a* (iii) | | *a* (iv) | | *a* (v) | | *c* (vii) | | *c* (viii) | |
|---|---|---|---|---|---|---|---|---|---|---|---|---|---|---|
| | *r* | *n* | *r* | *n* | *r* | *n* | *r* | *n* | *r* | *n* | *r* | *n* | *r* | *n* |
| 10,000–8,800 | 20 | 3 | 20 | 1 | — | 0 | 36 | 2 | — | 0 | — | 0 | — | 0 |
| 8,800–8,300 | 26 | 1 | 50 | 1 | 65 | 1 | 60 | 2 | — | 0 | — | 0 | — | 0 |
| 8,300–7,800 | 110 | 2 | — | 0 | 210 | 1 | 178 | 2 | 50 | 1 | — | 0 | — | 0 |
| 7,800–7,000 | 37 | 1 | 32 | 1 | 37 | 3 | 109 | 3 | 56 | 2 | 100 | 1 | 93 | 1 |
| 7,000–5,400 | 208 | 2 | 29 | 2 | 86 | 6 | 49 | 13 | 80 | 5 | 86 | 3 | 103 | 2 |
| 5,400–3,000 | 75 | 1 | — | 0 | 75 | 4 | 37 | 13 | 66 | 13 | — | 0 | 46 | 5 |
| 3,000–500 | — | 0 | — | 0 | — | 0 | 75 | 3 | 59 | 5 | 70 | 1 | 64 | 14 |

lakes. It is all the more interesting then that the samples of bog peat considered have a well-marked modal rate of 51–60 cm per 1000 years and that this is not very dissimilar from the modal rate of organic deposits, sedimentary as well as sedentary, considered together.

These data are remarkable for the overall similarity which they demonstrate between the distributions of accumulation rates for the different deposit types. They show that, although the rates achieved by a given deposit type vary between wide limits, the modal rates for all types are similar and do not change systematically from time to time or place to place; rather are differences from the mode to be attributed to local conditions of particular sites at particular times. There is certainly no evidence for a substantial and necessary gradient of increasing rates along the series nekron mud → fine detritus mud → coarse detritus mud which is often assumed nor any indication that fen peat accumulates relatively slowly. There is some indication, indeed, that fine detritus mud accumulates somewhat less rapidly than the other deposits considered.

It is likely that, given adequate nutrient supply, the productivity of open water is related to its depth, down to the limit of penetration of light. Moreover, chemical or biological breakdown of nekron during the earliest stages of its deposition is likely to be least intense under deep water where the temperature and oxygen availability are low. High production and low breakdown will lead to rapid accumulation in deep water whilst the converse may be true of the shallows. Areas central to a lake are therefore likely to fill more rapidly than those marginal pools in which the slowed accretion of nekron is not offset by a large contribution of detritus from higher plants. It is probable, too, that wave action has its most pronounced erosional effects in the very regions where fine detritus mud is first produced. In still more marginal regions, occupied by dense reedswamp or similar vegetation, the nekron fraction of the deposit is relatively small and the greater part of the coarse detritus mud derives from the massive parts of the phanerogams, which, in spite of the undoubtedly greater oxidative breakdown, guarantee an accumulation rate about equal to that of deep water nekron mud. Moreover, conditions beneath water level in a reedswamp are probably not conducive to extensive degradation of dead plant remains because the ex-

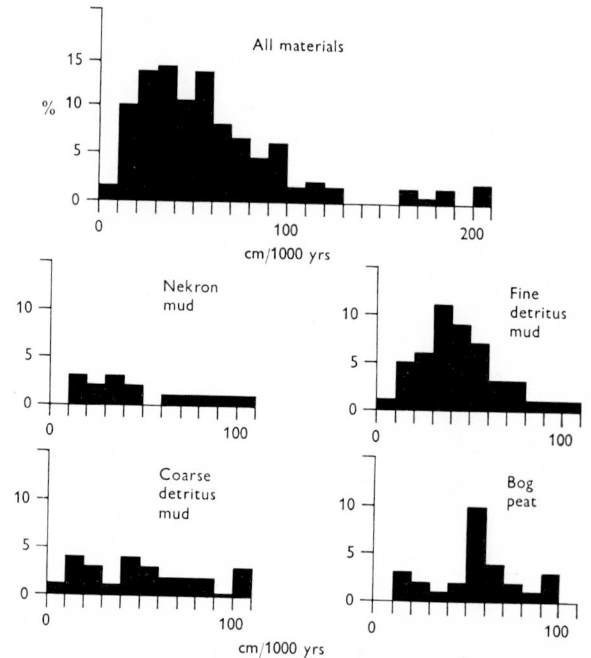


Figure 8. Frequency distributions of deposit accumulation rates. The ordinate measures percentage in class in the uppermost histogram and number in class in all others. A few high records have been omitted for convenience of presentation.

change of water with the main body of the lake is minimal and oxygen depletion considerable. These stagnant conditions are probably at their strongest in areas of extensive swamps such as occur behind newly emerging shorelines and in more-or-less evenly shallow lagoons; perhaps for these reasons, the higher rates of detritus mud accumulation are recorded from such periods and sites. Moreover, in such situations the reedswamp is growing directly over freshly deposited inorganic substrates which presumably give it a richer source of nutrients than does the organic mud over which it normally grows in more restricted habitats.

## DISCUSSION

The stratigraphic data show that, except in the deepest and most extensive lakes, vegetation change has predominated over vegetation stability throughout the Post-glacial. The analysis of a sample of these data, however, emphasizes the variety of the courses which have been followed from place to place and time to time. Since the information is drawn only from sites at which the hydroseres are judged to have been predominantly autogenic, change can only rarely lead to a reduction of accumulated deposit and since the vegetation stages are arranged broadly in order of decreasing tolerance of flooding (with the exception of stage 12), it is hardly surprising that a 'lower' stage follows a 'higher' stage in only 17 % of recorded transitions. Excluding these 'reversals', of the remaining 66 theoretically possible transitions between vegetation stages, 35 (53 %) are in fact recorded although only 23 (35 %) substantially so. If the theoretical possibilities are reduced more closely to reality by disregarding the truly open water stages (stages 1, 2 and 3) the comparable figures are: possible transitions 39, realized transitions 27 (69 %) and frequently realized transitions 17 (44 %). In a system which has often been considered one of the most unvarying in plant ecology, the variety of courses actually taken is remarkable and demands that, even in situations free from allochthonous influences, predictions be made only in terms of probability.

At any site where the water is initially too deep for phanerogam growth, the processes which distinguish the hydrosere in the strict sense cannot begin until sediment accumulation has so shallowed the water that floating-leaved macrophytes can grow in it. Exceptionally, in kettle lakes, the development of floating vegetation mats of aquatic *Sphagnum* overcomes this inhibition. Progress beyond the floating-leaved macrophyte stage depends on further shallowing of the water or the development of floating mats. There is evidently a tendency for accumulation, of fine detritus mud, to slow somewhat in the floating-leaved macrophyte stage. Nevertheless, this stage is one of the three shortest of all. The explanation of this apparent anomaly lies in the frequency with which the floating-leaved macrophytes are replaced by fast growing floating *Sphagnum* rafts, particularly in the small inland basins where the input of nutrients is small. Equally, in larger basins, where nutrients are not in short supply because of the greater input and circulation, direct transition from floating-leaved macrophytes to floating fen or to swamp carr takes place. Whilst at many sites the open water is obliterated by these processes, still the single most probable, and somewhat slower, course is for reedswamp to succeed floating-leaved macrophytes as the water becomes sufficiently shallow.

Rafts of vegetation, once established, are likely to be relatively long-lived because as peat accumulates the raft is depressed by the increased weight and living conditions at the surface do not change. Accession of new nutrients is virtually non-existent, however, which tends to reduce the biomass production and hence the rate of peat accumulation. Transitions from swamp carr, once it is established, are unlikely to occur rapidly for similar reasons. Bushes and trees growing precariously on insecurely rooted sedge tussocks periodically capsize, but this sporadic contribution to deposition does not result in an overall high accumulation rate. It does leave deep pools, however, and ensures the continued regeneration of swamp carr. In a sense it is the physical instability of floating mat vegetation and swamp carr which contributes most to their biological stability.

Reedswamp is the single vegetation type most frequently associated with hydroseres, except for the terminal stage of bog. More than any other vegetation it is responsible for transforming a fundamentally aquatic environment to a more-or-less terrestrial one. The success of the most important species (e.g. *Phragmites communis*, *Schoenoplectus lacustris*) which form reedswamps is in large part due to their possession of luxuriant above-water parts. Although reedswamps do not grow in consistently turbulent water, they can begin in water up to about 1 m deep and, once established, can withstand occasional flooding to considerably greater depths. Equally, dominant reedswamp species can persist on fen surfaces which are only periodically under water. In relatively shallow water, reedswamp has great competitive advantage over floating-leaved or submerged macrophytes because its over-water stems and leaves not only cast shade but induce an above-water atmosphere of high humidity and carbon dioxide content. On the other hand, by the impeding of water movement, inorganic silt and clay are deposited, so enriching the substratum with nutrients which, in the predominantly reducing environment, can readily be mobilized by the growing plants. In the face of this apparent luxuriance it is surprising that the characteristic deposit of reedswamp, namely coarse detritus mud, does not accumulate very much more rapidly than other organic materials. Part of the reason for this may be that the possession of large aerial photosynthesizing organs ensures a high energy availability much of which is used in the root-rich reducing surface muds to break down the newly dead detritus. Whether bulk depletion occurs by this means or simply by redistribution into deeper water, only a very small proportion of the total leaf and stem production of a reedswamp is ever found in a subfossil coarse detritus mud, common and identifiable though these remains are. In spite of the lack of acceleration of accumulation under reedswamp, the facts that it usually does not grow in water more than a metre deep and on unflooded surfaces must compete with better adapted plants, ensures that the reedswamp stage is not prolonged. Moreover, simply because it heralds and initiates terrestrial conditions in a formerly flooded place, reedswamp is the starting point for a greater variety of vegetation successions than could possibly begin in water.

In small inland basins, where most of the affluent drainage is in the form of seepage from the banks, the first establishment of a zone of reedswamp around the lake edge is probably the turning point in the hydroseral development of the basin as a whole. Until then any nutrients reaching the shallows could be distributed at least throughout the epilimnion of the whole lake. A marginal reedswamp not only inhibits exchange between peripheral and central water but provides an immediate metabolic sink for nutrients entering from the banks. The main body of water devoid of reedswamp receives no nutrient replenishment and quickly deteriorates to the state where aquatic Sphagna are the only likely invaders. For the same basic reasons, although Sphagna often invade reedswamps they rarely grow along their landward edges.

The transition from reedswamp to fen is less important, at least in the analysed sample of sites, than established hydrosere theory would credit. It seems rarely to happen in small steep-sided inland sites, particularly those in nutrient-poor glacial drift, but in larger basins, where margins are not so steep and nutrient enrichment is greater, it is less unusual. Floristically, the transition from reedswamp to fen is very gradual, reflecting an equally gradual change in environmental conditions imposed by the progressive raising of the surface of the organic substratum. As the latter rises to mean annual water level, the likelihood of the surface drying for prolonged periods in summer increases. This not only encourages the ecesis of species less tolerant of waterlogging than are reedswamp plants but, perhaps more importantly, reduces the mobility of nutrients in the rooting medium by the imposition of oxidizing conditions. A fen developed over a relatively incompressible substratum of mud, as distinct from one on a floating mat, is subject to erratic changes in surface conditions depending on the frequency of flooding more than anything else. Oxidative decomposition of the surface peat of fens mitigates against continuously rapid accretion, but the fact that practically nothing is borne away from the surface by water movement usually more than makes up for this. Nevertheless, the circumstances under which fens persist for a long time whilst rapidly accumulating peat must be very rare, except on floating mats; in the data analysed, Tregaron seems to represent these unusual sites. The relative longevity of most of the sampled fens must be attributed to the uncertainties of accumulation on physically stable fens and to the relative changelessness of the environment on a floating mat.

Swamp carr rivals fen as a successor of reedswamp. The shallower the water in which it initially grows the faster does its growth and death eliminate the open water. But whilst it persists, the pools between the *Carex* tussocks, overshadowed by bushes and trees, are ecologically daunting niches which usually become occupied by *Sphagnum* spp. Swamp carr is therefore yet another stage which encourages the development of bog without the intervention of fen. The transition from swamp carr to fen or fen carr is very rare indeed.

Once aquatic Sphagna are established at any stage of a hydrosere its future development is irrevocably determined if no strong allogenic factor supervenes. Even on the surface of a thin (*c.* 30 cm) floating mat of loose *S. cymbifolium* peat, species such as *S. papillosum* and *S. cuspidatum* begin to grow and before long a gently undulating lawn is established on which *Oxycoccus quadripetalus*, *Erica tetralix*, *Rhyncospora alba* and *Eriophorum angustifolium* form the vanguard of a large suite of bog angiosperms. The anatomical structure of *Sphagnum* endows this vegetation with near-independence of the lake water table. As the vegetation and the peat it produces grow upwards the living surface plants rely increasingly on direct precipitation for their water supply and an ombrogenous bog develops from a topogenous mire.

On a fen surface, *Sphagnum* spp. (e.g. *S. squarrosum*) may become widely established if the surface is only rarely flooded so that oxidation and acidification of the surface peat occur. Alternatively, where irregularities on the fen surface allow the persistence of small pools isolated for long periods from nutrient enrichment by flood waters, the sequence beginning with aquatic Sphagna may be followed on a small scale. This provides nucleii from which larger bogs spread as the mire surface depends more and more on rain for its water supply. Once a fen surface is isolated from periodic suffusion by lake water it must become a bog or maintain a tenuous stability by patchy, cyclical, breakdown and regeneration of the fen itself.

Many trees (e.g. *Alnus glutinosa*, *Salix atrocinerea* and *Betula pubescens*) are tolerant of a fair degree of waterlogging and are therefore able to colonize fens in which the water table is not perennially at the surface. The fen carrs so formed, although initially more productive than the fens they replace, contribute to the degradation of the surface peat by the aeration resulting from the penetration of tree roots. The concomitant acidification affects both the ground flora and the trees. Carrs initially dominated by *Salix atrocinerea*, *Alnus glutinosa*, or *Frangula alnus* with a rich understorey containing *Carex paniculata*, *Thelypteris palustris*, *Filipendula ulmaria*, *Eupatorium cannabinum* and even remnant *Phragmites communis* are replaced by *Betula pubescens* woodlands in which the shrub and herbaceous layers are relatively depauperate with *Molinia caerulea*, *Myrica gale*, *Sphagnum recurvum* and *S. papillosum* assuming importance. If the rainfall is sufficient to maintain *Sphagnum* growth a bog, from which even *Betula pubescens* may be excluded, is the result. Without such rainfall, a fen carr may persist, indeed may be enriched temporarily by a variety of woodland species including *Quercus robur*, but its longevity depends on the extent to which it can continue to tap and efficiently re-cycle the nutrients in the preformed peat and mud. The oxidizing conditions created in the soil down to the level at which perennial waterlogging prevents root growth inhibit effective nutrient mobilization and contribute to the ultimate failure of the woodland and the establishment of bog. There are insufficient data relevant to the longevity of fen carr and associated vegetation stages but it must be suspected that they only persist through many tree generations in those localities where the trees are able to tap directly the mineral soil beneath shallow peat and mud.

The hydrosere system described is, of course, an abstraction for it derives from frequencies of phenomena and rates of change drawn from many different sources. Can the complete histories of individual sites be accommodated within the system? This may be tested by briefly relating to it the actual records of three sites, none of which has contributed much to the system's construction and two of which are in the south of England from which few statistical data were available.

In the Meare Pool region of the Somerset Levels (Godwin, 1955*a*) newly exposed marine clay first bore *Phragmites* reedswamp which, as the influence of fresh water increased and drainage lines were impeded by plant growth, thinned to open water with floating-leaved macrophytes (e.g. *Nymphaea alba*) and only patches of *Phragmites*, *Cladium mariscus* and swamp carr in the shallows. After about 75 cm of organic mud had accumulated under these conditions the water was colonized by floating hypnoid mosses together with *Menyanthes trifoliata* and, on the loose floating mat so formed, bog Sphagna and ericoid plants established. In terms of the vegetation stage numbers used earlier in this paper, this Meare Pool sequence could be represented:

$$12 \rightarrow 5 \rightarrow 4 \rightarrow 10 \rightarrow 11$$

The transition from reedswamp to open water with *Nymphaea alba* ($5 \rightarrow 4$) may best be looked upon as a small allogenic perturbation of the otherwise autogenic sequence. In the data used for constructing the system as a whole the remaining transitions are common in hydroseres of coastal and estuarine localities. The transition from $4 \rightarrow 10$, here imposed by a preceding allogenic change but itself autogenic, is rare in these circumstances but common in kettles. The series of changes leading to the growth of the floating hypnoid mat was accomplished within Zone VIIa and can therefore have taken no more than 2000 years and probably less; this again is consistent with the slower of the rates determined elsewhere. On a rough estimate only, the transition from the beginning of raft formation to the establishment of an ombrogenous bog probably took less than 1000 years, slightly less than the appropriate modal rate. Elsewhere on the Somerset Levels the continued growth of bogs so impeded the drainage that the Meare Pool bog was flooded and returned almost cataclysmically to the open water with floating-leaved macrophyte stage. Subsequent development again followed a path to bog although this time through reedswamp (*Molinia caerulea*, *Phragmites communis* and *Cladium mariscus* stages) and aquatic Sphagna; in terms of the standard vegetation stages:

4 → 5 → 10 → 11

Accepting that irregularities in the drainage régime effectively protected much of this region from the base-rich run-off from the neighbouring Lias hills, it is nonetheless remarkable that the hydroseres had such a strong tendency to bog development without the intervention of fen or fen carr.

At Trundle Mere on the margin of the East Anglian Fenland (Godwin & Clifford, 1939), a rise in the regional water table early in Zone VII a created a stretch of open water which passed through floating-leaved macrophyte and reedswamp (*Cladium mariscus* and *Carex* spp.) stages followed by aquatic Sphagna and raised bog, all within less than 3000 years. In nearby areas (e.g. Ugg Mere) the Zone VII a water-logging of the soil created fen woodlands the pools of which were occupied by *Sphagnum* sec. Subsecunda. From these nucleii other Sphagna spread until the onset of the brackish water incursion about 2000 B.C. which deposited the fen clay. Although Trundle Mere was unaffected by this event, Ugg Mere was flooded and was afterwards occupied by floating-leaved macrophytes which gave way directly, and through reedswamp, to *Sphagnum* bog. In the notation adopted earlier, these successions can be represented:

Trundle Mere   2 → 4 → 5 → 10 → 11,
Ugg Mere   12 → 9 → 9(+11) → flooding → 4 → 11 → 5 → 4.

These are entirely consistent with sequences of the system with the exceptions that 12 → 9 is only once recorded elsewhere (presumably because wetted woodlands rarely fail to dry out again thereby destroying the record of their vicissitudes) and 4 → 11 occurs only once and then indicated by pollen only. As in the Somerset Levels, the tendency to terminal bog formation is evidently very strong. Further out in the Fenland 'basin' bases supplied by powerful affluent rivers during periods of heavy rainfall seem to have arrested hydrosere development at the fen carr stage, although even there (e.g. Wood Fen, Ely) records of *Sphagnum* amongst the *Alnus* and *Betula* indicate the probability of bog formation had the allogenic factors been removed.

At Carrowreagh, in the west of central Ireland (Jessen, 1949), early Post-glacial chalk- and shell-mud was covered by a brown sediment containing the remains of planktonic algae, water lillies, *Menyanthes trifoliata* and *Cladium mariscus* in early Zone VII a. This predominantly floating-leaved macrophyte stage was followed later in the same Zone by *Phragmites-Cladium* and later (early Zone VII b) by pure *Phragmites* reedswamp, at least near the centre of the basin. On the deposits of this, a fen flourished fleetingly during Zone VII b and was succeeded, in the later part of the Zone, by fen carr (*Alnus glutinosa*, *Betula* sp., *Fraxinus excelsior*). Much of the flora of this carr probably derived from parents in a swamp carr which had directly succeeded the reedswamp on the borders of the fen. Whilst the fen and carr had been developing in the central and northern areas, aquatic *Sphagnum* spp. had directly invaded reedswamp further south and the bog to which this process gave rise duly replaced the carr to the north at the beginning of Zone VIII. This complex of transitions can perhaps best be represented:

North   2 → 4 → 5 → 8 → 9 → 11,
Centre   2 → 4 → 5 → 7 → 9 → 11,
South   2 → 4 → 5 → 10 → 11 → 11.

It demonstrates that separate, if related, courses may be followed by the hydroseres at a single site during the same period of time (in this case the 6000 years after the opening of Zone VII a) to culminate in the same terminal stage.

## CONCLUSIONS

This analysis of the stratigraphic data clearly confirms the operation of hydroseral processes during the Post-glacial. Vegetation types such as reedswamp and fen can be identified in Post-glacial deposits and can be shown unequivocally to have been temporally related to one another and to other vegetation types at a wide variety of localities. Moreover, although rates varied from stage to stage and place to place and although the process was occasionally arrested or even locally and temporarily reversed, it resulted in the progressive reduction of free water depth in the aquatic stages and increased dependence on direct rainfall in the terrestrial stages. In a sense, therefore, the zonation of vegetation observed in lakes and mires at any one time is an analogue of the sequence of vegetation types which may occur at any place.

But the range of vegetation types in a single locality at any one time does not necessarily reflect the sequence which has led to the current pattern at that site, nor does it alone predict the future of the vegetation there even if allogenic influences can be excluded. The course of a particular hydrosere in the past or in the future can only be hypothecated in a probablistic manner from a consideration of all possible transitions to and from existing vegetation types weighted for environmental site conditions and species availability.

The essential nature of the autogenic sequence seems not to have changed throughout the Post-glacial. This is probably due to two related reasons. On the one hand the properties of water bodies which matter most to plants (e.g. aeration, turbulence, temperature, turbidity) are relatively little affected by general climatic changes of the order which have been postulated for the Post-glacial. On the other, water and mire plants generally have wide climatic tolerances and are therefore unlikely to react directly to small changes in the aerial environment which may radically affect many strictly terrestrial species. The study of autogenic hydroseres is therefore inherently unpromising as a source of information about climatic fluctuations. By the same token, however, any major disturbance of a hydrosere (e.g. a transition from bog to swamp carr) is best attributed to some, almost catastrophic, environmental change (e.g. rise in silty water) which may well be a result of climatic changes (e.g. increase in precipitation: evaporation ratio) taking effect in the catchment of the basin.

The data also show that, whilst judged in terms of human lifetimes the hydroseral process may be slow, by comparison with autogenic terrestrial ecological changes of similar magnitude (e.g. bare rock to forest) it is relatively fast. If it is accepted that reedswamp, although amongst the most short-lived of hydroseral stages, may commonly persist for 500 years it is understandable that Spence (1967) only rarely recorded changes during the last 50 years in Scottish lowland lochs.

Variety is the keynote of hydroseral sequences for so long as local ecological conditions exclude *Sphagnum* spp. Once the nutrient level of the water falls to a level which gives *Sphagnum* spp. competitive advantage over other plants from the floating-leaved macrophyte stage onwards, the system is severely constrained and bog development directly ensues. Indeed, in a hydrosere where autogenic processes dominate, the probable course from any existing stage is more strongly directed towards bog than previously published generalizations imply. It is not stretching the evidence too far to claim that *throughout the British Isles* the true 'climax' of an initially topogenous hydrosere is not a terrestrial woodland but an ombrogenous bog and that only allogenic factors such as complete penetration of shallow peats by tree roots, critical changes in water level and human interference deflect or arrest this sequence. How closely must Godwin & Turner (1933) have come to drawing the same conclusion!

## SUMMARY


Peat and mud stratigraphy, amplified by pollen diagrams, has been used to study the courses and rates of British Post-glacial hydroseres.

The transition probabilities between the vegetation stages of hydroseres are crudely assessed. It is shown that, although certain sequences of transitions are 'preferred' in certain site types, variety is the keynote of the hydroseral succession. In spite of this, the data clearly indicate that bog is the natural 'climax' of autogenic hydroseres throughout the British Isles and the transition from fen to oakwood is unsubstantiated.

The persistence of vegetation types has not changed very significantly during the Post-glacial. Once a hydrosere is initiated to the floating-leaved macrophyte stage, conversion to fen frequently takes less than 1000 years and rarely twice as long. Bog formation from *Sphagnum* invasion of aquatic stages (including reedswamp) may be equally rapid but, if fen intervenes, may take between 2500 and 4000 years.

Sedimentary and sedentary deposits have accumulated at a modal rate of 21–60 cm per 1000 years. Rate variation between deposit types and between periods of Post-glacial time is insignificant. The evidence for the compaction of undisturbed deposits after their formation is meagre.


## REFERENCES


Bartley, D. D. (1960). Rhosgoch Common, Radnorshire: stratigraphy and pollen analysis. *New Phytol.* **59**, 238–62.

Bartley, D. D. (1962). The stratigraphy and pollen analysis of lake deposits near Tadcaster, Yorkshire. *New Phytol.* **61**, 277–87.

Birks, H. J. B. (1963–4). Chat Moss, Lancashire. *Mem. Manchr Lit. Phil. Soc.* **106**, 1–24.

Birks, H. J. B. (1965). Late-glacial deposits at Bagmere, Cheshire and Chat Moss, Lancashire. *New Phytol.* **64**, 270–85.

Buttery, B. R. & Lambert, J. M. (1965). Competition between *Glyceria maxima* and *Phragmites communis* in the region of Surlingham Broad. I. The competition mechanism. *J. Ecol.* **53**, 163–81.

Buttery, B. R., Williams, W. T. & Lambert, J. M. (1965). Competition between *Glyceria maxima* and *Phragmites communis* in the region of Surlingham Broad. II. The fen gradient. *J. Ecol.* **53**, 183–95.

Chapman, V. J. (1941). Studies in salt-marsh ecology. Section VIII. *J. Ecol.* **29**, 69–82.

Clapham, A. R. & Clapham, B. N. (1939). The valley fen at Cothill, Berkshire. Data for the study of Post-glacial history II. *New Phytol.* **38**, 167–74.

Clapham, A. R. & Godwin, H. (1948). Studies in the Post-glacial history of British vegetation. VIII. Swamping surfaces in peats of the Somerset Levels. IX. Prehistoric trackways in the Somerset Levels. *Phil. Trans. R. Soc.* B **233**, 233–73.

Clements, F. E. (1916). *Plant succession: an analysis of the development of vegetation.* Carnegie Inst. Washington, Publication No. 242.

Donner, J. J. (1957). The geology and vegetation of Late-glacial retreat stages in Scotland. *Trans. Roy. Soc. Edinb.* **63**, 221–64.

Donner, J. J. (1962). On the Post-glacial history of the Grampian Highlands of Scotland. *Soc. Scient. Fennica, Comm. Biol.* **24**, no. 6, 1–29.

Gillham, M. E. (1957). Vegetation of the Exe estuary in relation to water salinity. *J. Ecol.* **45**, 735–56.

Godwin, H. (1931). Studies in the ecology of Wicken Fen. I. The ground water level of the fen. *J. Ecol.* **19**, 449–73.

Godwin, H. (1936). Studies in the ecology of Wicken Fen. III. The establishment and development of fen scrub (carr). *J. Ecol.* **24**, 82–116.

Godwin, H. (1941). Studies in the ecology of Wicken Fen. IV. Crop-taking experiments. *J. Ecol.* **29**, 83–106.

Godwin, H. (1955*a*). Studies in the post-glacial history of British vegetation. XIII. The Meare Pool region of the Somerset Levels. *Phil. Trans. R. Soc.* B **239**, 161–90.

Godwin, H. (1955*b*). Vegetational history at Cwm Idwal: a Welsh plant refuge. *Svensk bot. Tidskr.* **49**, 35–43.

Godwin, H. (1959). Studies of the Post-glacial history of British vegetation. XIV. Late-glacial deposits at Moss Lake, Liverpool. *Phil. Trans. R. Soc.* B **242**, 127–49.

Godwin, H. & Bharucha, F. R. (1932). Studies in the ecology of Wicken Fen. II. The fen water table and its control of plant communities. *J. Ecol.* **20**, 157–91.

Godwin, H. & Clifford, M. H. (1939). Studies of the Post-glacial history of British vegetation. I. Origin and stratigraphy of Fenland deposits near Woodwalton, Hunts. II. Origin and stratigraphy of deposits in southern Fenland. *Phil. Trans. R. Soc.* B **229**, 323–406.

Godwin, H. & Mitchell, G. F. (1938). Stratigraphy and development of two raised bogs near Tregaron, Cardiganshire. *New Phytol.* **37**, 425–54.

Godwin, H. & Tallantire, P. A. (1951). Studies in the post-glacial history of British vegetation. XII. Hockham Mere, Norfolk. *J. Ecol.* **39**, 285–307.

Godwin, H. & Turner, J. S. (1933). Soil acidity in relation to vegetational succession in Calthorpe Broad, Norfolk. *J. Ecol.* **21**, 235–62.

Godwin, H. & Willis, E. H. (1959). Radiocarbon dating of prehistoric wooden trackways. *Nature, Lond.* **184**, 490–1.

Gorham, E. (1953). Some early ideas concerning the nature, origin and development of peat lands. *J. Ecol.* **41**, 257–74.

Gough, J. (1793). Reasons for supposing that Lakes have been more numerous than they are at present; with an Attempt to assign the Causes whereby they have been defaced. *Mem. Manchr Lit. Phil. Soc.* **4**, 1–19.

Hardy, E. M. (1939). Studies of the Post-glacial history of British vegetation. V. The Shropshire and Flint Maelor mosses. *New Phytol.* **38**, 364–96.

Harley, J. K. & Yemm, E. W. (1942). Ecological aspects of peat accumulation. I. Thornton Mire, Yorkshire. *J. Ecol.* **30**, 17–56.

Haslam, S. M. (1965). Ecological studies in the Breck Fens. I. Vegetation in relation to habitat. *J. Ecol.* **53**, 599–619.

Holdgate, M. W. (1955). The vegetation of some British upland fens. *J. Ecol.* **43**, 389–403.

Jessen, K. (1949). Studies in late-Quaternary deposits and flora-history of Ireland. *Proc. R. Ir. Acad.* **52**, B, 6, 154.

King, W. (1685). On the bogs and loughs in Ireland. *Phil. Trans. R. Soc.* **15**, 949.

Lambert, J. M. (1946). The distribution and status of *Glyceria maxima* (Hartm.) Holmb. in the region of Surlingham and Rockland Broads, Norfolk. *J. Ecol.* **33**, 230–67.

Lambert, J. M. (1951). Alluvial stratigraphy and vegetational succession in the region of the Bure Valley Broads. III. Classification, status and distribution of communities. *J. Ecol.* **39**, 149–70.

Lambert, J. M. & Jennings, J. N. (1951). Alluvial stratigraphy and vegetational succession in the region of the Bure Valley Broads. II. Detailed vegetational-stratigraphical relationships. *J. Ecol.* **39**, 120–48.

Lind, E. M. (1948–9). The history and vegetation of some Cheshire Meres. *Mem. Manchr Lit. Phil. Soc.* **90**, 1–20.

Matthews, J. R. (1914). The White Moss Loch: a study in biotic succession. *New Phytol.* **13**, 134–48.

Mitchell, G. F. (1951). Studies in Irish Quaternary deposits: no. 7. *Proc. R. Ir. Acad.* **53**, B, 11, 162.

Mitchell, G. F. (1965). Littleton Bog, Tipperary: an Irish vegetational record. *Geol. Soc. America. Special Paper* **84**, 1–16.

Morrison, M. E. S. (1959). Evidence and interpretation of 'Landnam' in the north-east of Ireland. *Bot. Notiser* **112**, 185–204.

Nichols, H. (1967). Vegetational change, shoreline displacement and the human factor in the Late

Quaternary history of south-west Scotland. *Trans. Roy. Soc. Edinb.* **67**, 145–87.

Oldfield, F. (1960). Studies in the Post-glacial history of British vegetation: Lowland Lonsdale. *New Phytol.* **59**, 192–217.

Oldfield, F. & Statham, D. C. (1963). Pollen-analytical data from Urswick Tarn and Ellerside Moss, North Lancashire. *New Phytol.* **62**, 53–66.

Pallis, M. (1911). The river valleys of East Norfolk *in* Tansley, A. G. (ed.) *Types of British Vegetation*, pp. 214–45. Cambridge University Press.

Pearsall, W. H. (1917). The aquatic and marsh vegetation of Esthwaite Water, Part I. *J. Ecol.* **5**, 180–202.

Pearsall, W. H. (1918). The aquatic and marsh vegetation of Esthwaite Water, Part II. *J. Ecol.* **6**, 53–74.

Pearsall, W. H. (1920). The aquatic vegetation of the English Lakes. *J. Ecol.* **8**, 163–201.

Pearsall, W. H. (1921). The development of vegetation in the English Lakes considered in relation to the general evolution of glacial lakes and rock basins. *Proc. Roy. Soc.* B **92**, 259–84.

Pearson, M. C. (1960). Muckle Moss, Northumberland. I. Historical. *J. Ecol.* **48**, 647–66.

Pennington, W. (1947). Studies of the Post-glacial history of British vegetation. VII. Lake sediments: pollen diagrams from the bottom deposits of the north basin of Windermere. *Phil. Trans. R. Soc.* B **233**, 137–75.

Pennington, W. (1964). Pollen analyses from the deposits of six upland tarns in the Lake District. *Phil. Trans. R. Soc.* B **248**, 205–44.

Pigott, C. D. & Pigott, M. E. (1963). Late-glacial and Post-glacial deposits at Malham, Yorkshire. *New Phytol.* **62**, 317–34.

Poore, M. E. D. & Walker, D. (1958–59). Wybunbury Moss, Cheshire. *Mem. Manchr Lit. Phil. Soc.* **101**, 1–24.

Sculthorpe, C. D. (1967). *The biology of aquatic vascular plants*, p. 417. London: Arnold.

Seagrief, S. C. (1960). Pollen diagrams from southern England: Cranes Moor, Hampshire. *New Phytol.* **59**, 73–83.

Seagrief, S. C. & Godwin, H. (1960). Pollen diagrams from southern England: Elstead, Surrey. *New Phytol.* **59**, 84–91.

Seddon, B. (1962). Late-glacial deposits at Llyn Dwythwch and Nant Ffrancon, Caernarvonshire. *Phil. Trans. R. Soc.* B **244**, 459–81.

Smith, A. G. (1958*a*). Post-glacial deposits in south Yorkshire and north Lincolnshire. *New Phytol.* **57**, 19–49.

Smith, A. G. (1958*b*). Two lacustrine deposits in the south of the English Lake District. *New Phytol.* **57**, 363–86.

Smith, A. G. (1959). The mires of south-western Westmorland: stratigraphy and pollen analysis. *New Phytol.* **58**, 105–27.

Spence, D. H. N. (1964). The macrophytic vegetation of freshwater lochs, swamps and associated fens, *in* Burnett, J. H. (ed.) *The vegetation of Scotland*, pp. 306–425. Edinburgh: Oliver and Boyd.

Spence, D. H. N. (1967). Factors controlling the distribution of freshwater macrophytes with particular reference to the lochs of Scotland. *J. Ecol.* **55**, 147–70.

Tansley, A. G. (1939). *The British Islands and their Vegetation.* Cambridge University Press.

Tansley, A. G. (1941). Note on the status of salt-marsh vegetation and the concept of 'formation'. *J. Ecol.* **29**, 212–14.

Thomas, K. W. (1965). The stratigraphy and pollen analysis of a raised peat bog at Llanllwch, near Carmarthen. *New Phytol.* **64**, 101–17.

Turner, J. (1962). The *Tilia* decline: an anthropogenic interpretation. *New Phytol.* **61**, 328–41.

Turner, J. (1964). The anthropogenic factor in vegetational history. I. Tregaron and Whixall mosses. *New Phytol.* **63**, 73–90.

Turner, J. (1965). A contribution to the history of forest clearance. *Proc. Roy. Soc.* B **161**, 343–53.

Tutin, T. G. (1941). The hydrosere and current concepts of the climax. *J. Ecol.* **29**, 268–79.

Walker, D. (1955). Studies in the post-glacial history of British vegetation. XIV. Skelsmergh Tarn and Kentmere, Westmorland. *New Phytol.* **54**, 222–54.

Walker, D. (1965). The Post-glacial period in the Langdale Fells, English Lake District. *New Phytol.* **64**, 488–510.

Walker, D. (1966). The late-Quaternary history of the Cumberland Lowland. *Phil. Trans. R. Soc.* B **251**, 1–210.

Walker, D. & Godwin, H. (1954). Lake stratigraphy, pollen analysis and vegetational history, *in* Clark, J. G. D. *Excavations at Star Carr*, pp. 25–69. Cambridge University Press.

White, J. M. (1932). The fens of North Armagh. *Proc. R. Ir. Acad.* **40** B, 233–83.

# THE ECOLOGICAL HISTORY OF BLELHAM BOG NATIONAL NATURE RESERVE

by F. Oldfield

*School of Biological and Environmental Studies, The New University of Ulster, Coleraine, N. Ireland*

## INTRODUCTION

Most pollen-analytical and associated stratigraphic and plant macrofossil studies explore either palaeoecological or phytogeographical problems. The emphasis varies with the worker, the sites studied and the balance between the different techniques employed. Within recent years, in Britain at least, most Late- and Post-glacial studies have tended to be palaeoecological in approach, to some extent in recognition of the fact that Professor Godwin's major work (1956) went a long way towards resolving many of the problems which preoccupied an earlier generation of phytogeographers. Further stimuli have come from the realization that, for most critical groups of vascular plants, the Quaternary palaeobotanist can never hope to reach the level of identification required by the modern biosystematist and from the increasing range of taxa which can be recognized or subdivided in an ecologically useful if taxonomically inadequate manner on the basis of their fossil remains. Equally important has been the association of pollen-analytical studies with parallel techniques dealing, for example, with mollusca or sediment chemistry whilst the increasing availability and precision of radiocarbon dating techniques have served to emphasize this trend. Dr Pennington's critical review of late-Quaternary palaeoecological studies in north-west England in this volume gives an excellent indication of the fruits of this type of labour. From this and other works, it becomes immediately apparent that a high proportion of the recent effort within this field has gone into the study of the Late-glacial period and of that part of the Post-glacial record between *c.* 3500 B.C. and the beginning of the present millennium, the period of man's growing effect on British vegetation in prehistoric and early historic times. The early and mid-Post-glacial periods are less well known, most writers until recently having tended to accept with few reservations the postulates of the 1930s and '40s. Even less attention has been devoted to the record of the last few centuries despite the fact that, in most cases, the increasing volume of documentary evidence goes little or no way towards compensating for the progressively more fragmentary nature of the 'scientific' record.

The consequence of this neglect is revealed most clearly in relation to problems of habitat and community conservation. The *immediate* antecedents of the present-day plant cover, though usually more relevant to practical questions than the remote antecedents, are often much less well documented. Thus there are usually few *direct* points of contact between the general stratigraphic and pollen-analytical work at a site and ecological studies of present-day communities. More often than not, the former provides little more than a general context for the latter (e.g. Bartley, 1960; Pearson, 1960; Chapman, 1964). The problem is part of the broader difficulty experienced in any attempt, within ecology, to relate the relatively short term studies of processes and 'dynamics', such as those of Watt (1947), to environmental and 'sociological' changes in the longer term.

The present study was undertaken with these problems in mind, particularly those posed by the acquisition of and need to formulate long term management plans for a particular National Nature Reserve.

## THE SITE AND REASONS FOR ACQUISITION

Blelham Bog (54° 23′ N, 2° 58′ W; Nat. Grid Ref. NY 366006) lies at approximately 43 m (140 ft) O.D. in the south of the English Lake District. The bog, together with the Tarn of the same name which flanks it to the south, lies within a roughly east–west through valley between the head of Esthwaite Water and the northern basin of Windermere (figure 1). The area surrounding the Blelham basin comprises Upper Silurian Flags and Slates of the Coniston series, overlain by glacial drift the topographical

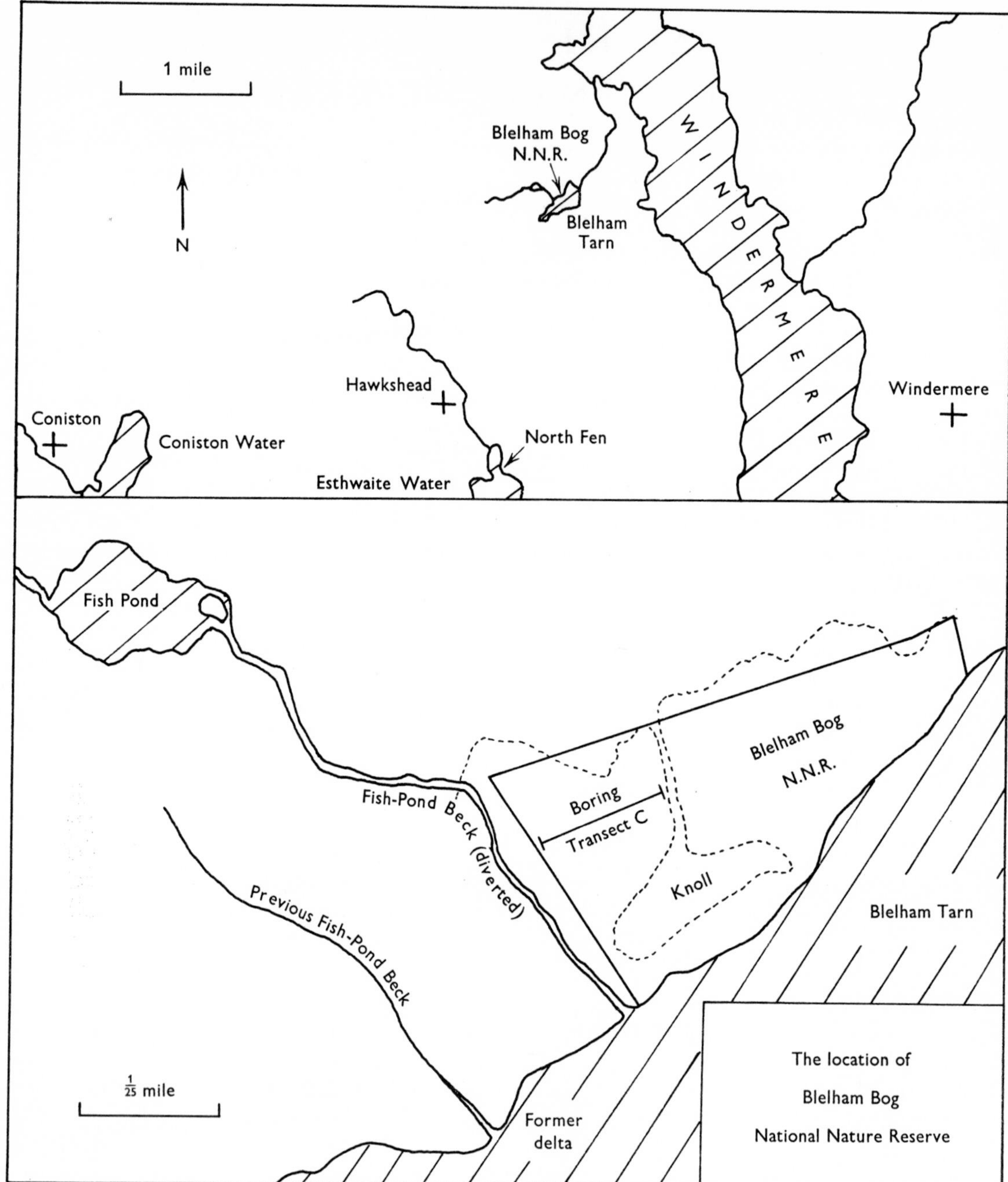


Figure 1. Location map of Blelham Bog. The approximate position of Boring Transect C is shown. Other Transects are located accurately in figure 2.

expression of which includes a number of conspicuous kettleholes.

The site was acquired by the Nature Conservancy in 1954 on the recommendation of the late Professor Pearsall. He regarded the site as 'an almost unique example of *Sphagnum* bog developing from wet willow woodland' (Pearsall, 1954). As such, it was thought to occupy an intermediate position between Esthwaite North Fen, where Pearsall (1917, 1918) had traced a hydroseral succession from open water through reedswamp and *Carex* fen to willow and alder carr, and Rusland Moss, a fully developed raised bog. At Blelham Bog, Pearsall claimed to find the lateral change from western birch, alder and willow carr to eastern *Sphagnum-Myrica-Molinia* 'bog' (plate 1) reflected in the stratigraphic development of the latter over woody carr peat. The evidence was taken to indicate the natural and progressive development of ombrogenous conditions over telmatic peat still forming where the carr woodland persisted. However, the exclusion of stock from the site since 1956 set in train a series of rapid vegetational changes which were difficult to interpret entirely in the light of the above hypothesis and accordingly a vegetation survey was carried out in 1959 which led in turn to the present investigation begun in 1961.

## THE PRESENT VEGETATION

The reserve falls naturally into two halves, separated by an artificial causeway linking the rising ground along the northern edge with the rocky knoll overlooking the Tarn. To the west of the Causeway, the land is fairly closely wooded, to the east it is more or less open (plate 1). Each half can be further subdivided.

### THE VEGETATION WEST OF THE CAUSEWAY

During the 1959 survey, no attempt was made to analyse the vegetation of this area in detail (Elliott, 1959), though a single comprehensive species list was compiled. A survey of the species composition along boring transect line C (figures 1 and 5) brings out the main contrasts between plant communities within this area (table 1). To the east of boring 3, and between borings 5 and 7, *Alnus glutinosa*, *Betula pubescens* and *Salix* spp. form a discontinuous and uneven aged canopy over a shrub and field layer comprising a mixture of fen, marsh and wet bog species. Around borings 3 and 4, the canopy comprises mainly mature trees, whilst *Myrica gale* and *Molinia caerulea* dominate the shrub and field layers respectively. To the west of boring 7, the canopy includes occasional *Fraxinus excelsior* and *Fagus sylvatica*; the most abundant species within the field layer is *Juncus effusus* and many woodland herbs are recorded.

TABLE 1. *Plants west of the Causeway recorded along the line of Transect C*

| | Borings 3 and 4 Ridge east of kettlehole C | Borings 5–11 kettlehole C | Borings 10 and 12 Platform west of kettlehole C |
|---|---|---|---|
| *Alnus glutinosa* | . | * | * |
| *Betula pubescens* | * | * | * |
| *Salix* spp. | * | * | * |
| *Fagus sylvatica* | . | . | *(S) |
| *Fraxinus excelsior* | . | . | *(R) |
| *Sphagnum* spp. (mainly cf. *papillosum*) | . | * | . |
| *Erica tetralix* | . | * | . |
| *Drosera rotundifolia* | . | * | . |
| *Potentilla erecta* | . | * | . |
| *Narthecium ossifragum* | . | * | . |
| *Molinia caerulea* | * | * | . |
| *Myrica gale* | * | * | . |
| *Juncus acutiflorus* | . | * | . |
| *Menyanthes trifoliata* | . | * | . |
| *Succisa pratensis* | . | * | . |
| *Iris pseudacorus* | . | * | . |
| *Hydrocotyle vulgaris* | . | * | . |
| *Cirsium palustre* | . | * | * |
| *Filipendula ulmaria* | . | * | * |
| *Galium palustre* | . | * | * |
| *Crataegus monogyna* | . | . | * |
| *Lonicera periclymenum* | . | . | * |
| *Rubus fruticosus* agg. | . | * | * |
| *Juncus effusus* | . | . | * |
| *Achillaea ptarmica* | . | . | * |
| *Angelica sylvestris* | . | . | * |
| *Lotus uliginosus* | . | . | * |
| *Mentha* sp. | . | . | * |
| *Valeriana officinalis* | . | . | * |
| *Thelypteris palustris* | . | . | * |
| *Hypnum cupressiforme* | . | . | *(W) |
| *Brachypodium sylvaticum* | . | . | * |
| *Anemone nemorosa* | . | . | * |
| *Lysimachia nemorum* | . | . | * |
| *Oxalis acetosella* | . | . | * |
| *Prunella vulgaris* | . | . | * |

(S) One single poorly developed specimen observed.
(R) Plentiful and conspicuous regeneration observed.
(W) Observed on damp rotted wood at ground level only.

### THE VEGETATION EAST OF THE CAUSEWAY (table 2)

In the report on the 1959 survey, Elliott distinguishes four main zones based on the analysis of the species composition in fixed metre quadrats forming a grid 20 yards × 10 yards (c. 18 m × 9 m).

(i) Fringing the lake is a narrow reedswamp mainly composed of *Scirpus lacustris*, *Typha latifolia* and *Equisetum limosum*.

(ii) Immediately landward of this, a *Carex* species with characteristics intermediate between *C. acuta* and *C. elata* is dominant, growing along with *Phragmites communis* and *Potentilla palustris*.

(iii) In the remainder of the 'far-eastern' section of the reserve, the vegetation bears some resemblance to the flushed bog types of McVean and Ratcliffe (1962), being marked by a much greater frequency of *Phragmites*, of *Carex* spp. and of *Sphagnum auriculatum* than is the area closer to the causeway (table 2). This far-eastern area is fed by a system of flushes (plate 2), described in detail by Elliott (1959).

(iv) The 'near-eastern' section of the reserve, adjacent to the Causeway, is the area which Professor Pearsall originally noted as incipient raised bog.

TABLE 2. *Analysis of vegetation east of the Causeway (from Elliott, 1959)*

| | Near-eastern 'incipient raised bog' | | | | | | Far-eastern 'flushed bog' | | | | | |
|---|---|---|---|---|---|---|---|---|---|---|---|---|
| Quadrat groups ... | 1 | | 2 | | 3 | | 4 | | 5 | | 6 | |
| No. of quadrats ... | 8 | | 5 | | 5 | | 8 | | 7 | | 5 | |
| Soil pH ... | 3·9 | | 3·7 | | 4·9 | | 4·0–5·0 | | 5·8 | | 5·5 | |
| D = Mean Domin value<br>C = Constancy class | D | C | D | C | D | C | D | C | D | C | D | C |
| *Myrica gale* | 4 | I | 7 | I | 6–7 | I | 6–7 | I | 4 | I | 4 | I |
| *Molinia caerulea* | 4 | I | 8 | I | 6 | I | 3–4 | I | 3 | I | 4 | I |
| *Narthecium ossifragum* | 8 | I | 3–4 | I | . | . | + | . | 2–3 | II | . | . |
| *Erica tetralix* | 3–4 | I | . | . | . | . | + | II | 1·3 | II | 1·2 | II |
| *Potentilla erecta* | 3 | I | . | . | . | . | . | . | . | . | . | . |
| *Carex panicea* | 3 | I | . | . | . | . | 3 | I | 3 | I | 3 | I |
| *C. rostrata* | . | . | . | . | . | . | 5–6 | I | 3 | II | 3 | II |
| *C. demissa* | . | . | . | . | . | . | . | . | 3 | I | * | . |
| *Drosera rotundifolia* | 3 | I | . | . | . | . | . | . | . | . | 2 | II |
| *D. anglica* | . | . | . | . | . | . | . | . | 3 | I | . | . |
| *Eriophorum angustifolium* | 3 | I | . | . | . | . | 2–3 | II | 3 | I | 3 | I |
| *Rhynchospora alba* | . | . | . | . | . | . | . | . | + | . | 3 | I |
| *Utricularia vulgaris* | . | . | . | . | . | . | . | . | 3–4 | I | . | . |
| *Juncus acutiflorus* | . | . | . | . | 5–6 | I | . | . | . | . | . | . |
| *J. kochii* | . | . | . | . | . | . | . | . | 3 | I | 2 | I |
| *Eleocharis multicaulis* | . | . | . | . | . | . | . | . | 2 | II | 4 | II |
| *Potamogeton polygonifolius* | . | . | . | . | . | . | . | . | 3 | I | 4 | I |
| *Phragmites communis* | . | . | . | . | . | . | 3–4 | I | 4 | I | 3 | I |
| *Sphagnum auriculatum* | . | . | . | . | 4 | II | 3–4 | I | 3–4 | I | 6–7 | I |
| *S. papillosum* | 7–8 | I | . | . | 3 | I | + | . | . | . | . | . |
| *S. recurvum* | . | . | . | . | 4–5 | I | . | . | . | . | . | . |
| *Hypnum cupressiforme* | . | . | 2–3 | I | . | . | . | . | . | . | . | . |
| *H. revolvens* | . | . | . | . | . | . | . | . | 2–3 | II | . | . |
| *Scorpidium scorpidioides* | . | . | . | . | . | . | . | . | 3 | I | . | . |
| *Calypogeia trichomanis* | . | . | . | . | 2 | I | . | . | . | . | . | . |
| *Lophocolea bidentata* | . | . | . | . | 2 | II | . | . | . | . | . | . |
| *Campylium stellatum* | . | . | . | . | 2 | II | . | . | . | . | . | . |
| Open water | . | . | . | . | . | . | . | . | 3–4 | 100% freq. | 5–7 | 100% freq. |

Group 1. Quadrats typical of the near-eastern section of the reserve.

Group 2. Quadrats representing the driest parts of the near-eastern section of the reserve.

Group 3. Quadrats representing lower, marginal, flushed areas in the near-eastern section of the reserve.

Group 4. Quadrats representing typical 'flushed bog' areas in the far-eastern section of the reserve away from the direct influence of flushes.

Groups 5 and 6. Quadrats from separate flushes crossing the far-eastern part of the reserve.

Table 2 summarizes the main elements in the vegetation of this area in 1959, the high cover-abundance of *Narthecium ossifragum* and *Sphagnum papillosum* being especially noteworthy then, though much less so now. Elliott also notes variations from this, either where slightly drier conditions prevail or where, in marginal situations, the bog surface receives some mineral supply by way of drainage from the nearby slopes.

## BORING TRANSECTS AND GENERAL STRATIGRAPHY

Four transects of borings were made and their positions are shown in figures 1 and 2, where they are labelled A to D. The stratigraphy established by these boring transects is shown in figures 3, 4 and 5. The parts of the reserve which they cross can be divided into five sections, each characterized by particular sub-surface configuration and a certain degree of unity in stratigraphic succession.

### East of the Causeway

(i) Kettlehole AB. This is crossed north–south* by transect A and east–west by transect B, borings 1 to 5. It extends westwards under the Causeway for *c.* 25 m into the western carr-covered part of the reserve, where it is sampled in the first two borings of transect C.

(ii) Lake margin. Crossed east–west by the eastern end of transect B and north–south by the shallow transect D, this is separated from (i) by a sub-surface

* For the sake of brevity and convenience, approximate bearings are given as though the long landward boundary fence of the reserve ran due east–west.

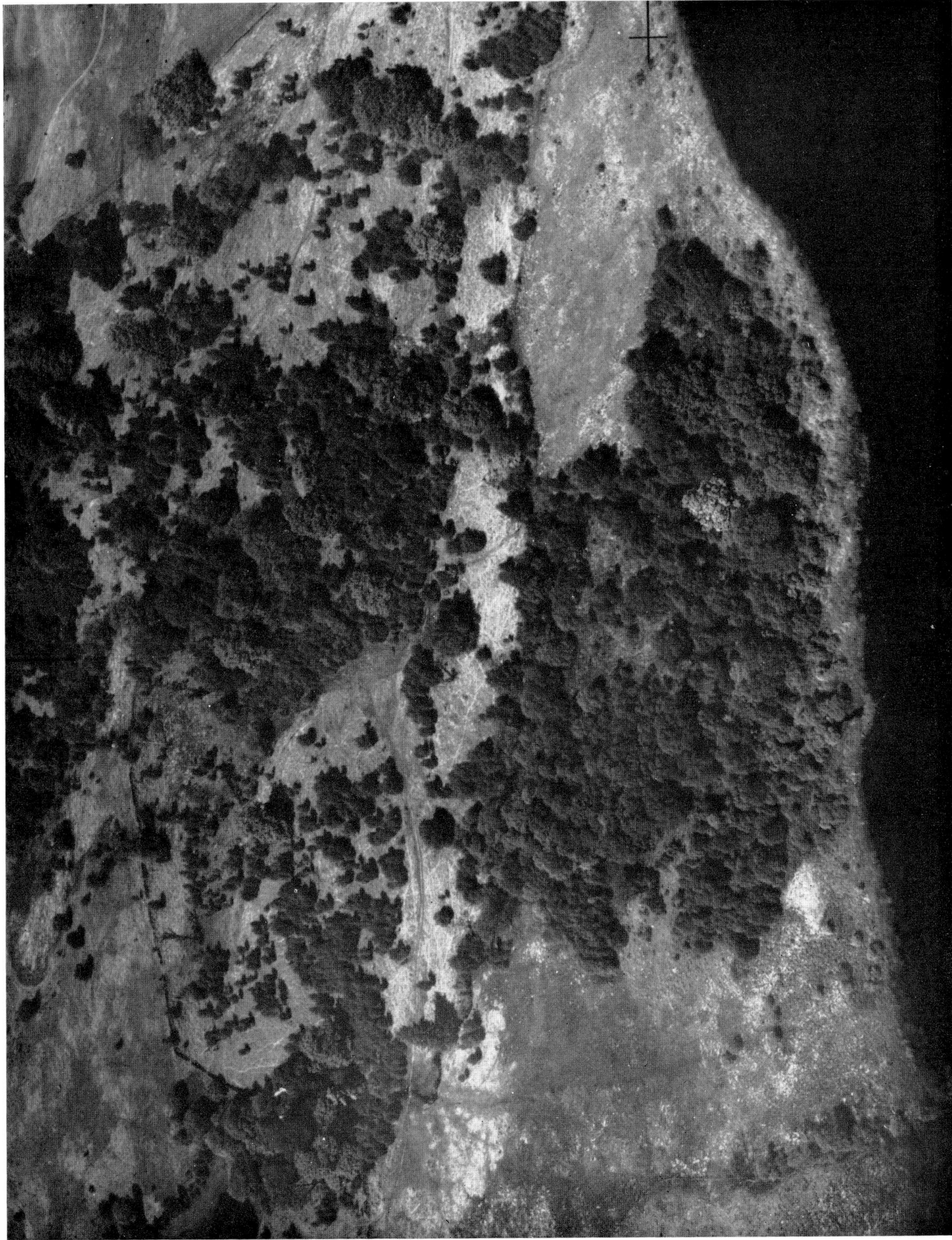

*Oblique aerial photograph of Blelham Bog National Nature reserve and adjacent areas*

The open eastern area of the bog lies towards the top right-hand corner of the photograph. Below and to the right of this lies the wooded knoll. The smaller more or less unwooded patch below the eastern bog area covers kettlehole C. The former delta of the Fish-Pond Beck lies in the bottom right-hand corner of the plate; its present outlet, close to the western reserve boundary, lies some 7 cm above this. The Fish-pond itself appears as a dark patch on the bottom edge of the plate, 2·5 cm from the left-hand corner. The prominent lighter-leaved deciduous trees on the knoll and in the wooded area to the left of the reserve are mature standards of *Quercus petraea* from the original Randy Pike Plantation. (*Photo from* Cambridge University Collection: *copyright reserved.*)

(*facing p.* 144)

(*a*) *Vertical photograph of the main areas of the Reserve* Lines of flushing can be clearly seen across the eastern (top) half of the photograph.

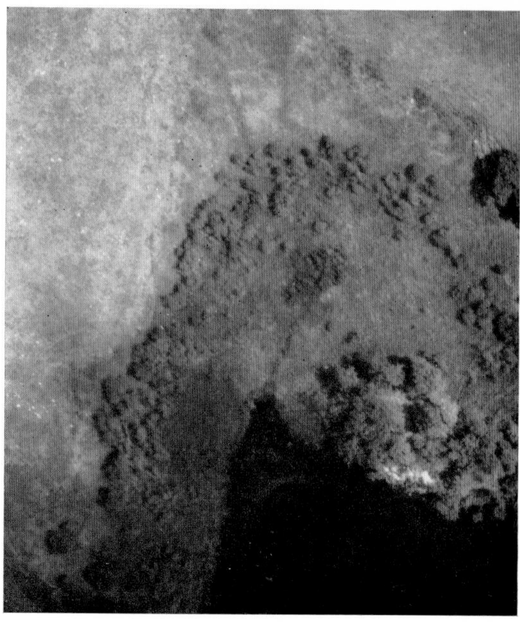

(*b*) *Vertical photograph of the north-east corner of Blelham Tarn* The eastern reserve boundary lies along the bottom edge of the plate and the area shown does not form part of the protected area. The main vegetation boundary, lying roughly parallel to the water's edge is related to the effects of an early nineteenth century lowering in the lake level referred to in the text.

(*c*) *Vertical photograph of part of the Randy Pike Plantation area* The bottom edge of the photograph cuts across the area of the drained hillside kettlehole between 1 and 3 cm from the left hand corner. The course of the *Juncus* flush marking the line of the early nineteenth century drainage ditch referred to in the text lies roughly parallel to and 2 cm from the left-hand edge of the photograph alongside the curving edge of the wooded ground.

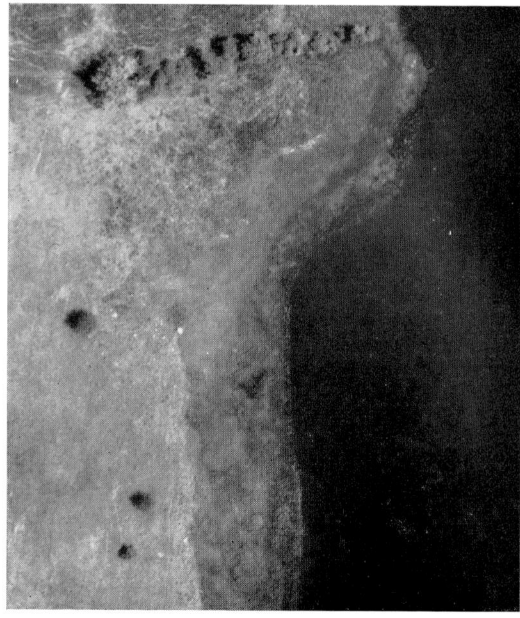

(*d*) *Vertical photograph of the North Western shores of Blelham Tarn*
The former delta and lower course of the Fish-Pond Beck can be seen near the top of the plate whilst parallel to the water's edge lies the vegetation boundary associated with the nineteenth century lowering of lake level. (*Photos from* Cambridge University Collection: *Copyright reserved.*)

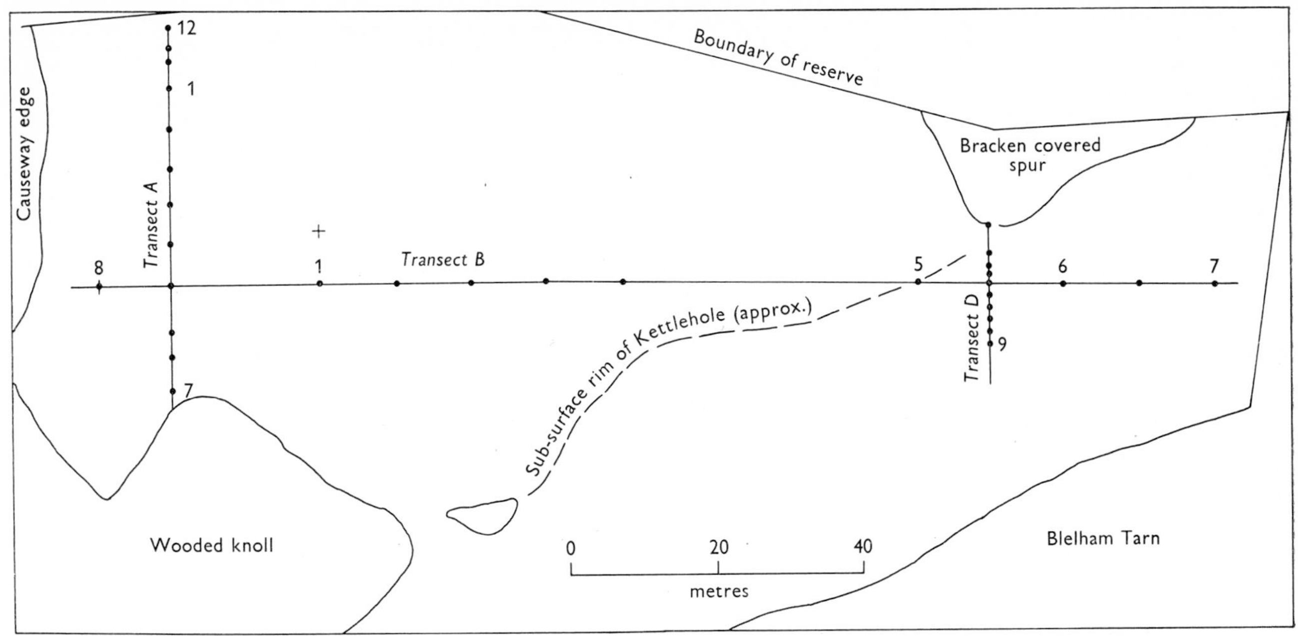


Figure 2. Location of Boring Transects across Blelham Bog east of the Causeway. I, Location of Profile I (figure 6); II, Location of Profile II (figure 6).

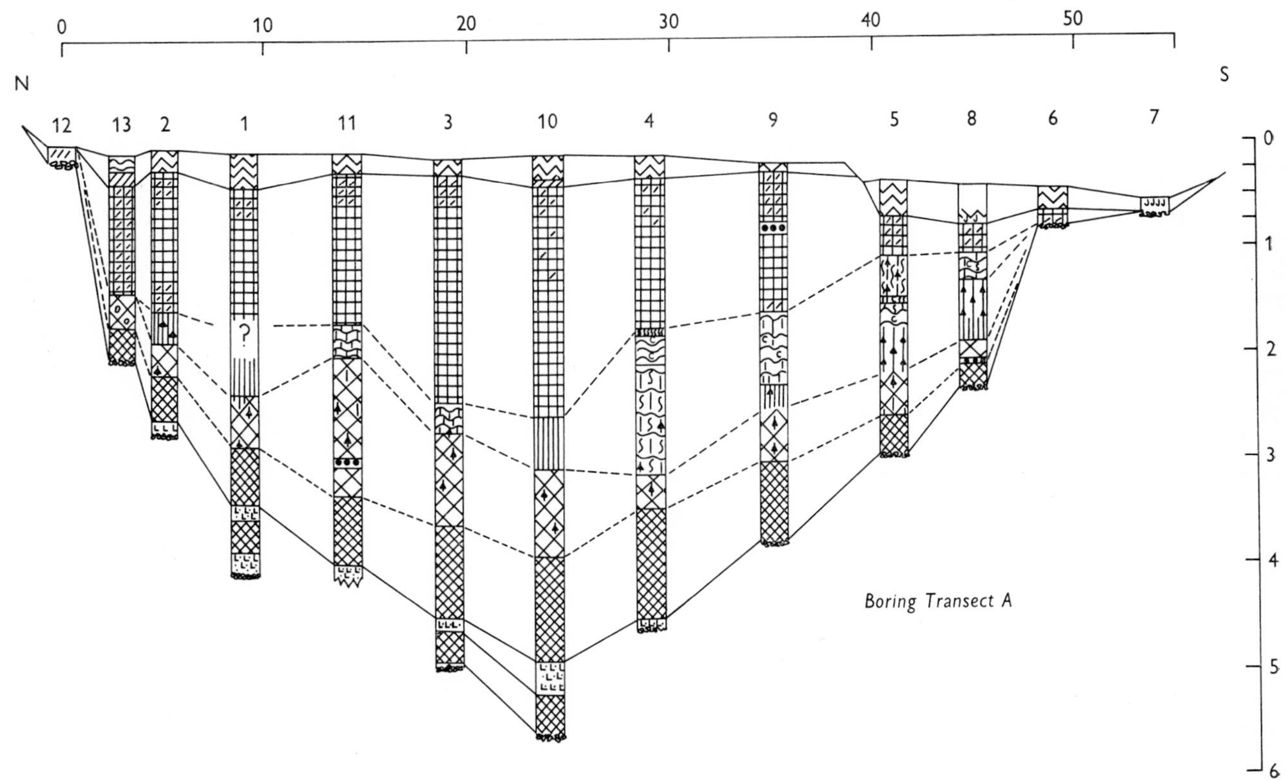


Figure 3. Boring Transect A across Kettlehole AB. Both vertical and horizontal scales are in metres. Stratigraphic symbols are as for figure 5.

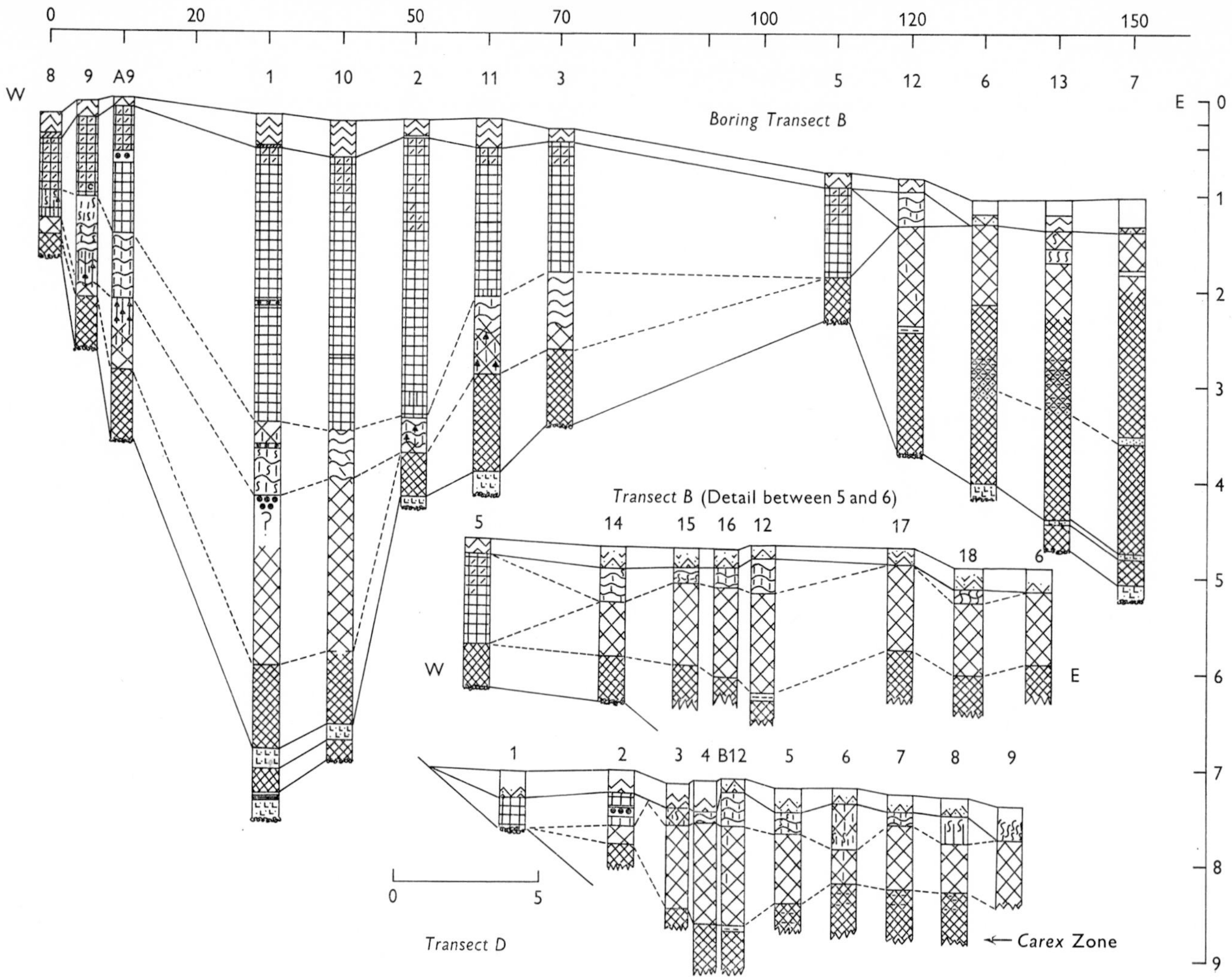

Figure 4. Boring Transects B and D, east of the Causeway. Stratigraphic symbols are as for figure 5. Scales in metres.

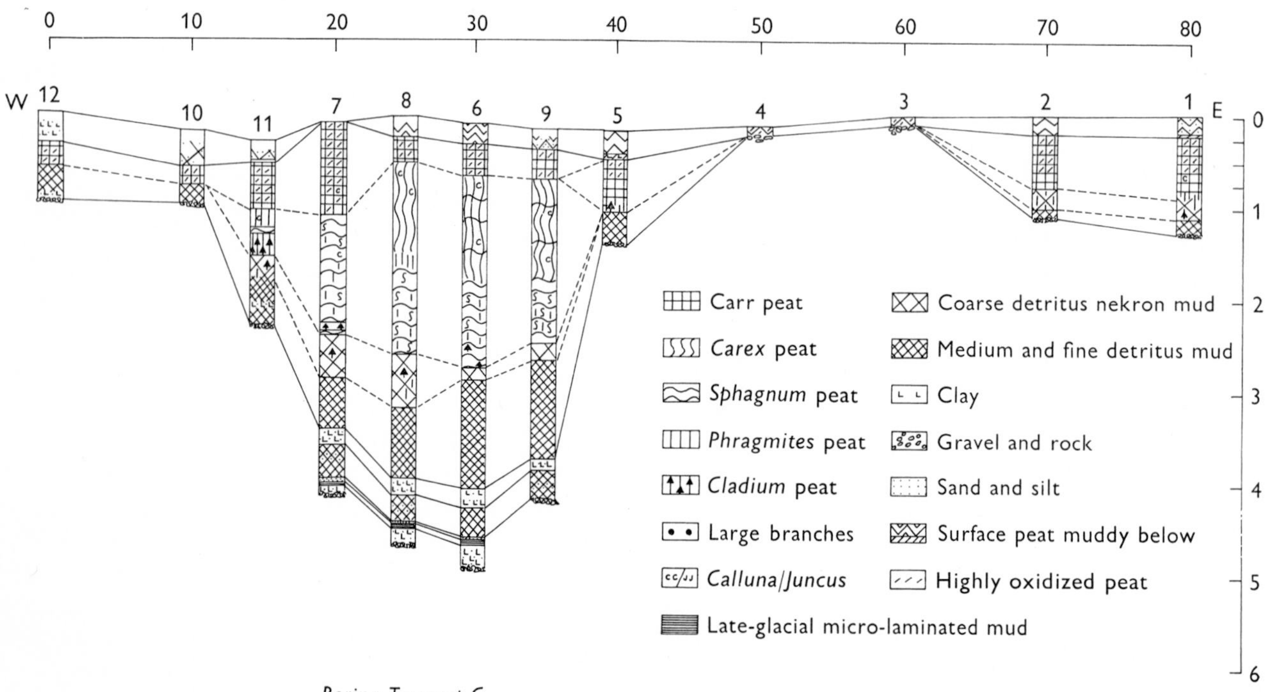


Figure 5. Boring Transect C, west of the Causeway, and key to stratigraphic symbols. Scales in metres.

TABLE 3. *Field records of macroscopic remains from each main type of deposit*

| | Open water sediments | | | | Reedswamp, fen and 'schwingmoor' deposits | | | | Carr peats | | | 'Surface peats' | | | |
|---|---|---|---|---|---|---|---|---|---|---|---|---|---|---|---|
| Cf. *Alnus*—wood | | | (C) | (D) | (A) | | | | **A** | (B) | | | | | |
| Cf. *Betula*—wood | (A) | (B) | | **D** | (A) | B | | D | (A) | B | (C) | (A) | | | |
| *Betula*—fruits | (A) | | | (D) | | | | | | | | | | | |
| *Salix*—wood | **A** | B | | **D** | **A** | B | | | **A** | **B** | (C) | | | | |
| *Salix*—leaves | (A) | (B) | (C) | | | | | | (A) | | | | | | |
| *Frangula alnus*—wood | A | | | | **A** | **B** | **C** | | (A) | (B) | (C) | | | | |
| *Corylus avellana*—nuts | | | | (D) | | | | | (A) | | | | | | |
| *Calluna vulgaris*—twigs | | | | | A | B | **C** | | | (B) | (C) | | | | |
| *Erica tetralix*—twigs | | | | | | | | (D) | | | | | | | |
| *Myrica gale*—twigs | | | | | | | C | (D) | | (B) | | (A) | | | (D) |
| *Molinia caerulea*—rootlets | | | | | | | | | | | | **A** | **B** | **C** | **D** |
| *Potentilla* cf. *erecta*—achenes | | | | | (A) | (B) | C | | | (B) | | | | | |
| *Eriophorum angustifolium*—rootlets | | | | (D) | | | (C) | | | | | | (B) | | |
| *Sphagnum* spp.—leaves | | | | | **A** | **B** | **C** | **D** | | | | (A) | | | |
| *Polytrichum s. lat.*—leaves and stems | | | | | | (B) | (C) | | | | | | | | |
| *Equisetum* cf. *limosum*—stems and nodes | | | (C) | **D** | | (B) | (C) | **D** | | | | | | | |
| *Carex* spp.—tissue | | | (C) | | **A** | B | **C** | **D** | (A) | (B) | (C) | | | | (D) |
| *Carex* spp.—fruits | (A) | (B) | | D 1 | | | (C) | (D) | | | | | | | |
| *Phragmites communis*—leaves | A | (B) | (C) | **D** | **A** | **B** | **C** | **D** | (A) | (B) | (C) | | | | **D** |
| *Cladium mariscus*—rhizomes | A | (B) | (C) | | **A** | B | (C) | | | (B) | (C) | | | | |
| *Cladium mariscus*—fruits | **A** | (B) | C | | **A** | (B) | (C) | | (A) | | | | | | |
| *Potentilla palustris*—achenes | **A** | (B) | | (D) | (A) | (B) | (C) | | | | | | | | |
| *Menyanthes trifoliata*—seeds | | | | **D** | | | | D | | | | | | | |
| *Juncus* spp.—stem bases | | | | | | | | | | | | (A) | | | |
| *Hydrocotyle vulgaris*—fruits | | | | | | | (C) | | | | | | | | |
| *Potamogeton* spp.—fruitstones | **A** | **B** | C | (D) | | | | | | | | | | | |
| *Nymphaea alba*—seeds | | | | **D** | | | | | | | | | | | |
| *Myriophyllum* spp.—seeds | (A) | | | | | | | | | | | | | | |

A = records from Transect A, over kettlehole AB (section i);
B = records from the western part of Transect B, over kettlehole AB (section i);
C = records from the middle section of Transect C, over kettlehole C (section iv);
D = records from Transect D and the eastern part of Transect B, over the lake margins (section ii).
Letters in brackets refer to infrequent and isolated records; bold face letters refer to the most abundant records.

solid or morainic divide, linking the spur projecting into the reserve from the north with the eastern end of the wooded knoll to which the Causeway runs.

WEST OF THE CAUSEWAY

(iii) Sub-surface ridge, separating kettleholes AB and C and recorded in borings 3 and 4 of transect C.

(iv) Kettlehole C, crossed east–west by transect C between borings 5 and 11.

(v) Sub-surface platform, extending from the western rim of kettlehole C to the western boundary fence of the reserve (transect C, borings 10 to 12).

The stratigraphical succession in each section is summarized below.

(i) The basal deposits in kettlehole AB are gravels, then clays alternating with greenish open-water nekron muds. These basal deposits are succeeded by coarser shallow-water lake muds which are overlain by a complex of reed-swamp, fen and, finally, carr peats which persist to within less than 80 cm of the present-day surface. The plant remains and peat types from these levels are very varied (table 3). The top layers of these deposits become progressively more oxidized, giving rise to a peat in which the only recognizable plant remains are rotted pieces of wood and occasional resistant seeds which occur in a crumbly, amorphous, black or brown, aerobically decayed matrix. Both *Calluna* twigs and achenes of a *Potentilla* species (not *P. palustris* but almost certainly *P. erecta*) occur sparsely at this level. This material comes to within 15–30 cm of the surface, at which level there is a very sharp transition to the recent deposits formed by the present-day vegetation and its immediate antecedents: a pale to medium brown muddy peat containing the living rootlets and rhizomes of the present-day plants, which is discussed in more detail below.

(ii) South and east of the sub-surface kettlehole rim and towards the edge of the open water, the rock and gravel floor below the penetrable deposits slopes down quite steeply. Basal clays and nekron muds are recorded as in the kettlehole and, above the highest clay, lake mud usually persists to within no more than 100 cm of the present-day surface, interrupted by occasional, narrow, silt horizons, and bands of diatomite.

Near the lake margins, within the zone of present-day *Carex acuta/elata* dominance and beyond it towards the open water, nekron mud with plant detritus coarsening towards the surface, forms the oozy substratum for the felted roots and rhizomes of the present-day reedswamp species. Landwards of this zone, where the present-day surface is dominated by a mixture of *Myrica, Molinia, Phragmites and Sphagnum auriculatum*, the nekron muds are overlain mainly by very coarse shallow water detritus mud containing *Phragmites* remains, seeds of *Nymphaea alba* and pieces of unidentified wood. Lenses of *Sphagnum* peat, including remains of *Myrica*, *Erica tetralix*, *Eriophorum angustifolium* and *Potentilla* cf. *erecta*, also occur close to the surface in this part. The surface layer resembles that found over the kettlehole although it is generally thinner, more silty, less consolidated and very watery.

(iii) Over the sub-surface ridge between the two kettleholes just west of the causeway, 10–20 cm of black amorphous peaty soil forms the only organic accumulation.

(iv) In kettlehole C, the general sequence of gravels, alternating clays and muds, pure muds, swamp, fen and carr peats, and surface, fresh rootlet, peat resembles that in kettlehole AB. Two significant differences may be noted.

(*a*) Whereas in the eastern kettlehole, the basal stratigraphy in every boring save the deepest (B1) shows a simple threefold division, namely clay/mud/clay, at least five borings in this second basin (C6, 7 and 8 as well as two later ones not shown) have revealed the presence of a greasy, fissile, microlaminated mud layer interrupting the lower clay just below its upper contact. Samples taken from these basal sediments in kettlehole C have yielded the Late-glacial pollen diagram from Blelham Bog, referred to elsewhere in this volume (Pennington, pp. 43–58). The following field notes record the lowest part of boring C7:

| | |
|---|---|
| 257–311 cm | Greenish medium detritus mud |
| 311–328 cm | Sandy silty clay |
| 328–355 cm | Greenish-brown medium detritus mud with silty band between 337 and 344 cm |
| 355–361 cm | Blue-grey, sandy, clay-silt |
| 361–364 cm | Greasy, olive-brown, fissile microlaminated mud |
| 364–370 cm | Impenetrable, blue-grey, stony clay. |

In kettlehole AB, only boring B1 from the deep centre of the basin revealed the intercalation of laminated clay-mud in the lowest clay.

(*b*) The stage between the lower open-water nekron muds and the immediately sub-surface oxidized carr peat is characterized by much more *Sphagnum* and rather more *Myrica* and *Potentilla* cf. *erecta* than is the case in the eastern kettlehole. There is also rather less *Alnus*, *Salix* and *Betula* wood and fewer *Cladium* rhizomes and nutlets. Only within *c.* 70 cm of the surface do true wood peats begin to predominate.

(v) The two borings down to the sub-surface platform on the western edge of the reserve (C10 and 12) record greenish-brown open-water nekron muds above impenetrable gravel. These muds are overlain abruptly by a band of peaty soil above which there is a very sudden transition to clay-silts which persist to the surface.

Table 4 compares the main vegetational subdivisions of the reserve with the evidence of subsurface topography and peat succession revealed in the stratigraphy.

TABLE 4. *A comparison between vegetational zones and stratigraphic contexts over the whole bog*

| Vegetational zone | Stratigraphic context |
|---|---|
| *East of Causeway* | |
| (i) Reedswamp | |
| (ii) *Carex* zone | Lake margin |
| (iii) Flushed bog zone | |
| (iv) 'Incipient raised bog' | Most of kettlehole AB |
| *West of Causeway* | |
| (v) Wettest 'carr' areas over mixed ground flora (borings C1–C2 and C5–C11) | All of kettlehole C and western edge of kettlehole AB |
| (vi) Mature carr woodland over *Myrica* and *Molinia* (borings C3 and C4) | Kettlehole rim |
| (vii) Far-western *Juncus effusus* marsh intermixed with woodland elements (borings C10–C12) | Sub-surface platform |

## THE SHALLOW POLLEN PROFILES

Four pollen diagrams (figure 6) were prepared from the subsurface peat at the site with a view to tracing the most recent stages in vegetational development. The samples used for diagram II were collected individually in 1961 but the rest were taken from monoliths collected in tins in 1965 and 1966. All evidence from macroscopic analyses is included in the stratigraphic descriptions given below.

PROFILE I (located on figure 2)

| | |
|---|---|
| 0–9 cm | Fibrous *Molinia* turf with numerous *Betula* fruits, occasional seeds of *Erica tetralix*, *Potentilla* cf. *erecta* and *Juncus* sp. and a few leaves of *Sphagnum papillosum* |
| 9–14 cm | As above but less fibrous |
| 14–16 cm | *Myrica—Molinia—Sphagnum* peat with a few seeds of *Potentilla* cf. *erecta*, *Carex* sp. and *Erica tetralix*; conspicuously blackened |
| 16–21 cm | Rather humified *Molinia* peat with occasional *Myrica* stems |
| 21–28 cm | Slightly silty *Sphagnum papillosum—Eriophorum angustifolium—Molinia* peat with *Myrica* stems and *Equisetum* cf. *limosum* node and stem fragments; occasional *Juncus* sp. and *Potentilla* cf. *erecta* seeds; bounded below by a clear cut contact sloping at *c.* 45° |
| 28–50 cm | Woody, amorphous, carr peat with abundant sedge remains and occasional seeds of *Rubus fruticosus* agg. |

PROFILE II (Boring 8, Transect B)

| | |
|---|---|
| 0–10 cm | Fibrous *Molinia* turf |
| 10–18 cm | As above but less fibrous |
| 18–21 cm | Blackened *Myrica—Molinia—Sphagnum* peat |
| 21–27 cm | Pale brown humified *Molinia* peat |
| 27–50 cm | Black, oxidized carr peat. |

PROFILE III (located on figure 5)

| | |
|---|---|
| 0–20 cm | Tough, fibrous *Molinia* turf |
| 20–36 cm | As above but less fibrous |
| 36–48 cm | More humified *Molinia* peat, silty above 42 cm |
| 48–66 cm | Highly humified sedge peat. |

PROFILE IV (3 m north of Boring 10, Transect C, figure 5)

| | |
|---|---|
| 0–18 cm | Grey clay—silt |
| 18–23 cm | Brown clay-mud becoming more organic at base |
| 23–30 cm | Coarse stony-clay (moraine). |

In the subdivision of the pollen diagrams shown on figure 6, the following points may be noted.

(i) In the diagrams from Profiles I, II and III, taken from the kettleholes, the stratigraphic boundary between the underlying 'carr' peats and the *Molinia–Myrica–Sphagnum* peat is marked by a sharp change in pollen content.

(ii) The pollen spectra from the underlying carr peats may be compared with Post-glacial diagrams by Pennington (1965) and Evans (unpub.) from the Tarn centre and its northern edge respectively. In diagrams I and II, the spectra from the carr peats consistently show high *Quercus*/*Ulmus* ratios which indicate a date later than the VII a/b boundary. In both cases the total absence of *Plantago lanceolata* pollen points to a Zone VII b rather than a Zone VIII date, in terms of Pennington's zonation, whilst the higher *Pinus*/*Ulmus* ratio and the absence of *Fraxinus* pollen from the carr peats in diagram II suggest an earlier date than do the spectra from diagram I over the deeper part of the kettlehole. In diagram III the carr peat spectra, with their declining *Pinus* and increasing *Alnus* frequencies, clearly date from the Zone VI/VII a boundary.

(iii) The whole of diagram IV and the upper analyses in diagrams I, II and III may be roughly subdivided and correlated as shown, mainly on the basis of the *Betula* and *Alnus* frequencies. The Zones defined are briefly described below.

*Zone A.* Well represented only in diagram II and marked by exceptionally high *Quercus* and *Alnus* with low *Betula* and *Pinus* frequencies. A single sample (28–30 cm) in diagram I straddles the stratigraphic boundary with the carr peats, and from its exceptionally high *Quercus* frequencies it appears in part to represent Zone A.

*Zone B.* High *Betula* and reduced *Quercus* and *Alnus* frequencies. The two diagrams from profiles to the west of the Causeway (III and IV) show increased *Pinus* values, whilst those to the east show high *Plantago lanceolata* frequencies.

*Zone C.* High *Alnus* and low *Betula* frequencies.

*Zone D.* High *Betula* and low *Alnus* frequencies. The three diagrams representing this Zone all show *Pinus* and *Plantago lanceolata* maxima whilst diagrams I and III include a *Fagus* maximum in this Zone.

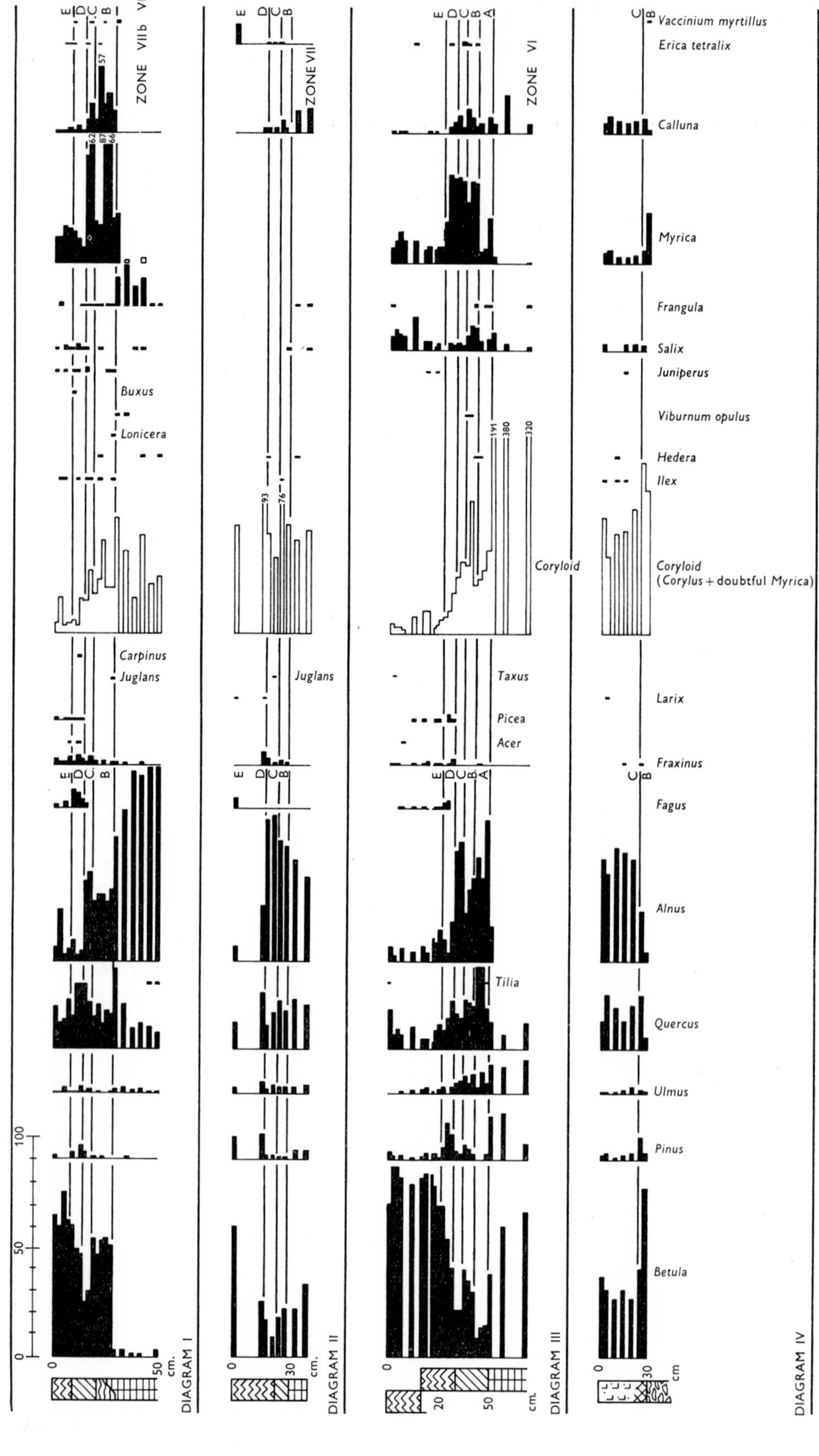


Figure 6. Pollen diagrams from shallow peat profiles. A pollen sum of 150 or more tree pollen (excluding *Corylus* and *Salix*) is used throughout. Samples of the top 20 cm of Profile III were taken from a cut sod, the remainder from a monolith tin.

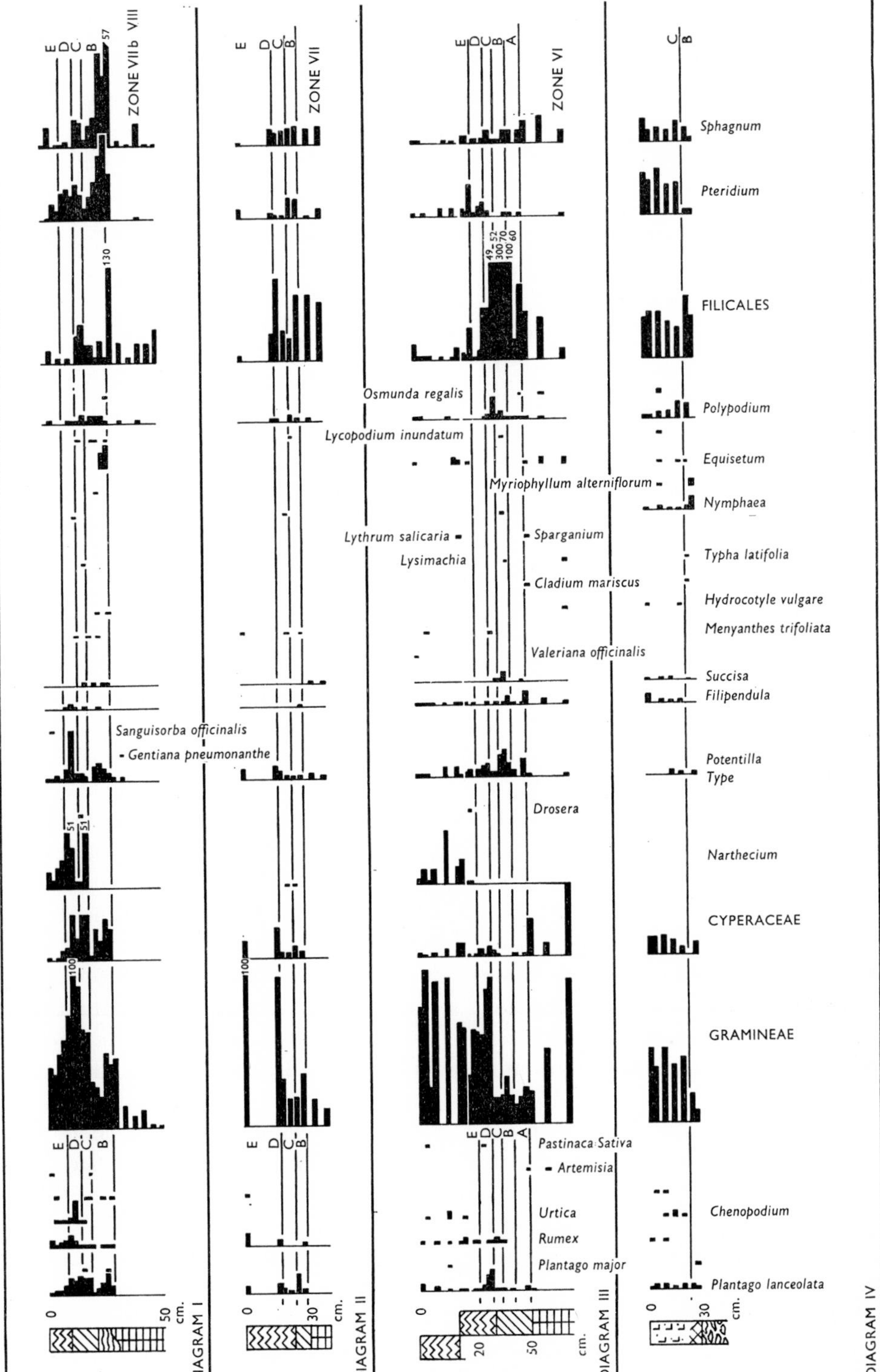


Figure 6. (*cont.*)

*Zone E. Betula* values high throughout. *Fagus* and *Plantago lanceolata* frequencies are reduced.

The correlation proposed would clearly place the whole of diagram IV in Zones B and C as identified in the other three diagrams. The fluctuating, but at times relatively high, weed pollen values in the upper peats, but more especially the presence of the pollen of planted tree genera (e.g. *Picea* and *Larix*), refer the upper peats to a very recent time interval, almost certainly within the last two centuries.

(iv) In view of the above, the stratigraphic and pollen-analytical change between the carr peats and the surface deposits marks a considerable hiatus especially near the edge of kettlehole C. However, despite the demonstration of a substantial lapse in time between the accumulation of the carr peat and the surface peat, there is no significant evidence of pollen destruction at or immediately below the contact in any of the analyses.

## DOCUMENTARY EVIDENCE

A survey of all available maps and literature points to a number of anthropogenic factors which may have influenced the development of the vegetation at the site.

(i) Cowper & Collingwood (1899) refer to a charcoal smelting hearth (bloomery) almost certainly located on the knoll overlooking Blelham Tarn. West (1774) also refers to such a bloomery but neither source establishes the dates when it was operative. However, it is not likely to have persisted beyond the seventeenth century at the very latest.

(ii) A conveyance preserved in the Wray Castle Estate Records (W.C.E.R.) and dated to 1836 shows a map of the bog and the surrounding area. The bog itself is referred to as Moss Intake. Later conveyances in 1887, 1899, 1906 and 1925 (W.C.E.R.) refer to this same area as 'A dale or parcel of Peat Moss or turbary ground, containing 103 roods and 30 perches situate near Blelham Tarn'. The term 'dale' in this context is a word of Scandinavian origin widely used in north-west England to indicate a strip of peat over which an individual had cutting rights. This, together with the reference to turbary and the ubiquity of peat cutting and burning throughout the area indicates previous cutting of peat at the site. Topographical traces of this are limited to shallow rectilinear irregularities on the bog surface, especially in the part of kettlehole AB immediately to the north of the knoll. Transect A crosses one of the shallow steps resulting from uneven cutting between borings 5 and 9. A similar feature is crossed by Transect C to the west of the Causeway between borings 7 and 11. The contrast between quadrat groups 2 and 3 (table 2) also reflects the changes in vegetation type related to topographical variations over kettlehole AB.

(iii) The course of the stream draining Blelham Tarn eastwards into Windermere is known to have been straightened with some consequent lowering of water level in the Tarn (plate 2*b* and *d*). Greenwood's map of 1818 shows the stream in its original unstraightened state, whilst a conveyance of 1824 (W.C.E.R.) makes no reference to straightening or its consequences, despite other comments on the stream (e.g. eel fishing). A later conveyance in 1834 refers to 'the recent partial drainage of Blelham Tarn' and a further one in 1836 draws attention both to some transfer of lands recently negotiated as a result of the diversion and to the creation of new fields alongside the Tarn. This particular activity can thus be securely dated to *c.* 1830.

(iv) The first 6-in Ordnance Survey map of the area (1847–8) refers to the wooded hillsides to the north of the Tarn as 'Randy Pike Plantation', whereas Greenwood (1818), who locates other and smaller patches of woodland, shows this area as unforested. This would suggest a date for the establishment of the wood somewhere between *c.* 1820 and 1840. The only surviving trees from the original plantation are the scattered *Quercus petraea* standards located on plate 2. Most of the remaining area of the plantation is covered by rather discontinuous secondary woodland of *Betula pubescens* over an acidophilous ground flora including extensive patches of *Pteridium*.

(v) The same first Ordnance Survey map locates and labels a drain running from a damp hollow on the afforested hillside out into the middle of the present reserve, ending near the eastern half of Transect C. In view of the other evidence for local land improvement around 1830, this would be the likely approximate date for the construction of the drain. It can still be identified as a long damp *Sphagnum–Juncus* flush leading from the swampy kettlehole which it drains (plates 1 and 2). The flush crosses the reserve boundary and at flood times carried a lot of water into the western half. Shorter, but otherwise very similar, flushes referred to by Elliott (1959) lead into the far-eastern part of the reserve and these may also be successors to artificial drains.

(vi) On the 1847–8 map and on previous plans, the Fish Pond Beck is shown as flowing to the Tarn on a more westerly course than its present one, to emerge at what is now a relict delta feature *c.* 100 m west of the reserve boundary (plates 1 and 2*d*). The 25-in survey of 1888 shows the present-day course and, for the first time, clearly marks the Causeway which divides the reserve. This points to a second phase of local land improvements which as yet cannot be accurately dated within the 40-year time span between the two surveys. Local tradition has it that the Causeway was used for coach parties riding across to the knoll for picnics in Victorian times. Since its construction, numerous birch trees have grown on it.

(vii) Correspondence between local land-owners and the Nature Conservancy refers to the extent to which the reserve was grazed by sheep in the open parts until 1956 when it was fenced. Since then, changes in the eastern half of the reserve include a spectacular extension of *Phragmites* in the flushed bog area and the much taller growth of *Myrica* throughout (Elliott, 1959). In the area close to the Causeway, many seedlings of *Betula pubescens* and some occasional *Quercus petraea* have spread onto kettlehole AB.

## DISCUSSION

### Early- and mid-Post-glacial successions in the kettleholes

From the base of the Post-glacial muds which rest on the Late-glacial clays, the stratigraphy records the infilling of the basins by what appears to be a natural and more or less uninterrupted hydrosere passing from open water through fen and carr stages and culminating in the oxidized peat just below the present surface. Above the open water stage, various aspects of the stratigraphy can be to some extent related to differing topographic contexts. In both kettleholes, *Cladium* fruits and rhizomes are absent from the deepest central parts where the thickness of underlying lake muds is greatest. This parallels evidence from the Bure Valley Broads where Lambert & Jennings (1951) noted that *Cladium* tended to dominate the fen stage of the hydroseres mainly on the firmer peats outside the deepest Broads basins and to be replaced by other reedswamp and fen species where the substratum was less solid. In the deepest boring in the eastern kettlehole there are almost 5 m of peat and mud with wood remains, the lowest being mainly *Salix*. Such a depth of swamp and fen peats recalls the 'swamp-carr' communities established as a hydroseral stage over the deepest and least firm parts of the Bure Valley Broads, in which the development of woody vegetation on unstable, semi-floating, fen communities depresses the surface and partially reduplicates the succession. The pollen spectra from 500 cm and 305 cm in boring B 1 (table 5) show that the level of the lowest branches dates from the beginning of Zone VII a (*c.* 4000–5000 B.C.) and the transition from fen to true carr peats lies somewhat later in the same Zone, before *c.* 3000 B.C.

In the western kettlehole C, the predominance of *Sphagnum* peats with frequent *Calluna* and occasional *Myrica* remains, lying immediately above the open water nekron muds, suggests the development of a 'schwingmoor' over the partly infilled middle of the basin at quite an early stage in the succession there.

TABLE 5. *Tree and shrub pollen frequencies from 305 cm and 500 cm, boring 1, Transect B, expressed as a percentage of total tree pollen less Corylus and Salix.*

| Genus | 305 cm | 500 cm |
|---|---|---|
| *Betula* | 5 | 45 |
| *Pinus* | — | 10 |
| *Ulmus* | 15 | 14 |
| *Quercus* | 21 | 23 |
| *Alnus* | 60 | 9 |
| *Corylus* | 29 | 33 |
| *Frangula* | 6 | — |
| *Salix* | — | 4 |

In both basins, the oxidized sub-surface peats record the culmination of the various hydroseres in a carr community within or even mainly above the seasonal variations of the water-table. The presence of *Calluna* twigs and *Potentilla* cf. *erecta* achenes in parts of this peat may reflect some transition towards increasingly base-poor and possibly ombrogenous conditions.

### The antecedents of the present plant cover

(i) The *Juncus effusus* marsh over the platform west of kettlehole C. The stratigraphic section (figure 5) shows that the layer of inorganic silt which forms the surface deposit over this area is localized within the zone flanking the diverted Fish Pond Beck. The silt overlies the westward continuation of the kettlehole C carr peats and grades into more organic sediments and finally into the surface peats over the kettlehole itself.

Pollen-diagram IV from the silts comes from a point to the north of the Transect, where the silts lie directly over moraine. The lowest pollen sample is rich in *Nymphaea alba* and *Myriophyllum alterniflorum* pollen which suggests a still, open-water source for some of the deposit. The change within the silt accumulation, from more to less organic, suggests progressive erosion of soil surfaces above the point of deposition. The present flora of this part of the reserve (table 1) shows that the flooding associated with the spread of silt took place over a rather damp deciduous woodland, sufficiently unevenly to have allowed the persistence of many species from the woodland ground flora. The single *Fagus* pollen grain within the silt layer presumably predates the flooding and suggests a recent date.

All the points noted above accord with the deduction that the silts were laid down by the diverted Fish Pond Beck at an early stage after its diversion. The Fish Pond itself would thus be the source of the aquatic pollen types noted in Pollen-diagram IV. This interpretation places Zone B (figure 6) within the nineteenth century, between 1848 and 1888.

(ii) THE CARR-PEAT SURFACE OVER THE KETTLEHOLES. The previously noted references to documentary and topographical evidence for peat cutting together with the pollen-analytical indications of a variable but lengthy hiatus above the carr peats and good pollen preservation within their uppermost levels, all point to the artificial origin of the immediate substratum upon which the recent surface peats have accumulated. It seems reasonable to assume that, before peat cutting began, the area over both kettleholes had been developed into a small raised-bog complex. This was eventually cut fairly evenly down to the upper surface of the carr peats.

(iii) THE RECENT PEAT OVER KETTLEHOLE C. The surface peat over kettlehole C shows relatively little stratigraphic change except for a tendency towards siltiness near the base and the increasingly fibrous nature of the peat near the surface. Pollen diagram III, taken close to the eastern edge of the kettlehole where the surface peats are thickest, shows that the silt can be dated to Zone B (figure 6). The pollen and spore frequencies suggest that earlier stages of local dominance by *Myrica* and a fern species (presumably *Thelypteris palustris* which is still present) have changed towards much greater abundance of *Molinia* and *Narthecium*.

The pollen correlations noted earlier suggest that the time interval covered by the spread of silt over the western platform area is represented between levels 32 and 42 cm in diagram III (Zones B and C). Bearing in mind the deductions in the previous sections, the siltiness of the Zone B peat in diagram III may be interpreted as a further result of the diversion of the Fish Pond Beck, leading to a brief phase of more widespread flushing and increased mineral supply over the kettlehole surface itself.

However, the stratigraphic and pollen-analytical correlations with the Fish Pond Beck silts show that accumulation of recent peat over the kettlehole began shortly before the stream diversion. The likeliest cause of earlier flooding is the drainage channel noted in the 1848 survey and referred to in an earlier section where it is tentatively dated to about 1830. This disgorges into the northern edge of both kettleholes but in a zone to the west of the present Causeway (plate 2). A pollen-analytical point tending to confirm this is the *Quercus* maximum in Zone A. The present state of the early nineteenth century Randy Pike Plantation confirms that *Quercus petraea* standards were a major component in the original woodland and these probably gave rise to the high relative frequencies of *Quercus* pollen prior to the secondary development of *Betula* throughout the woodland and of *Alnus* more locally in and around the reserve.

(iv) THE RECENT PEAT OVER KETTLEHOLE AB, EAST OF THE CAUSEWAY. The surface peat succession over this kettlehole shows much more variation than that over kettlehole C. Both the macroscopic and pollen analytical records from the earliest surface peats in Profile I (28–21 cm) point to sudden and sustained flooding over the uneven carr peat surface. This is followed in both Profiles (I and II) by mainly *Molinia* peats which are conspicuously blackened at intermediate levels. These become progressively drier and more fibrous towards the present surface.

The sample at 28–30 cm in Profile I, spanning the lower contact of the surface peats, includes high *Myrica*, Gramineae, Filicales and *Sphagnum* frequencies which relate to the surface vegetation developed after the flooding. The proposed correlation, on the basis of the *Quercus* peak, between this sample and Zone A as defined in Pollen-diagram III, coupled with the fact that the samples were taken from the side rather than the bottom of an infilled hollow, suggest that the earliest flooding over kettlehole AB was contemporary with the first flooding of kettlehole C. Both are thus probably related to the early nineteenth-century drainage ditch previously noted. Also, as over kettlehole C, the Zone B peats are slightly silty suggesting even more widespread flushing after the diversion of the Fish Pond Beck. The local successions over the kettleholes are thus very closely parallel up to this point. Beyond it, whereas 36 cm of fresh, wet *Molinia* peat accumulated

TABLE 6. *The main factors affecting the different vegetational zones of Blelham Bog since 1800*

| Pollen zones | Far-western platform area | Kettlehole C | Kettlehole AB east of Causeway | Eastern lake margin: Flushed bog | Eastern lake margin: *Carex* zone | Eastern lake margin: Reed swamp | Dates |
|---|---|---|---|---|---|---|---|
| E | | *Salix–Alnus–Betula* Carr (C) | Growth of *Myrica* and tree seedlings | *Sphagnum–Molinia* peat | *Carex* peat | Reedswamp peat | 1956 |
| E | | | 'Sphagnum–Myrica–Molinia bog' | | | | |
| D | *Juncus* marsh and residual woodland flora | Flushing; Rapid peat accumulation | No further flushing; Grazing; Retarded accumulation of acid peat | | | | |
| C, B | Silt accumulation | (B) ii; (B) i; Silt | Causeway; Silt | | | | Between 1848 and 1888 |
| A | Damp mixed deciduous woodland | Flushing; Rapid peat accumulation | Flushing; Rapid peat accumulation | | | | |
| A | | (A) ii Flooding of formerly cut surface | | (A) i | | | *c.* 1830 |
| | | HIATUS | | | | | |
| VII a and b | | Former raised-bog area cut down to mid-Post-glacial carr peats | | Open-water | | | Pre-1800 |

(A) i Partial drainage of Blelham Tarn.
ii Cutting of early nineteenth-century drain and flooding of kettleholes.

(B) i Diversion of Fish-Pond Beck.
ii Construction of Causeway

(C) Fencing of reserve and exclusion of grazing animals.

in Profile III, 19 cm and 21 cm of partially oxidized and much drier peat accumulated in Profiles I and II respectively. The contrast in scale of deposition becomes progressively greater towards the present surface, Zone E being represented by 26 cm in Profile III and only 8 cm in Profile I.

The recent period leading to an increasing contrast in rates of accumulation and degree of waterlogging between the areas on each side of the Causeway begins with the increase of *Betula* pollen frequencies from the brief minimum defining Zone C. The first stage of the *Betula* increase coincides with weed pollen maxima in all three kettlehole pollen diagrams. It was deduced from the nineteenth-century cartographic evidence that the Causeway dividing the two halves of the reserve was built within the same time interval as the diversion of the Fish Pond Beck (1848–88). Its construction must therefore have come at approximately the point in time from which progressive divergence in succession between the western and eastern kettlehole areas is recorded.The weed pollen maxima were probably the direct result of the local disturbance involving, as the nature of the Causeway reveals, the importation onto the site of a great deal of rock and earth from elsewhere. Furthermore, the Causeway is now entirely overgrown by birch trees which form the nearest parents for the numerous *Betula* disseminules found in the top 10 cm of peat over the eastern kettlehole. The Causeway lies between the outfall of the early nineteenth-century drainage ditch (now the main flush feeding the reserve) and the eastern kettlehole. At the time of its construction it must therefore have cut off the area immediately to the east from its major source of ground water and mineral supply since the other flushes, noted by Elliott (1959), by-pass the whole area between the Causeway and Boring 11 on Transect B (plate 2). The exceptionally low pH values over this area, the retarded accumulation and the tendency towards drying out, become immediately comprehensible in relation to the building of the Causeway.

More recently, grazing of this drier area, perhaps coupled with the low pH values, has emphasized the contrast on either side of the Causeway by eliminating tree regeneration in the eastern part. It is only within the last decade that the fencing of the reserve has allowed *Quercus petraea* and *Betula pubescens* to establish seedlings in this zone.

(v) THE FAR-EASTERN LAKE MARGINS. Borings 14 to 7 on Transect B and 1 to 9 on Transect D (figure 2) are all from the lake margin area beyond the subsurface rim of kettlehole AB, which links the Knoll to a bracken-covered spur projecting into the northern edge of the reserve. Within this area, around the fringes of open water, both reedswamp and *Carex* peats lie directly over open-water nekron muds. In most of the area from the *Carex* zone the surface *Sphagnum* and *Molinia* peats also lie directly over shallow water nekron muds with, in 11 borings out of 16, no intervening reedswamp or *Carex* peats. The lateral zonation over the present-day surface, i.e. open-water—reedswamp—*Carex* fen—'flushed bog', is not reflected in the vertical sections. Despite appearances, it is not the result of a previous gradual and progressive hydroseral succession, since each of the present vegetation types has developed, for the most part, directly over shallow water muds presumably as a result of a sudden fall in lake level. The direct continuity between the surface peats in this area and those over the kettleholes indicates that roughly the same time interval is involved. This suggests an early nineteenth-century date for the inferred fall in water level. Documentary evidence noted previously identifies a period of lake-level lowering resulting from the straightening of the outflow in approximately 1830.

(vi) SUMMARY OF RECENT SUCCESSIONS. The above discussion relates the course of ecological change in each part of the reserve to recent anthropogenic factors which are well documented and which can, in most cases, be linked directly to particular aspects of the stratigraphic and pollen-analytical record at the site. Table 6 summarizes the main changes and their effects in each part of the reserve.

## CONCLUSIONS

Detailed study of the recent ecological changes at the site revealed in the surface peats, together with an analysis of all relevant documentary evidence, clearly shows that the original *raison-d'etre* of the reserve cannot be upheld. The far western area is a recently flooded deciduous woodland, the far eastern section is mainly dried out lake-margin, whilst the kettleholes between are waterlogged peat cuttings. The contrast between the vegetation to east and west of the Causeway is not the result of ombrogenous development over carr woodland but of the location of the Causeway itself in relation to drainage from the adjacent hillside and of the recent pattern of grazing at the site. Not only the present status of the various parts of the reserve but also future conservation policy must spring from these deductions. Such value as the reserve has would appear to lie in the variety of habitats and the resulting floristic diversity which it encompasses and the potential significance of long term ecological studies related to the events, processes and environmental factors revealed in the work completed to date. An extension of the grid of fixed quadrats into the western half of the reserve and a more thorough ecological and hydrological study of the wooded hillsides above the bog are the most obvious and urgent preliminaries to further work. A study of long term ecological changes over flooded and unreclaimed peat cuttings in response to carefully evaluated environmental parameters should ultimately be of some general value beyond the individual site since the type of habitat is a common one throughout upland Britain.

## SUMMARY

Pollen-analytical and stratigraphic studies at Blelham Bog National Nature Reserve in north-west England have revealed the detailed ecological history of the site during the last two centuries. The succession of ecological changes so revealed can be related directly to human activity on and around the bog as reconstructed from documentary sources. The present vegetation over the site as a whole is extremely varied with woodland, swamp, carr, fen and bog communities all represented. The pattern of differentiation as well as the present ecological trends have been interpreted in terms of the historical evidence for local human interference. Peat cutting at the site was followed, during the nineteenth century, by flooding from artificial drains dug across the nearby hillside, the construction of a Causeway across the reserve, the diversion of a stream bordering the site to the west and the lowering of level of Blelham Tarn. All these changes, together with selective grazing in more recent times, gave rise to a vegetation pattern part of which was formerly interpreted as the transitional stages of a hydrosere leading from fen woodland to bog. This hypothesis has not been confirmed by the present evidence.

I am especially grateful to the Nature Conservancy who financed the field work at the site, to Dr R. J. Elliott, for help and encouragement at an early stage in the work and for permission to use the results of his unpublished vegetational survey; also to Dr H. Frankland, for much helpful information, to Drs I. G. Simmons and H. Nichols for help with field work, to Mrs J. Walsh and Miss E. Critchley for help in the laboratory and drawing office respectively and to Mr K. E. Barber for identification of sub-fossil *Sphagnum* remains.

## REFERENCES

Bartley, D. D. (1960). Rosgoch Common, Radnorshire, stratigraphy and pollen analysis. *New Phytol.* **59**, 238–62.

Chapman, S. B. (1964). The ecology of Coom Rigg Moss, Northumberland. *J. Ecol.* **52**, 299–322.

Cowper, H. S. & Collingwood, W. G. (1899). Reports on excavations of Springs Bloomery near Coniston Hall, Lancs. *Trans. Cumb. West. Ant. Arch. Soc.* **15**, 211–36.

Elliott, R. J. (1959). *Blelham Bog National Nature Reserve.* Unpublished Survey. The Nature Conservancy, London.

Evans, G. H. (unpub.). *Pollen Diagram from Blelham Tarn, Northern Margin.*

Godwin, H. (1956). *The History of the British Flora.* Cambridge.

Greenwood, C. (1818). *A Map of the County Palatine of Lancaster from an Actual Survey.* Wakefield.

Lambert, J. M. & Jennings, J. N. (1951). Alluvial stratigraphy and vegetational succession in the region of the Bure Valley Broads. II. Detailed vegetational-stratigraphic relationships. *J. Ecol.* **39**, 120–48.

McVean, D. N. & Ratcliffe, D. A. (1962). *Plant Communities of the Scottish Highlands. A Study of Scottish Mountain & Forest Vegetation.* H. M. Stationery Office.

Pearsall, W. H. (1917). The aquatic and marsh vegetation of Esthwaite Water. Part I. *J. Ecol.* **5**, 180–202.

Pearsall, W. H. (1918). The aquatic and marsh vegetation of Esthwaite Water. Part II. *J. Ecol.* **6**, 53–74.

Pearsall, W. H. (1954). Unpublished letter on Blelham Bog to Miss F. Whelans, Nature Conservancy. The Nature Conservancy London.

Pearson, M. C. (1960). Muckle Moss, Northumberland. I. Historical. *J. Ecol.* **48**, 647–66.

Pennington, W. (1965). The interpretation of some Post-glacial vegetation diversities at different Lake District sites. *Proc. Roy. Soc.* B **161**, 310–23.

Watt, A. S. (1947). Pattern and process in the plant community. *J. Ecol.* **35**, 1–22.

West, T. (1774). *Antiquities of Furness.* Ulverston.

Wray Castle Estate Records (1824, 1834, 1836, 1887, 1899, 1906 and 1925). Lancashire County Record Office, Preston.

# MAXIMUM SUMMER TEMPERATURE IN RELATION TO THE MODERN AND QUATERNARY DISTRIBUTIONS OF CERTAIN ARCTIC-MONTANE SPECIES IN THE BRITISH ISLES

## PART 1. THE MODERN RELATIONSHIPS

by Ann P. Conolly

*Department of Botany, University of Leicester*

and Eilif Dahl

*Botanical Institute, the Agricultural College of Norway, Vollebekk*

### INTRODUCTION

The general acceptance of climate as an important determinant of plant distribution, spatially and temporally, on large and small scales, prompts the search for correlations between climatic data and the areas inhabited by particular species or vegetation types. This search dates back to the Ancient Greeks (Jessen, 1948) and has now generated an extensive literature. Meteorological observations from an ever-increasing number of stations have improved the climatic component of these correlations during recent years just as the compilation of detailed species' distribution maps has increased the validity of the botanical basis.

Amongst the earlier correlations to be recognized was that between some measure of summer temperature and the upper and poleward limits of species and vegetation types. In particular it was thought that the alpine and polar timberlines coincided with the 10 °C mean July temperature (Dengler, 1930, with references). An anomaly was soon discovered in that the mean July temperature at timberline proved to be higher in oceanic than in continental areas (Hagem, 1917; Brockmann-Jerosch, 1919), a relationship which applies also to the mean temperature of the four warmest months of the year. This latter formulation, introduced by Mayr (1909), has been preferred by most Scandinavian authors. Dahl's (unpublished) correlation ($r = >0{\cdot}98$) between the altitude of the climatic timberline and mean temperature of the three warmest months of the year for the western United States was improved by Aass (1965) in relation to Norwegian data by the introduction of continentality (expressed as annual temperature amplitude) as a modifying factor.

That oceanic vascular plants have distributions related to winter temperatures was recognized by de Candolle (1855) and has since been considerably elaborated for particular species, e.g. the northern and eastern limit of *Ilex aquifolium* has been correlated with the −0·5 °C isotherm of the coldest month of the year (Holmboe, 1913, 1925; Iversen, 1944).

Observing that alpine plants grown in lowland gardens suffered during the hottest spells of summer, Dahl (1951) suggested that alpine and northern species are restricted in their distributions to areas of low summer temperatures. Expressed as the average of the highest annual temperatures recorded over a period of years, extrapolated to the highest summits, good correlations were established with the distributions of many alpine or northern plants in Scandinavia (Dahl, 1951), the British Isles (Conolly, 1961) and North America (Dahl, 1963, 1964, 1966).

Numerous other climatic parameters may be correlated with the ranges of plant species amongst which have been suggested, for example, accumulated temperatures above a certain basal temperature, length of growing season, number of days with maximum temperatures above or minimum temperatures below certain values (Enquist, 1933), amount of rainfall (Dansereau, 1957, with references), air humidity and wind velocity. It would be naïve to expect a species to be limited by a single climatic factor all round its range, so multidimensional representation of factors has been devised (Iversen, 1944; Hintikka, 1963), or several climatic parameters have been combined to form indices, as those for oceanicity proposed by Kotilainen (1933), Amann (1929), de Martonne (1927) and Gams (1931–2). Nevertheless, single temperature correlations are often the most striking and suggestive, as those between *Linnaea borealis* and minimum February temperatures below 34 °F (1·1 °C), *Erica mediterranea* and minimum February temperatures above 38 °F (3·3 °C), *Phyteuma tenerum* and average July temperatures exceeding 62 °F (16·6 °C) and *Blysmus rufus* and July temperatures below 60 °F (15·5 °C) (Perring & Walters, 1962).

Although a close correlation between a limit to the distribution of a species and some climatic factor may hint at the causal factors involved, the attribution of these with any certainty is fraught with difficulties. Spurious correlations may result from the linkage between climatic factors. In some cases, however, analysis proves possible as in some areas of south-west Norway where direct species relationships with high winter temperatures or high rainfall can be distinguished from apparent correlations with the linked parameters (Lye, 1968). Attempts to attain correlations by combining a number of climatic variables in a single index may achieve their immediate aim but, by their very sophistication, obscure any possible causal relationships. Climatic indices can be concocted to delimit almost any area on a map, but their significance for the plants growing within that area may be extremely dubious.

A climatic factor which truly affects the distribution or growth of a species must in some way act on the physiological processes in the individual plants which constitute that species. Climatic distributional correlations must seek explanation in physiological mechanisms revealed as significant by autecological studies (Dahl & Mork, 1959; Dahl, 1964; Pigott, 1958; Mooney & Billings, 1961). Moreover, all stages in the life-cycle of the species must be considered, particular account being taken of different optimum and minimum conditions for seed setting, germination and for vegetative growth (cf. *Stratiotes aloides*, Samuelsson, 1934) and also of biotypic variation within a species.

Species whose modern distribution can confidently be related to particular climatic factors and whose distribution in the past is known to differ from that at present have often been used as indices of climatic change (Iversen, 1944; Godwin, 1956, with references). Naturally, because the fossil record is never complete, such estimates are only of minimum differences and moreover they do not take account of possible biotype differentiation and of consequent changed climatic tolerances. In spite of these difficulties the method, judiciously used, has already yielded important results be it applied to the changed distribution of a single species recorded from many sites or to rich assemblages of species from a single site. The purpose of the second part of this paper is to attempt a critical assessment of temperature changes within the British Isles between defined Quaternary periods and the present day based on changes in the distribution of arctic-montane species. Most of these belong to Matthews' (1937, 1955) Arctic–Alpine or Arctic–Subarctic elements and a few to his Northern Montane, Alpine or other groups. From what has been written above it is evident that such considerations must rest on substantially authenticated correlations between modern distribution patterns and particular climatic parameters derived from current meteorological data, and it is to the investigation of these that the first part of this paper is devoted.

## THE ISOTHERM MAPS

Isotherm maps showing (figure 1) the mean annual maximum temperature (referred to later as maximum summer temperature) reduced to sea level and (figure 2) the estimated mean annual maximum temperature of the highest points in the British Isles were compiled from the published records of the highest temperatures recorded from all British meteorological stations for each year of the period 1921–50. For stations with 10 or more relevant observations during this thirty-year period, all figures were averaged and the result used as the mean annual maximum for station level. The distribution of stations with usable data was very uneven, particularly few occurring in the crucial north-west of Scotland. Because of this, incomplete data from Strathy (8 years), Cape Wrath (7 years) and Benbecula (6 years) were used and mean annual maximum temperatures calculated for them by the difference method using Stornoway as the reference station.

Reduction of station data to sea level equivalent was achieved by applying a temperature lapse rate of 0·6 °C per 100 m altitude interval to the station data and constructing isotherms at 1 °C intervals from the results (figure 1).

The positions of the highest points in the landscape were plotted on a map and their altitudes multiplied by the lapse rate to derive expected temperature depressions between sea level and summit. Superimposition of this map on the sea-level isotherm map allowed the ready estimation of mean annual maximum temperatures at these high points and the drawing of a second isotherm map based on these (figure 2). On this map, all points on the lower side of an isotherm experience a lower mean annual maximum summit temperature than that represented by the isotherm; the reverse relationship applies to points on the higher side of the isotherm. However, whereas it is unlikely that 'cooler' areas will occur on the 'warmer' side of an isotherm since the summits are likely to represent the coldest areas, there will be innumerable 'warmer' areas on the 'cooler' side at altitudes below those of the summits. Where it would have been difficult to delineate the isotherms around isolated high mountains only the estimated mean annual maximum temperature of the summit is shown on the map.

There are three major limitations to the use of these maps.

1. A climatological map is no better than the observations on which it is based. The paucity of adequate records for crucial areas such as northern Scotland, northern Ireland and the English Lake District impart uncertainties to the courses of the isotherms in these areas.

2. Although a temperature lapse rate of 0·6 °C per 100 m of altitude difference has been used (cf. Dahl, 1951), Conolly (1961) believes that 0·68 °C per 100 m might be closer to the truth in Britain during July, a contention supported by Johannessen's (1956) use of 0·7 °C per 100 m in Norway for the same month. Dahl (unpublished) found that lapse rates in mean monthly summer temperatures determined from radiosonde observations over the western United States were greater inland than on the coast. This is probably attributable to the influence of water content on air mass stability and suggests that a lower lapse rate should probably be used in areas of high atmospheric humidity than in those with lower air moisture contents. Errors in the selection of a lapse rate are of little consequence in the reduction of station observations to sea level, since most stations are not very high, but estimates of summit temperatures, on the other hand, may be substantially affected. A lapse rate underestimated by 0·1 °C per 100 m applied through an altitude difference of 1000 m results in the attribution of a temperature 1·0 °C higher than the true figure to the point in question. This size and sign of error may have been made for the highest mountains in Wales and Scotland.

3. The isotherm maps are constructed from the observations of a specific period of years, viz. 1921–50. Plant distribution maps, on the other hand, often include records made over a long period of time and may not be correct for the climatologically selected period, particularly in their more isolated outliers from which the species concerned may well have disappeared. Anomalies in the congruence of plant and isotherm distributions of this kind may reflect climatic change in the very recent past but could equally well be attributed to other causes, e.g. human interference and changes in land use. In those cases where the distribution patterns based on 'all records' differ markedly from those based on 'recent records' (i.e. since 1930), e.g. *Lycopodium alpinum* and *Cryptogramma crispa*, separate correlations for the two sets of data have been determined.

These methodological limitations suggest that precise correspondences between plant and isotherm distribution are not to be expected however real the effect of maximum summer temperature may be in determining the range of particular taxa. Nevertheless, if variations of less than 1 °C in apparently limiting isotherms are accepted as error inherent in the method, a number of striking correlations can be demonstrated.

## ANNUAL MAXIMUM SUMMIT ISOTHERMS AND THE DISTRIBUTION OF BRITISH ARCTIC–MONTANE SPECIES

The summit isotherm map (figure 2) was compared with the distribution maps of all British species for which adequate data were available (Perring & Walters, 1962). Conolly's (1961) experience that the limit of a single species conformed best with different isotherms in different areas prompted the search for correlations on a regional basis: northern England with southern Scotland; Wales and neighbouring England; Highland Scotland (i.e. Scotland north of the Antonine wall); and Ireland. Additional differences between northern England and southern Scotland (including the Northumbrian Cheviots) were noted. England and Wales were treated as one region for species whose area was such that a separation could not be justified.

'Northern' and 'montane' species were selected for further study, but the distribution patterns of some of these did not remotely correspond in shape with that of the summit isotherms, or were too restricted, and these species were excluded. Correlation of some others was impeded by the occurrence of many, apparently significant, outliers. Individual plant records which conflicted with an otherwise adequate correlation were checked back at source and some found to be spurious. For some of the most interesting species the most complete distribution data available (Nature Conservancy Biological Records Centre, and other published and unpublished sources) were incorporated before correlations were sought.

Where the distribution of a species within a region nearly fills the area encompassed by a particular isotherm, the distribution and the isotherm are deemed correlated. The same is claimed for those cases where, though a species is absent from parts of the relevant area, the authors judge the correlation to be none-the-less substantial, e.g. *Betula nana* in Highland Scotland (figure 9) or *Cirsium heterophyllum* in England and Wales (figure 21). Correlations with isotherms between those shown on the map are described as follows:

$n+^{\circ}$ indicates correlation with an isotherm somewhat outside that for $n^{\circ}$, numerically defined as $n+0{\cdot}2^{\circ}$;

$n-^{\circ}$ indicates correlation with an isotherm somewhat inside that for $n^{\circ}$, numerically defined as $n-0{\cdot}2^{\circ}$;

$n$ to $(n+1)^{\circ}$ indicates correlation with an isotherm about midway between those for $n^{\circ}$ and $n+1^{\circ}$, numerically defined as $n+0{\cdot}5^{\circ}$.

The ease with which a correlation can be established is related to the degree to which other environmental or historical factors limit the distribution of a species within its temperature tolerance range. Thus correlations for species with strong edaphic preferences, e.g. the limestone plants *Dryas octopetala* and *Draba incana*, may be expected to be inherently less certain than those for more ecologically tolerant species. Table 1 (see appendix, p. 215) shows correlations based on relatively slight data in parentheses whilst those of greater dubiety, on account of paucity of records are marked thus (+). Species for which a correlation is established in at least one region but which is widespread (i.e. apparently unlimited) in another is marked '.' in the latter; absence from a region is denoted by '–'.

Correlations for 105 British and one introduced species are summarized in table 1 together with the corresponding isotherms for Scandinavia where these are available (Dahl, 1951). Many species lack a Scandinavian correlation because they grow in comparatively warm regions there (>29 to 30°), because they are rare or absent there or because their distribution patterns are in some way aberrant. The table also shows Scandinavian limiting isotherm figures for some species which are now very rare in Britain or for which only subfossil records are known. England and Wales were treated as one region for species whose area either extended into central or southern England or where the distance between Pennine localities and the nearest ones in Wales and the Border was 100 km or less. In these cases no correlation is shown under northern England. Figures 3 to 22 (see appendix, pp. 179–198) demonstrate examples of some British correlations cartographically and, *inter alia*, also demonstrate some discontinuities related to edaphic conditions (e.g. *Draba incana*, figure 17; *Salix reticulata*, figure 5) and a commonly encountered limitation to the south-west of species' area (e.g. *Cirsium heterophyllum*, figure 21; *Festuca altissima*, figure 22).

There are evident systematic differences in the relation between the limitation of species distribution by annual maximum summit isotherms from region to region. The investigation of these differences was extended by including Scandinavia as an additional region, for comparison with which only British data based on all records (i.e. including those made before 1930) were used, since the Scandinavian maps (cf. Hultén, 1950) do not distinguish records made 'since 1930'. Moreover, species with limiting isotherms of 27° or higher in Wales and (northern) England were excluded from the comparisons because their areas are such that this regional distinction no longer holds. The differences between the limiting isotherm values of all reasonably widespread species (i.e. those with isotherms not in parentheses in table 1) in all pairs of regions were calculated. The average differences for all species shared by each pair of regions were then compared (by $t$-test) with the null hypothesis of zero difference and the statistical significance of each deviation from zero determined (table 2, see appendix, p. 217).

From these considerations it appears that, for the given groups of species severally shared by particular pairs of regions, the limiting isotherms in Scandinavia tend to lie about 4 °C higher than in Highland Scotland, rather more than 3 °C higher than in Ireland and about 1·5 °C higher than in northern England. There is no significant difference between northern England and Wales, although in both regions the limiting isotherms tend to be higher than in Highland Scotland and Ireland; the difference between Highland Scotland and Ireland is not significant. Of the 14 species with limiting isotherms between 27 and 29 °C, excluded from this analysis because their extension of area into central or southern England obscures the regional distinctions used, eight have reliable isotherms in Ireland. The mean difference between these central and/or southern English limiting isotherm values and those for Ireland for these 8 species has the highly significant difference from zero of 2·91 °C. The great apparent difference between the values of limiting isotherms in Scandinavia and the British Isles might be reduced were correlations appropriate only to the south-west of Scandinavia to be used as there is some evidence (Dahl, 1963, 1966) that many species are limited by lower temperatures there than in the east and north-east.

The extent to which individual species deviate from the mean differences in limiting temperatures between pairs of regions may be used to hypothesize the action of other constraining factors, e.g. other climatic factors, edaphic conditions and biotype variability.

### Regional variation in correlation very different from the average

Species, the limiting isotherm values of which in northern England and Highland Scotland differ by 3 °C or more (including some which are rare in the sense of table 1 (see appendix, p. 215), constitute a well-defined element: *Asplenium viride*, *Draba incana* (figure 17), *Potentilla crantzii*, *Vaccinium uliginosum*, *Bartsia alpina*, *Juncus alpinoarticulatus* and *Kobresia simpliciuscula* (figure 6). To these might perhaps be added *Chamaepericlymenum suecicum* and *Tofieldia pusilla* which are too rare in northern England to be related to any isotherm yet which certainly grow there at maximum summer temperatures considerably higher than they will tolerate in Highland Scotland. With the exception of *Vaccinium uliginosum* and *Chamaepericlymenum suecicum* (the distribution of which is somewhat bizarre) the species of this element tend to be calcicolous, to be rare or absent from southern Scotland (except *Asplenium viride*) and to be geographically restricted in the north of England, all but *Chamaepericlymenum suecicum* growing in or near Upper Teesdale.

*Dryas octopetala* (figure 16) shows 2 °C difference in the isotherms which limit its distribution in Ireland and Highland Scotland although those relating to Ireland and northern England are virtually the same. Such a deviation from the general pattern may be due to biotypic variability within the species (Porsild, 1959; Pigott, 1956).

### Low correlation difference between Scandinavia and British regions

Unlike all other species, *Sedum villosum*, *Kobresia simpliciuscula* (figure 6) and *Minuartia stricta* grow in areas with higher maximum summer temperatures in northern England than in Scandinavia. In the latter, *S. villosum* grows in wet flushes in the lower alpine and upper forest belts, oddly scattered geographically about a centre in the Jotunheimen mountains; in France and Spain it is widespread in the lowlands (Hultén, 1958). *K. simpliciuscula* and *M. stricta* are typical alpine species in Scandinavia, hardly penetrating the forest belt, but in central Europe the first descends to 1300–1400 m altitude (Hegi, 1939) whilst the second formerly grew even lower in subalpine mires north of the Alps proper (Hegi, 1962). The contrasts in climatic tolerances demonstrated by these species between northern Europe on the one hand and central and south-western Europe on the other argue biotypic differences between the geographically separated populations of each species. However, the temperature relationships established between the northern England (Teesdale) and Scandinavian populations of *K. simpliciuscula* and *M. stricta* hint that the Teesdale plants might be more closely related to those of central Europe, a suggestion in conformity with the occurrence of other species (e.g. *Gentiana verna*, *Myosotis alpestris*, *Primula farinosa*, *Sesleria albicans* Kit. ex Schult.) in the two areas. On the other hand the great difference in limiting isotherm values of *K. simpliciuscula* between northern England and Highland Scotland suggests that the Scottish plants may be of the Scandinavian biotype. The requirements of the British *S. villosum* seem to lie between those of the Scandinavian and south-west European populations.

*Cryptogramma crispa* (figure 20), *Sedum rosea*, *Saxifraga stellaris* (figure 18) and *Cerastium arcticum* differ in their British and Scandinavian limiting isotherms by 1 °C or less. The comparison for *C. arcticum* is somewhat dubious since the taxonomic position of the Scottish plant is insufficiently certain, whilst its ecological requirements there (McVean & Ratcliffe, 1962) are known to be different from the strongly calcicolous related taxon in Scandinavia.

### High correlation difference between Scandinavia and British regions

In addition to 22 species which show a difference in value of their limiting isotherms between Scandinavia and Highland Scotland of 5 °C or more (table 1), twelve more which have limiting isotherms below 24 °C in Highland Scotland grow in areas with maximum temperatures exceeding 29 °C in southern Scandinavia.

Although these 34 species have such high limiting isotherms in Scandinavia, they nevertheless tend to be rare or absent from cool oceanic parts of south-west Norway, only 11 descending to low altitudes there, viz: *Asplenium viride*, *Draba incana*, *Potentilla crantzii*, *Chamaepericlymenum suecicum*, *Betula nana*, *Vaccinium uliginosum*, *Bartsia alpina*, *Saussurea alpina*, *Tofieldia pusilla*, *Carex capillaris* and *Poa alpina*. Seven of these eleven have already been noted as behaving irregularly in their regional distribution in the British Isles and of these, 6 species grow in Teesdale as also do *Betula nana*, *Carex capillaris* and *Poa alpina*. Moreover, all but three of the eleven are calcicoles and of those which are not (*Chamaepericlymenum suecicum*, *Betula nana* and *Vaccinium uliginosum*) *Vaccinium uliginosum* in Britain is probably biotypically different from its southern Scandinavian counterpart. In the Scandinavian mountains the 8 calcicoles are widespread and have scattered outlying areas on limestone in the south, notably Östergötland, Västergötland, Öland and Götland in Sweden and in Esthonia. The total floras of these southern areas are peculiar for their intermixture of northern, xerothermic southern and local endemic elements and contain species which were more widespread during the Late-glacial (e.g. *Helianthemum oelandicum*; Iversen, 1954). Like the Teesdale assemblage (Godwin, 1956; Godwin & Walters, 1967) they may have survived the Post-glacial roughly where they are now found, and they may belong to biotypes different from conspecific populations growing elsewhere (cf. Petterson, 1965).

Contrasting with this distribution pattern is that of the 23 species which do not descend to low altitudes in coastal south-west Norway. Of these, only one (*Epilobium alsinifolium*) occurs in Ireland. This, together with the fact that only four others of the 34 species presently being considered extend westward into Ireland in spite of the availability there of maximum summer temperatures which do not exceed those limiting the distribution of these species in Highland Scotland, suggests that some other climatic factor may interfere with the south-westerly extension of these species in both Scandinavia and the British Isles. If the comparison of limiting isotherms between Scandinavia and Highland Scotland is made to include only those 16 species which are also present in Ireland, the average difference between values of limiting isotherms is 3·22 °C, almost 1 °C lower than the corresponding figure calculated from all species common to the first mentioned regions. This is close to the average difference between Scandinavian and Irish limiting isotherms and obliterates any systematic difference between Highland Scotland and Ireland seen in relation to Scandinavia. Again, excluding species absent from Ireland, the average difference between Scandinavian and north English–south Scottish limiting isotherms becomes 1·64 °C (14 species) and that between north English–south Scottish and Highland Scottish becomes 1·56 °C (23 species). The sum of these two, 3·2 °C, is close to the average difference between Scandinavia and Highland Scotland, a consistency which suggests a real climatic cause.

## SCANDINAVIAN SPECIES ABSENT FROM THE BRITISH ISLES

Of the species with low limiting isotherms in Scandinavia (Dahl, 1951), only a few occur in the British Isles as well (table 3, see appendix, p. 217).

The representation in the British Isles of Scandinavian species in the different Scandinavian limiting isotherm classes clearly increases around the 25 °C isotherm. Of the species in Scandinavia with limiting isotherms there below 25 °C, only 24 % occur in the British Isles; of those limited by isotherms of 25 °C and higher, 62 % reach the British Isles.

Amongst the 11 British species with limiting isotherms in Scandinavia below 25 °C are *Cerastium arcticum*, *Kobresia simpliciuscula*, *Minuartia stricta* and *Sedum villosum*. The unusual distribution of the latter three species has already been attributed to biotype differences between populations. The group also includes *Artemisia norvegica* whose status in Scandinavia has now been questioned by a new find at low altitude in south-west Norway (Ryvarden & Kaland, 1968).

The 39 species with limiting isotherms of 25 °C or higher in Scandinavia which are not found in the British Isles comprise several which grow in tall herb meadows on brown earths (e.g. *Aconitum septentrionale*, *Ranunculus platanifolius*, *Viola biflora*, *Astragalus frigidus*, *Epilobium hornemanni*, *E. lactiflorum*, *Gentiana purpurea*, *Myosotis silvatica ssp. frigida*, *Luzula parviflora*) and others which grow on more acid soils (*Cassiope hypnoides*, *Deschampsia atropurpurea*, *Pedicularis lapponica*). None are known to occur at low levels along the south-west coast of Norway, although *Saxifraga cotyledon* is found at low levels in the inner fjords, and twenty-four are entirely missing from Rogaland in south-west Norway in spite of apparently suitable habitats. On the other hand, of the 64 species with limiting isotherms of 25 °C or higher which *do* occur in the British Isles, only three are missing from Rogaland and many descend to low altitudes near the coast. This is yet another indication of a limitation imposed on the south-western extension of many species by an unknown, but probably climatic, factor.

If the data in table 3 are modified for the effects of this unknown factor and for inconsistencies attributed to biotype differences, it is evident that few species with limiting isotherms lower than 25 °C in Scandinavia also occur in the British Isles whilst a majority of species with limiting isotherms above 25 °C in Scandinavia do occur in the British Isles. In terms of average relationships of the relevant plants (p. 164) a limiting isotherm of 25 °C in Scandinavia corresponds with one of about 22 °C in Scotland and Ireland. If the maximum summer temperature during some part of the Post-glacial were only 2 °C higher than at present, species limited by the 25 °C or lower isotherms in Scandinavia could only have survived in Britain in areas which now have mean maximum temperatures of 20 °C or less. These are very restricted (figures 2, 3) and may have proved inadequate refuges for plants with such exacting demands for cool summer climates.

## DISCUSSION

The data presented establish, beyond reasonable doubt, that the distribution limits of most British arctic and montane species are correlated, towards warmer regions, with maximum summer temperatures. For most species, close correlation with a particular limiting isotherm is maintained only within a limited geographical region, but the values attached to the isotherms fall progressively in the following order: Scandinavia and central and/or southern England, northern England and Wales, Highland Scotland and Ireland. These features of the correlations suggest that, as was supposed for the poleward and altitudinal timberline, continentality or oceanicity exercises a modifying influence on the control of plant distribution by maximum summer temperature.

Although these conclusions seem unassailable as generalizations it is none-the-less important to examine exactly what they mean in terms of the local distributions of plants in the field. The correlations are based on mean annual maximum temperatures for the highest summits in a landscape. It is true that many arctic and montane species do in fact grow at or near summits, not descending to low levels, particularly towards the warmest extremity of their ranges, e.g. *Athyrium alpestre*, *Sibbaldia procumbens*, *Saxifraga nivalis*, *S. rivularis*, *Salix reticulata* (figure 5), *Loiseleuria procumbens* (figure 7), *Gentiana nivalis*, *Gnaphalium supinum*, *Juncus trifidus* (figure 8), *J. biglumis*, *Luzula arcuata*, *Carex saxatilis*, *C. vaginata*, *C. rariflora*, *C. atrata*, *C. norvegica*, *C. atrofusca* (figure 3), *C. lachenalii*, *Poa alpina* and *P. flexuosa* (cf. Wilson, 1956; Ratcliffe, 1960; McVean & Ratcliffe, 1962), yet this is not true for all species that correlate well with the summit isotherms. Many grow on the north and east faces of high mountains some way below the summits and it is on these slopes that, for meteorological reasons, local climates cooler than might be anticipated from their altitudes alone are of common occurrence. Species limited by high maximum temperatures would naturally inhabit such cool niches and, as a result, their distributions would still correlate well with summit temperature data.

It sometimes happens that species which demonstrate strong correlation with summit isotherms nevertheless occur well below the summits and in localities which in all probability do not experience summers unusually cool for their altitudes. Some of these, at least, owe their origins to propagules washed down from higher altitudes and the constant renewal of open habitats along the banks of streams down which they travel. Some populations and individual records of this kind are ephemeral, but others are more or less constantly maintained. Many of the aberrant outliers of otherwise high altitude species considered in this paper are of one of these kinds or the other, e.g. *Alchemilla alpina*, *Cardaminopsis petraea*, *Epilobium anagallidifolium*, *Saxifraga stellaris* (figure 18), *S. aizoides*, *Oxyria digyna* (figure 12), *Polygonum viviparum*, *Luzula spicata* (figure 11), *Deschampsia alpina*, *Festuca vivipara*, *Cherleria sedoides* (Burgess, 1935; Trail, 1923; Corstorphine, MS).

There are low-level occurrences of arctic and montane species which cannot be explained in these ways however. Often they are very close to sea level on coastal cliffs or sand dunes, especially along western and northwestern coasts, e.g. *Draba incana* (figure 17), *Saxifraga oppositifolia*, *S. aizoides*, *Sedum rosea*, *Silene acaulis* and *Thalictrum alpinum* (figure 13) (cf. Praeger, 1934; Wilson, 1956; Conolly, 1961; McVean & Ratcliffe, 1962) where maximum summer temperatures *at sea level* tend to be lower than elsewhere. Moreover, in addition to this general trend in sea level isotherms, an additional depression of maximum temperature occurs in immediate proximity to the sea (Dahl, 1951).

Given that summer-cool areas exist at low altitudes and are occupied by some otherwise montane species, it is reasonable to ask why others, growing at high altitudes within the same region, do not also grow there. The question is particularly cogent to the cases of those species which are limited by isotherms 4 or 5 °C higher in Scandinavia than in Highland Scotland, and raises once more the problem of the modification of temperature limitation towards the south-west.

Amongst those species which show south-western limits poorly related to annual maximum temperatures, are some whose distribution areas in Europe are more or less complementary to those of oceanic south-western plants, e.g. *Picea abies*: *Ilex aquifolium*, *Luzula parviflora*: *Narthecium ossifragum* (Böcher, 1951), *Aconitum septentrionale*: *Digitalis purpurea*. It is generally accepted that the frost-sensitivity of oceanic south-western plants restricts them to areas with relatively high winter temperatures; conversely eastern and north-eastern plants inhabit areas of relatively low winter temperatures. The selective pressure of low winter temperatures may be exercised in many ways. The seeds of *Picea abies*, for instance, do not germinate after storage in moist soil below 10 °C and are also harmed by intermittent frost and thaw (Mork, 1938, 1952). In oceanic south-western Norway the autumn shed seed is unprotected by snow cover from intermittent wetting and drying, freezing and thawing and as a result the viability of the seed is reduced and the south-westward extension of the spruce limited (cf. Robak, 1960). *Pinus sylvestris*, on the other hand, is self-reproducing along the south-west coast of Norway probably due to the fact that its less sensitive seeds are shed in the spring and not exposed to the environmental rigours which afflict the spruce.

A requirement for low winter temperatures may also explain some cases of unusually big differences of correlation with annual maximum summit isotherms between oceanic and continental areas. Winter temperatures, expressed as monthly means, decrease with altitude within the area under consideration. Species demanding cold winters would therefore be restricted to high altitudes and, in a relatively oceanic region such as Highland Scotland would therefore show a spurious correlation with maximum summer temperature. In Scandinavia, with colder winters, they might even grow in the southern lowlands (e.g. *Betula nana*, *Rubus chamaemorus*), or be genuinely limited by maximum summer temperature. Such species would not be expected to descend to low altitudes along the western and south-western coasts. As a result of the pressure of similar forces, many Scandinavian species might not be able to grow in the British Isles and many which grow in Highland Scotland and northern England might fail to extend to Ireland.

There are, of course, many other possible explanations of aberrant correlation patterns between species limits and annual maximum summit temperatures. Most of the species which exhibit more than 5 °C difference in limiting isotherms between Scandinavia and Highland Scotland are widespread, common and edaphically tolerant in the former but rare or restricted to calciferous soils in the latter, and it may be that an interaction between climate and soil conditions results in correlations with unusually low isotherms in Scotland. Moreover, large populations may have an ecological aggressiveness which is not shared by small, genotypically depleted, relict populations many of which may still occupy only the areas to which they were restricted during the Post-glacial climatic amelioration. The outcome is that their present limits may be correlated with lower isotherms than the actual temperature tolerances of the species demand (Conolly, 1961).

These proposals for explanations of aberrations from the general pattern of correlations between species distribution and annual maximum summit isotherms leave two major phenomena for which physiological explanations must be sought, namely the way in which high summer temperatures may deleteriously affect plant growth, reproduction or both and how this may vary along a gradient of oceanicity so that a single species is limited by isotherms as much as 3 °C apart at opposite ends of its range.

Lange (1959, 1961, 1962, 1967), Lange & Lange (1962, 1963), Biebl (1965) and Russian authors have investigated temperature conditions lethal to plants usually by immersing whole plants or parts of them in water at known temperatures and subsequently determining their survival. The measure of lethality usually used is that temperature, 30 minutes exposure to which kills 50 % of the experimental individuals. Lethal temperatures so defined usually exceed 40 °C for higher plants but are lower for marine algae. Death is thought to result from the denaturing of proteins and the associated inactivation of important enzymes.

The temperatures attained by plants, or some of their parts, in the field may differ considerably from the air temperatures recorded in nearby meteorological shelters. Lange (1959) observed that some plants of the Mauritanian desert maintained a leaf temperature below that of ambient air by great loss of water whilst the leaves of other species became hotter than the surrounding air. The lethal temperatures for 'cool' species proved to be consistently lower than those of 'hot' species. If deprived of an adequate water supply on hot days the leaves of 'cool' species became overheated and the plants died.

The temperature of a plant may rise above that of the ambient air as a result of absorption of solar radiation and heat radiated by other bodies in the vicinity. The temperature actually reached depends, of course, on many factors which influence exposure to sun (height, degree of shading, etc.) and the dissipation of heat to the environment (transpiration efficiency, size and shape of leaves, humidity, wind velocity, etc.) (Gates, 1962).

The effect of the heat balance on the survival of one arctic-montane species may be developed from the data available for *Koenigia islandica* (Dahl, 1963). The lethal temperature for this species is about 45 °C (Löve & Sarkar, 1957) and its distribution in continental areas is limited by the 24 °C annual maximum isotherm. Because *K. islandica* is a very small herb its temperature may be equated to that of the soil surface on which it grows. In spite of uncertainties about the magnitudes of some of the physical constants involved, it seems likely that *K. islandica* under wind-still air at 25 °C with a relative humidity of 50%, and completely water-saturated soil would reach 43 °C whilst a partial drying of the soil would result in temperatures higher, but not above 53 °C. It is therefore understandable that this species might be excluded from regions with maximum summer air temperatures exceeding 24 °C.

Where air temperatures are the same, plants may be expected to reach higher temperatures in moist oceanic air than in dry continental air, and this might prove to be the underlying cause of the different correlations with isotherms shown by individual species from one region to another. For low plants of open-habitats such as *Salix herbacea*, *Silene acaulis* and *Sibbaldia procumbens*, the expected differences due to regional variations in atmospheric moisture content would be of the sign and magnitude actually observed. For shade-tolerant species as well as erect and woody plants, however, calculations are more difficult to make.

The distribution limits of many British arctic-montane species are certainly related to the annual maximum temperatures experienced at the places where they grow. The meteorological and physiological mechanisms which determine these correlations are still obscure, nor should it be expected that a single explanation will suffice for all cases. Nevertheless, it is amply obvious that the general concepts which Dahl (1951) propounded for parts of the Scandinavian flora can equally well be applied in a more sophisticated manner to the distribution of British arctic–montane species.

## SUMMARY

The distributions of 106 arctic–montane species of the British Isles are correlated with annual maximum summit isotherms, ranging from 20 °C for *Carex atrofusca* to 29 – °C for *Rubus saxatilis*.

A significant and systematic difference of 3 °C is established between the correlations proposed by Dahl (1951) for Scandinavia and those now elaborated for Scotland and Ireland. Regional differences also exist in the British Isles, the same species, on average, being limited by isotherms 1·6 °C higher in northern England than in Ireland and Highland Scotland.

Many species which are absent from Ireland and other south-western areas despite the availability of sufficiently cool localities, also show greater than average differences between their limiting isotherms in Britain and Scandinavia and tend to be restricted to high altitudes in Scotland. These anomalous distribution patterns may be related to a requirement for low winter temperatures, an explanation which could also account for parallel phenomena within Scandinavia and the absence of some Scandinavian species from the British Isles. Other deviations from the general pattern of correlations may be due to edaphic requirements, biotype variation or loss of genetically determined aggressiveness. The Teesdale populations of some plants which are rare there seem to have greater affinities, in relation to summer temperature tolerance, with Alpine than with Scandinavian biotypes of the same species. The rarity of suitably cool localities during the Post-glacial warm period in the British Isles may account for the absence from Scotland and Ireland of many species which, in Scandinavia, have limiting isotherms of 25 °C or lower. The differences in limiting isotherms of species between Scandinavia and Britain and regional variations within the British Isles may be attributable to the modifying effects of oceanicity.

Possible physiological reasons for the ecological behaviour described are considered.

# PART 2. THE FOSSIL RECORD IN RELATION TO MAXIMUM SUMMER TEMPERATURE

by Ann P. Conolly

*Department of Botany, University of Leicester*

## INTRODUCTION

The desire to deduce precise climatic parameters from the fossil floras of any period has prompted the intensified search for the factors determining the limits of distribution of species at the present day. Selected individual species for which relevant, preferably experimental autecological, data are available may be used to establish the climatic conditions of the past and particularly the temperature component. Iversen's (1944) elegant demonstration of the present-day climate's control on the distribution of fertile *Ilex*, *Hedera* and *Viscum* served to illuminate the climatic conditions of particular periods during the Post-glacial when some or all of these plants were producing pollen in Denmark.

Deductions based on single species, however, are open to the criticism that biotype evolution may have occurred between the prehistoric period and the present which, whilst not affecting the morphological characters from which the fossils are identified, nevertheless may have altered the species' ecological tolerances. Although this problem is directly insoluble, the criticism loses much of its force if a suite of species are concerned, all of which point consistently to a single conclusion, for it is inherently unlikely that closely parallel evolution will have taken place in all the species during the one period of time. Biotypic variety apart, a plant's reaction to its climatic environment is so complex that the correlation of its distribution with a single attribute of the climate is likely to vary from place to place because of the modifying effect of other components. The data presented in Part 1 of this paper demonstrate that this is the case with the response of very many species to the level of annual maximum temperature from region to region within the British Isles and between these and Scandinavia. To be relevant for the investigation of past climates, therefore, living plants and their fossil antecedents must belong to the same geographical region or the probable deviation in tolerance limits between the region inhabited by the living plant and that from which the fossil derives must be confidently assessable.

The present contribution is concerned with only one climatic attribute, mean annual maximum temperature, and uses as its basic data the Quaternary fossil records in the British Isles of *c.* 53 of the 106 arctic–montane species for which this parameter has been shown to be of significance. The regional variation in summer temperature tolerance exhibited by single species in different geographical regions (Part 1 of this paper) has required that only the temperature correlation relevant to the region of the fossil find be used; this has to be estimated if a present-day value is not listed. Furthermore, where the present-day information strongly suggests that, for a particular species, some other factor is modifying the temperature-area correlation in part of its range, then the primary indications of the fossil finds need to be taken with reserve or a 'correction' applied. Wherever possible, groups of species and regional groups of sites have been used to overcome some of the problems already discussed.

Fossil distributions are based on the published and unpublished records of many authors and those of living plants are taken from the *Atlas of the British Flora* (Perring & Walters, 1962), revised when necessary as described in Part 1 of this paper.

It is well to stress at the outset that all the uncertainties relating to the correlation of modern plant distribution limits with mean annual maximum isotherms are even more abundantly relevant to the extrapolation of these correlations to the prehistoric scene. All must be tentative if it is to avoid misleading.

## THE QUATERNARY DISTRIBUTION OF THE ARCTIC–MONTANE FLORA IN THE BRITISH ISLES

Fifty-three of the 106 arctic–montane species discussed in Part 1 of this paper are represented by fossil records and to these have been added a few species which are rare or extinct in the British Isles but for which Scandinavian limiting isotherms are available. Records based on pollen or macroscopic remains have been used wherever a determination refers the material to a single species or to not more than two possible species.

The data as presented are grouped according to geographical region (of which there are 10: figures 23–26) and age. Table 4 lists sites of records from the Late-Weichselian (Zones I–III), Early Post-glacial (Zones IV, V, VI) and Later Post-glacial (Zone VII onwards). Table 5 gives the Weichselian and pre-Weichselian sites; late- and early-glacial periods are generally included with the relevant interglacial, but Early-Weichselian material is attributed to the Weichselian glacial. Individual sites are located by National Grid Reference (figures 23 to 26 and tables 4 and 5 (see appendix, p. 218–20)); the latter also show the present-day mean annual maximum isotherm value appropriate to each site (cf. Part 1 of this paper, and figure 2). Sites of controversial age are either listed in an *addendum* to table 4, and lettered instead of numbered (as also are a number of late additions), or are distinguished by an asterisk in the normal numbered sequence. Sites from which unpublished data have been used are marked †.

A selection of the maps which were prepared relating the occurrence of a species at different prehistoric periods to its present distribution (Perring & Walters, 1962), is illustrated in figures 3, 5, 7 and 9 to 21. These do not show the nature of the fossil material identified nor do they distinguish uncertainties of identification from those of dating, information which is included (in part) in figures 27 to 32 (see appendix, p. 203–214).

The total distribution of a species at a particular time in the past can only be hazarded if the fossil records are numerous and widespread, conditions which apply, in the British Isles, only to *Salix herbacea* and *Betula nana*. The fossil distributions of these two species were compared with modern mean annual summit maximum isotherms in the manner already used for present-day distributions (Part 1 of this paper) and the results compared (figures 14 and 15, 9 and 10). The Late-Weichselian sites of *S. herbacea*, excepting that from Brook, Kent, are embraced by the present-day 28 °C mean annual maximum summit isotherm whereas the Weichselian and pre-Weichselian sites fall, for the most part, in the 'over 30 °C' area. Using the present-day correlation of *Salix herbacea* distribution with the 25 °C isotherm (northern England), and assuming the same to have applied formerly, the Late-Weichselian *sens. lat.* most probably experienced mean annual maximum temperatures about 3 °C below those of today, whilst earlier periods may have been up to 5 °C cooler. The distribution of *Betula nana* fossil sites, however, is so widespread as to be impossible to correlate with any useful isotherm.

Other solutions to the problem involve the use of the cumulative evidence of all species recorded from a fossil site, whilst sites may also be grouped into regions corresponding, at least in part, with the regional divergences discovered in Part 1 of this paper. By comparison between the present-day mean annual maximum summit isotherm closest to the site and the similar isotherm which currently limits the distribution of each species concerned within the relevant region, the minimum temperature depressions indicated by the species' fossil occurrences are calculated. These assessments of temperature 'depressions' are shown on a regional basis in figures 27 to 32 for each subdivision of the Quaternary previously defined. The variation from region to region within any time period is easily seen although the presentation somewhat obscures changes in a single region through time. In these diagrams, the 53 species (to which is added *Koenigia islandica* because of its importance) are tabulated in order of increasing value of present-day limiting isotherms. Scandinavian isotherm values are also shown, particularly to draw attention to large discrepancies which would arise with certain species were calculations to be based on Scandinavian rather than appropriate, regional, British data; where these differ from table 1 they indicate the 'warmest' areas in which the species grows in Scandinavia (derived from Hultén, 1950 in conjunction with Dahl, 1951 (figure 2), and Dahl, personal communication); they do not represent a correlation and are therefore placed in brackets. A few non-British or very rare species are included at the base of the diagrams for reference purposes. The temperature depression implied by the fossil occurrence of each species at any given locality is indicated by a bar of appropriate length on a scale of degrees Centigrade or by a dot or special symbol if none is implied. A dashed bar indicates dubious or approximate identification; a dotted extension to a bar records the indication of the more exacting species where an alternative identification is offered. Uncertainties of dating are indicated by wavy bars, broken when the determination is dubious and repeated through the likely time range. No indications of frequency per site of fossil materials are given and separated entries for pollen and macroscopic records are shown for only 3 species.

The temperature 'depressions', indicated by the lengths of the bars in figures 27 to 32, are derived as described above. Where an appropriate regional isotherm correlation is not available, that from another region is used and a correction factor applied which is based on average differences between regions (Part 1, and table 2 with the adjustments from p. 164). When a northern England figure is used for application to an East Anglian (or south-eastern) fossil occurrence, 1 °C is added to it; a Highland Scottish figure applied to these same regions is increased by 3 °C. Northern Midlands records and species with northern English distributions reaching southern Pennines merit proportionately lower additions. Sites in Devon and Cornwall, for which no local isotherms are available, are assigned figures 1 °C lower than those appropriate in Wales. *Regional isotherm equivalents* derived in this way are entered in special columns of the figures; they are clearly somewhat speculative, probably more so for some species than for others, but particularly where more than one species is concerned they probably lead to estimates within 1 °C of the truth.

The use of the Highland Scotland isotherm corrected by 3 °C for East Anglian assessments needs special justification. Arctic–montane species growing closest to East Anglia and south-east England at the present day (i.e. those within the 27 and 28 °C limiting isotherms) and which also grow in Ireland are there correlated with limiting isotherms about 3 °C lower than in the English regions. On average, Irish and Highland Scottish limiting isotherms for a given species are not significantly different and the 3 °C correction may therefore reasonably be applied directly to the Scottish figures. This argument can be extended by way of Ireland and Highland Scotland to Scandinavia, but for species with a big regional difference in isotherm value between Highland Scotland and Scandinavia very different results will be obtained depending on the region from which the basic data are derived. Where figures derived from Scandinavia and Highland Scotland are presented for a single fossil record, the former is indicated by a vertical line across the horizontal bar. Although only marked for East Anglia—the region most likely to equate with Scandinavia—it should perhaps be applied to the other British regions also: it is likely to represent the upper limit of possible discrepancy.

The species whose regional variation in relationship to annual maximum temperature isotherms reflects the modifying effect of other climatic criteria (Part 1 of this paper) will be less reliable as indicators of summer temperature depression than those species more directly related. For this reason species which show a greater than average (5 °C or more) difference between limiting annual maximum isotherms in Highland Scotland and Scandinavia and which are also absent from Ireland have been marked with an asterisk in figures 27 to 32.

The summit isotherm map (figure 2), owing to paucity of data, may be inadequate for extrapolating to lowland sites when these lie in a region of high summits but away from their immediate vicinity (e.g. Solway Firth and Cumberland coast). This could lead to under-estimation both of the isotherm values appropriate to such fossil sites (tables 4, 5) and of the implied temperature depressions. More detailed meteorological data are required.

## CLIMATIC ESTIMATES BASED ON THE FOSSIL RECORD

The depression of mean annual maximum temperature for any prehistoric period in each of the ten geographical regions may be judged from figures 27 to 32. For any period the fossil record provides a number of possible temperature maxima since the tolerances of species vary but the accumulation of all the data for a region usually allows an 'acceptable' maximum depression to be selected which does violence only to those data which are based on the most uncertain foundations. Such acceptable estimates are tabulated in tables 6 to 8 (see appendix, pp. 221–22). It shou'd b estressed that these estimates are of the minimum climatic difference needed to account for the records.

The validity of even these 'acceptable' figures clearly depends on the extent to which some of the primary assumptions are justified. Inaccurate identification may arise from faulty determinations by particular workers although where this is known to be the case the records have been corrected or rejected. On the other hand, although many species appear to retain their characteristics unchanged over long periods, any changes, such as physiological reaction to climatic conditions, which were not paralleled by morphological changes will always defy recognition in the fossil forms. The chance of errors derived from this kind of evolutionary change obviously increases with the age of the fossils and must be particularly likely in complex aggregate 'species'.

There is no *a priori* reason for supposing that the pattern of mean annual maximum summit isotherms remains constant through time, only the values of the isotherms changing. It is an assumption which may well be reasonable for the period between the Late-

Weichselian and the present but for more remote periods it is inherently suspect when the effects of distribution of land and sea on temperature and other climatic attributes are considered. Sites which are today coastal in East Anglia, for instance, will have been inland, and perhaps even close to an ice front, during some glacial periods. The probable modifications of climatic patterns by changes in the configuration of the land such as these, argue caution in the too-ready acceptance of these results.

Except for the Late-Weichselian and Post-glacial, the pollen analytical zonation of which is more or less confidently determinable, the time periods used are relatively long. Nevertheless, it is not beyond conception that errors have arisen as a result of faulty dating of the fossil material either by the uncritical interpretation of pollen diagrams or by real inconsistencies in vegetation pattern, and therefore recovered pollen spectra, between regions or even within them. The importance of this possible source of error can only be assessed when the synchroneity of British pollen analytical zones has been more critically studied and the relative chronological validities of pollen analysis, stratigraphy and radiocarbon dating revealed.

Regional comparison of the temperature depressions summarized in tables 6 to 8 may be misleading if account is not taken of the inherent differences due to the varying geographical situation; the potential maximum depression for a region will, by virtue of its contained fossil temperature-indicators, depend on the degree to which the isotherm value of the 'warmest' site exceeds that of the 'coldest' species found as a fossil there. Thus the Isle of Man, with no present-day maximum summer isotherm much above 23 °C, could not demonstrate more than 2·2 °C depression at any period, even if the most exacting of the arctic–montane species recorded fossil in the British Isles were to be found there. The maximum depressions of (maximum) summer temperatures realizable for each region are summarized in table 9 (see appendix, p. 223). In this table regional maximum depressions (column 2) are calculated on the highest maximum summer temperature isotherms (figure 2) located in each region (column 1)—i.e. the values for the 'warmest' potential site—and are based on the theoretical possibility of the occurrence as a fossil of the 'coldest' of the 54 species with a reliable determination (figures 27 to 32). Column 2 thus gives the absolute maxima that could occur were a site to be discovered with the most exacting species in the 'warmest' area of each region. Maximum depressions calculated on the highest maximum summer isotherm realized by existing sites (column 3) are shown in column 5. In addition the number of sites in each period which have isotherm values not more than 1 °C below the maximum possible for their region is shown in column 4, together with the total number of sites. This emphasizes those areas where paucity of sites lessens the chance of the most exacting species being found in the 'warmest' areas. It will be realized from the above that regional comparisons become of less value where the temperature assessments (tables 6 to 8) lie near the (regional) potential maxima, especially where these are low.

## DISCUSSION

### Late-Weichselian

It has long been known (e.g. Godwin, 1949) that many species occurred well outside their present limits during the Late-Weichselian. But although the extensions covered relatively great distances they were not indiscriminately widespread and at least a hint of a relationship to specific annual maximum isotherms can be discerned. In the case of *Salix herbacea*, for which many data are available, it is even possible to recognize a limiting isotherm which bounds the Late-Weichselian find sites. Even where sites are 150 to 300 km from the nearest extant stations of the recovered species, as is the case for *Thalictrum alpinum*, *Draba incana*, *Cirsium* cf. *heterophyllum* and *Saxifraga* cf. *hypnoides* in their fossil occurrences at Nazeing, Essex (Allison, Godwin & Warren, 1952), the implied depression of mean annual maximum temperature for Zone III (or earlier) is about 3 to 4 °C, close to the value of 3 °C deduced from the overall pattern of *Salix herbacea* finds.

Lack of data make it impossible to judge whether the increased summer 'warmth' of Zone II led to the elimination or areal reduction of species already present in Zone I.

The general impression gained from the regional results (table 6), but discounting *Betula nana*, is that the Zone I record is too meagre for meaningful comparison, and that for Zone III (or the undifferentiated Late-Weichselian) temperature depressions of *c.* 3 °C (*c.* 4·5 °C if *Koenigia* pollen is taken into account) were experienced in northern England (with south Scotland) and in Ireland. Depressions were considerably less in the South-west and apparently also in Scotland, Wales and the Isle of Man. But for neither of the latter two regions nor for the majority of Scottish sites could more than very low figures be expected, owing to the juxtaposition of sites with low (present-day) isotherms (cf. table 9). There are insufficient data for the Midlands, East Anglia and the South-east, though there is a hint of temperature depressions at least as great as for northern England and Ireland. So far as the data are available they show no very striking difference in mean annual maximum temperature from zone to zone in any region although Zone III, other than in Ireland, is consistently from half to one degree 'colder' than Zone II. The higher values that would be registered using *Betula nana* would be discounted, at least for East Anglia and the South-east, were the calculations to be based on its Scandinavian behaviour in relation to the annual maximum isotherms.

### Post-glacial

Post-glacial records, particularly those dating from Zone VII and later, generally come from sites close to the present stations of the species involved. Where the recent history of a species is one of decline during the last 50 to 100 years, the fossil sites from which it is recorded with few exceptions lie within the recently vacated area. The species recorded from Zone IV at Nazeing, some 150 km or more from their modern localities, are mostly only tentatively identified or rest on scant evidence.

Post-glacial *Betula nana* (figure 10) is exceptional in the remoteness of many of its localities from those it presently inhabits. Calculations based on its modern Scottish temperature relationships point to a mean annual maximum temperature depression considerably greater than those based on any other species: e.g. one of nearly 5 °C for Zone IV in the South-east, although this would not apply were the Scandinavian distribution data to be used as a basis. Indeed, the Scottish correlation may well be discrepant; the species occurs in Teesdale both at the present-day and in the Post-glacial (Hutchinson, 1966; Hewetson, personal communication), its central European populations are sub-alpine and in Scandinavia it grows in the lowlands. For this reason estimates based on *Betula nana* have been kept separate in tables 6 to 8.

The effect of the supposed increase in temperature at the onset of the Post-glacial may have been to eliminate *Salix herbacea* from central and eastern Ireland in Zone IV (Godwin, 1956). Assuming that *Salix herbacea* was limited by the 23 °C mean annual maximum isotherm then as now, Zone IV would have required to be 2 °C colder than today (by this parameter) in order for the plant to have occurred in its known sites but colder than this to have grown in areas exceeding the present-day 25 °C isotherm; the absence of Zone IV records from the latter area implies this was not the case, although known here for the Late-Weichselian (figures 14 and 15). This suggests an increase in mean annual maximum temperature up to about 2 °C colder than today in the early Post-glacial from *c.* 3 °C colder (or more) in the Late-Weichselian.

### The post-glacial thermal maximum

It has been supposed that relatively high summer, and other, temperatures during the Post-glacial, might have contributed to the extinction of species which do not grow in some localities which seem thermally suited to them at the present day. A small area of Devon and Cornwall today experiences a mean annual summit maximum below 25 °C yet is free of species like *Salix herbacea* and *Saxifraga stellaris* the distributions of which are correlated with this isotherm in Wales. However, were the suggested regional correction of −1 °C between Wales and south-west England to apply then *Salix herbacea* might require mean annual summit maxima not higher than 24 °C. Such areas are not recorded in south-west England, and plants such as those already mentioned together with *Subularia aquatica* and *Betula nana*, all recorded from the Late-Weichselian there, would certainly not have survived through a post-glacial period with higher temperatures than now.

For those species which, in Scandinavia, have limiting isotherms of 25 °C or less, it was argued (Part 1 of this paper) that in the British Isles areas with present-day mean annual maximum summit temperatures of 20 °C or less would be a requisite of survival during a post-glacial increase in temperature of 2 °C or more above those of today. Only few areas with appropriate temperatures even as low as 20 °C exist in Scotland and Ireland (figures 2 and 3), and extinctions and reductions to near extinction would be expected. *Salix polaris* (with a limiting isotherm of 25 °C in Scandinavia), recorded from supposedly Late-Weichselian deposits in the Edinburgh area (Reid, 1899), was probably one such victim. *Koenigia islandica* and *Artemisia norvegica* both with Scandinavian correlations with the 24 °C isotherm, and both with widely scattered Late-Weichselian pollen finds (including some from Wales, and for the former also northern Ireland), have narrowly escaped to survive in Scotland alone.

The occurrence of a Post-glacial thermal maximum would place large mountain massifs or particularly high mountains at an advantage compared with isolated hills of only moderate height as refugia for plants intolerant of high summer temperatures, by ensuring continuous corridors of migration to climatically tolerable localities. Even small populations which might manage to survive in great isolation would nevertheless be susceptible to genotype depletion and lowering of dispersal potential, and consequently be especially prone to later extermination and result in correlations (today) with lower (maximum) isotherms than would otherwise be expected. For these reasons, the very mass of the Scottish Highlands and the great heights of Snowdonia are more likely to harbour arctic–montane plants at the present day than are the isolated peaks of Ireland, adequately low temperatures notwithstanding, though the selection of species surviving at a site must often have been fortuitous.

### Earlier periods

Fossils of the Weichselian and earlier periods are perforce restricted to East Anglia and the Midlands. They lie remote from the present localities of their modern counterparts, many of them in the region bounded by the 30 °C mean annual maximum summit isotherm, thus implying a considerable fall in temperature to accommodate the more exacting species.

The indications based on all species from all sites suggest that at some time during the Weichselian, the depression of the mean annual maximum temperature was as much as 5 to 6 °C for East Anglia and about 4 °C for the Midlands. For pre-Weichselian periods there is little evidence of temperature reductions as great as this: some 4 °C in East Anglia and the South-east for the Gipping glacial, perhaps *c.* 3 °C for the Hoxnian in Ireland and *c.* 4·5 °C for Late- or Early-glacial stages of the Hoxnian in East Anglia (increased to *c.* 5 °C on records of *Salix polaris*). There are no pointers to depressions of much more than 4 °C at any stage of the Lower Middle Pleistocene in East Anglia, especially if the Scandinavian base for the assessments are followed; but here again fewness of reliable entries vitiates the drawing of conclusions.

## CONCLUSIONS

The tentative nature of all these deductions is increased by the paucity of data from many periods or regions; moreover the fossils preserved can only represent a small and unknown proportion of the former total flora for any period. Nevertheless, in so far as the temperature indications of these species are valid at all, they should contribute to minimum estimates of mean annual maximum temperature depressions. No inferences about the amplitude of the yearly temperature regime can be drawn, nor any deductions about relative continentality or oceanicity. For this, as for the problem of those species whose relationship with maximum summer temperature are apparently modified by some other climatic factor, specific requirements related to winter temperatures are needed. The problem of assessing the hypothetical behaviour of arctic–montane species in East Anglia would be aided were data available on their relationship with maximum summer temperature in the central European Alps, to counter the comparison with Scandinavia.

With all their shortcomings, however, these data provide the best available botanical complement to the estimates of Manley (1951, 1959), Wright (1961), Dahl (1964) and Lamb, Lewis & Woodroffe (1966). When the amounts of independent but grossly subjective estimation which contribute to all these methods are considered it is as surprising as it is reassuring that the results compare so well.

## SUMMARY

The fossil record of arctic–montane plants is interpreted in the light of their present-day distribution correlations with mean annual maximum summit temperatures. The fossil distributions, maps of several of which are given, are used to attempt the estimation of this climatic attribute during several Quaternary intervals. In terms of this temperature parameter it is concluded:

(*a*) that the Weichselian was at least 5 °C (or even 6°) colder than now in East Anglia, and perhaps 3·5 or 4 °C colder in the Midlands,

(*b*) that the Late-Weichselian *sens. lat.* was probably at least 3 °C colder than now on average in Great Britain and Ireland but probably less so in south-west England,

(*c*) that Zone IV in Ireland may have been up to about 2 °C colder than now,

(*d*) that a postulated Post-glacial thermal maximum 2–3 °C warmer than at present accords well with the pattern of extinction and survival of arctic–montane species.

We wish to acknowledge the Botanical Society of the British Isles Distribution Maps Scheme and especially the help accorded to us by Dr F. Perring and the staff of the Biological Records Centre of the Nature Conservancy for the data on present-day distribution on which the maps are based and for making these available. We are grateful to our many correspondents for information about local records, particularly Dr D. A. Ratcliffe, Dr G. Roger, Miss Ursula Duncan and Mr R. D. Meikle, to Sir George Taylor for an extract from the MS Corstorphine Flora of Angus and to the National Museum of Wales and the Holmesdale Natural History Club for the loan of specimens.

During the preparation of Part 2 of this paper multifarious help was accorded to me (A.P.C.) by members of the Cambridge University Sub-department of Quaternary Research and particularly by Professor H. Godwin, F.R.S. and Dr R. G. West, F.R.S.; for this I am most grateful. For advice and access to hitherto unpublished data I am indebted to Professor G. F. Mitchell, Mrs C. A. and Dr J. H. Dickson, Mrs H. H. and Mr J. Birks, Miss Frances Bell, Miss R. Andrew, Dr F. A. Hibbert, Dr G. Evans, Miss Valerie Hewetson, Dr N. T. Moar, Dr P. Moore, Dr B. Seddon, Mr P. A. Tallantire, Dr C. Turner, Mrs W. Tutin and Mrs G. Wilson. I also thank Professor E. Dahl for many suggestions.

Dr D. A. Ratcliffe and Mr H. H. Lamb kindly discussed parts of our work with us and Dr G. Halliday, Dr D. M. Moore and Dr Brenda Turner read and improved the manuscript. We owe much to Dr D. Walker for his generous help as editor and to Miss Miriam Perkins for typing the manuscript.

## REFERENCES

Aass, B. (1965). Skoggrenser i Norge. Hovedoppgave i geografi. University of Oslo. (Unpublished.)

Allison, J., Godwin, H. & Warren, S. H. (1952). Late-glacial Deposits at Nazeing in the Lea Valley, North London. *Phil. Trans. R. Soc.* B **236**, 169–240.

Amann, J. (1929). L'hygrothermie du climat, facteur déterminant de la répartition des espèces atlantiques. *Revue bryol.* New Series **2**, 126–33.

Biebl, R. (1965). Temperaturresistenz tropischer Pflanzen auf Puerto Rico. *Protoplasma,* **59**, 133–56.

Böcher, T. W. (1951). Distributions of Plants in the Circumpolar Area in relation to Ecological and Historical Factors. *J. Ecol.* **39**, 376–95.

Brockmann-Jerosch, H. (1919). Baumgrenze und Klimacharakter. *Beitr. geobot. Landesaufn. Schweiz,* **6**, 1–255.

Burgess, J. J. (1935). *Flora of Moray.* Elgin: Courant and Courier.

Conolly, A. P. (1961). Some Climatic and Edaphic Indications from the Late-glacial Flora. *Proc. Linn. Soc. Lond.* **172**, 56–62.

Corstorphine, R. H. & Corstorphine, M. *Flora of Angus* (unpublished). Manuscript in possession of Sir George Taylor.

Dahl, E. (1951). On the relation between summer temperature and the distribution of alpine vascular plants in the lowlands of Fennoscandia. *Oikos,* **3**, 22–52.

Dahl, E. (1963). On the Heat Exchange of a Wet Vegetation Surface and the Ecology of *Koenigia islandica. Oikos,* **14**, 190–211.

Dahl, E. (1964). Present-Day Distribution of Plants and Past Climate, in *The Reconstruction of Past Environments* (ed. J. H. Hester & J. Schoenwetter), no. 3, 52–9. Fort Burgwin Research Center.

Dahl, E. (1966). The Heat Exchange of Plants and its Importance to Plant Morphology and Distribution. *Blyttia* **24**, 105–29.

Dahl, E. & Mork, E. (1959). On the Relationships between Temperature, Respiration and Growth in Norway Spruce. *Meddr norske Skogsfors Ves.* **53**, 81–93.

Dansereau, P. (1957). *Biogeography.* New York: Ronald Press.

de Candolle, A. (1855). *Géographie Botanique Raissonnée.* Paris: Masson.

Dengler, A. (1930). *Waldbau auf ökologischer Grundlage.* Berlin: Springer.

Enquist, F. (1933). Baumgrenzuntersuchungen. *Svenska Skogsv. För. Tidskr.* **31**, 145–214.

Gams, H. (1931–2). Die klimatische Begrenzung von Pflanzenarealen und die Verteilung der hygrischen kontinentalität in den Alpen. *Z. Ges. Erdk. Berl.* 1931, 321–56; 1932, 52–68, 178–98.

Gates, D. M. (1962). *Energy Exchange in the Biosphere.* New York: Harper and Row.

Godwin, H. (1949). The Spreading of the British Flora. *J. Ecol.* **37**, 140–7.

Godwin, H. (1956). *The History of the British Flora.* Cambridge University Press.

Godwin, H. & Clapham, A. R. (1951). Peat Deposits on Cross Fell, Cumberland. *New Phytol.* **50**, 167–71.

Godwin, H. & Walters, S. M. (1967). The Scientific Importance of Upper Teesdale. *Proc. bot. Soc. Br. Isl.* **6**, 348–51.

Hagem, O. (1917). Furuens og granens frøsaetting i Norge. *Meddr Vestland. forst. Fors. Stn.* **2**, 1–188.

Hegi, G. (1939). *Illustrierte Flora von Mittel-Europa,* **2**, 57. 2nd ed. Munich.

Hegi, G. (1962). *Illustrierte Flora von Mittel-Europa,* **3**: 2, 815–16. (2nd ed.) Munich: Hanser.

Hintikka, V. (1963). Über das Grossklima einiger Pflanzenareale in zwei Klimakoordinatensystemen dargestellt. *Suomal. eläin-ja kasvit. Seur. van kasvit. Julk.* **34**, no. 5, 1–64.

Holmboe, J. (1913). Kristtornen i Norge. *Bergens Mus. Årb.* no. 7, 1–91.

Holmboe, J. (1925). Einige Grundzüge von der Pflanzengeographie Norwegens. *Bergens Mus. Årb.* Naturvidensk. Raekke no. 3, 1–54.

Hultén, E. (1950). *Atlas of the Distribution of Vascular Plants in N.W. Europe.* Stockholm: Generalstabens Litografiska Anstalts.

Hultén, E. (1958). The Amphi-Atlantic Plants and their

Phytogeographical Connections. *K. Svenska Vetensk-Akad. Handl.* 4th Ser. **7**, no. 1, 1–340.

Hutchinson, T. C. (1966). The Occurrence of Living and Sub-Fossil Remains of *Betula nana* L. in Upper Teesdale. *New Phytol.* **65**, 351–7.

Iversen, J. (1944). *Viscum, Hedera* and *Ilex* as Climate Indicators. *Geol. För. Stockh. Förh.* **66**, 463–83.

Iversen, J. (1954). The Late-glacial Flora of Denmark and its Relation to Climate and Soil. *Danm. geol. Unders.* (II) 80, 87–119.

Jessen, K. F. W. (1948). *Botanik der Gegenwart und Vorzeit.* Waltham, Mass.: Chronica Botanica.

Johannessen, T. W. (1956). Varmevekslinger i bygninger og klimaet. *Rapp. Norg. Byggforsk. Inst.* no. 21, 1–258.

Kelly, M. R. (1964). The Middle Pleistocene of North Birmingham. *Phil. Trans. R. Soc.* B **247**, 533–92.

Kotilainen, M. J. (1933). Zur Frage der Verbreitung des atlantischen Florenelementes Fennoskandias. *Suomal eläin-jä kasvit. Seur. van. kasvit. Julk.* **4**, no. 1, 1–75.

Lamb, H. H., Lewis, R. P. W. & Woodroffe, A. (1966). Atmospheric Circulation 8000 to 0 B.C. In *World Climate from 8000 to 0 B.C.* (ed. J. S. Sawyer), 174–215. Roy. Meteorological Society, London.

Lange, O. L. (1959). Untersuchungen über Wärmehaushalt und Hitzeresistenz mauritanischer Wüsten- und Savannenpflanzen. *Flora, Jena* **147**, 595–651.

Lange, O. L. (1961). Die Hitzeresistenz einheimischer immerund wintergrüner Pflanzen im Jahreslauf. *Planta* **56**, 666–83.

Lange, O. L. (1962). Über die Beziehungen zwischen Wasser- und Wärmehaushalt von Wüstenpflanzen. *Veröff. geobot. Inst., Zürich* **37**, 155–68.

Lange, O. L. (1967). Investigations on the Variability of Heat-resistance in Plants. In *The Cell and Environmental Temperature* (ed. A. S. Troshin), 131–41. Oxford: Pergamon.

Lange, O. L. & Lange, R. (1962). Die Hitzresistenz einiger mediterraner Pflanzen in Abhängigkeit von der Höhenlage ihrer Standorte. *Flora, Jena* **152**, 707–10.

Lange, O. L. & Lange, R. (1963). Untersuchungen über Blattetemperaturen, Transpiration und Hitzresistenz und Pflanzen mediterraner Standorte (Costa Brava, Spanien). *Flora, Jena,* **153**, 387–425.

Lewis, F. J. (1904). Interglacial and Postglacial Beds of the Cross Fell District. *Rep. Br. Ass. Advmt Sci.* 74th Meeting, 798–9.

Lewis, F. J. (1911). The Plant Remains in the Scottish Peat Mosses, Part IV. *Trans. R. Soc. Edinb.* **47**, 793–833.

Löve, A. & Sarkar, P. (1957). Heat Tolerances of *Koenigia islandica. Bot. Notiser* **110**, 478–81.

Lye, K. A. (1968). The Horizontal and Vertical Distribution of Oceanic Plants in South-west Norway and their Relation to the Environment. *Nytt Mag. Bot.* **15** (in the press).

McVean, D. N. & Ratcliffe, D. A. (1962). *Plant Communities of the Scottish Highlands.* London: H.M. Stationery Office.

Manley, G. (1951). The Range of Variation of the British Climate. *Geogrl J.* **117**, 43–68.

Manley, G. (1959). The Late-glacial Climate of Northwest England. *Lpool Manchr geol. J.* **2**, 188–215.

Martonne, E. de (1927). *Traité de Géographie Physique,* **3**. *Biogéographie.* Paris: Armand Colin.

Matthews, J. R. (1937). Geographical Relationships of the British Flora. *J. Ecol.* **25**, 1–90.

Matthews, J. R. (1955). *Origin and Distribution of the British Flora.* London: Hutchinson.

Mayr, H. (1909). *Waldbau auf naturgesetzlicher Grundlage.* Berlin: Parey.

Mitchell, G. F. (1960). The Pleistocene History of the Irish Sea. *Advmt Sci., Lond.* 313–25.

Mitchell, G. F. (1965). The St Erth Beds—an Alternative explanation. *Proc. Geol. Ass.* **76**, 345–66.

Mitchell, G. F. & Orme, A. R. (1967). The Pleistocene Deposits of the Isles of Scilly. *Q. Jl geol. Soc. Lond.* **123**, 59–92.

Mooney, H. A. & Billings, W. D. (1961). Comparative Physiological Ecology of Arctic and Alpine Populations of *Oxyria digyna. Ecol. Monogr.* **31**, 1–29.

Mork, E. (1938). Gran og furufrøets spiring ved forskjellig temperatur og fuktighet. *Meddr norske Skogsfors. Ves.* **6**, 225–49.

Mork, E. (1952). Faktorer som virker på spireevnen hos furugran- og bjørkefrø. *Meddr norske Skogsfors. Ves.* **9**, 159–72.

Perring, F. H. & Walters, S. M. (1962). *Atlas of the British Flora.* London: Nelson.

Petterson, B. (1965). Götland and Öland. Two limestone Islands Compared. *Acta phytogeogr. suec.* **50**, 131–40.

Pigott, C. D. (1956). The Vegetation of Upper Teesdale in the North Pennines. *J. Ecol.* **44**, 545–86.

Pigott, C. D. (1958). Biological Flora of the British Isles: *Polemonium caeruleum* L. *J. Ecol.* **46**, 507–25.

Porsild, A. E. (1959). *Dryas Babingtoniana* nom. nov. An Overlooked Species of the British Isles and Western Norway. *Bull. natn. Mus. Can.* **160**, 133–48.

Praeger, R. Ll. (1934). *The Botanist in Ireland.* Dublin: Hodges, Figgis.

Ratcliffe, D. A. (1960). The Mountain Flora of Lakeland. *Proc. bot. Soc. Br. Isl.* **4**, 1–25.

Reid, C. (1899). *The Origin of the British Flora.* London: Dulau.

Robak, H. (1960). Spontaneous and Planted Forest in West Norway. *Skr. Rekk. geogr. Avh.* no. 7, 17–34.

Ryvarden, L. & Kaland, P. E. (1968). *Artemisia norvegica* Fr. funnet i Rogaland. *Blyttia* **26**, 75–84.

Samuelsson, G. (1934). Die Verbreitung der höheren Wasserpflanzen in Nordeuropa. *Acta phytogeogr. suec.* **6**, 1–211.

Trail, J. W. H. (1923). *A Memorial Volume.* Aberdeen University Press.

Wilson, A. (1956). *The Altitudinal Range of British Plants.* (2nd ed.) Arbroath, Scotland: Buncle.

Wright, H. E. (1961). Late Pleistocene Climate of Europe: A Review. *Bull. geol. Soc. Am.* **72**, 933–84.

The most important unpublished material consulted is:

The Card Index of Fossil records in the Sub-department of Quaternary Research, Cambridge.

Dickson, J. H. (1965). *Historical Biogeography of the British Moss Flora*. Ph.D. dissertation, University of Cambridge.

Moar, N. T. (1964). *The History of the Late-Weichselian and Flandrian Vegetation in Scotland*. Ph.D. dissertation, University of Cambridge.

Moore, P. D. (1966). *Stratigraphical and Palynological Investigations of Upland Peats in Central Wales*. Ph.D. thesis, University of Wales (Aberystwyth).

Trotman, D. M. (1963). *Data for Late Glacial and Post Glacial History in South Wales*. Ph.D. thesis, University of Wales (Swansea).

# Appendix A: Figures

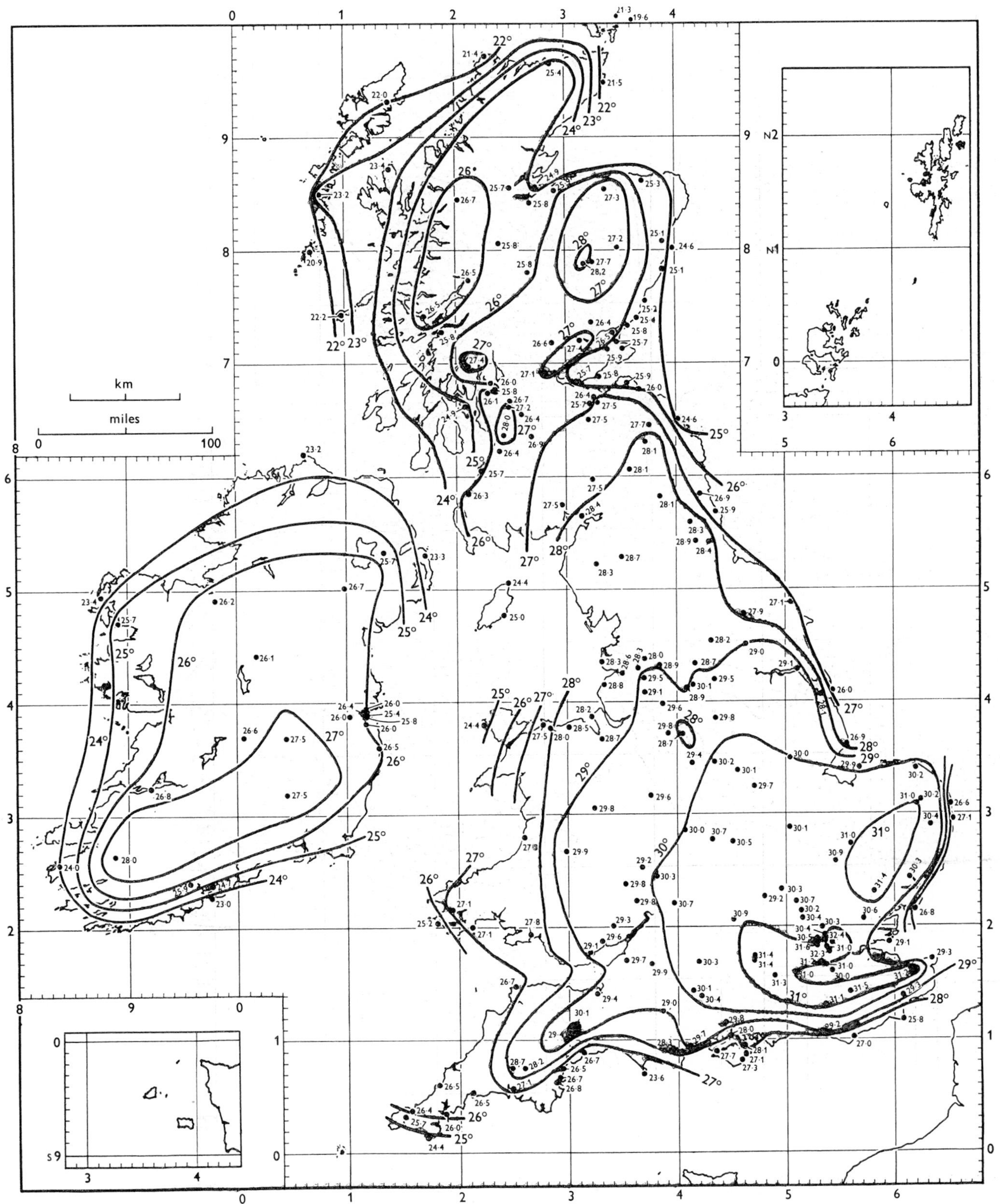


Figure 1. Isotherm map (°C) of mean annual maximum temperature reduced to sea-level. Dots mark the position of the meteorological stations.

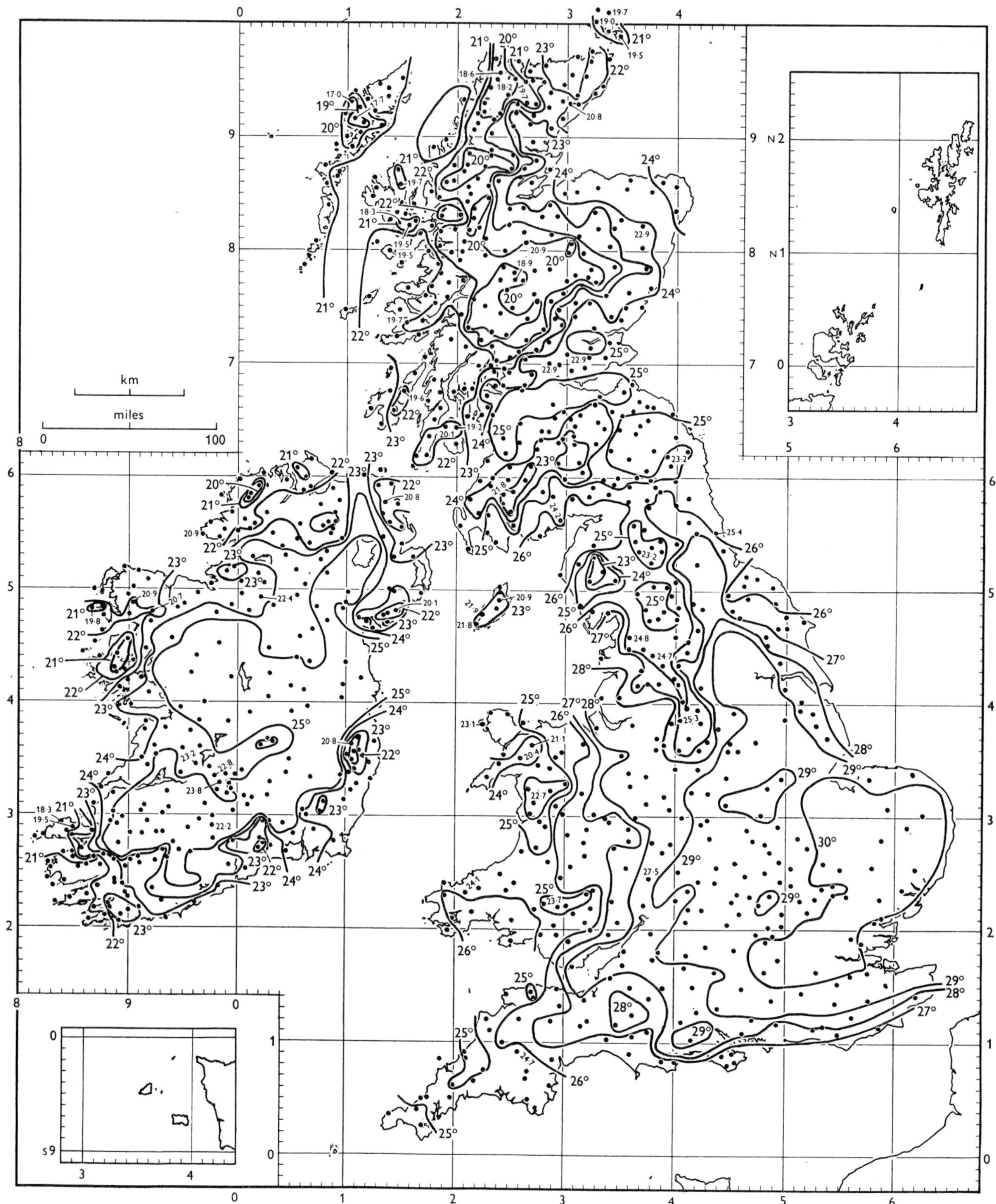

Figure 2. Isotherm map (°C) of mean annual maximum temperature for the highest places in the landscape. Dots mark the positions of the summits used in the calculations. Spot heights are shown for all summits more than 1 °C below the drawn enclosing isotherm.

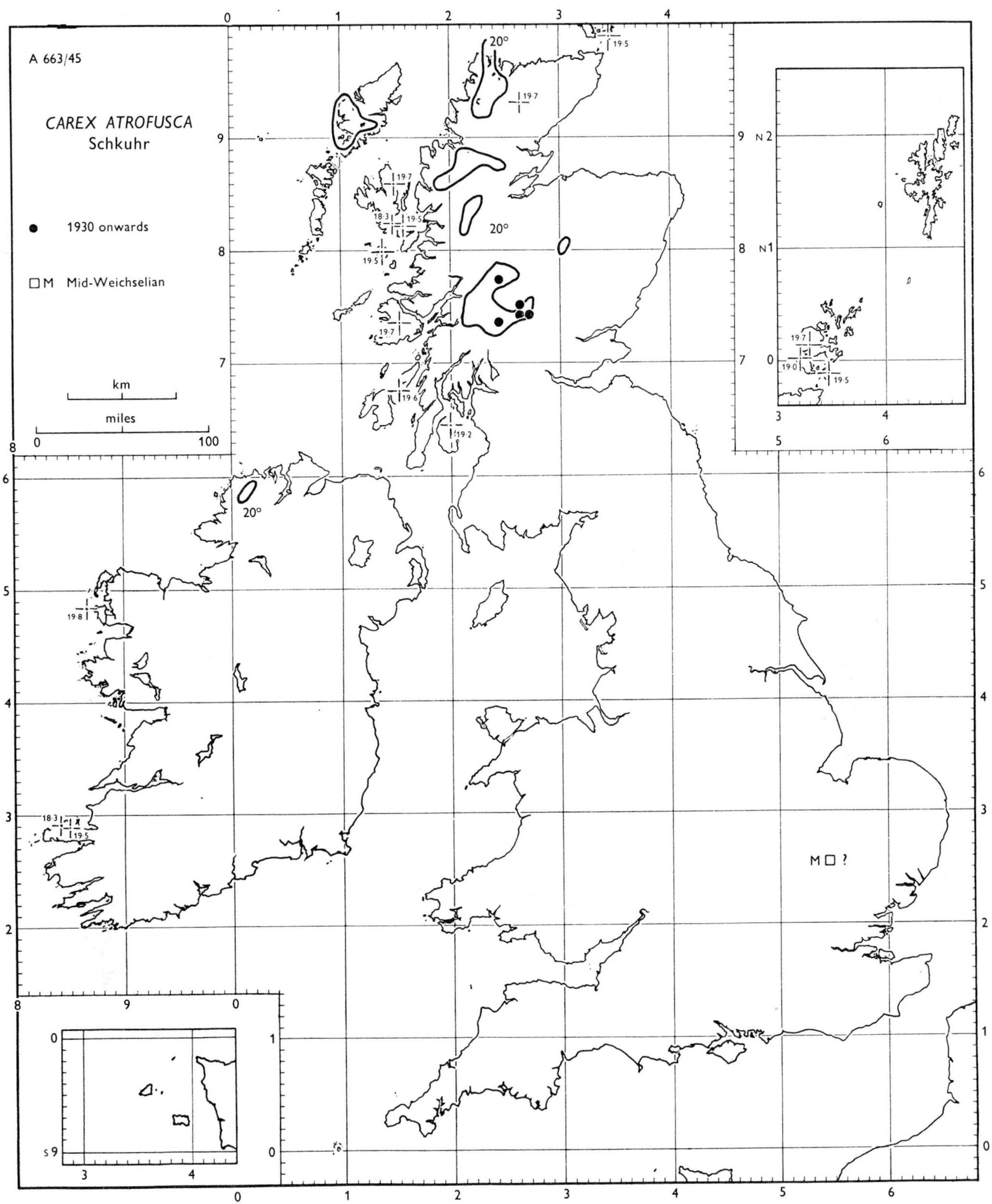


Botanical Society of the British Isles Distribution Maps Scheme

Figure 3. Map showing the present-day distribution of *Carex atrofusca* Schkuhr with the 20 °C maximum summer temperature summit isotherm and the location of a Weichselian fossil record (of doubtful determination). On this and subsequent maps the sign ┼ indicates spot heights for outlying summits with the relevant mean annual maximum temperature value.

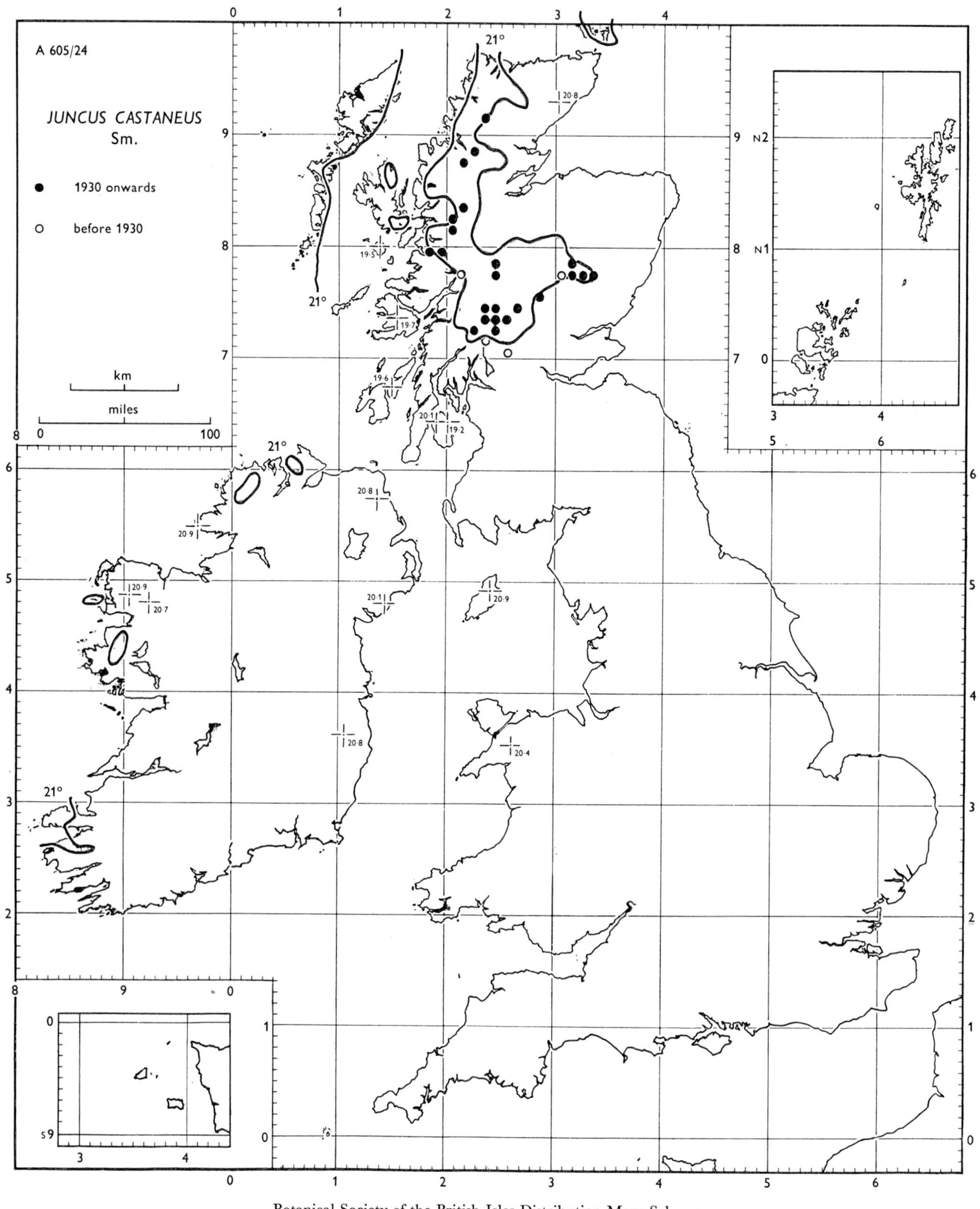


Botanical Society of the British Isles Distribution Maps Scheme

Figure 4. Map showing the present-day distribution of *Juncus castaneus* Sm. with the 21 °C maximum summer temperature summit isotherm.

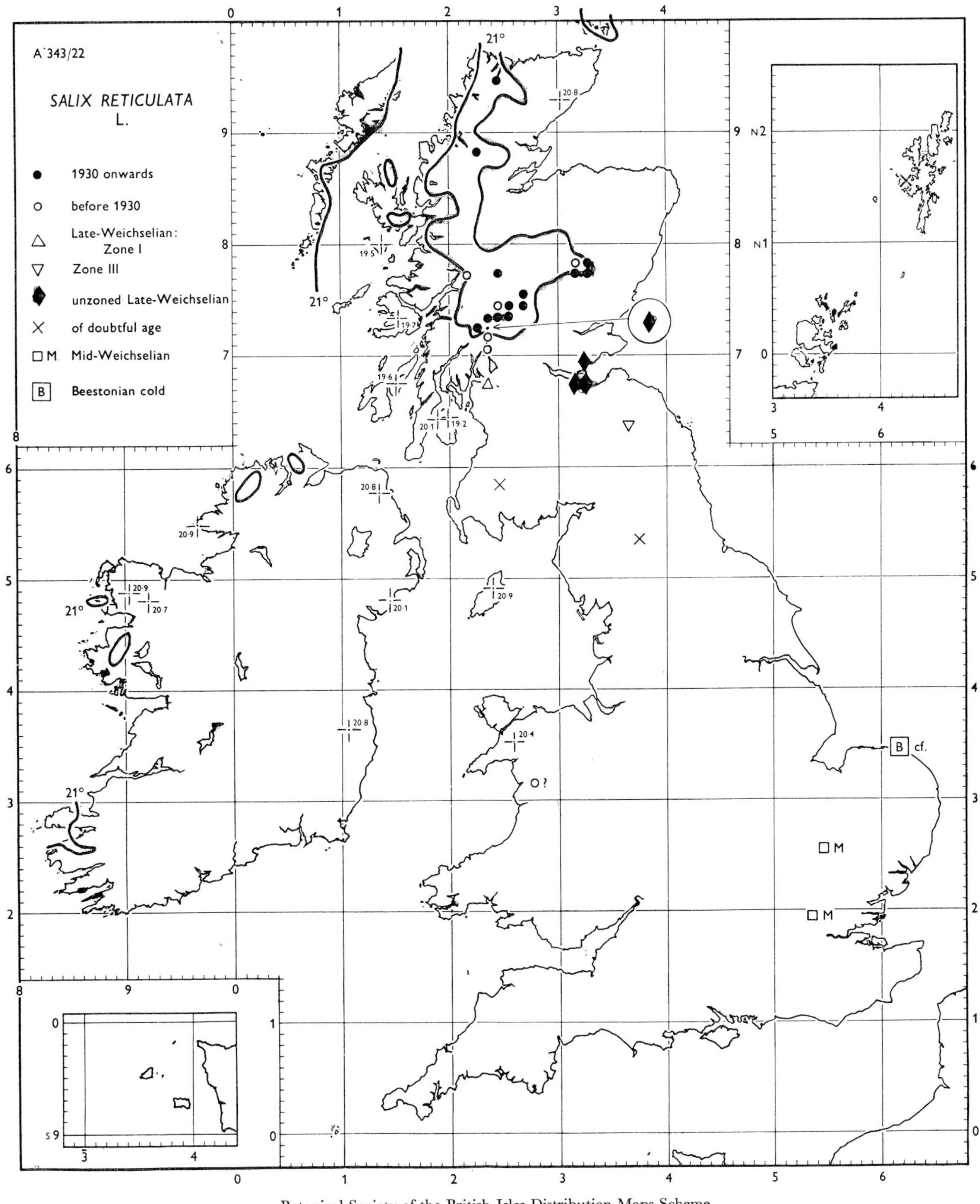


Botanical Society of the British Isles Distribution Maps Scheme

Figure 5. Map showing the present-day distribution of *Salix reticulata* L. with the 21 °C maximum summer temperature summit isotherm, and the location of fossil records for Late-Weichselian, Weichselian and Beestonian periods. Sites marked X are from Lewis (1904, 1911).

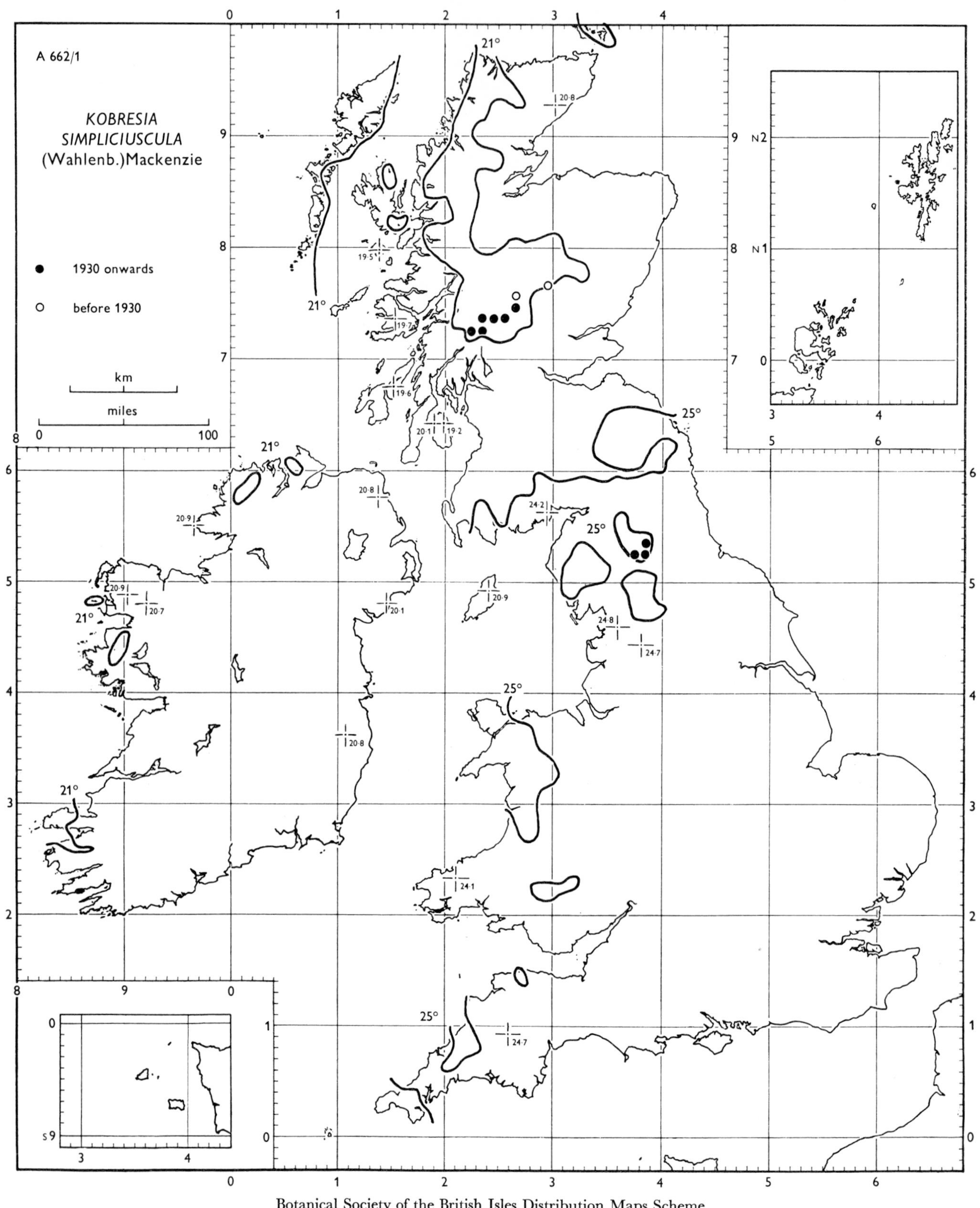


Figure 6. Map showing the present-day distribution of *Kobresia simpliciuscula* (Wahlenb.) Mackenzie with the 21 °C maximum summer temperature summit isotherm for Ireland, Highland Scotland and the Isle of Man, and the 25 °C isotherm for Wales, England and adjacent Scotland.

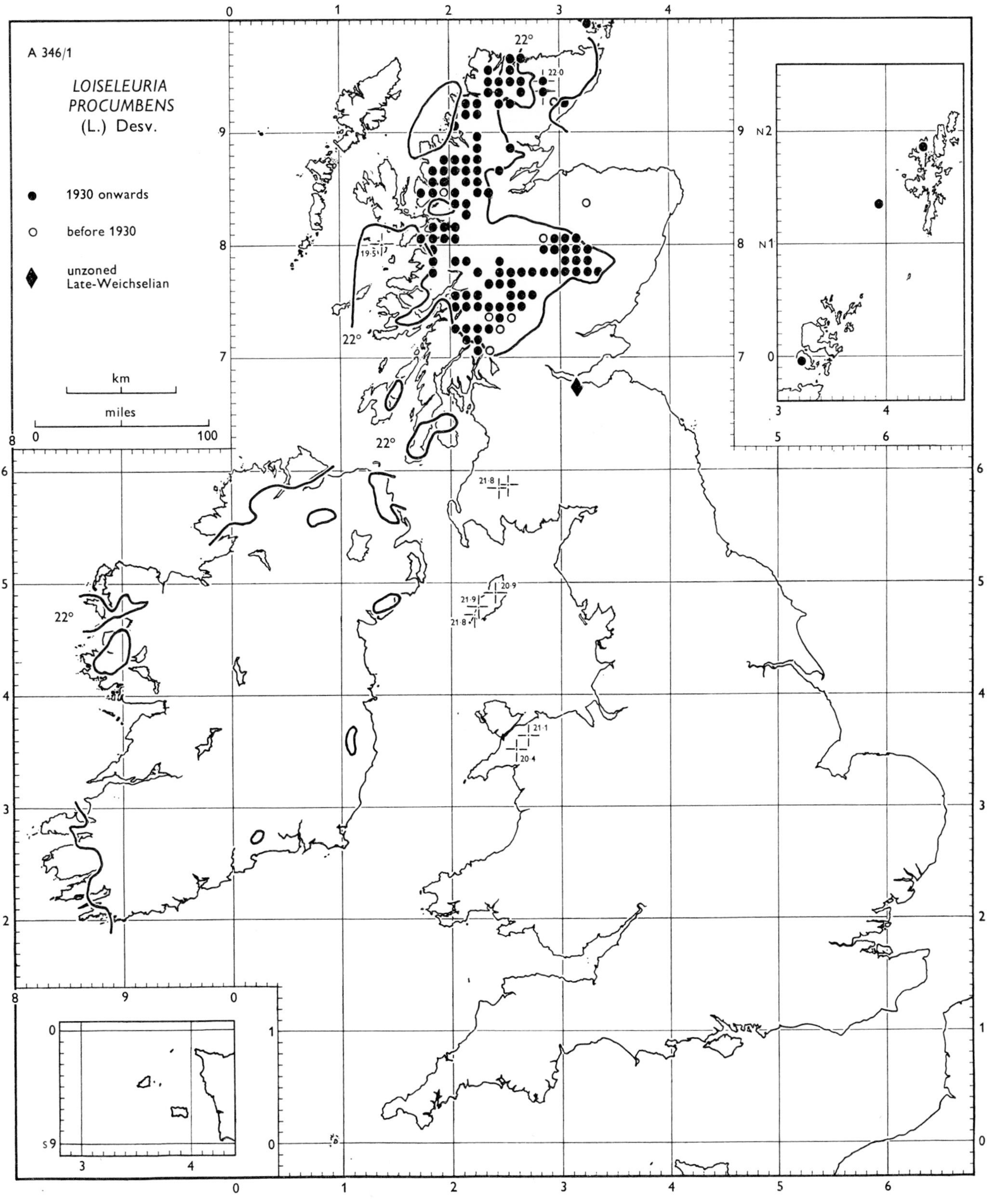


Botanical Society of the British Isles Distribution Maps Scheme

Figure 7. Map showing the present-day distribution of *Loiseleuria procumbens* (L.) Desv. with the 22 °C maximum summer temperature summit isotherm, and the location of an unzoned Late-Weichselian fossil record.

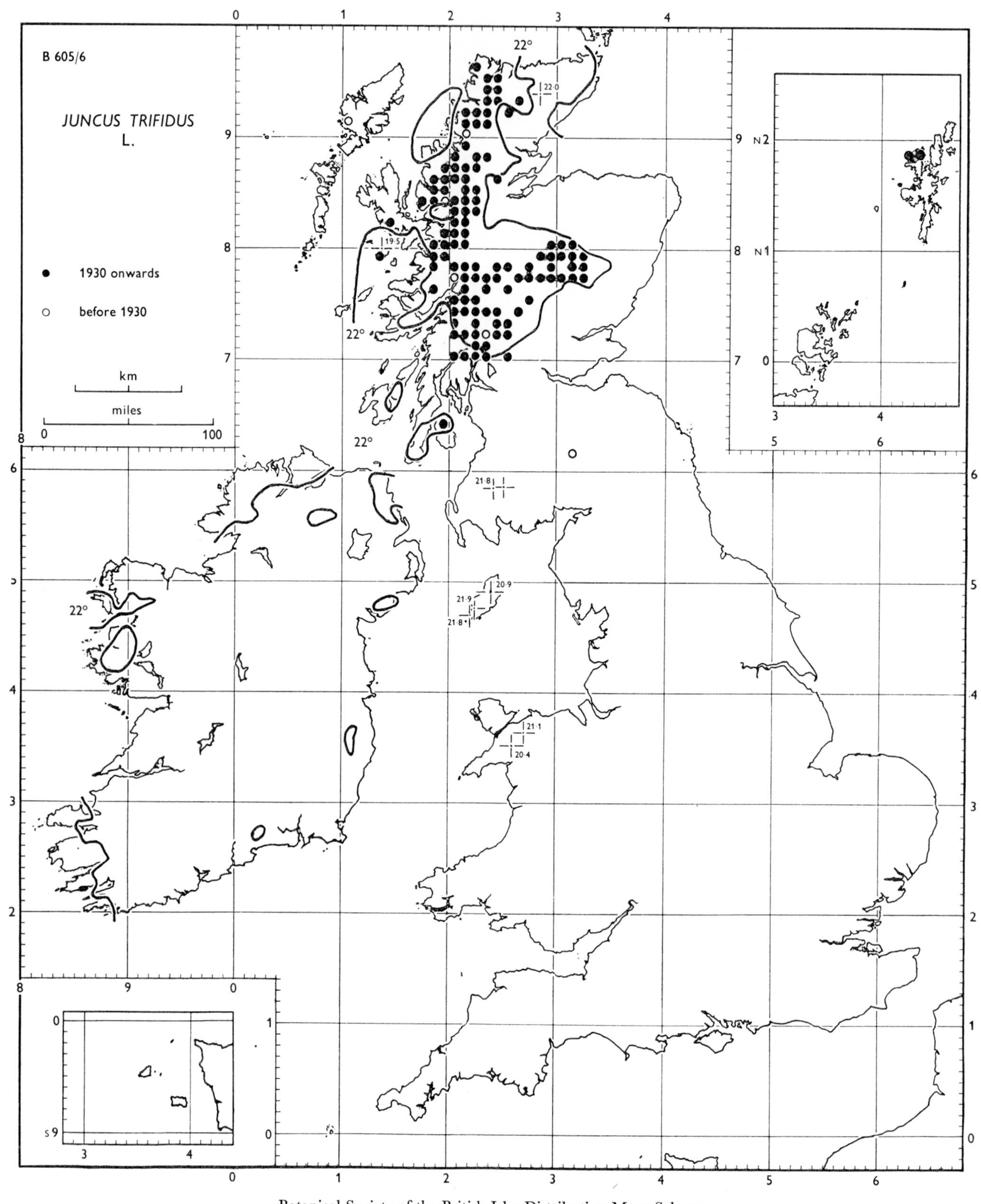


Botanical Society of the British Isles Distribution Maps Scheme

Figure 8. Map showing the present-day distribution of *Juncus trifidus* L. with the 22 °C maximum summer temperature summit isotherm.

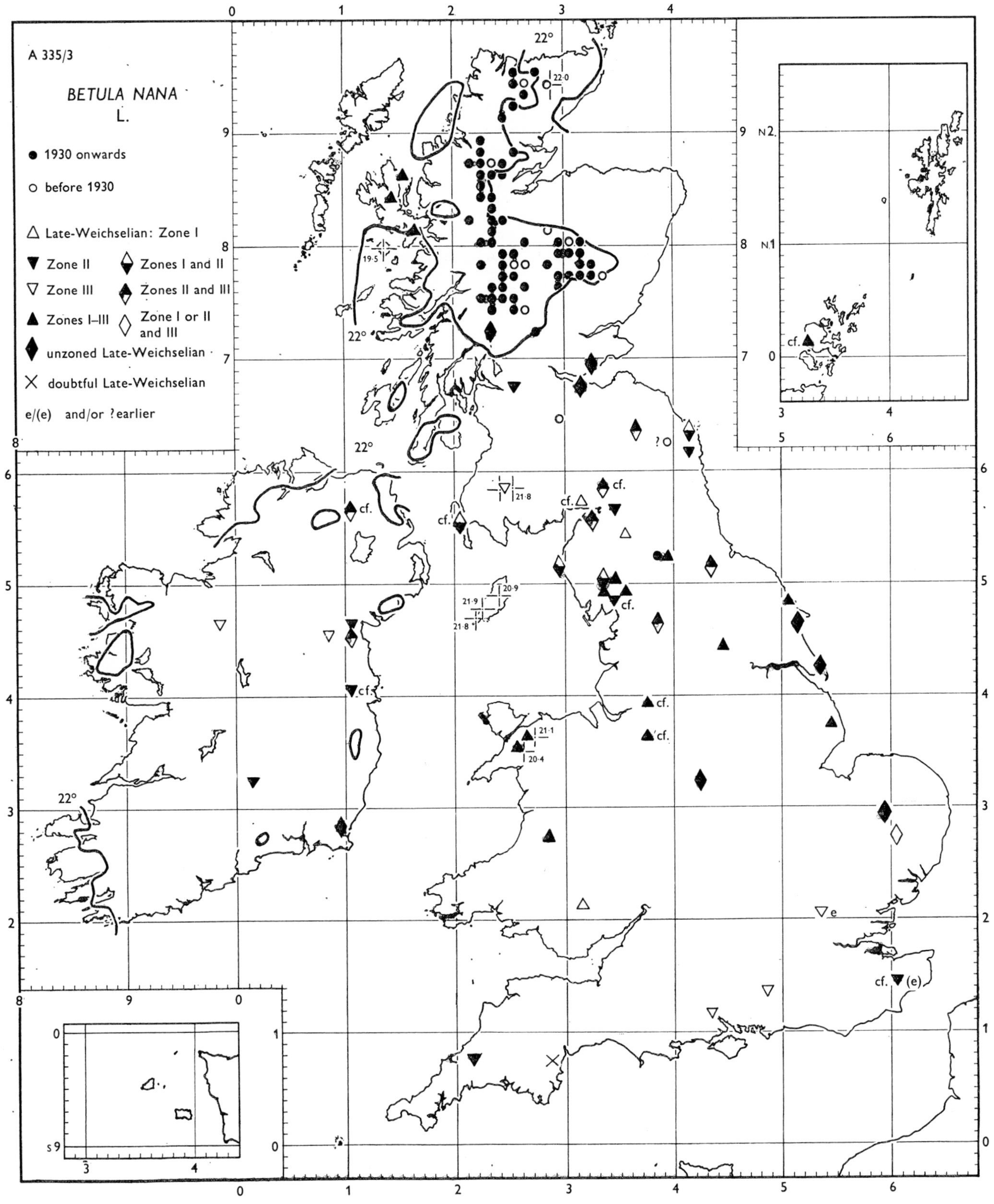


Botanical Society of the British Isles Distribution Maps Scheme

Figure 9. Map showing the present-day distribution of *Betula nana* L. with the 22 °C maximum summer temperature summit isotherm, and the location of Late-Weichselian fossil records.

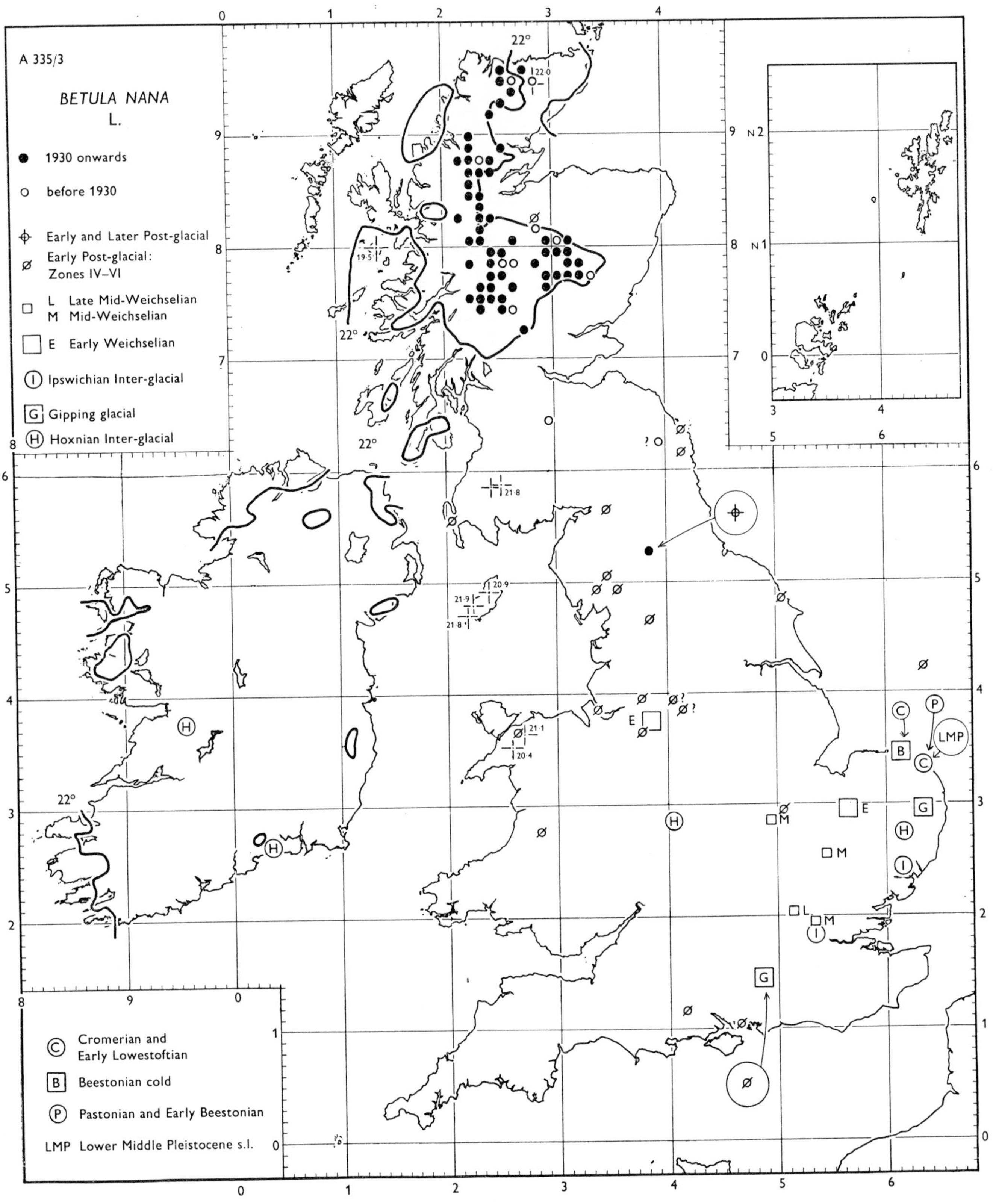


Botanical Society of the British Isles Distribution Maps Scheme

Figure 10. Map showing the present-day distribution of *Betula nana* L. with the 22 °C maximum summer temperature summit isotherm, and the location of fossil records from the Post-glacial, Weichselian and pre-Weichselian.

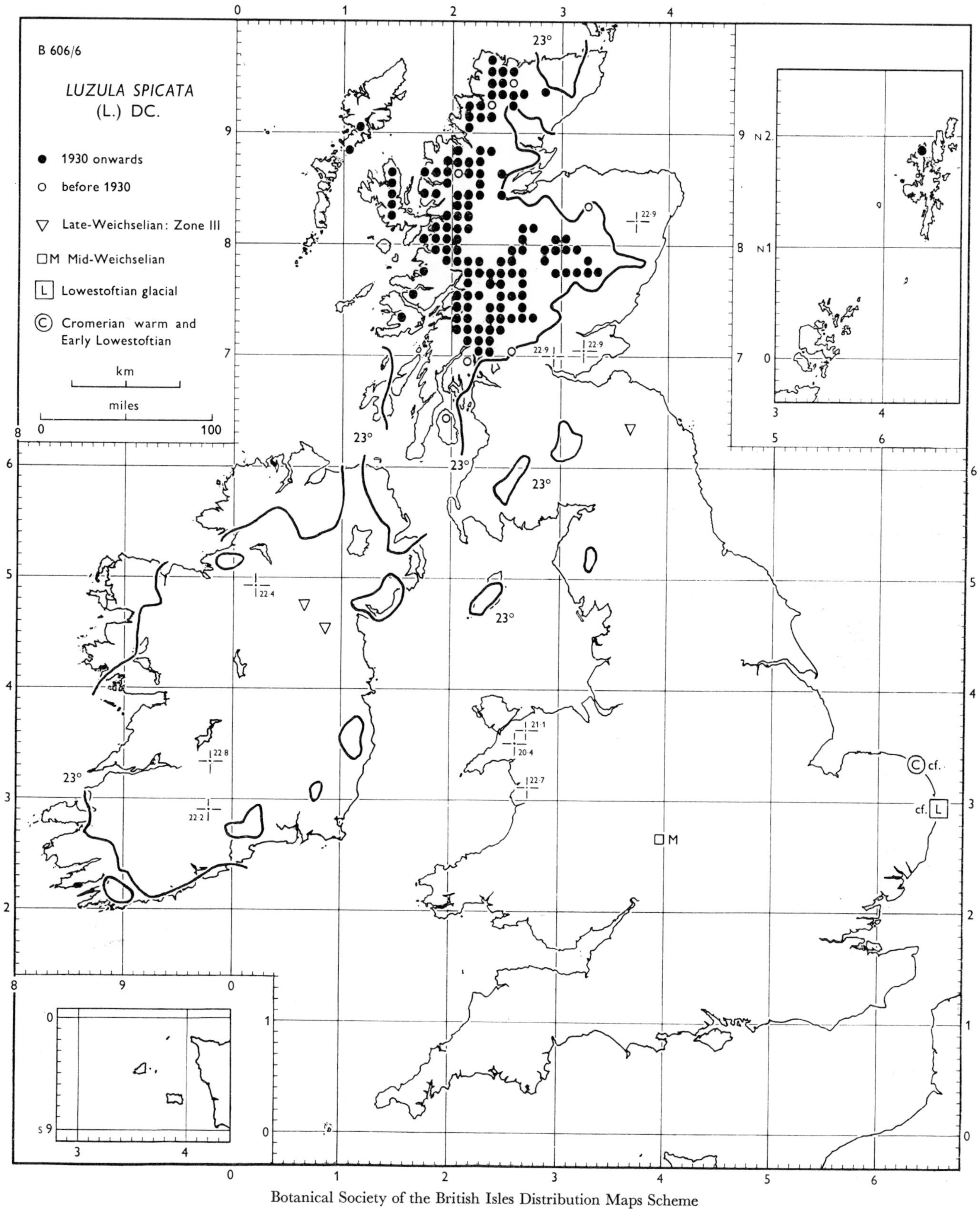


Botanical Society of the British Isles Distribution Maps Scheme

Figure 11. Map showing the present-day distribution of *Luzula spicata* (L.) DC. with the 23 °C maximum summer temperature summit isotherm and the location of fossil records from the Late-Weichselian, Weichselian and pre-Weichselian.

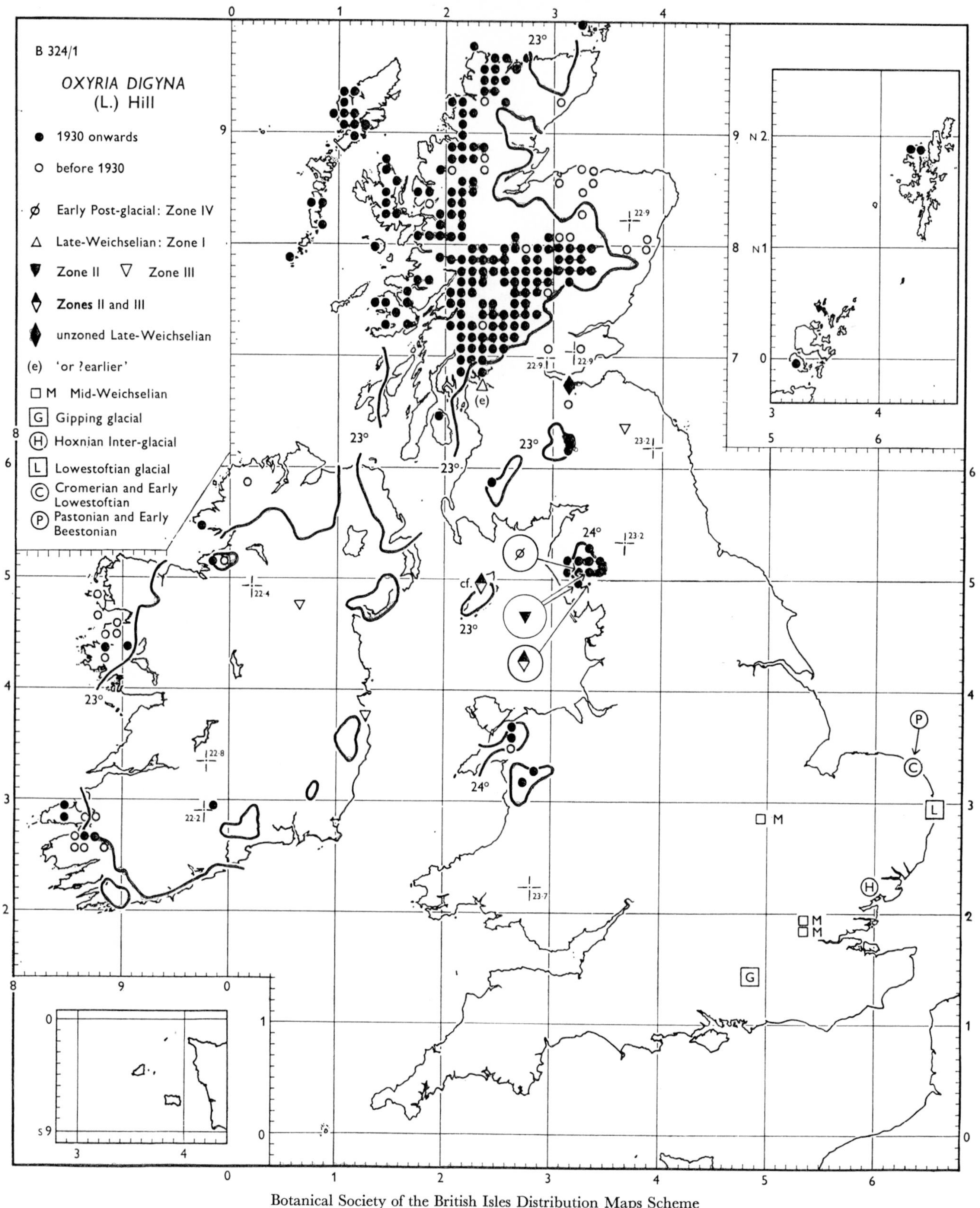


Figure 12. Map showing the present-day distribution of *Oxyria digyna* (L.) Hill with the 23 °C maximum summer temperature summit isotherm for Ireland, Highland Scotland, South Scotland and the Isle of Man, and the 24 °C isotherm for Wales and England. Also shown is the location of fossil records from the Post-glacial, Late-Weichselian, Weichselian and certain pre-Weichselian periods.

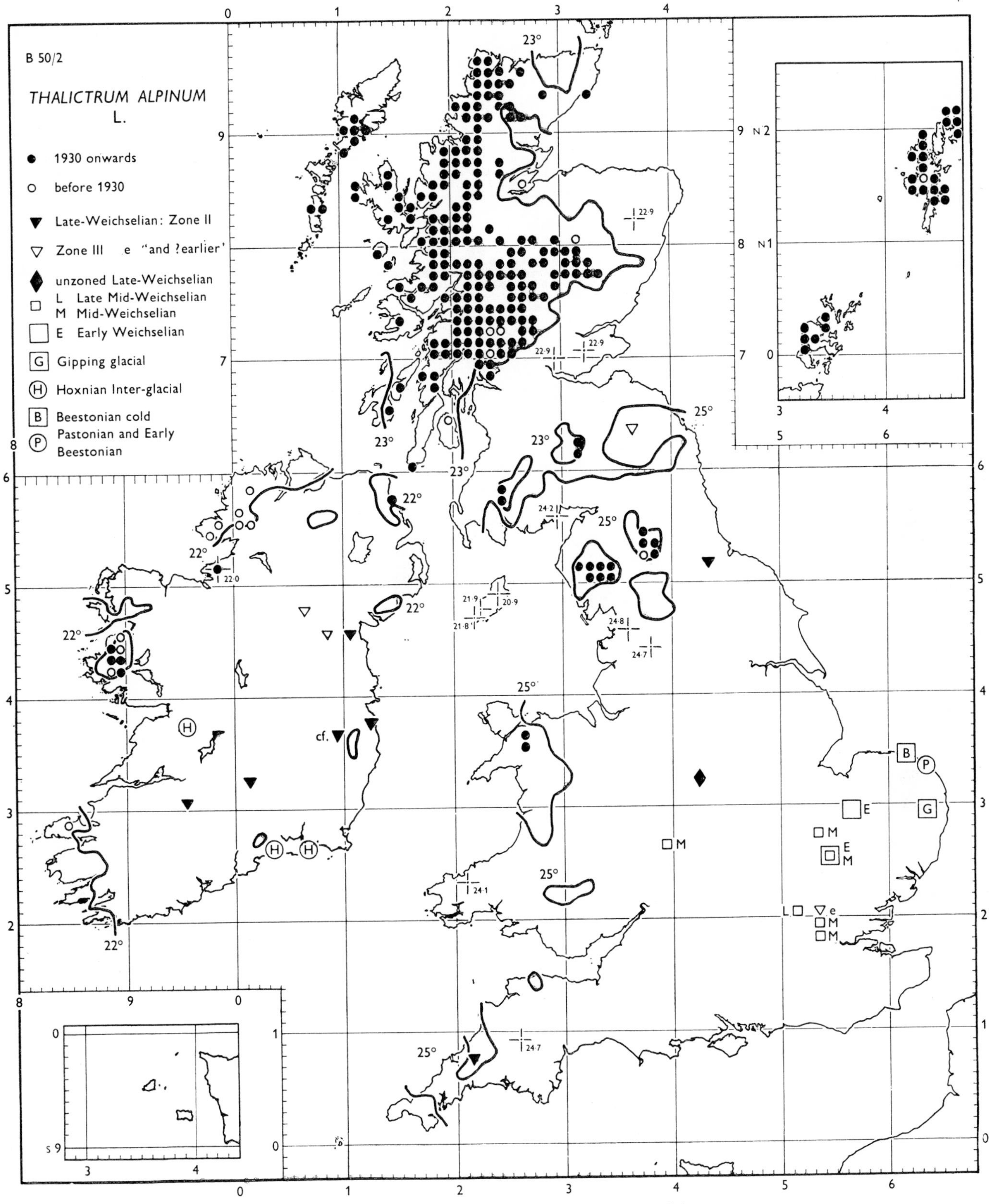


Figure 13. Map showing the present-day distribution of *Thalictrum alpinum* L. with the 23 °C maximum summer temperature summit isotherm for Highland Scotland and South Scotland, the 22 °C isotherm for Ireland and the Isle of Man, and the 25 °C isotherm for Wales, England and adjacent Scotland. Also shown is the location of fossil records from Late-Weichselian, Weichselian and certain pre-Weichselian periods.

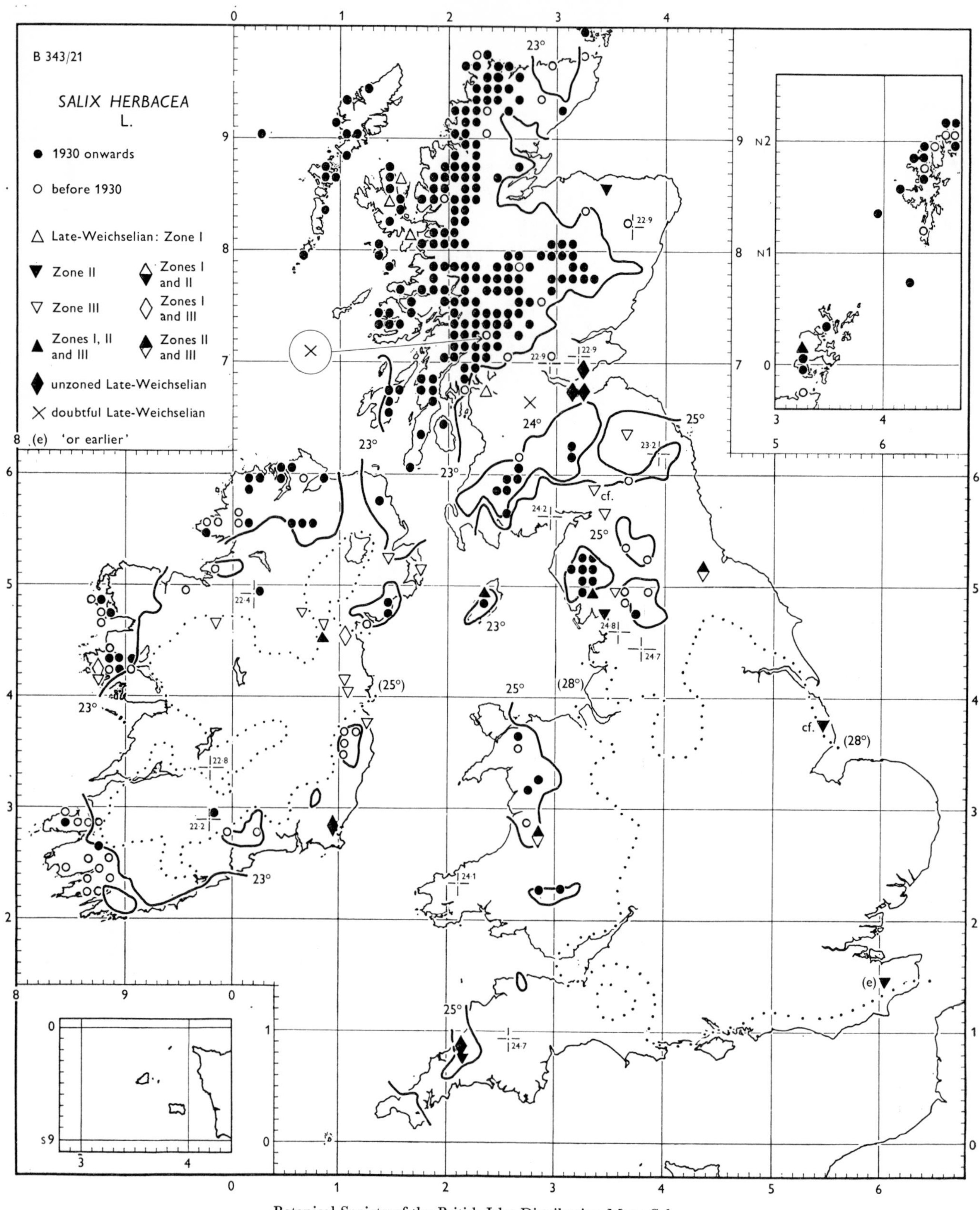


Botanical Society of the British Isles Distribution Maps Scheme

Figure 14. Map showing the present-day distribution of *Salix herbacea* L. with the 23 °C maximum summer temperature summit isotherm for Ireland, Highland Scotland and the Isle of Man, the 24 °C isotherm for South Scotland, and the 25 °C isotherm for Wales, England and adjacent Scotland. The location is also shown of fossil records from the Late-Weichselian and, for comparison, the 25 °C isotherm (dotted) in Ireland and the 28 °C isotherm (dotted) for England and Wales. Further explanation in text.

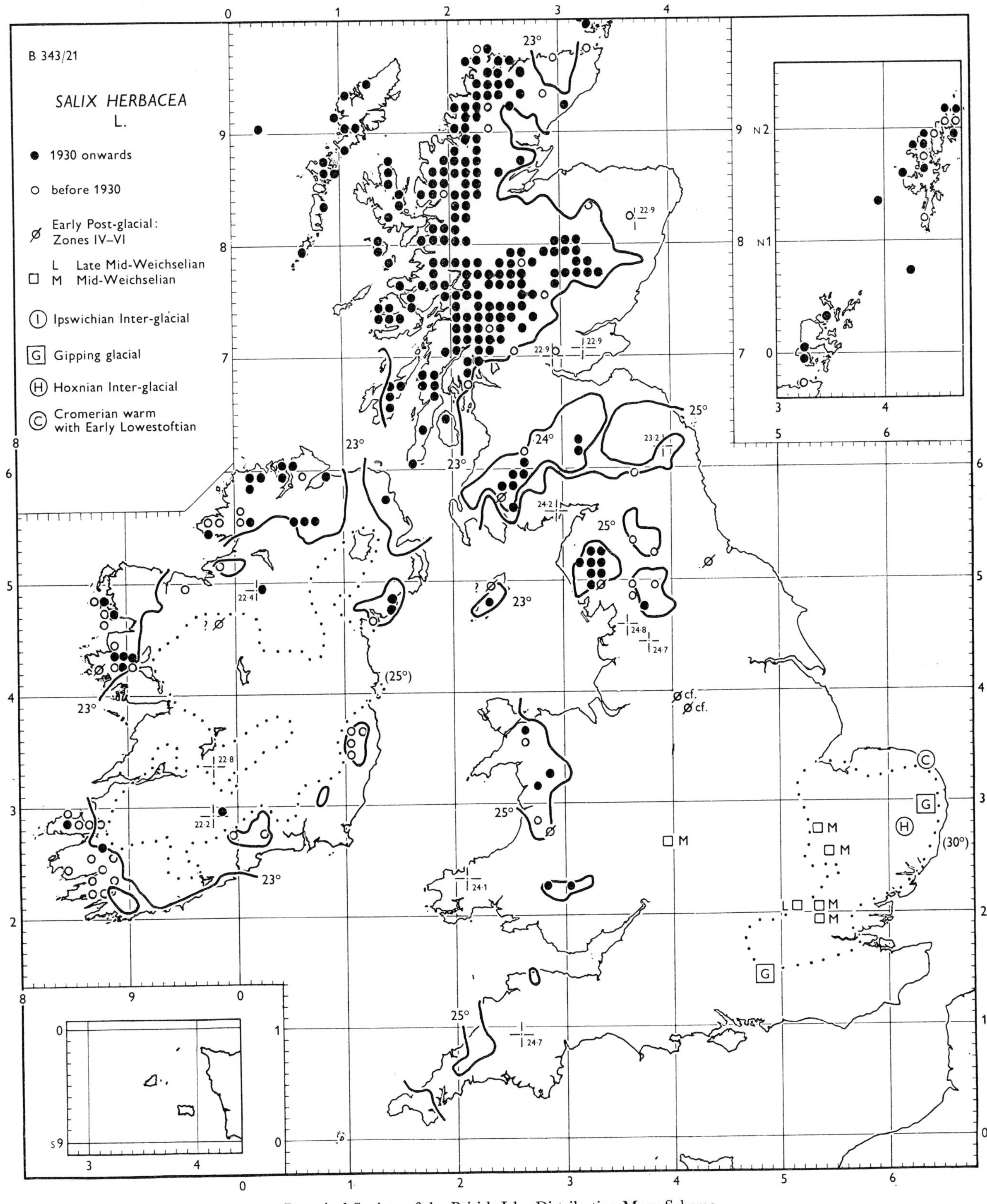


Figure 15. Map showing the present-day distribution of *Salix herbacea* L. with the 23 °C maximum summer temperature summit isotherm for Ireland, Highland Scotland and the Isle of Man, the 24 °C isotherm for southern Scotland, and the 25 °C isotherm for Wales, England and adjacent Scotland. Also shown is the location of fossil records from the Post-glacial and from the Weichselian and pre-Weichselian, and for comparison the 25 °C isotherm (dotted) in Ireland and the 30 °C isotherm (dotted) for England. Further explanation in text.

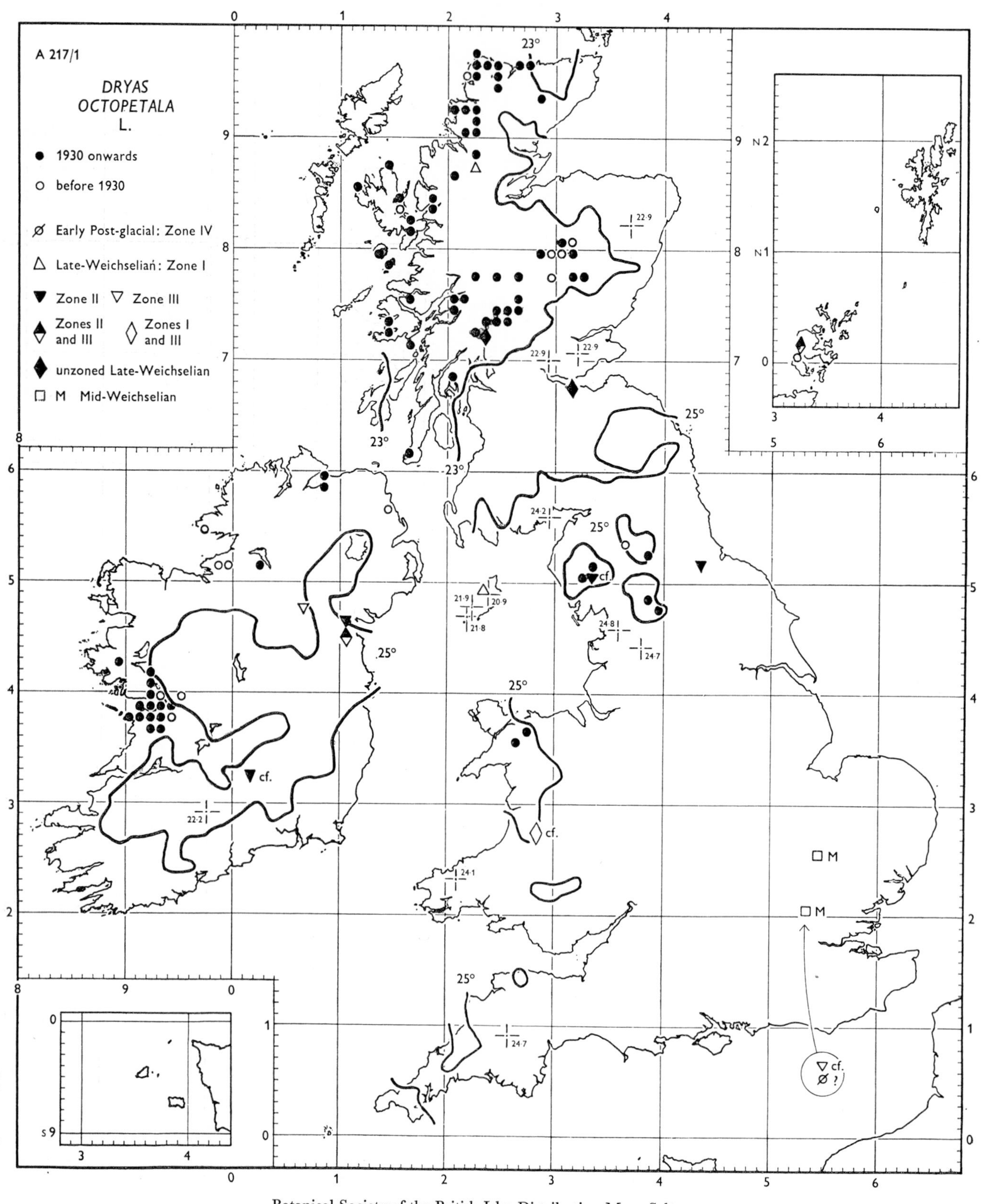


Botanical Society of the British Isles Distribution Maps Scheme

Figure 16. Map showing the present-day distribution of *Dryas octopetala* L. with the 23 °C maximum summer temperature summit isotherm for Highland Scotland and the 25 °C isotherm for Ireland, Wales, England and adjacent Scotland. Also shown is the location of fossil records from the Post-glacial, Late-Weichselian and Weichselian.

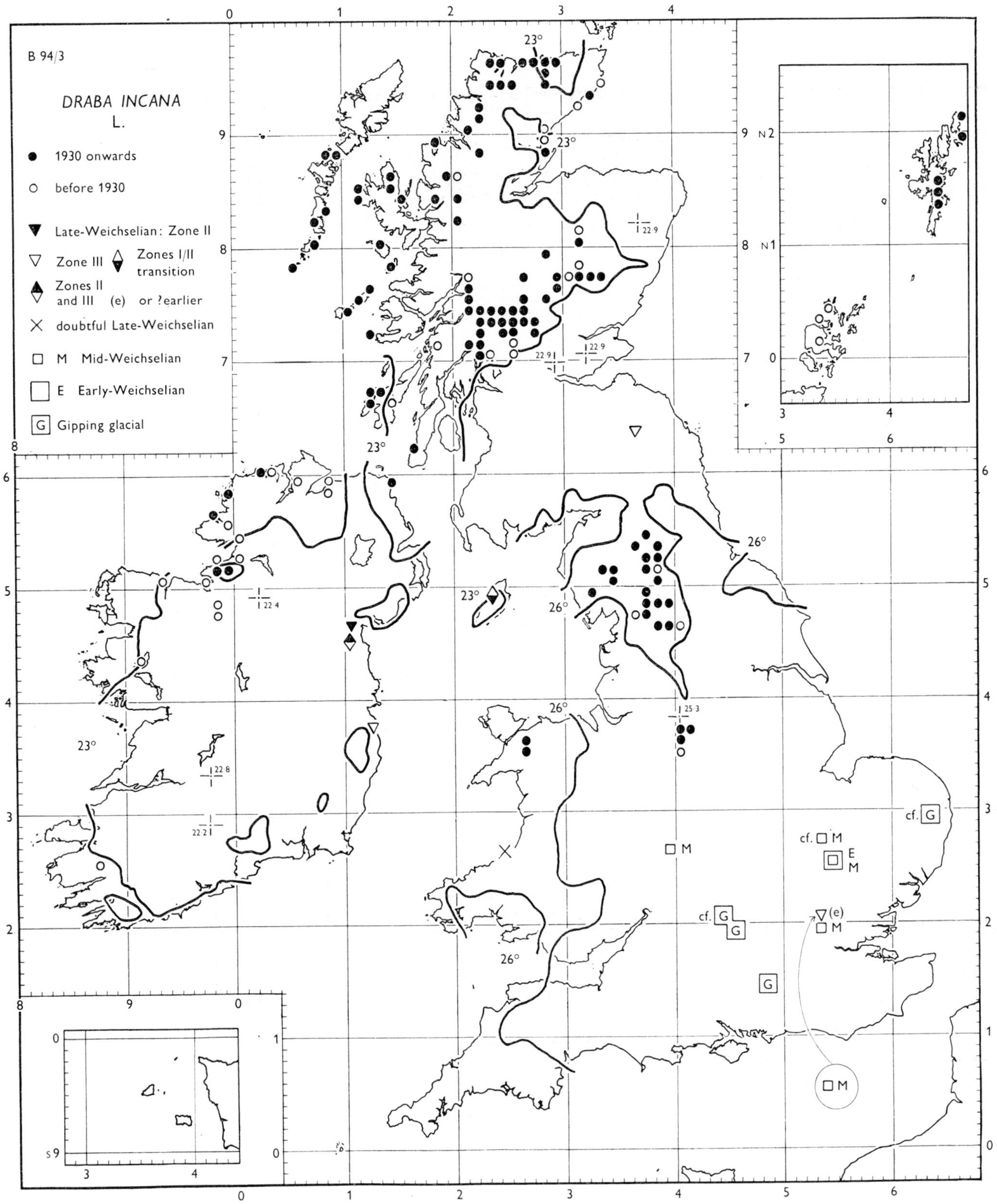


Botanical Society of the British Isles Distribution Maps Scheme

Figure 17. Map showing the present-day distribution of *Draba incana* L. with the 23 °C maximum summer temperature summit isotherm for Ireland, Highland Scotland and the Isle of Man and the 26 °C isotherm for Wales, England and adjacent Scotland. Also shown is the location of fossil records from the Late-Weichselian, Weichselian and Gipping.

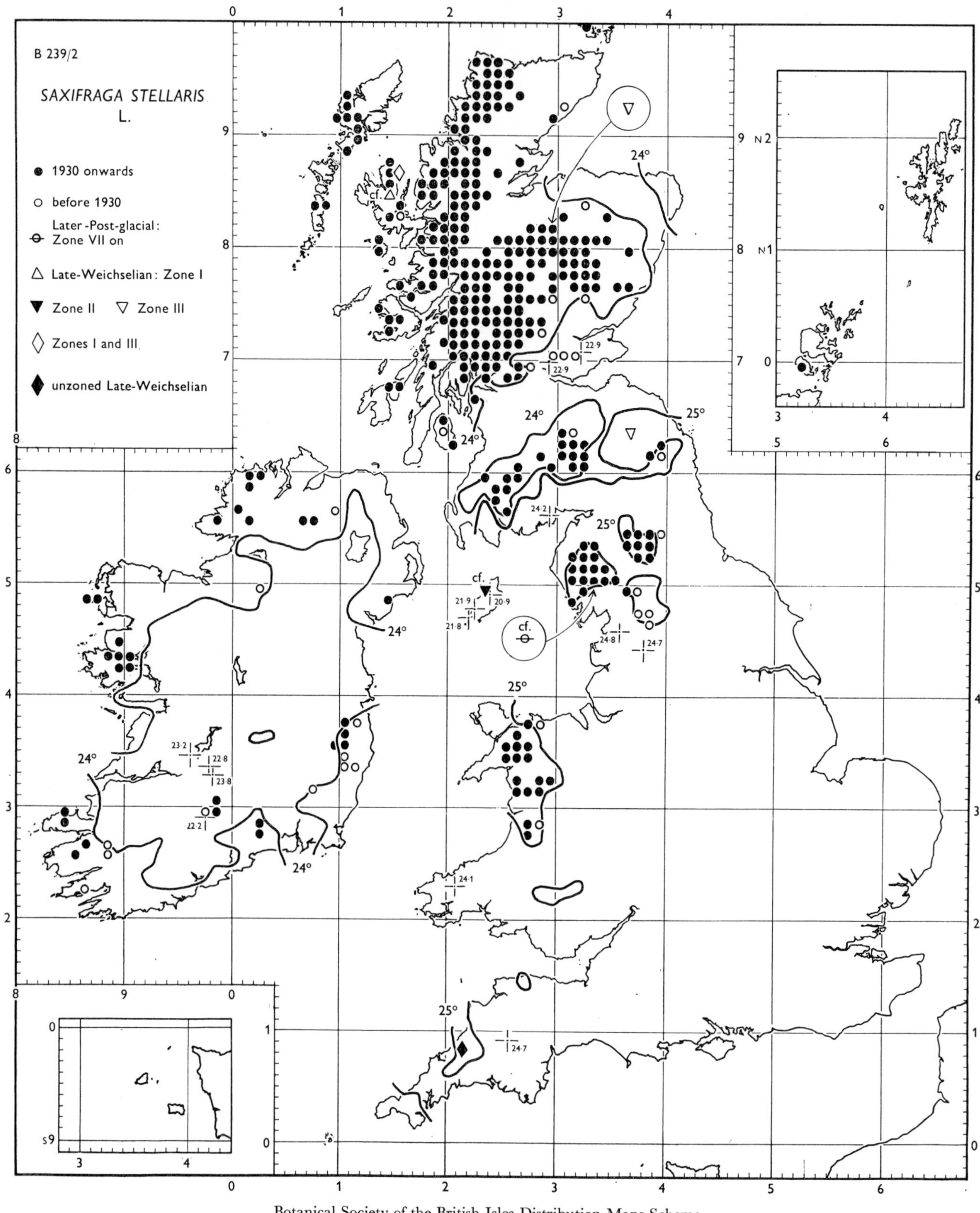


Figure 18. Map showing the present-day distribution of *Saxifraga stellaris* L. with the 24 °C maximum summer temperature summit isotherm for Ireland, Highland Scotland and South Scotland and the 25 °C isotherm for Wales, England and adjacent Scotland. Also shown is the location of fossil records from the Post-glacial and Late-Weichselian.

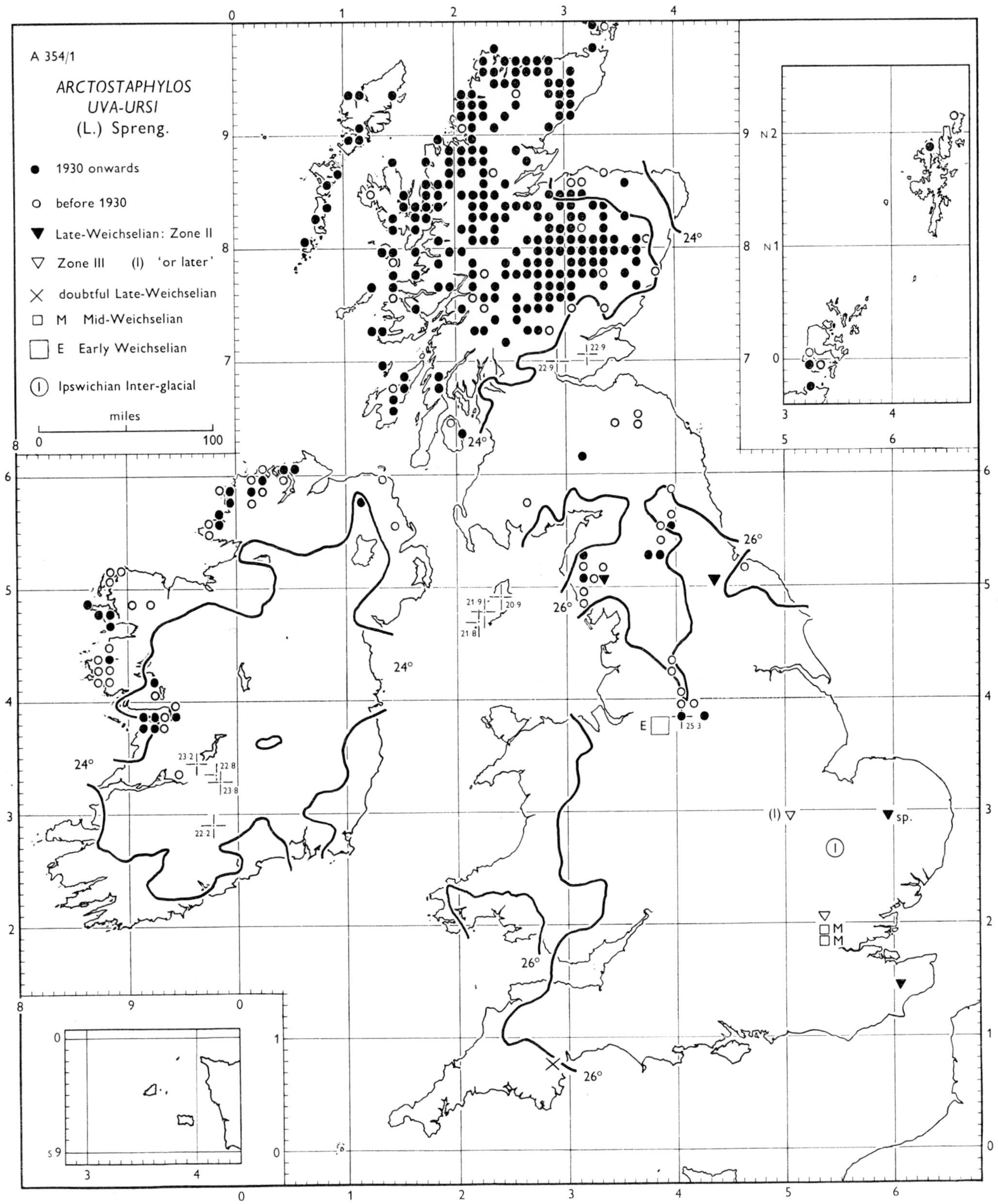


Botanical Society of the British Isles Distribution Maps Scheme

Figure 19. Map showing the present-day distribution of *Arctostaphylos uva-ursi* (L.) Spreng. with the 24 °C maximum summer temperature summit isotherm for Ireland and Highland Scotland and the 26 °C isotherm for Wales, England and adjacent Scotland. Also shown is the location of fossil records from the Late-Weichselian, Weichselian and Ipswichian. The record marked 'sp.' represents pollen determined as '*Arctostaphylos* sp.'

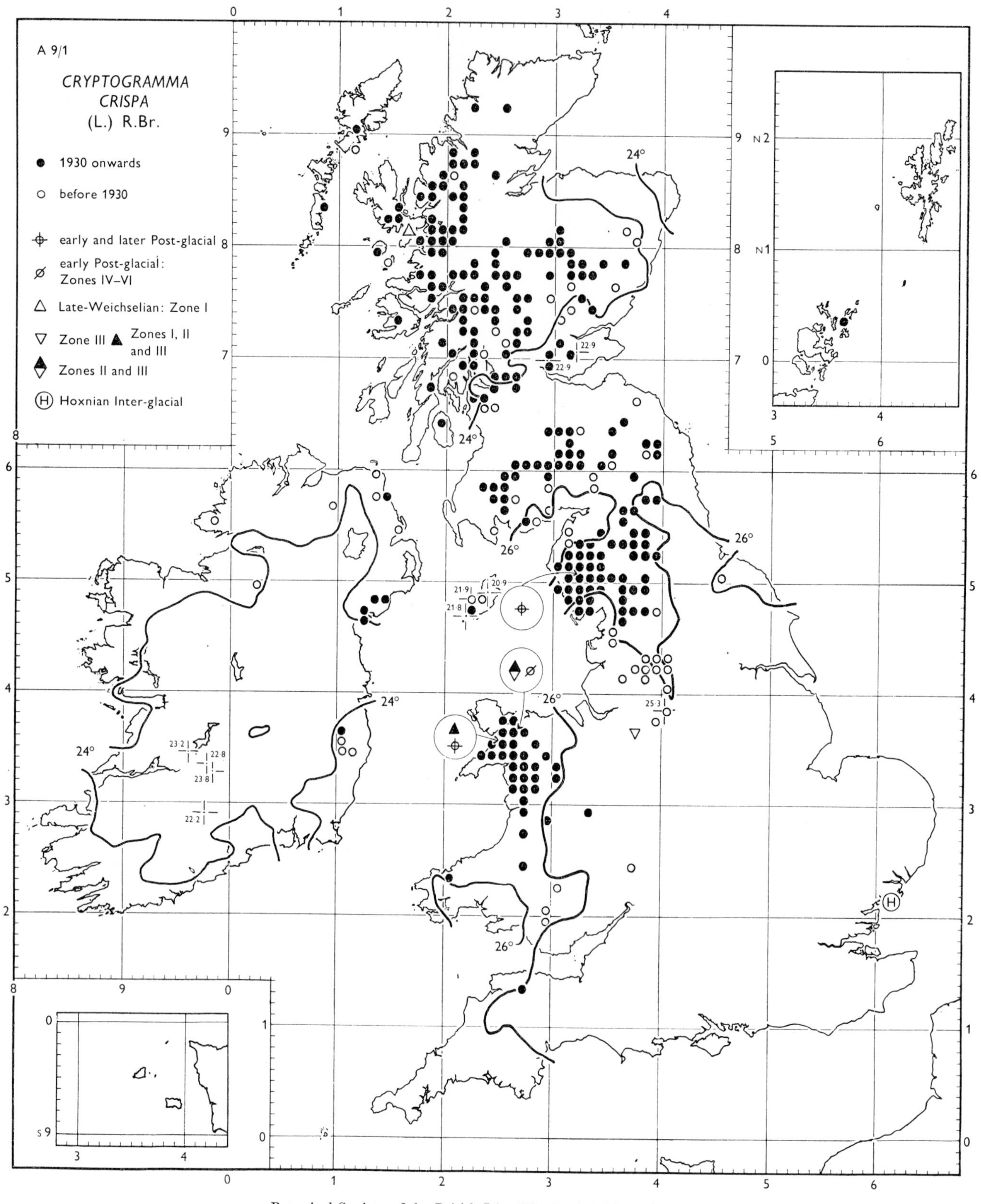


Botanical Society of the British Isles Distribution Maps Scheme

Figure 20. Map showing the present-day distribution of *Cryptogramma crispa* (L.) R.Br. with the 24 °C maximum summer temperature summit isotherm for Ireland and Highland Scotland and the 26 °C isotherm for Wales, England and adjacent Scotland. Also shown is the location of fossil records from the Post-glacial, Late-Weichselian and Hoxnian.

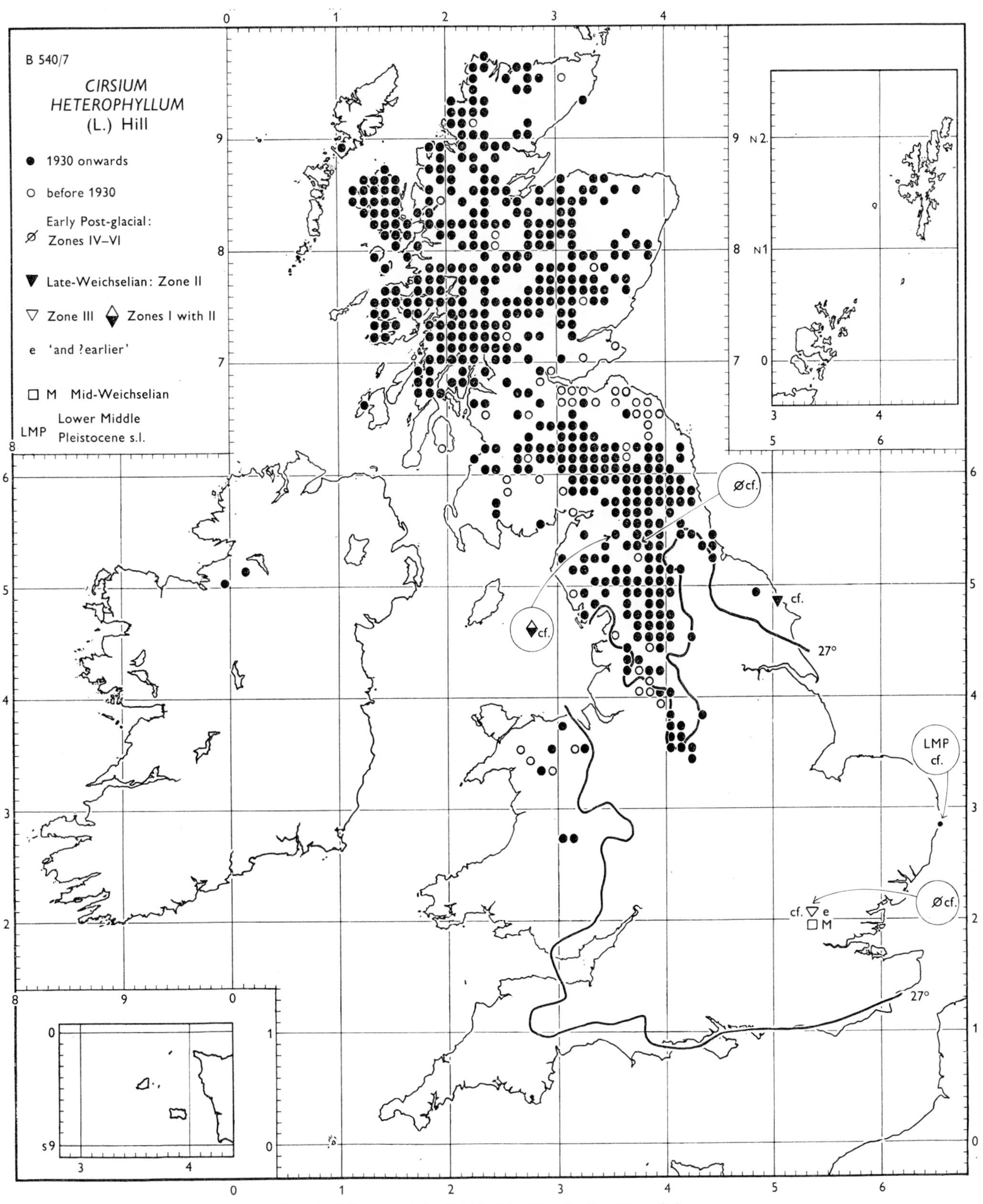


Botanical Society of the British Isles Distribution Maps Scheme

Figure 21. Map showing the present-day distribution of *Cirsium heterophyllum* (L.) Hill (known introductions omitted) with the 27 °C maximum summer temperature summit isotherm, and the location of fossil records from the Post-glacial, Late-Weichselian, Weichselian and Lower Middle Pleistocene. Most of these records are given as 'cf.' or as 'possibly *C. palustre*'.

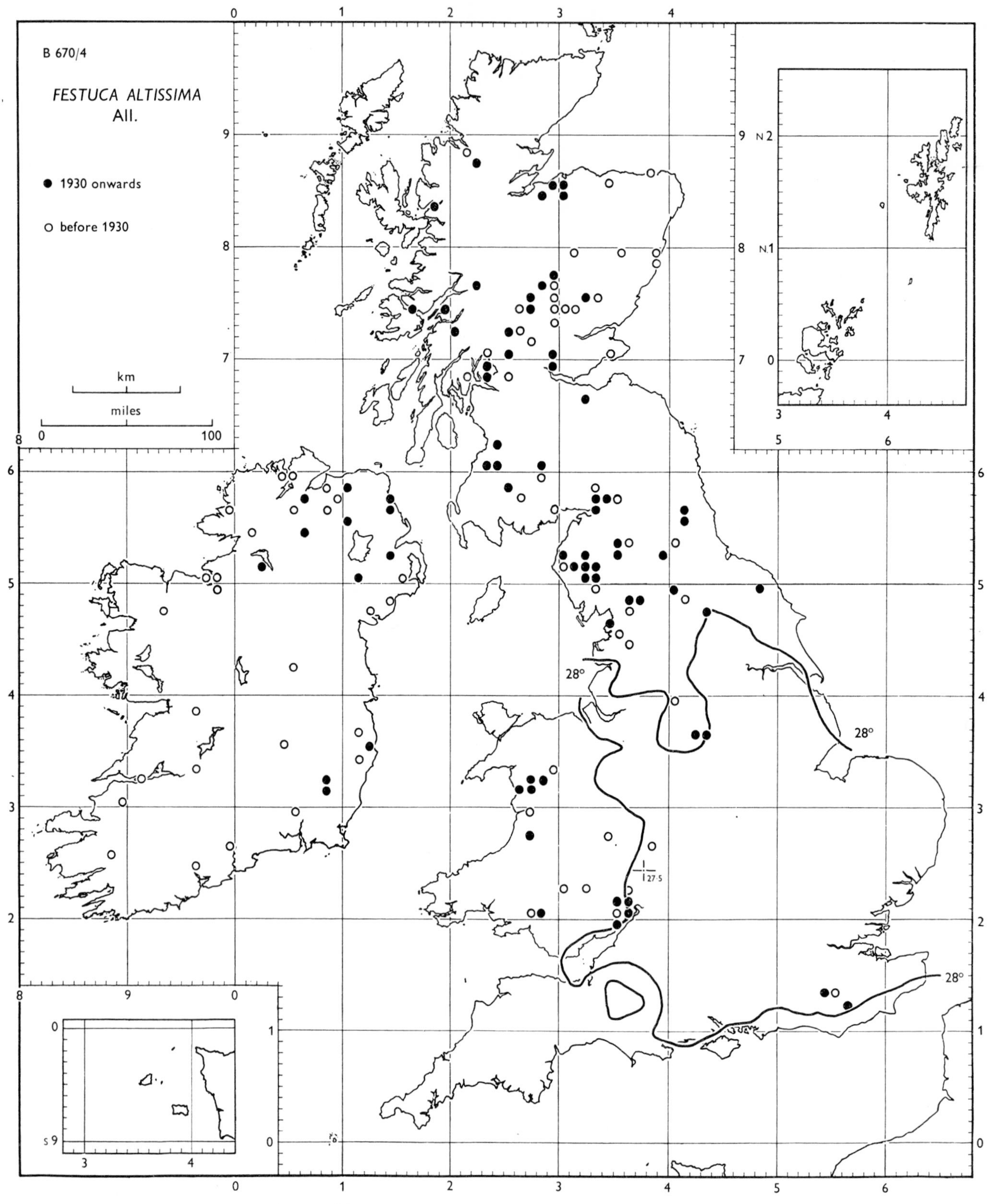


Botanical Society of the British Isles Distribution Maps Scheme

Figure 22. Map showing the present-day distribution of *Festuca altissima* All. with the 28 °C maximum summer temperature summit isotherm.

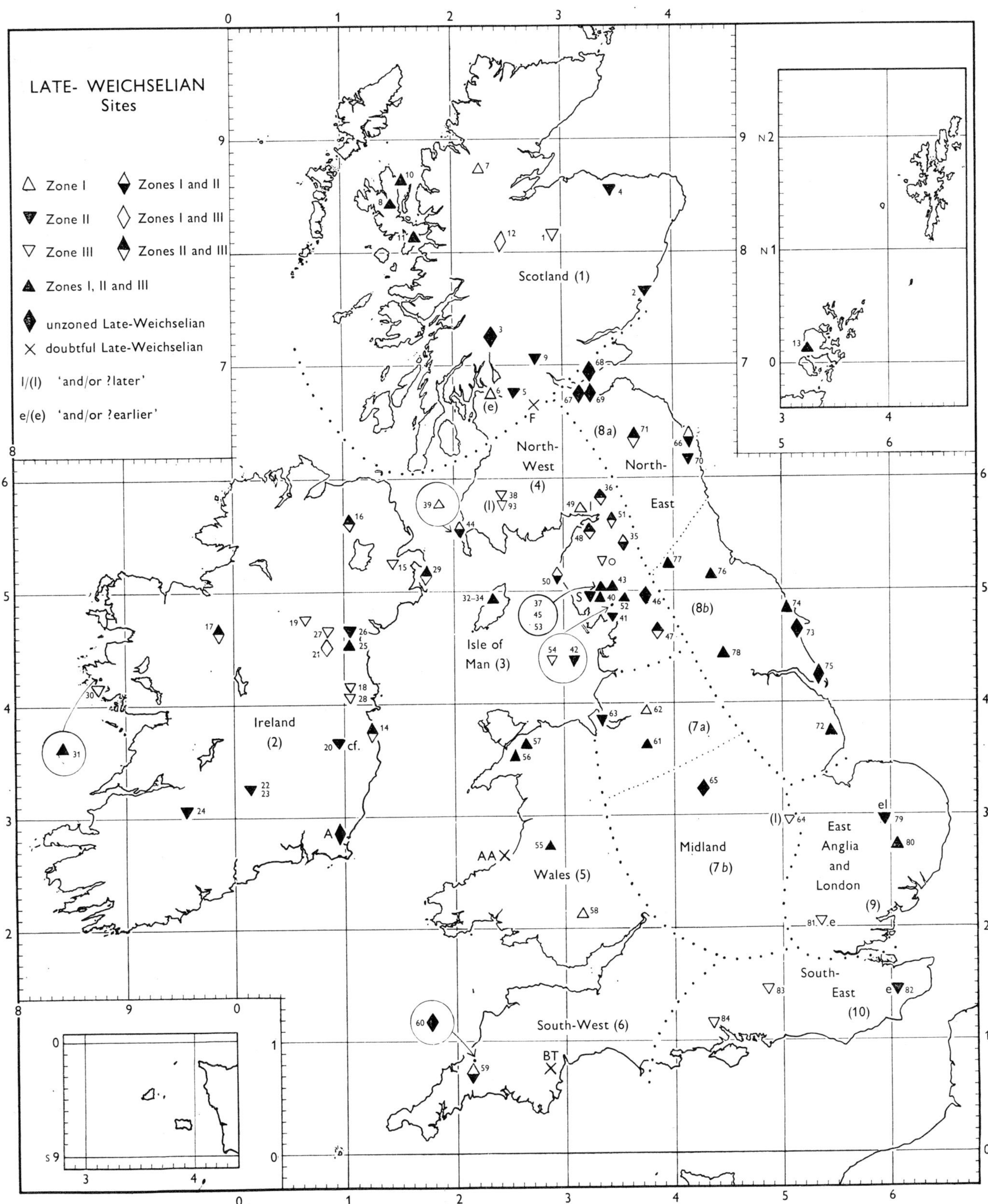


Figure 23. Map showing location of *Late-Weichselian* sites for which fossils of the arctic-montane species are recorded and the regional and sub-regional boundaries. Sites are numbered consecutively and are alphabetical within each region; they are located to the nearest 10 km National Grid reference. An 'e' or an 'l' alongside a symbol signifies *either* that there is a record from a period somewhat earlier: 'e', or later: 'l', *in addition* to that indicated by the symbol, *or* that the dating is possibly somewhat older: '(e)', or younger: '(l)' than that shown by the symbol.

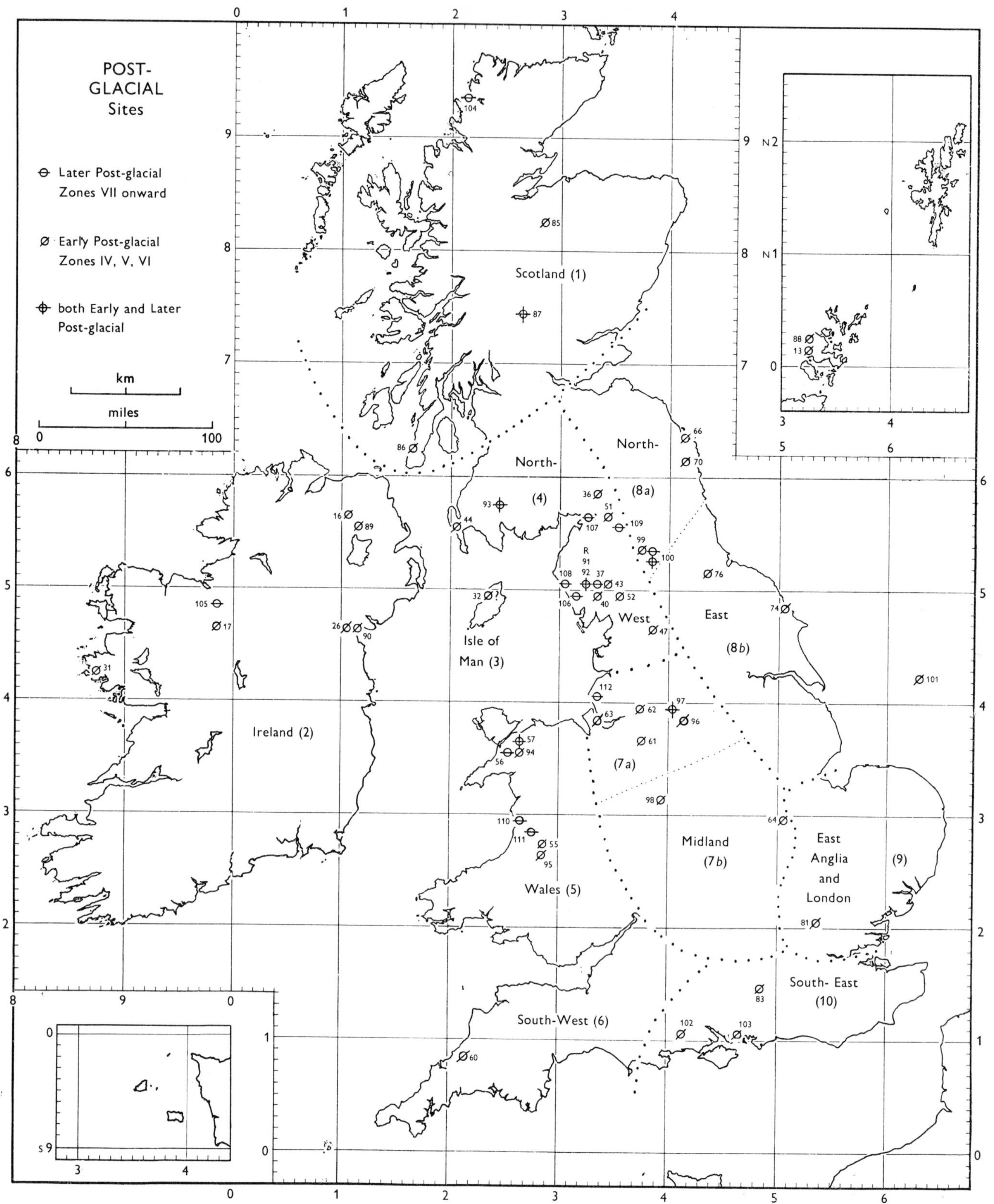


Figure 24. Map showing location of *Post-glacial* sites for which fossils of the arctic-montane species are recorded and the regional and sub-regional boundaries. Sites are numbered consecutively with the Late-Weichselian, the number retained where sites recur; they are alphabetical within each region. 'e', 'l' and (e), (l) as in figure 23.

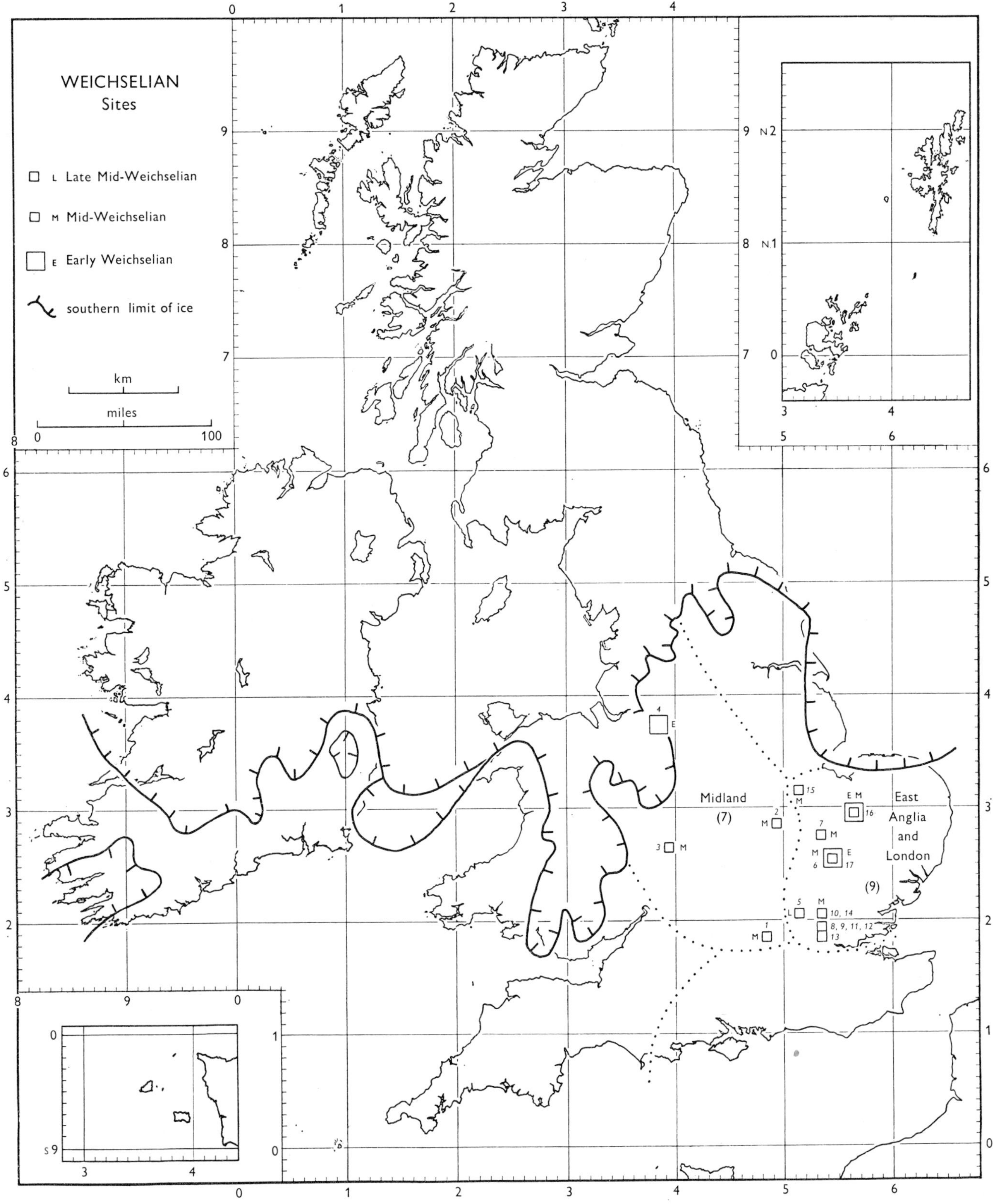


Figure 25. Map showing location of *Weichselian* sites for which fossils of the arctic-montane species are recorded, the regional boundaries and the southern limit of ice after Mitchell (1960). Sites are numbered (in italic) consecutively and are alphabetical within each region.

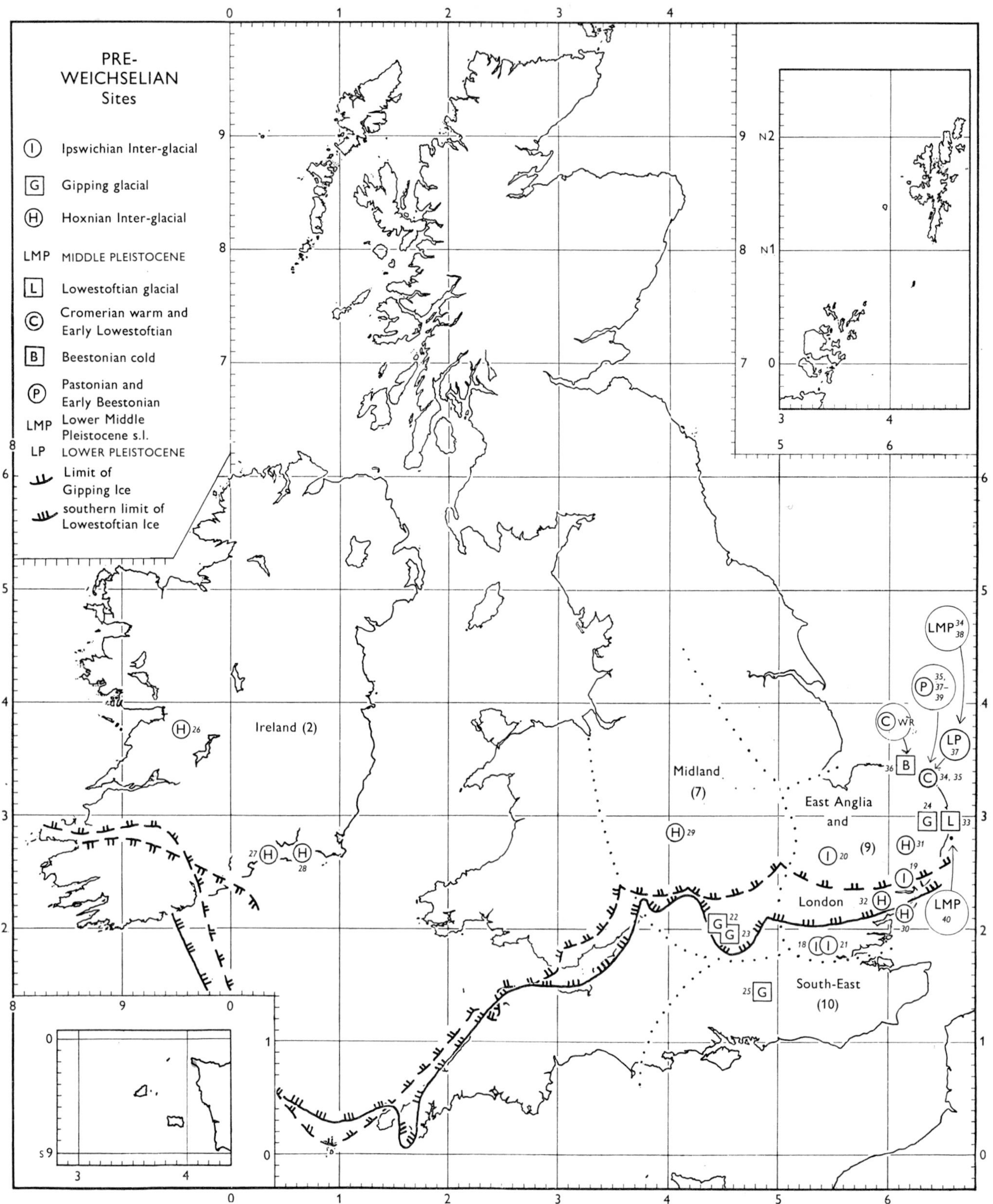


Figure 26. Map showing location of *pre-Weichselian* sites for which fossils of the arctic-montane species are recorded, the regional boundaries and the southern limit of ice for the Gipping and the Lowestoftian periods after Mitchell (1960) and his later amendments (1965, personal communication, 1967; Mitchell & Orme, 1967). Sites are numbered (in italic) consecutively with the Weichselian, the number retained where sites recur; they are alphabetical within each region.

## FIGURES 27–32

Diagrams to show the reduction in maximum summer temperature implied by the fossil occurrence of arctic-montane species at different periods from sites in each geographical region. The species are grouped according to their limiting isotherm values (table 1, p. 215 of this paper), those correlated with the lowest temperatures shown first, using the Scottish correlations up to 25 °C and the English thereafter, with the exception of *Arctostaphylos uva-ursi* which is grouped under its English value. Species are alphabetical within each temperature group. The Scandinavian limiting isotherms are also given, for comparison, and to indicate instances with differences between calculations based on Scottish and on Scandinavian values; where these differ from table 1 they are figures suggested by Dahl (pers. comm.) or obtained from Hultén (1950) and Dahl (1951) to signify the 'warmest' areas in which the species grows in Scandinavia; they are placed in brackets as they do not indicate a correlation with that isotherm. The calculated *regional isotherm equivalents* (Reg. iso. equ.) are shown in a separate column. Horizontal bars and symbols show, according to length, the temperature depression, in °C, implied by the occurrence of each species at any given site. Each site entry is shown by a symbol representing one occurrence.

## Key to Symbols

*Record of unqualified identification*

Length measures implied temperature depression.

Temperature indication *nil.*

Age uncertain[1]; length measures implied temperature depression.

Age uncertain[1]; temperature indication *nil.*

Cross-bar marks reduction in length if Scandinavian data were to be used for calculations.

Dotted extension indicates length if Scandinavian data were to be used for calculations.

*Record of dubious or approximate identification*

Length measures implied temperature depression.

Temperature indication *nil.*

Age uncertain[1]; length measures implied temperature depression.

Age uncertain[1]; temperature indication *nil.*

Dotted extension indicates length for the more exacting of two alternative identifications.

Cross-bar marks reduction in length for the more exacting of two alternative identifications if Scandinavian data were to be used for the calculations.

Reg. iso. equ. = Regional isotherm equivalent
(figures in brackets = species present in region, but no value shown in table 1).
* = Species less reliable as indicators of mean annual maximum temperature (see text).

M = Macro identification.
P = Pollen identification.
S = Spore identification.
Temperatures are in degrees centigrade (°C). Scale in °C.

[1] Entries of uncertain date are repeated through the likely time periods.

## Notes to figure 27

1. Or earlier.
2. Lewis' records (1911) have been excluded owing to doubt as to age.
3. ● *S. oppositifolia* s.s. ○ pollen-type not always distinguishable from *S. aizoides*.
4. Zone III/IV transition.
5. *S. hypnoides* pollen-type.
6. *S. hypnoides* agg. including *S. rosacea*. M.
7. *S. oppositifolia* pollen-type includes *S. aizoides*.
8. Would be reduced to *nil* temperature depression on basis of central European distribution.
9. Or *Rumex crispus* (pollen).
10. 'Bistorta'-type pollen including *P. viviparum*.
11. Pollen-type
12. Doubtful Late-Weichselian.
13. *S. hypnoides* agg. M.
14. End of Zone III or Zone IV.
15. Lewis' (1904) records of *S. reticulata*, *S. herbacea*, *S. lapponum*, *S. myrsinites* and *S. arbuscula* from Cross Fell have been excluded owing to doubt as to age.
16. Or *S. cernua*.
17. *S. hypnoides* ss. M.
18. *Arctostaphylos* sp.
19. & ? before Zone III.
20. Or ? before Zone III.

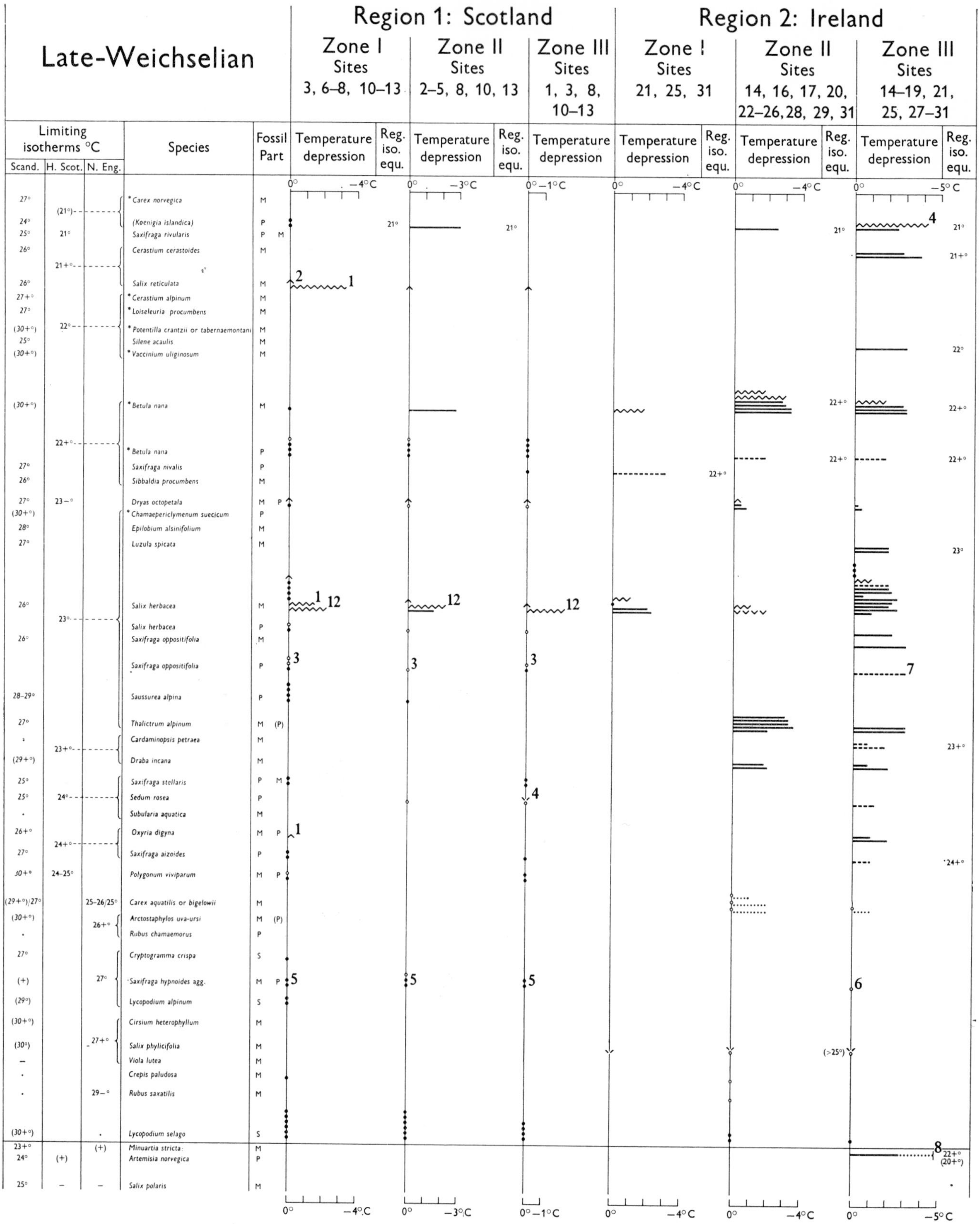

## Late-Weichselian

| Region 1: Scotland | | | Region 2: Ireland | | |
|---|---|---|---|---|---|
| Zone I<br>Sites<br>3, 6–8, 10–13 | Zone II<br>Sites<br>2–5, 8, 10, 13 | Zone III<br>Sites<br>1, 3, 8, 10–13 | Zone I<br>Sites<br>21, 25, 31 | Zone II<br>Sites<br>14, 16, 17, 20, 22–26, 28, 29, 31 | Zone III<br>Sites<br>14–19, 21, 25, 27–31 |
| Temperature depression 0° to −4°C | Temperature depression 0° to −3°C | Temperature depression 0° to −1°C | Temperature depression 0° to −4°C | Temperature depression 0° to −4°C | Temperature depression 0° to −5°C |

| Limiting isotherms °C: Scand. | H. Scot. | N. Eng. | Species | Fossil Part | Scotland Zone I Reg. iso. equ. | Scotland Zone II Reg. iso. equ. | Ireland Zone I Reg. iso. equ. | Ireland Zone II Reg. iso. equ. | Ireland Zone III Reg. iso. equ. |
|---|---|---|---|---|---|---|---|---|---|
| 27° | | | *Carex norvegica | M | | | | | |
| | (21°) | | | | | | | | |
| 24° | | | (Koenigia islandica) | P | | | | | |
| 25° | 21° | | Saxifraga rivularis | P M | 21° | 21° | | 21° | 21° |
| 26° | | | Cerastium cerastoides | M | | | | | 21+° |
| | 21+° | | | | | | | | |
| 26° | | | Salix reticulata | M | | | | | |
| 27+° | | | *Cerastium alpinum | M | | | | | |
| 27° | | | *Loiseleuria procumbens | M | | | | | |
| (30+°) | 22° | | *Potentilla crantzii or tabernaemontani | M | | | | | |
| 25° | | | Silene acaulis | M | | | | | |
| (30+°) | | | *Vaccinium uliginosum | M | | | | | 22° |
| (30+°) | | | *Betula nana | M | | | | 22+° | 22+° |
| | 22+° | | | | | | | | |
| | | | *Betula nana | P | | | | 22+° | 22+° |
| 27° | | | Saxifraga nivalis | P | | | 22+° | | |
| 26° | | | Sibbaldia procumbens | M | | | | | |
| 27° | 23–° | | Dryas octopetala | M P | | | | | |
| (30+°) | | | *Chamaepericlymenum suecicum | P | | | | | |
| 28° | | | Epilobium alsinifolium | M | | | | | |
| 27° | | | Luzula spicata | M | | | | | 23° |
| 26° | | | Salix herbacea | M | | | | | |
| | 23° | | | | | | | | |
| | | | Salix herbacea | P | | | | | |
| 26° | | | Saxifraga oppositifolia | M | | | | | |
| | | | Saxifraga oppositifolia | P | | | | | |
| 28–29° | | | Saussurea alpina | P | | | | | |
| 27° | | | Thalictrum alpinum | M (P) | | | | | |
| · | | | Cardaminopsis petraea | M | | | | | |
| | 23+° | | | | | | | | 23+° |
| (29+°) | | | Draba incana | M | | | | | |
| 25° | | | Saxifraga stellaris | P M | | | | | |
| 25° | 24° | | Sedum rosea | P | | | | | |
| · | | | Subularia aquatica | M | | | | | |
| 26+° | | | Oxyria digyna | M P | | | | | |
| | 24+° | | | | | | | | |
| 27° | | | Saxifraga aizoides | P | | | | | 24+° |
| 30+° | 24–25° | | Polygonum viviparum | M P | | | | | |
| (29+°)/27° | | 25–26/25° | Carex aquatilis or bigelowii | M | | | | | |
| (30+°) | | | Arctostaphylos uva-ursi | M (P) | | | | | |
| | | 26+° | | | | | | | |
| · | | | Rubus chamaemorus | P | | | | | |
| 27° | | | Cryptogramma crispa | S | | | | | |
| (+) | | 27° | Saxifraga hypnoides agg. | M P | | | | | |
| (29°) | | | Lycopodium alpinum | S | | | | | |
| (30+°) | | | Cirsium heterophyllum | M | | | | | |
| | | 27+° | | | | | | | |
| (30°) | | | Salix phylicifolia | M | | | | | (>25°) |
| — | | | Viola lutea | M | | | | | |
| · | | | Crepis paludosa | M | | | | | |
| · | | 29–° | Rubus saxatilis | M | | | | | |
| (30+°) | | · | Lycopodium selago | S | | | | | |
| 23+° | | (+) | Minuartia stricta | M | | | | | 22+° |
| 24° | (+) | | Artemisia norvegica | P | | | | | (20+°) |
| 25° | — | — | Salix polaris | M | | | | | |

Figure 27

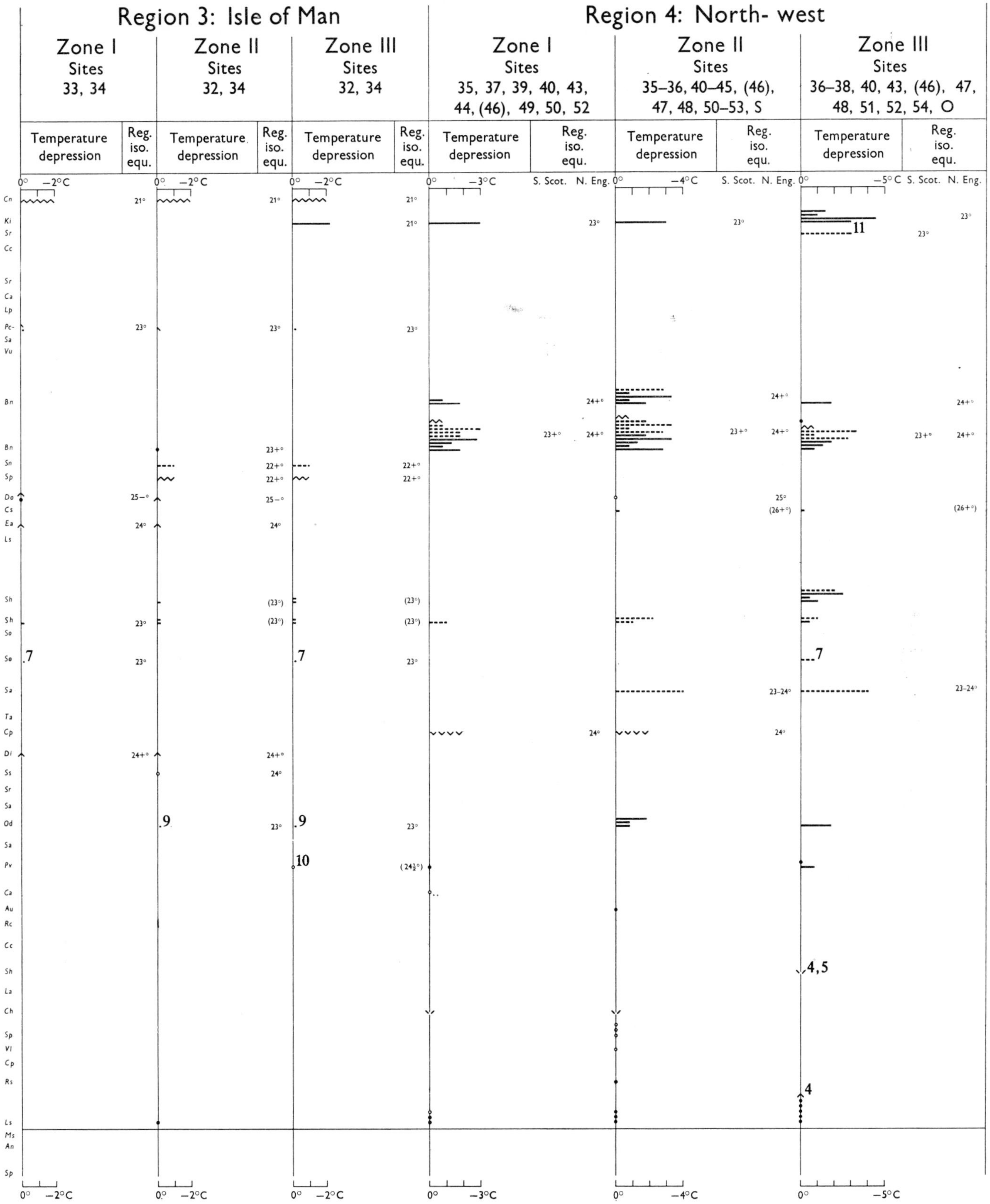

| Region 3: Isle of Man | | | | | | Region 4: North-west | | | | | |
|---|---|---|---|---|---|---|---|---|---|---|---|
| Zone I Sites 33, 34 | | Zone II Sites 32, 34 | | Zone III Sites 32, 34 | | Zone I Sites 35, 37, 39, 40, 43, 44, (46), 49, 50, 52 | | Zone II Sites 35–36, 40–45, (46), 47, 48, 50–53, S | | Zone III Sites 36–38, 40, 43, (46), 47, 48, 51, 52, 54, O | |
| Temperature depression | Reg. iso. equ. | Temperature depression | Reg. iso. equ. | Temperature depression | Reg. iso. equ. | Temperature depression | Reg. iso. equ. | Temperature depression | Reg. iso. equ. | Temperature depression | Reg. iso. equ. |
| 0° −2°C | | 0° −2°C | | 0° −2°C | | 0° −3°C | S. Scot. N. Eng. | 0° −4°C | S. Scot. N. Eng. | 0° −5°C | S. Scot. N. Eng. |

Figure 27 (*cont.*)

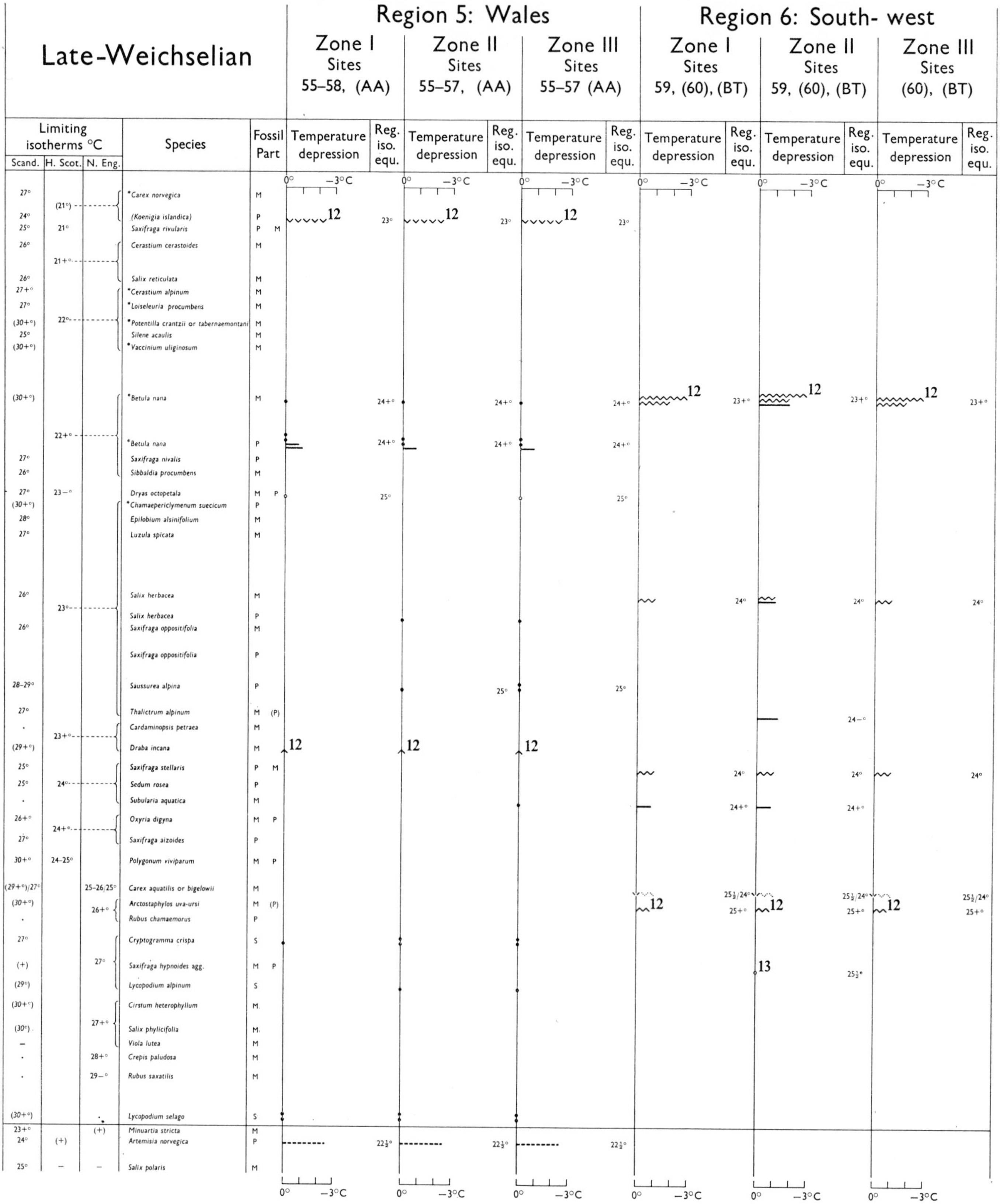

| Late-Weichselian | | | | | Region 5: Wales | | | | | | Region 6: South-west | | | | | |
|---|---|---|---|---|---|---|---|---|---|---|---|---|---|---|---|---|
| | | | | | Zone I Sites 55–58, (AA) | | Zone II Sites 55–57, (AA) | | Zone III Sites 55–57 (AA) | | Zone I Sites 59, (60), (BT) | | Zone II Sites 59, (60), (BT) | | Zone III Sites (60), (BT) | |
| Limiting isotherms °C | | | Species | Fossil Part | Temperature depression | Reg. iso. equ. | Temperature depression | Reg. iso. equ. | Temperature depression | Reg. iso. equ. | Temperature depression | Reg. iso. equ. | Temperature depression | Reg. iso. equ. | Temperature depression | Reg. iso. equ. |
| Scand. | H. Scot. | N. Eng. | | | 0° −3°C | | 0° −3°C | | 0° −3°C | | 0° −3°C | | 0° −3°C | | 0° −3°C | |
| 27° | (21°) | | *Carex norvegica | M | | | | | | | | | | | | |
| 24° | | | (Koenigia islandica) | P | 12 | 23° | 12 | 23° | 12 | 23° | | | | | | |
| 25° | 21° | | Saxifraga rivularis | P M | | | | | | | | | | | | |
| 26° | 21+° | | Cerastium cerastoides | M | | | | | | | | | | | | |
| 26° | | | Salix reticulata | M | | | | | | | | | | | | |
| 27+° | | | *Cerastium alpinum | M | | | | | | | | | | | | |
| 27° | | | *Loiseleuria procumbens | M | | | | | | | | | | | | |
| (30+°) | 22° | | *Potentilla crantzii or tabernaemontani | M | | | | | | | | | | | | |
| 25° | | | Silene acaulis | M | | | | | | | | | | | | |
| (30+°) | | | *Vaccinium uliginosum | M | | | | | | | | | | | | |
| (30+°) | | | *Betula nana | M | | 24+° | | 24+° | | 24+° | 12 | 23+° | 12 | 23+° | 12 | 23+° |
| | 22+° | | *Betula nana | P | | 24+° | | 24+° | | 24+° | | | | | | |
| 27° | | | Saxifraga nivalis | P | | | | | | | | | | | | |
| 26° | | | Sibbaldia procumbens | M | | | | | | | | | | | | |
| 27° | 23−° | | Dryas octopetala | M P | | 25° | | | | 25° | | | | | | |
| (30+°) | | | *Chamaepericlymenum suecicum | P | | | | | | | | | | | | |
| 28° | | | Epilobium alsinifolium | M | | | | | | | | | | | | |
| 27° | | | Luzula spicata | M | | | | | | | | | | | | |
| 26° | 23° | | Salix herbacea | M | | | | | | | | 24° | | 24° | | 24° |
| | | | Salix herbacea | P | | | | | | | | | | | | |
| 26° | | | Saxifraga oppositifolia | M | | | | | | | | | | | | |
| | | | Saxifraga oppositifolia | P | | | | | | | | | | | | |
| 28–29° | | | Saussurea alpina | P | | | | 25° | | 25° | | | | | | |
| 27° | | | Thalictrum alpinum | M (P) | | | | | | | | | | 24−° | | |
| · | 23+° | | Cardaminopsis petraea | M | | | | | | | | | | | | |
| (29+°) | | | Draba incana | M | 12 | | 12 | | 12 | | | | | | | |
| 25° | | | Saxifraga stellaris | P M | | | | | | | | 24° | | 24° | | 24° |
| 25° | 24° | | Sedum rosea | P | | | | | | | | | | | | |
| · | | | Subularia aquatica | M | | | | | | | | 24+° | | 24+° | | |
| 26+° | 24+° | | Oxyria digyna | M P | | | | | | | | | | | | |
| 27° | | | Saxifraga aizoides | P | | | | | | | | | | | | |
| 30+° | 24–25° | | Polygonum viviparum | M P | | | | | | | | | | | | |
| (29+°)/27° | | 25–26/25° | Carex aquatilis or bigelowii | M | | | | | | | | 25½/24° | | 25½/24° | | 25½/24° |
| (30+°) | | 26+° | Arctostaphylos uva-ursi | M (P) | | | | | | | 12 | 25+° | 12 | 25+° | 12 | 25+° |
| · | | | Rubus chamaemorus | P | | | | | | | | | | | | |
| 27° | | 27° | Cryptogramma crispa | S | | | | | | | | | | | | |
| (+) | | | Saxifraga hypnoides agg. | M P | | | | | | | | | 13 | 25½° | | |
| (29°) | | | Lycopodium alpinum | S | | | | | | | | | | | | |
| (30+°) | | 27+° | Cirsium heterophyllum | M | | | | | | | | | | | | |
| (30°) | | | Salix phylicifolia | M | | | | | | | | | | | | |
| — | | | Viola lutea | M | | | | | | | | | | | | |
| · | | 28+° | Crepis paludosa | M | | | | | | | | | | | | |
| · | | 29−° | Rubus saxatilis | M | | | | | | | | | | | | |
| (30+°) | | · | Lycopodium selago | S | | | | | | | | | | | | |
| 23+° | | (+) | Minuartia stricta | M | | | | | | | | | | | | |
| 24° | (+) | | Artemisia norvegica | P | | 22½° | | 22½° | | 22½° | | | | | | |
| 25° | — | — | Salix polaris | M | | | | | | | | | | | | |
| | | | | | 0° −3°C | | 0° −3°C | | 0° −3°C | | 0° −3°C | | 0° −3°C | | 0° −3°C | |

Figure 27 (*cont.*)

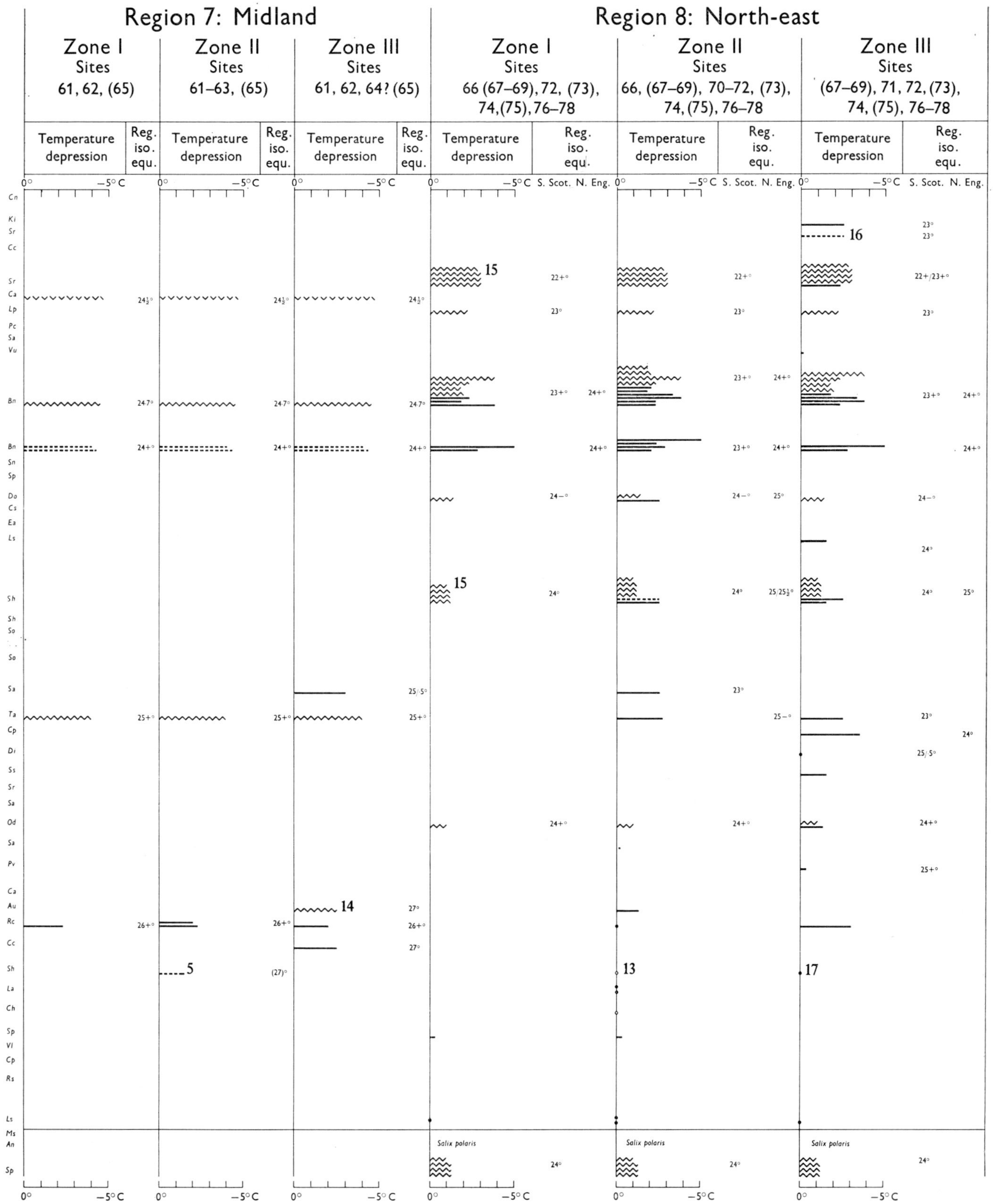

| Region 7: Midland | | | Region 8: North-east | | |
|---|---|---|---|---|---|
| Zone I<br>Sites<br>61, 62, (65) | Zone II<br>Sites<br>61–63, (65) | Zone III<br>Sites<br>61, 62, 64? (65) | Zone I<br>Sites<br>66 (67–69), 72, (73), 74, (75), 76–78 | Zone II<br>Sites<br>66, (67–69), 70–72, (73), 74, (75), 76–78 | Zone III<br>Sites<br>(67–69), 71, 72, (73), 74, (75), 76–78 |
| Temperature depression / Reg. iso. equ. | Temperature depression / Reg. iso. equ. | Temperature depression / Reg. iso. equ. | Temperature depression / Reg. iso. equ. (S. Scot. N. Eng.) | Temperature depression / Reg. iso. equ. (S. Scot. N. Eng.) | Temperature depression / Reg. iso. equ. (S. Scot. N. Eng.) |
| 0° –5° C | 0° –5° C | 0° –5° C | 0° –5° C | 0° –5° C | 0° –5° C |

| | Reg. 7 Zone I | Reg. 7 Zone II | Reg. 7 Zone III | Reg. 8 Zone I | Reg. 8 Zone II | Reg. 8 Zone III |
|---|---|---|---|---|---|---|
| Cn | | | | | | |
| Ki | | | | | | 23° |
| Sr | | | | | | 16 23° |
| Cc | | | | | | |
| Sr | | | | 15 22+° | 22+° | 22+/23+° |
| Ca | $24\frac{1}{2}$° | $24\frac{1}{2}$° | $24\frac{1}{2}$° | | | |
| Lp | | | | 23° | 23° | 23° |
| Pc | | | | | | |
| Sa | | | | | | |
| Vu | | | | | | |
| Bn | 24.7° | 24.7° | 24.7° | 23+° 24+° | 23+° 24+° | 23+° 24+° |
| Bn | 24+° | 24+° | 24+° | 24+° | 23+° 24+° | 24+° |
| Sn | | | | | | |
| Sp | | | | | | |
| Do | | | | 24–° | 24–° 25° | 24–° |
| Cs | | | | | | |
| Ea | | | | | | |
| Ls | | | | | | 24° |
| Sh | | | | 15 24° | 24° $25/25\frac{1}{2}$° | 24° 25° |
| Sh | | | | | | |
| So | | | | | | |
| So | | | | | | |
| Sa | | | 25/5° | | 23° | |
| Ta | 25+° | 25+° | 25+° | | 25–° | 23° |
| Cp | | | | | | 24° |
| Di | | | | | | 25/5° |
| Ss | | | | | | |
| Sr | | | | | | |
| Sa | | | | | | |
| Od | | | | 24+° | 24+° | 24+° |
| Sa | | | | | | |
| Pv | | | | | | 25+° |
| Ca | | | | | | |
| Au | | | 14 27° | | | |
| Rc | 26+° | 26+° | 26+° | | | |
| Cc | | | 27° | | | |
| Sh | | 5 | (27)° | | 13 | 17 |
| La | | | | | | |
| Ch | | | | | | |
| Sp | | | | | | |
| Vl | | | | | | |
| Cp | | | | | | |
| Rs | | | | | | |
| Ls | | | | | | |
| Ms | | | | Salix polaris | Salix polaris | Salix polaris |
| An | | | | | | |
| Sp | | | | 24° | 24° | 24° |
| | 0° –5° C | 0° –5° C | 0° –5° C | 0° –5° C | 0° –5° C | 0° –5° C |

Figure 27 (*cont.*)

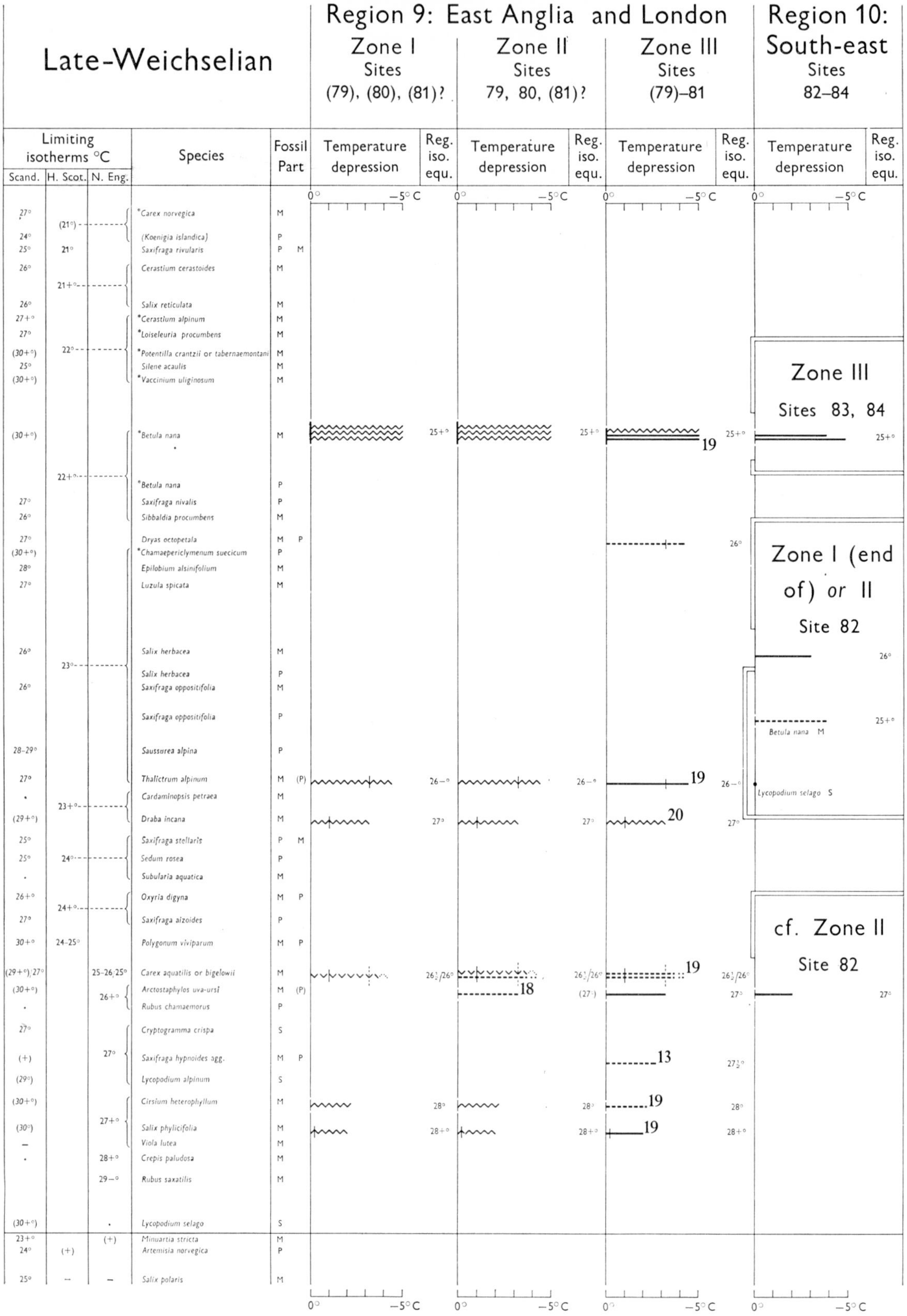

| Late-Weichselian | | | | | Region 9: East Anglia and London — Zone I, Sites (79), (80), (81)? | Region 9: Zone II, Sites 79, 80, (81)? | Region 9: Zone III, Sites (79)–81 | Region 10: South-east, Sites 82–84 |
|---|---|---|---|---|---|---|---|---|
| Limiting isotherms °C: Scand. | H. Scot. | N. Eng. | Species | Fossil Part | Temperature depression 0°–5° C / Reg. iso. equ. | Temperature depression 0°–5° C / Reg. iso. equ. | Temperature depression 0°–5° C / Reg. iso. equ. | Temperature depression 0°–5° C / Reg. iso. equ. |
| 27° | | | *Carex norvegica | M | | | | |
| | (21°) | | | | | | | |
| 24° | | | (Koenigia islandica) | P | | | | |
| 25° | 21° | | Saxifraga rivularis | P M | | | | |
| 26° | | | Cerastium cerastoides | M | | | | |
| | 21+° | | | | | | | |
| 26° | | | Salix reticulata | M | | | | |
| 27+° | | | *Cerastium alpinum | M | | | | |
| 27° | | | *Loiseleuria procumbens | M | | | | |
| (30+°) | 22° | | *Potentilla crantzii or tabernaemontani | M | | | | Zone III, Sites 83, 84 |
| 25° | | | Silene acaulis | M | | | | |
| (30+°) | | | *Vaccinium uliginosum | M | | | | |
| (30+°) | | | *Betula nana | M | 25+° | 25+° | 19 — 25+° | 25+° |
| | 22+° | | | | | | | |
| | | | *Betula nana | P | | | | |
| 27° | | | Saxifraga nivalis | P | | | | |
| 26° | | | Sibbaldia procumbens | M | | | | |
| 27° | | | Dryas octopetala | M P | | | 26° | Zone I (end of) or II, Site 82 |
| (30+°) | | | *Chamaepericlymenum suecicum | P | | | | |
| 28° | | | Epilobium alsinifolium | M | | | | |
| 27° | | | Luzula spicata | M | | | | |
| 26° | | | Salix herbacea | M | | | | 26° |
| | 23° | | | | | | | |
| | | | Salix herbacea | P | | | | |
| 26° | | | Saxifraga oppositifolia | M | | | | |
| | | | Saxifraga oppositifolia | P | | | | Betula nana M — 25+° |
| 28–29° | | | Saussurea alpina | P | | | | |
| 27° | | | Thalictrum alpinum | M (P) | 26−° | 26−° | 19 — 26−° | Lycopodium selago S |
| · | | | Cardaminopsis petraea | M | | | | |
| | 23+° | | | | | | | |
| (29+°) | | | Draba incana | M | 27° | 27° | 20 — 27° | |
| 25° | | | Saxifraga stellaris | P M | | | | |
| 25° | 24° | | Sedum rosea | P | | | | |
| · | | | Subularia aquatica | M | | | | |
| 26+° | | | Oxyria digyna | M P | | | | cf. Zone II, Site 82 |
| | 24+° | | | | | | | |
| 27° | | | Saxifraga aizoides | P | | | | |
| 30+° | 24–25° | | Polygonum viviparum | M P | | | | |
| (29+°)/27° | | 25–26;25° | Carex aquatilis or bigelowii | M | 26½/26° | 26½/26° | 19 — 26½/26° | |
| (30+°) | | 26+° | Arctostaphylos uva-ursi | M (P) | | 18 — (27°) | 27° | 27° |
| · | | | Rubus chamaemorus | P | | | | |
| 27° | | | Cryptogramma crispa | S | | | | |
| (+) | | 27° | Saxifraga hypnoides agg. | M P | | | 13 — 27½° | |
| (29°) | | | Lycopodium alpinum | S | | | | |
| (30+°) | | | Cirsium heterophyllum | M | 28° | 28° | 19 — 28° | |
| | | 27+° | | | | | | |
| (30°) | | | Salix phylicifolia | M | 28+° | 28+° | 19 — 28+° | |
| – | | | Viola lutea | M | | | | |
| · | | 28+° | Crepis paludosa | M | | | | |
| | | 29−° | Rubus saxatilis | M | | | | |
| (30+°) | | · | Lycopodium selago | S | | | | |
| 23+° | | (+) | Minuartia stricta | M | | | | |
| 24° | (+) | | Artemisia norvegica | P | | | | |
| 25° | – | – | Salix polaris | M | | | | |

Figure 27 (*cont.*)

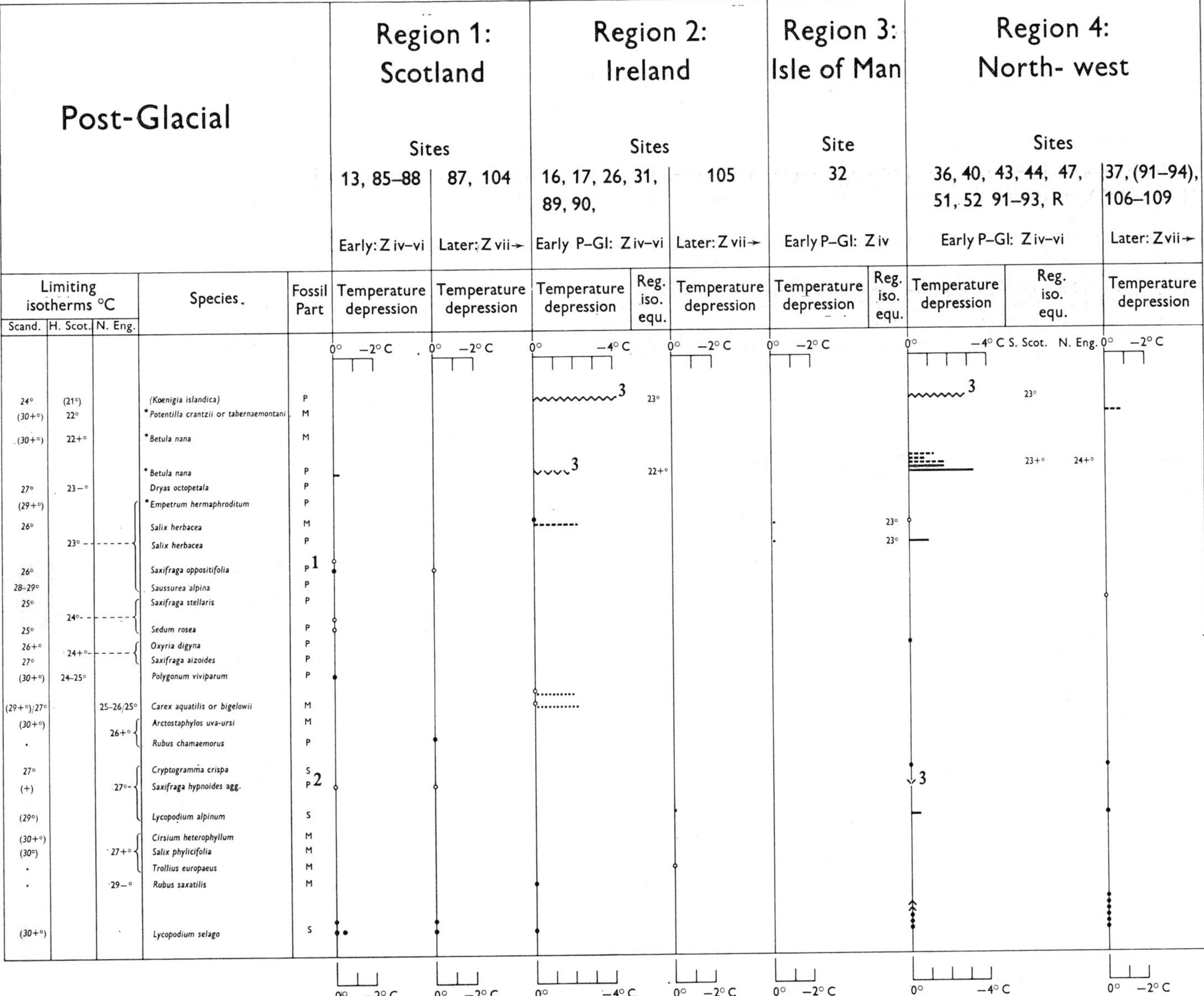


Figure 28

## Notes to Figure 28

1. ● *S. oppositifolia* s.s. ○ pollen-type including *S. aizoides*. --- cf. *S. oppositifolia*.
2. *S. hypnoides*-type.
3. Zone III/IV transition.
4. Zone III (end) or IV.
5. Includes estimate for Dogger Bank.
6. Records of macroscopic *Salix herbacea* as well as *S. reticulata*, *S. lapponum*, *S. myrsinites* and *S. arbuscula* from Cross Fell (Lewis, 1904) have been excluded owing to doubt (Godwin & Clapham, 1951)

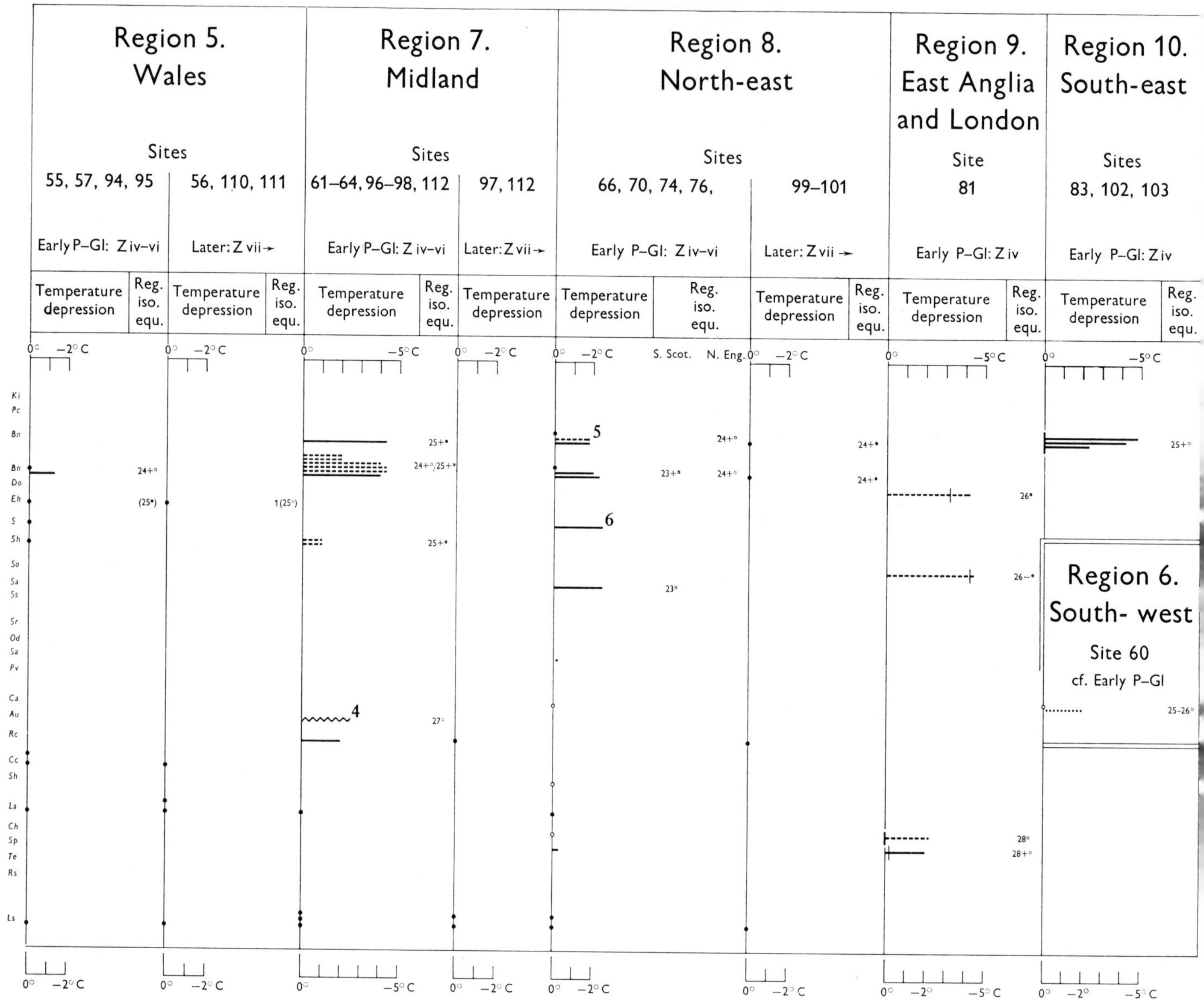

| Region 5. Wales | | | | Region 7. Midland | | | Region 8. North-east | | | | Region 9. East Anglia and London | | Region 10. South-east | |
|---|---|---|---|---|---|---|---|---|---|---|---|---|---|---|
| Sites 55, 57, 94, 95 | | Sites 56, 110, 111 | | Sites 61–64, 96–98, 112 | | 97, 112 | Sites 66, 70, 74, 76, | | 99–101 | | Site 81 | | Sites 83, 102, 103 | |
| Early P–Gl: Z iv–vi | | Later: Z vii → | | Early P–Gl: Z iv–vi | | Later: Z vii → | Early P–Gl: Z iv–vi | | Later: Z vii → | | Early P–Gl: Z iv | | Early P–Gl: Z iv | |
| Temperature depression | Reg. iso. equ. | Temperature depression | Reg. iso. equ. | Temperature depression | Reg. iso. equ. | Temperature depression | Temperature depression | Reg. iso. equ. | Temperature depression | Reg. iso. equ. | Temperature depression | Reg. iso. equ. | Temperature depression | Reg. iso. equ. |
| 0° −2° C | | 0° −2° C | | 0° −5° C | | 0° −2° C | 0° −2° C | S. Scot. N. Eng. | 0° −2° C | | 0° −5° C | | 0° −5° C | |

Ki
Pc
Bn
Bn
Do
Eh
S
Sh
So
Sa
Ss
Sr
Od
Sa
Pv
Ca
Au
Rc
Cc
Sh
La
Ch
Sp
Te
Rs
Ls

24+° (25°) 1(25°) 25+° 24+°/25+° 25+° 27° 5 6 4 23+° 24+° 24+° 23° 24+° 24+° 26° 26–° 28° 28+° 25+°

Region 6. South-west
Site 60
cf. Early P–Gl
25-26°

0° −2° C | 0° −2° C | 0° −5° C | 0° −2° C | 0° −2° C | 0° −2° C | 0° −2° −5° C | 0° −2° −5° C

Figure 28 (*cont.*)

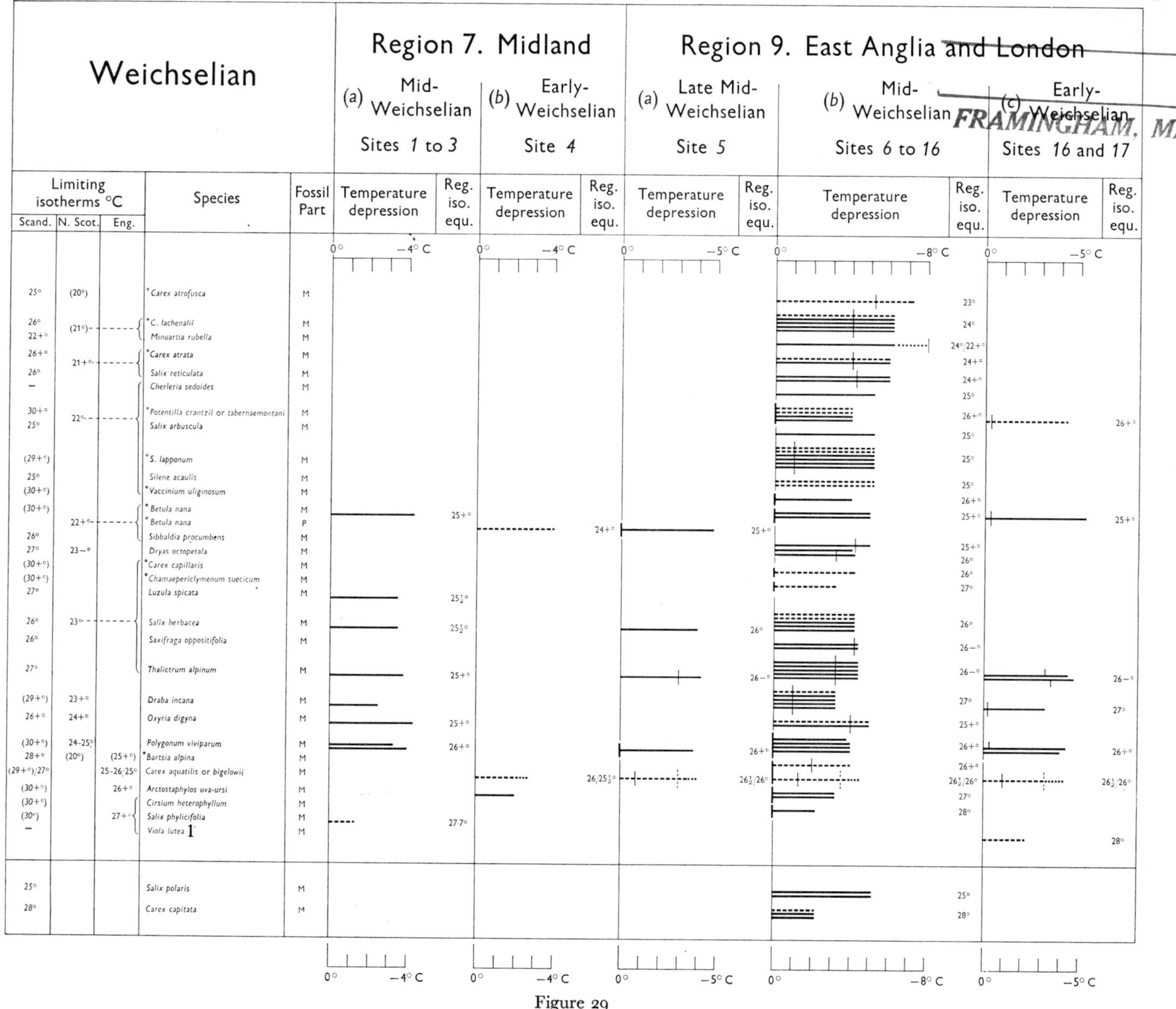


Figure 29

1, as subgenus Melanium

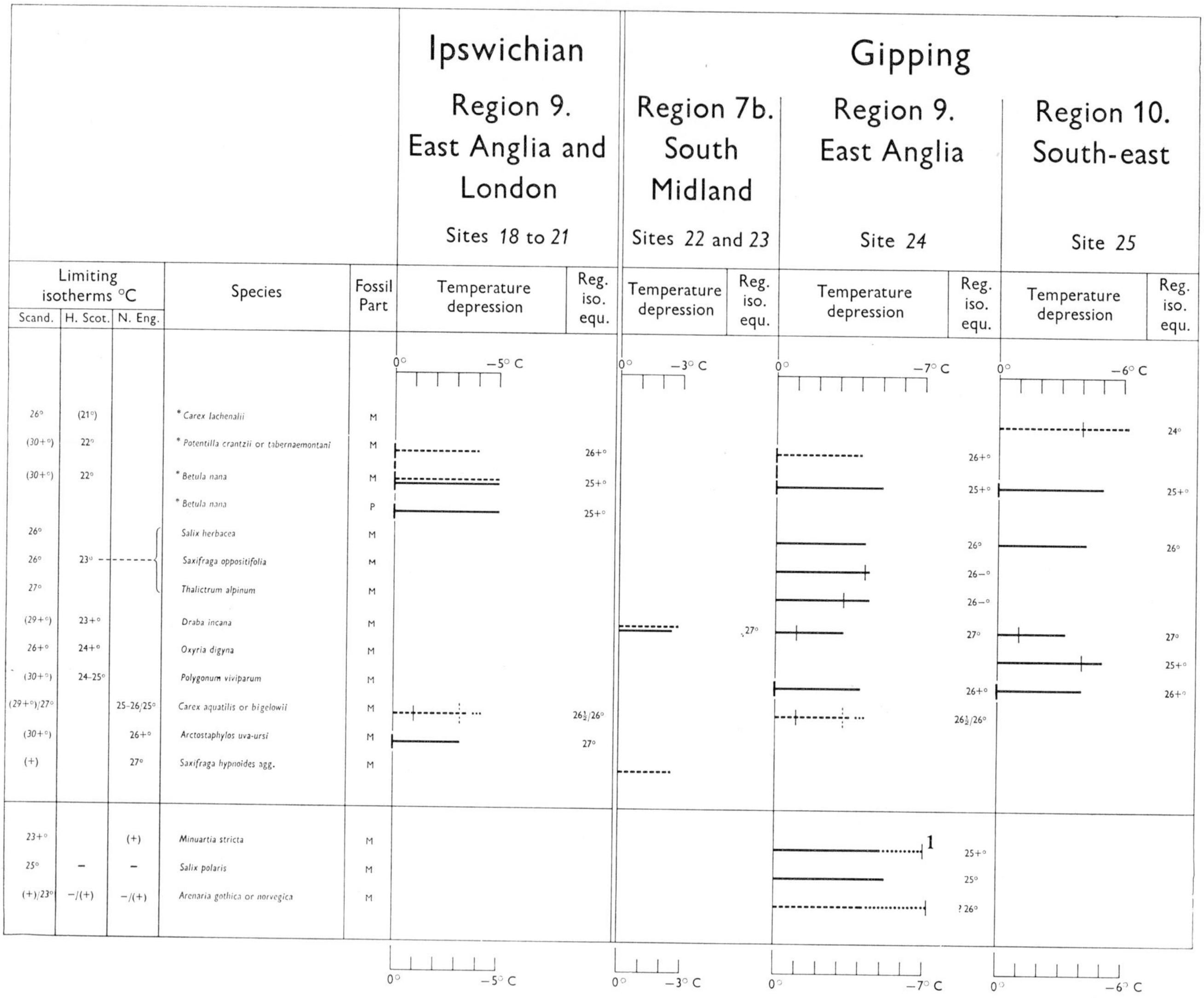

| Limiting isotherms °C: Scand. | H. Scot. | N. Eng. | Species | Fossil Part | Ipswichian — Region 9. East Anglia and London — Sites *18* to *21*: Temperature depression | Reg. iso. equ. | Gipping — Region 7b. South Midland — Sites *22* and *23*: Temperature depression | Reg. iso. equ. | Gipping — Region 9. East Anglia — Site *24*: Temperature depression | Reg. iso. equ. | Gipping — Region 10. South-east — Site *25*: Temperature depression | Reg. iso. equ. |
|---|---|---|---|---|---|---|---|---|---|---|---|---|
| 26° | (21°) | | * *Carex lachenalii* | M | | | | | | | | 24° |
| (30+°) | 22° | | * *Potentilla crantzii or tabernaemontani* | M | | 26+° | | | | 26+° | | |
| (30+°) | 22° | | * *Betula nana* | M | | 25+° | | | | 25+° | | 25+° |
| | | | * *Betula nana* | P | | 25+° | | | | | | |
| 26° | 23° (for *Salix herbacea*, *Saxifraga oppositifolia* and *Thalictrum alpinum* together) | | *Salix herbacea* | M | | | | | | 26° | | 26° |
| 26° | | | *Saxifraga oppositifolia* | M | | | | | | 26−° | | |
| 27° | | | *Thalictrum alpinum* | M | | | | | | 26−° | | |
| (29+°) | 23+° | | *Draba incana* | M | | | | 27° | | 27° | | 27° |
| 26+° | 24+° | | *Oxyria digyna* | M | | | | | | | | 25+° |
| (30+°) | 24–25° | | *Polygonum viviparum* | M | | | | | | 26+° | | 26+° |
| (29+°)/27° | | 25–26/25° | *Carex aquatilis or bigelowii* | M | | 26½/26° | | | | 26½/26° | | |
| (30+°) | | 26+° | *Arctostaphylos uva-ursi* | M | | 27° | | | | | | |
| (+) | | 27° | *Saxifraga hypnoides agg.* | M | | | | | | | | |
| 23+° | | (+) | *Minuartia stricta* | M | | | | | 1 | 25+° | | |
| 25° | – | – | *Salix polaris* | M | | | | | | 25° | | |
| (+)/23° | –/(+) | –/(+) | *Arenaria gothica or norvegica* | M | | | | | | ? 26° | | |

Scales: 0° to −5° C (Sites 18 to 21); 0° to −3° C (Sites 22 and 23); 0° to −7° C (Site 24); 0° to −6° C (Site 25).

1, on bases of C. European distribution would be reduced to *nil* temperature depression.

Figure 30

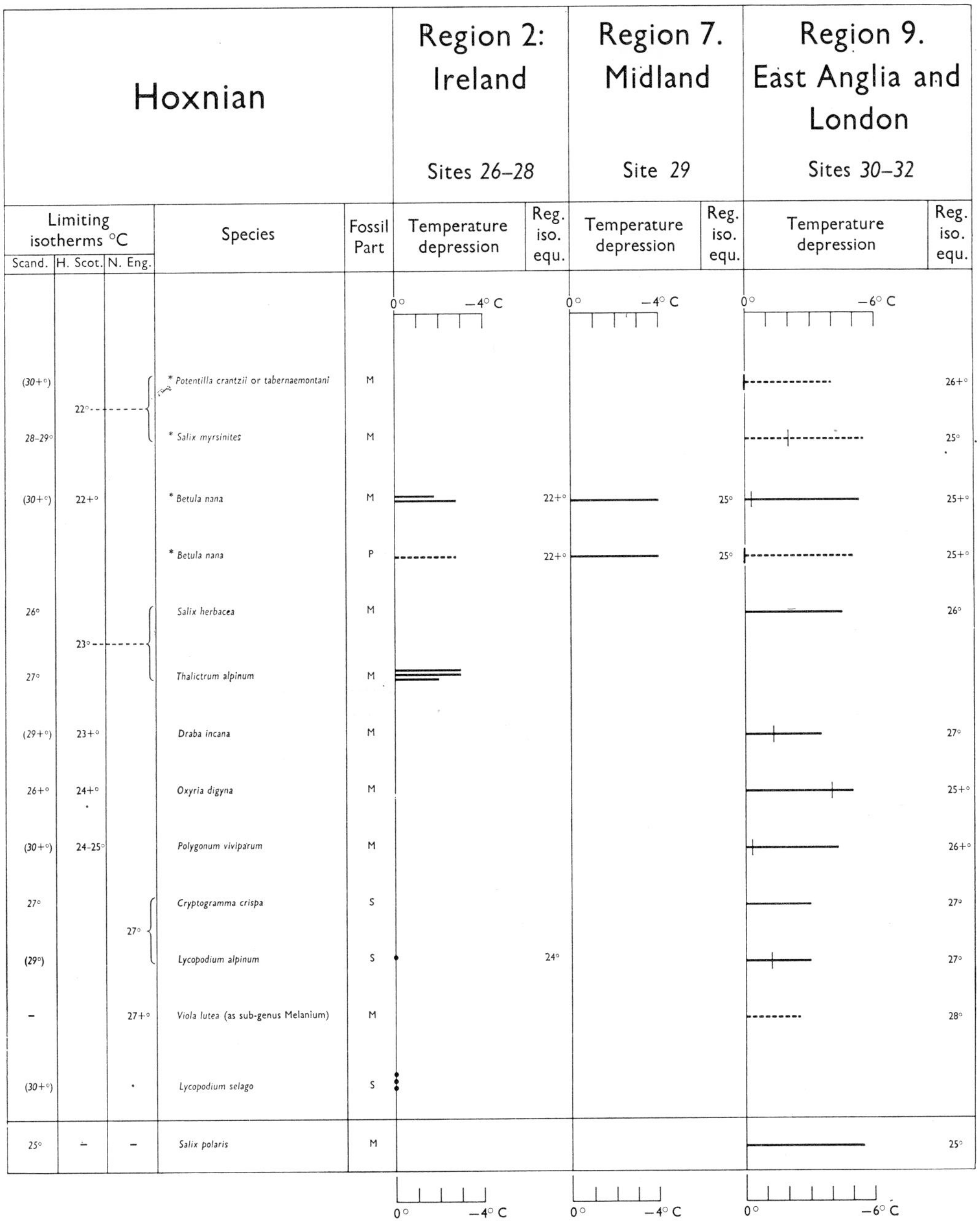

| Hoxnian | | | | | Region 2: Ireland<br>Sites 26–28 | | Region 7. Midland<br>Site 29 | | Region 9. East Anglia and London<br>Sites 30–32 | |
|---|---|---|---|---|---|---|---|---|---|---|
| Limiting isotherms °C<br>Scand. | H. Scot. | N. Eng. | Species | Fossil Part | Temperature depression<br>0° −4° C | Reg. iso. equ. | Temperature depression<br>0° −4° C | Reg. iso. equ. | Temperature depression<br>0° −6° C | Reg. iso. equ. |
| (30+°) | 22° | | * Potentilla crantzii or tabernaemontani | M | | | | | | 26+° |
| 28–29° | | | * Salix myrsinites | M | | | | | | 25° |
| (30+°) | 22+° | | * Betula nana | M | | 22+° | | 25° | | 25+° |
| | | | * Betula nana | P | | 22+° | | 25° | | 25+° |
| 26° | 23° | | Salix herbacea | M | | | | | | 26° |
| 27° | | | Thalictrum alpinum | M | | | | | | |
| (29+°) | 23+° | | Draba incana | M | | | | | | 27° |
| 26+° | 24+° | | Oxyria digyna | M | | | | | | 25+° |
| (30+°) | 24–25° | | Polygonum viviparum | M | | | | | | 26+° |
| 27° | | 27° | Cryptogramma crispa | S | | | | | | 27° |
| (29°) | | | Lycopodium alpinum | S | | 24° | | | | 27° |
| – | | 27+° | Viola lutea (as sub-genus Melanium) | M | | | | | | 28° |
| (30+°) | | · | Lycopodium selago | S | | | | | | |
| 25° | – | – | Salix polaris | M | | | | | | 25° |
| | | | | | 0° −4° C | | 0° −4° C | | 0° −6° C | |

Figure 31

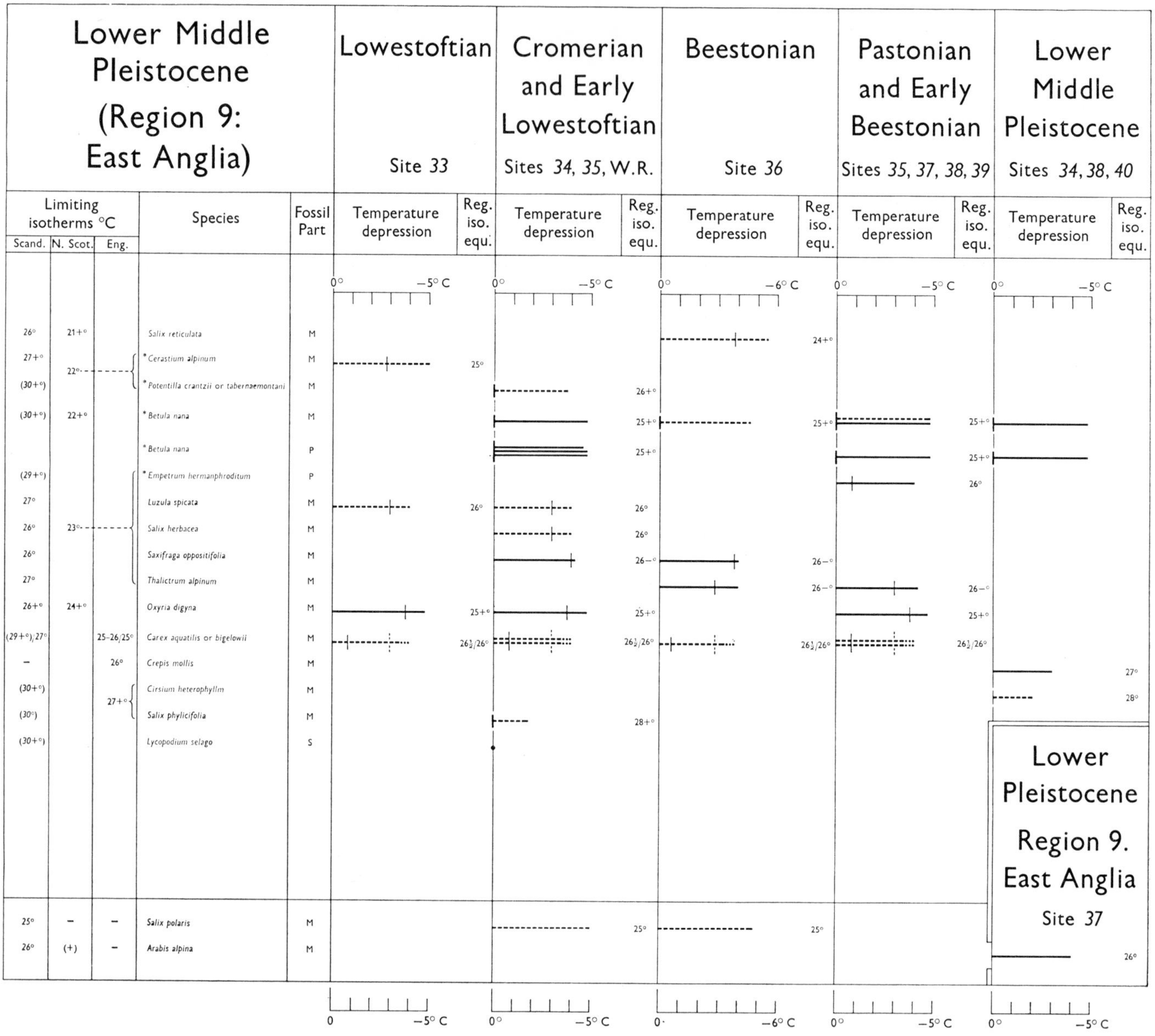

| Limiting isotherms °C: Scand. | N. Scot. | Eng. | Species | Fossil Part | Lowestoftian, Site 33: Reg. iso. equ. | Cromerian and Early Lowestoftian, Sites 34, 35, W.R.: Reg. iso. equ. | Beestonian, Site 36: Reg. iso. equ. | Pastonian and Early Beestonian, Sites 35, 37, 38, 39: Reg. iso. equ. | Lower Middle Pleistocene, Sites 34, 38, 40: Reg. iso. equ. |
|---|---|---|---|---|---|---|---|---|---|
| **Lower Middle Pleistocene (Region 9: East Anglia)** | | | | | | | | | |
| 26° | 21+° | | *Salix reticulata* | M | | | 24+° | | |
| 27+° | 22° | | **Cerastium alpinum* | M | 25° | | | | |
| (30+°) | | | **Potentilla crantzii or tabernaemontani* | M | | 26+° | | | |
| (30+°) | 22+° | | **Betula nana* | M | | 25+° | 25+° | 25+° | |
| | | | **Betula nana* | P | | 25+° | | 25+° | |
| (29+°) | | | **Empetrum hermanphroditum* | P | | | | 26° | |
| 27° | | | *Luzula spicata* | M | 26° | 26° | | | |
| 26° | 23° | | *Salix herbacea* | M | | 26° | | | |
| 26° | | | *Saxifraga oppositifolia* | M | | 26−° | 26−° | | |
| 27° | | | *Thalictrum alpinum* | M | | | 26−° | 26−° | |
| 26+° | 24+° | | *Oxyria digyna* | M | 25+° | 25+° | | 25+° | |
| (29+°)/27° | | 25–26/25° | *Carex aquatilis or bigelowii* | M | 26½/26° | 26½/26° | 26½/26° | 26½/26° | |
| – | | 26° | *Crepis mollis* | M | | | | | 27° |
| (30+°) | | 27+° | *Cirsium heterophyllm* | M | | | | | 28° |
| (30°) | | | *Salix phylicifolia* | M | | 28+° | | | |
| (30+°) | | | *Lycopodium selago* | S | | | | | |
| **Lower Pleistocene Region 9. East Anglia, Site 37** | | | | | | | | | |
| 25° | – | – | *Salix polaris* | M | | 25° | 25° | | |
| 26° | (+) | – | *Arabis alpina* | M | | | | | 26° |

Temperature depression scales: 0° to −5° C (Lowestoftian; Cromerian and Early Lowestoftian; Pastonian and Early Beestonian; Lower Middle Pleistocene; Lower Pleistocene); 0° to −6° C (Beestonian).

Figure 32

# Appendix B: Tables

## TABLE 1. *Limiting isotherms for British arctic-montane species*

All temperatures in °C. Scandinavian limiting isotherms where applicable based on Dahl (1951) are included for comparison. A point '.' signifies that the species within the area is too widespread to achieve a correlation. A plus (+) signifies that the species is too restricted within the area to achieve a correlation. An en rule '–' signifies that the species is unknown in the area. † introduced. Further explanation in text.

| | Northern England and southern Scotland | Wales (and central and southern England) | Highland Scotland | Ireland | Scandinavia | Comments |
|---|---|---|---|---|---|---|
| *Agropyron donianum* F. B. White | – | – | (20) | – | – | |
| *Alchemilla alpina* L. | 25 | – | 24+ | (23) | 27+ | |
| *Alopecurus alpinus* Sm. | (24+) | – | 22 | – | – | |
| *Arctostaphylos uva-ursi* (L.) Spreng. (figure 19) | 26+ | – | 24–25 | 24–25 | . | |
| *Arctous alpina* (L.) Niedenzu | – | – | 23+ | – | 27 | |
| *Asplenium viride* Huds. | . | 27 | 24 | 24 | . | |
| *Athyrium alpestre* Clairv. | – | – | 22 | – | 27 | |
| *Bartsia alpina* L. | (25+) | – | (20+) | – | 28+ | |
| *Betula nana* L. (figures 9 and 10) | (+) | – | 22+ | – | . | |
| *Cardaminopsis petraea* (L.) Hiit. | – | (24) | 23+ | (+) | . | Baltic shore to 27 to 28° |
| *Carex aquatilis* Wahlenb. | 25–26 | 25–26 | . | . | . | |
| *Carex atrata* L. | (+) | (+) | 21+ | – | 26+ | |
| *Carex atrofusca* Schkuhr (figure 3) | – | – | (20) | – | 25 | |
| *Carex bigelowii* Torr. | 25 | (23) | 23 | 23 | 27 | Southern Scotland 24 |
| *Carex capillaris* L. | (25) | (+) | 23 | – | . | Southern Scotland (+) |
| *Carex lachenalii* Schkuhr | – | – | (21) | – | 26 | |
| *Carex norvegica* Retz. | – | – | (21) | – | 27 | |
| *Carex rariflora* (Wahlenb.) Sm. | – | – | (21) | – | 26 | |
| *Carex rupestris* All. | – | – | (21+) | – | 26 | |
| *Carex saxatilis* L. | – | – | 21 | – | 25 | |
| *Carex vaginata* Tausch | (+) | – | 21 | – | . | Not in England |
| *Cerastium alpinum* L. | (23+) | (+) | 22 | – | 27+ | |
| *Cerastium arcticum* Lange | – | (+) | 21 | – | 22 | |
| *Cerastium cerastoides* (L.) Britton | – | – | 21+ | – | 26 | |
| *Chamaepericlymenum suecicum* (L.) Aschers. & Graebn. | (+) | – | 23 | – | . | |
| *Cherleria sedoides* L. | – | – | 22 | – | – | |
| *Cirsium heterophyllum* (L.) Hill (figure 21) | . | 27+ | . | (+) | . | |
| *Crepis mollis* (Jacq.) Aschers. | 26 | – | . | – | – | |
| *Crepis paludosa* (L.) Moench | . | 28+ | . | . | . | |
| *Cryptogramma crispa* (L.) Hook. (since 1930) | 26+ | 26 | 24–25 | (+) | | |
| *Cryptogramma crispa* (L.) Hook. (all records) (figure 20) | . | 27 | 24–25 | (24) | 27 | |
| *Dactylorchis purpurella* (T. & T. A. Steph.) Vermeul. | . | 28+ | . | 25 | (+) | |
| *Deschampsia alpina* (L.) Roem. & Schult. | – | (+) | 22 | (+) | 25 | |
| *Draba incana* L. (figure 11) | 26–27 | (+) | 23+ | 23+ | . | Not in southern Scotland |
| *Draba norvegica* Gunn. | – | – | 21 | – | 26 | |
| *Dryas octopetala* L. (figure 16) | 25 | (+) | 23– | 25– | 27 | Western Ireland peculiar; not southern Scotland |
| *Empetrum hermaphroditum* Hagerup | (23) | (+) | 23 | – | . | |
| *Epilobium alsinifolium* Vill. | 25 | (+) | 23 | (+) | 28 | |
| *Epilobium anagallidifolium* Lam. | 25+ | – | 23 | – | 26 | |
| †*Epilobium nerterioides* A. Cunn. | . | 28+ | . | 25+ | – | |
| *Equisetum variegatum* Web. & Mohr | 26 | 26 | . | . | . | Except maritime localities |
| *Erigeron borealis* (Vierh.) Simmons | – | – | (21) | – | 26 | |
| *Festuca altissima* All. (figure 22) | . | 28+ | . | 25 | . | |
| *Festuca vivipara* (L.) Sm. | 25 | 25 | 24+ | 25+ | . | Southern Scotland 24; no distribution map for Scandinavia |
| *Galium boreale* L. | 26+ | (24+) | . | . | . | |
| *Gentiana nivalis* L. | – | – | (21) | – | 28 | |
| *Gnaphalium norvegicum* Gunn. | – | – | (21) | – | 28 | |

TABLE I (*cont.*)

| | Northern England and southern Scotland | Wales (and central and southern England) | Highland Scotland | Ireland | Scandinavia | Comments |
|---|---|---|---|---|---|---|
| *Gnaphalium supinum* L. | – | – | 23 | – | 26 | |
| *Juncus alpinoarticulatus* Chaix | (25) | – | (22) | – | . | No distribution map for Scandinavia |
| *Juncus biglumis* L. | – | – | 21 | – | 26 | |
| *Juncus castaneus* Sm. (figure 4) | – | – | 21+ | – | 25 | |
| *Juncus trifidus* L. (figure 8) | (+) | – | 22+ | – | 27 | Not in England |
| *Juncus triglumis* L. | 25 | (+) | 23 | – | 27–28 | |
| *Kobresia simpliciuscula* (Wahlenb.) Mackenzie (figure 6) | (25) | – | (21) | – | 23–24 | |
| *Listera cordata* (L.) R.Br. | . | 27+ | . | 25 | . | |
| *Loiseleuria procumbens* (L.) Desv. (figure 7) | – | – | 22 | – | 27 | |
| *Luzula arcuata* Sw. | – | – | 21 | – | 25+ | |
| *Luzula spicata* (L.) DC. (figure 11) | – | – | 23 | – | 27 | |
| *Lycopodium alpinum* L. (since 1930) | 25 | 25+ | 24 | 24 | . | |
| *Lycopodium alpinum* L. (all records) | . | 27 | 24–25 | 24 | . | Scandinavia *c.* 28 |
| *Lycopodium selago* L. (since 1930) | 25–26 | 25+ | 24+ | . | . | |
| *Melampyrum sylvaticum* L. | (+) | – | 23+ | (+) | . | |
| *Meum athamanticum* Jacq. | 26 | (25) | . | – | (+) | |
| *Minuartia rubella* (Wahlenb.) Hiern | – | – | (21) | – | 22+ | |
| *Oxyria digyna* (L.) Hill (all records) (figure 12) | 24+ | 24 | 24+ | 23 | 26+ | Since 1930: Highland Scotland 23+; southern Scotland 23 |
| *Phleum alpinum* L. | (+) | – | 21+ | – | 29 | |
| *Poa alpina* L. | 25 | (+) | 22+ | (+) | . | Not in southern Scotland |
| *Poa flexuosa* Sm. | – | – | (21) | – | 24 | |
| *Poa glauca* Vahl | (+) | (+) | 22+ | – | . | |
| *Polygonum viviparum* L. | 25+ | (+) | 24–25 | (+) | . | |
| *Polystichum lonchitis* (L.) Roth | 25 | (+) | 23 | 23 | 27 | |
| *Potentilla crantzii* (Crantz) G. Beck | 25+ | (+) | 22 | – | . | Southern Scotland 23 |
| *Rubus chamaemorus* L. | 26+ | (+) | 24+ | (+) | . | |
| *Rubus saxatilis* L. | . | 29– | . | . | . | |
| *Sagina intermedia* Fenzl | – | – | 20 | – | 23+ | |
| *Sagina normaniana* Lagerh. | – | – | 21 | – | (+) | |
| *Sagina saginoides* (L.) Karst. | – | – | 21+ | – | 27+ | |
| *Salix arbuscula* L. | (23+) | – | 22 | – | 25 | Not in England |
| *Salix herbacea* L. (figures 14 and 15) | 25 | 25 | 23 | 23 | 26 | Southern Scotland 24 |
| *Salix lanata* L. | – | – | 21 | – | 27 | |
| *Salix lapponum* L. | (23) | – | 22 | – | . | |
| *Salix myrsinites* L. | (23) | – | 22 | – | 28–29 | Not in England |
| *Salix nigricans* Sm. | 27 | – | . | (+) | . | |
| *Salix phylicifolia* L. | 27+ | – | . | (+) | . | |
| *Salix reticulata* L. (figure 5) | – | – | 21+ | – | 26 | |
| *Saussurea alpina* (L.) DC. | (24) | (+) | 23 | 23 | 28–29 | Southern Scotland (23) |
| *Saxifraga aizoides* L. | 25 | – | 24+ | (24+) | 27 | |
| *Saxifraga cernua* L. | – | – | (21) | – | 25–26 | |
| *Saxifraga hypnoides* L. | . | 27 | 24+ | 24–25 | (+) | |
| *Saxifraga rivularis* L. | – | – | 21 | – | 25 | |
| *Saxifraga nivalis* L. | 24 | (+) | 22+ | (+) | 27 | Southern Scotland( +) |
| *Saxifraga oppositifolia* L. | 25– | 25 | 23 | 22+ | 26 | |
| *Saxifraga stellaris* L. (figure 18) | 25 | 25 | 24 | 24 | 25 | Southern Scotland 24 |
| *Sedum rosea* (L.) Scop. | 25+ | 25+ | 24 | 24 | 25 | |
| *Sedum villosum* L. | 25–26 | – | . | – | 24 | |
| *Sibbaldia procumbens* L. | (+) | – | 22+ | – | 26 | |
| *Silene acaulis* (L.) Jacq. | 24 | (23–24) | 22 | (22) | 25 | |
| *Subularia aquatica* L. | 25+ | 25 | 24 | 24+ | . | |
| *Thalictrum alpinum* L. (figure 13) | 25– | (+) | 23 | 22 | 27 | Southern Scotland 23 |
| *Thelypteris phegopteris* (L.) Slosson | . | 28+ | . | 25 | . | |
| *Tofieldia pusilla* (Michx.) Pers. | (+) | – | 22+ | – | 29 | |
| *Trollius europaeus* L. | . | 27+ | . | (+) | . | |
| *Vaccinium microcarpum* (Rupr.) Hook. f. | – | – | 24 | – | . | |
| *Vaccinium uliginosum* L. | (25+) | – | 22 | – | . | |
| *Veronica alpina* L. | – | – | (21) | – | 25+ | |
| *Veronica fruticans* Jacq. | – | – | (21+) | – | 25 | |

TABLE 1 *(cont.)*

| | Northern England and southern Scotland | Wales (and central and southern England) | Highland Scotland | Ireland | Scandinavia | Comments |
|---|---|---|---|---|---|---|
| *Viola lutea* Huds. | · | 27+ | · | (+) | – | |
| *Woodsia alpina* (Bolton) S. F. Gray | – | (+) | 21 | – | 28 | |
| *Aconitum septentrionale* Koelle | – | – | – | – | 28 | |
| *Arabis alpina* L. | – | – | (+) | – | 26 | |
| *Arenaria norvegica* Gunn. | – | – | (+) | – | 23 | |
| *Artemisia norvegica* Fr. | – | – | (+) | – | 22 | |
| *Campanula uniflora* Georgi | – | – | – | – | 22 | |
| *Carex capitata* L. | – | – | – | – | 28 | |
| *Koenigia islandica* L. | – | – | (+) | – | 24 | |
| *Minuartia stricta* (Sw.) Hiern | (+) | – | – | – | 23+ | |
| *Salix polaris* Wahlenb. | – | – | – | – | 25 | |

TABLE 2. *Comparison of limiting isotherms between regions within the British Isles and with Scandinavia*

The mean difference in °C and the number of comparisons included in each entry (in brackets) are given to the upper right of the diagonal and to the lower left the corresponding statistical significance. Reading horizontally, regions are taken as positive in relation to those in vertical columns. N.S. means not significant, × × means significant at the 1 per cent level, × × × means significant at the 0·1 per cent level. Further explanation in text.

The figures marked by asterisks are later modified (see p. 164)

* 1·64 (14) ** 3·22 (16) *** 1·56 (23)

| | Scandinavia | Northern England and southern Scotland | Wales | Highland Scotland | Ireland |
|---|---|---|---|---|---|
| Scandinavia | — | 1·46* (17) | 0·80 (5) | 4·13** (45) | 3·27 (10) |
| Northern England and southern Scotland | × × × | — | 0·06 (12) | 1·68*** (27) | 1·56 (14) |
| Wales | N.S. | N.S. | — | 1·15 (10) | 1·23 (8) |
| Highland Scotland | × × × | × × × | × × × | — | −0·19 (18) |
| Ireland | × × × | × × × | × × | N.S. | — |

TABLE 3. *Scandinavian species grouped according to the values of their limiting isotherms with the number and percentage of the elements reaching the British Isles*

| Scandinavian limiting isotherms, °C | 22 to 23– | 23 to 24– | 24 to 25– | 25 to 26– | 26 to 27– | 27 to 28– | 28 to 29– |
|---|---|---|---|---|---|---|---|
| Number of species in Scandinavia | 17 | 18 | 11 | 23 | 28 | 32 | 20 |
| Number of the above species in British Isles | 3 | 4 | 4 | 14 | 16 | 22 | 12 |
| Percentage in the British Isles | 18 | 22 | 36 | 61 | 57 | 69 | 60 |

TABLE 4. *List of sites for which Late-Weichselian and Post-glacial fossil records of the arctic-montane species enumerated in figures 27–32 are known*

† Includes unpublished data and data from theses.
‡ Includes or refers to Zone III/IV transition.
* Of somewhat uncertain age.

| Late-Weichselian sites | Nat. Grid | Present-day isotherm value of site (°C) |
|---|---|---|
| *Region 1: Scotland* | | |
| 1. Abernethy, Inv.† | 28/91 | 22 |
| 2. Burn of Benholm, Kinc. | 37/76 | 24 |
| 3. Crianlarich, Perths.* | 27/32 | 21 |
| 4. Garral Hill, Keith, Banffs. | 38/45 | 24–25 |
| 5. Garscadden Mains, Glasgow | 26/57 | 25 |
| 6. Garvel Park and New Dry Dock Greenock, Renf.† | 26/37 | 24–25 |
| 7. Loch Droma, Ross & Crom. | 28/27 | <20 |
| 8. Loch Fada, Skye† | 18/44 | 21–22 |
| 9. Loch Mahaik, Perths. | 27/70 | 24+ |
| 10. Loch Mealt, Skye† | 18/56 | 21 |
| 11. Loch Meodal, Skye† | 18/61 | 22 |
| 12. Loch Tarff, Inv.† | 28/40 | 21–22 |
| 13. Yesnaby, Orkney† | N30/21 | 21 |
| *Region 2: Ireland* | | |
| 14. Ballybetagh, Co. Dublin | 13/27 | 24 |
| 15. Belfast, Ormeau Bridge | 15/42 | 23 |
| 16. Cannons Lough, Co. Derry | 15/06 | 24 |
| 17. Carrowreagh, Co. Roscommon | 94/86 | 25+ |
| 18. Dunshaughlin, Co. Meath | 14/01 | 25–26 |
| 19. Drumurcher, Co. Monaghan | 04/67 | 25 |
| 20. Johnstown, Co. Wicklow | 03/96 | 25 |
| 21. Knocknacran, Co. Monaghan | 04/85 | 25 |
| 22. Littleton Bog, Leigh, Co. Tipperary | 03/12 | 25–26 |
| 23. Longford Pass, Co. Tipperary | 03/12 | 25+ |
| 24. Lough Gur, Co. Limerick | 93/50 | 25+ |
| 25. Mapastown, Co. Louth | 14/05 | 25+ |
| 26. Newtownbabe, Co. Louth | 14/06 | 25 |
| 27. Ralaghan, Co. Cavan | 04/86 | 25 |
| 28. Ratoath, Co. Meath | 14/01 | 25–26 |
| 29. Roddans Port, Co. Down | 15/71 | 23–24 |
| 30. Roundstone I, Co. Galway | 84/71 | 23 |
| 31. Roundstone II, Co. Galway | 84/72 | 22 |
| A. Ardcavan, Co. Wexford | 02/98 | 24 |
| *Region 3: Isle of Man* | | |
| 32. Ballaugh† | 24/39 | 23+ |
| 33. Ballyre† | 24/39 | 23+ |
| 34. Kirkmichael-Wyllin† | 24/39 | 23+ |
| *Region 4: North-west Britain* | | |
| 35. Abbots Moss, Cumb. | 35/54 | 26 |
| 36. Bigholm Burn, Dumfries.† | 35/38 | 26 |
| 37. Blelham Tarn Moss, Lancs.† | 35/30 | 25 |
| 38. Cooran Lane, Kirk† | 25/48 | 23 |
| 39. Culhorn Mains, Wigtowns.† | 25/05 | 25 |
| 40. Esthwaite, Lancs. | 34/39 | 26 |
| 41. Hawes Water, Lonsdale | 34/47 | 27+ |
| 42. Helton Tarn, Westmorland | 34/48 | 27 |
| 43. Kentmere, Westmorland | 35/40 | 25 |
| 44. Little Lochans, Wigtowns.† | 25/05 | 25 |
| 45. Low Wray Bay, Windermere† | 35/30 | 25 |
| 46. Lunds, Yorks. | 34/79 | <25 |
| 47. Malham Tarn & Moss, Yorks. | 34/86 | 25 |
| 93. Nick of Curleywee, Kirk.‡† | 25/47 | 24 |
| 48. Oulton Moor, Cumb. | 35/25 | 27·5 |
| 49. Robert Hill, Dumfries. | 35/17 | 26+ |
| 50. St Bees, Cumb. | 25/91 | 27 |
| 51. Scaleby, Cumb. | 35/46 | 27·5 |
| 52. Skelmergh, Westmorland | 34/59 | 25–26 |
| 53. Windermere, north | 35/30 | 25 |
| 54. Witherslack Hall, Westmorland | 34/48 | 27 |
| O. Overwater, Cumb.† | 35/32 | 24 |
| S. Seathwaite Tarn, Lancs. | 34/29 | 25 |
| *Region 5: Wales* | | |
| 55. Elan Valley, Cards./Montgom.† | 22/87 | 25 |
| 56. Llyn Dwythwch, Caerns. | 23/55 | 24+ |
| 57. Nant Ffrancon, Caerns. | 23/66 | <24 |
| 58. Waen Ddû, Brecon.† | 32/11 | 25+ |
| *Region 6: South-west England* | | |
| 59. Hawks Tor, Bodmin | 20/17 | 25 |
| 60. Stannon, Bodmin† | 20/18 | 25 |
| *Region 7: Midland England* | | |
| (*a*) *North* | | |
| 61. Bagmere, Cheshire | 33/76 | 28–29 |
| 62. Chat Moss, Cheshire | 33/79 | 28+ |
| 63. Moss Lake, Liverpool | 33/38 | 28+ |
| (*b*) South | | |
| 64. Apethorpe, Northants. | 52/09 | 29–30 |
| 65. Branston, Staffs.*† | 43/22 | 29+ |
| *Region 8: North-east Britain* | | |
| (*a*) *North* | | |
| 66. Bradford Kaim, Bamburgh, Northumb. | 46/13 | 25–26 |
| 67. Corstorphine, Edinburgh | 36/17 | 25+ |
| 68. Dronachy, Fife. | 36/29 | 25 |
| 69. Hailes (& ?Gayfield), Edinburgh | 36/27 | 25+ |
| 70. Longlee Moor, Northumb. | 46/11 | 25+ |
| 71. Whitrig Bog, Berwicks.† | 36/63 | 25–26 |
| (*b*) *South* | | |
| 72. Aby Grange, Lincs. | 53/47 | 28 |
| 73. Bridlington, Yorks.* | 54/16 | 26–27 |
| 74. Flixton (& Seamer), Yorks. | 54/08 | 26 |
| 75. Holmpton, Yorks.* | 54/32 | 28 |
| 76. Neasham, Co. Durham | 45/31 | 27–28 |
| 77. Romaldkirk, Teesdale | 35/92 | 27 |
| 78. Tadcaster, Yorks. | 44/44 | 29+ |
| *Region 9: East Anglia and London* | | |
| 79. Hockham Mere, Norfolk | 52/99 | 30+ |
| 80. Lopham Little Fen, Diss | 62/07 | 30+ |
| 81. Nazeing, Lea Valley | 52/30 | 30+ |
| *Region 10: South-east England* | | |
| 82. Brook, Kent | 61/04 | 29 |
| 83. Elstead, Surrey | 41/84 | 30 |
| 84. Nursling, Hants. | 41/31 | 29 |
| *Addendum* | | |
| AA. Aberaeron–Aberarth, Cards.† | 22/46 | 25–26 |
| BT. Bovey Tracey, Devon. | 20/87 | 26 |
| F. Faskine, Airdrie, Lanark. | 26/76 | 25+ |

TABLE 4 (*cont.*)

| Post-glacial sites | Nat. Grid | Present-day isotherm value of site (°C) |
|---|---|---|
| **Early Post-glacial: Zones IV, V and VI** | | |
| *Region 1: Scotland* | | |
| 85. Allt na Feithe Smelich, Inv.† | 28/82 | 22–23 |
| 86. Aros Moss, Argyll. | 16/62 | 22 |
| 87. Lochan-nan-cat, Lawers, Perths. | 27/64 | 20 |
| 88. The Loons, Orkney† | N30/22 | 21 estimated |
| 13. Yesnaby, Orkney† | N30/21 | 21 |
| *Region 2: Ireland* | | |
| 89. Ballyscullion, Co. Antrim | 15/15 | 25 |
| 16. Cannons Lough, Co. Derry | 15/06 | 24 |
| 17. Carrowreagh, Co. Roscommon | 94/86 | 25+ |
| 90. Jenkinstown, Co. Louth | 14/16 | 25 |
| 26. Newtownbabe, Co. Louth | 14/06 | 25 |
| 31. Roundstone II, Co. Galway | 84/72 | 22 |
| *Region 3: Isle of Man* | | |
| 32. Ballaugh | 24/39 | 23+ |
| *Region 4: North-west Britain* | | |
| 36. Bigholm Burn, Dumfries.†‡ | 35/38 | 26 |
| 40. Esthwaite, Lancs. | 34/39 | 26 |
| 43. Kentmere, Westmorland | 35/40 | 25 |
| 91. Langdale Coombe, Westmorland | 35/20 | 23 |
| 44. Little Lochans, Wigtowns.† | 25/05 | 25 |
| 47. Malham Tarn, Yorks. | 34/86 | 25 |
| 92. Mickleden, Langdale, Westmorland | 35/20 | 23+ |
| 93. Nick of Curleywee, Kirk.‡† | 25/47 | 24 |
| 51. Scaleby, Cumb. | 35/46 | 27·5 |
| 52. Skelmergh, Westmorland | 34/59 | 25–26 |
| R. Red Tarn, Langdale, Westmorland | 35/20 | 24 |
| *Region 5: Wales* | | |
| 94. Cwm Idwal, Caerns. | 23/65 | <24 |
| 55. Elan Valley, Cards./Montgom.† | 22/87 | 25 |
| 95 Llyn Gynon, Cards.† | 22/86 | 25+ |
| 57. Nant Ffrancon, Caerns. | 23/66 | <24 |
| *Region 6: South-west England* | | |
| 60. Stannon, Bodmin† | 20/18 | 25 |
| *Region 7: Midland England* | | |
| (*a*) North | | |
| 61. Bagmere, Cheshire | 33/76 | 28–29 |
| 62. Chat Moss, Cheshire | 33/79 | 28+ |
| 96. Kinder, Derbys.* | 43/18 | 26+ |
| 63. Moss Lake, Liverpool | 33/38 | 28+ |
| 97. Snake Pass, Derbys.* | 43/09 | 26+ |
| (*b*) South | | |
| 64. Apethorpe, Northants.‡ | 52/09 | 29–30 |
| 98. Rodbaston, Staffs. | 33/91 | 28–29 |

| Post-glacial sites (*cont.*) | Nat. Grid | Present-day isotherm value of site (°C) |
|---|---|---|
| *Region 8: North-east Britain* | | |
| (*a*) *North* | | |
| 66. Bradford Kaim, Bamburgh, Northumb. | 46/13 | 25–26 |
| 99. Cross Fell, Cumb. | 35/73 | <25 |
| 70. Longlee Moor, Northumb. | 46/11 | 25+ |
| 100. Widdybank (A), Upper Teesdale† | 35/82 | <25 |
| (*b*) *South* | | |
| 101. Dogger Bank | ?64/— | 26± estimated |
| 74. Flixton, Yorks. | 54/08 | 26 |
| 76. Neasham, Co. Durham | 45/31 | 27–28 |
| *Region 9: East Anglia and London* | | |
| 81. Nazeing, Lea Valley | 52/30 | 30+ |
| *Region 10: South-east* | | |
| 102. Crane's Moor, Hants. | 41/10 | 29–30 |
| 83. Elstead, Surrey | 41/84 | 30 |
| 103. Weevil Lake, Gosport, Hants. | 41/60 | 27–28 |
| **Later Post-glacial: Zones VII onward** | | |
| *Region 1: Scotland* | | |
| 87. Lochan-nan-cat, Lawers, Perths. | 27/64 | 20 |
| 104. Duartbeg, Sutherland† | 29/13 | 21–22 |
| *Region 2: Ireland* | | |
| 105. Treanscrabbah, Co. Sligo | 94/88 | 24+ |
| *Region 4: North-west Britain* | | |
| 37. Blelham Tarn Moss, Lancs. | 35/30 | 25 |
| 106. Barnscar, Cumb. | 34/19 | 25 |
| 107. Bowness Common, Cumb. | 35/26 | 26+ |
| 108. Ehenside Tarn, Cumb. | 35/00 | 26 |
| 91. Langdale Coombe, Westmorland | 35/20 | 23 |
| 92. Mickleden, Langdale, Westmorland | 35/20 | 23+ |
| 109. Moorthwaite, Cumb. | 35/55 | 26 |
| 93. Nick of Curleywee, Kirk.† | 25/47 | 24 |
| *Region 5: Wales* | | |
| 110. Borth Bog, Cards. | 22/69 | 25 |
| 56. Llyn Dwythwch, Caerns. | 23/55 | 24+ |
| 57. Nant Ffrancon, Caerns. | 23/66 | <24 |
| 111. Plynlimon, Cards.† | 22/78 | 24–25 |
| *Region 7: Midland England* | | |
| 97. Snake Pass incl. Feather Bed Moss, Derbys. | 43/09 | 27 |
| 112. South-west Lancs. coast | 34/30 | 28+ |
| *Region 8: North-east Britain* | | |
| 100. Widdybank (A, C), Upper Teesdale† | 35/82 | <25 |
| Widdybank (B), Upper Teesdale† | 35/83 | <25 |

TABLE 5. *List of sites for which Weichselian and earlier fossil records of the arctic-montane species enumerated in figures 27–32 are known*

† Includes unpublished data and data from theses

Sites *34–39* inclusive contain deposits of different ages. They are only entered in the table for the age appropriate to their fossil record of species enumerated in figures 27–32.

WEICHSELIAN AND PRE-WEICHSELIAN SITES

| | Nat. Grid | Present-day isotherm value of site (°C) |
|---|---|---|
| **WEICHSELIAN GLACIAL** | | |
| *Region 7: Midland England* | | |
| (*a*) Mid-Weichselian | | |
| *1.* Marlow, Bucks.† | 41/88 | 30+ |
| *2.* Thrapston, Northants.† | 42/98 | 29–30 |
| *3.* Upton Warren, Worcs. | 32/96 | 29 |
| (*b*) Early-Weichselian | | |
| *4.* Chelford, Cheshire | 33/87 | 28+ |
| *Region 9: East Anglia and London* | | |
| (*a*) Late Mid-Weichselian | | |
| *5.* Colney Heath, Herts. | 52/10 | 30 |
| (*b*) Mid-Weichselian | | |
| *6.* Barnwell Station, Cambs. | 52/45 | 30+ |
| *7.* Earith, Hunts.† | 52/37 | 30+ |
| Lea Valley: | | |
| *8.* Angel Road | 51/39 | 30+ |
| *9.* Barrowell Green | 51/39 | 30+ |
| *10.* Broxbourne | 52/30 | 30+ |
| *11.* Hedge Lane | 51/39 | 30+ |
| *12.* Ponders End | 51/39 | 30+ |
| *13.* Temple Mills | 51/38 | 30+ |
| *14.* Waltham Cross | 52/30 | 30+ |
| *15.* Market Deeping, Lincs.† | 53/11 | 30 |
| *16.* Wretton, Norfolk† | 52/69 | 30–31 |
| (*c*) Early-Weichselian | | |
| *16.* Wretton, Norfolk† | 52/69 | 30–31 |
| *17.* Sidgwick Avenue, Cambridge | 52/45 | 30+ |
| **IPSWICHIAN INTERGLACIAL AND ASSOCIATED LATE-GLACIAL** | | |
| *Region 9: East Anglia and London* | | |
| *18.* Admiralty Office, London | 51/28 | 30+ |
| *19.* Bobbitshole, Suffolk | 62/14 | 30+ |
| *20.* Histon Road, Cambridge | 52/46 | 30+ |
| *21.* Ilford, Essex | 51/48 | 30+ |
| **GIPPING GLACIAL** | | |
| *Region 7: Midland England* | | |
| *22.* Dorchester, Oxon. | 41/59 | 29–30 |
| *23.* Wolvercote, Oxford | 42/40 | 30− |
| *Region 9: East Anglia and London* | | |
| *24.* Broome Heath, Norfolk† | 62/39 | 30+ |
| *Region 10: South-east England* | | |
| *25.* Farnham, Surrey | 41/84 | 30+ |

| | Nat. Grid | Present-day isotherm value of site (°C) |
|---|---|---|
| **HOXNIAN INTERGLACIAL AND ASSOCIATED EARLY-GLACIAL AND LATE-GLACIAL** | | |
| *Region 2: Ireland* | | |
| *26.* Gort, Co. Galway | 93/57 | 25 |
| *27.* Kilbeg, Co. Waterford | 02/36 | 24 |
| *28.* Newtown, Co. Waterford | 02/66 | 25 |
| *Region 7: Midland England* | | |
| *29.* Nechells, Birmingham† | 42/08 | 29 |
| *Region 9: East Anglia and London* | | |
| *30.* Clacton-on-Sea, Essex | 62/11 | 30 |
| *31.* Hoxne, Suffolk | 62/17 | 30·5 |
| *32.* Marks Tey, Essex† | 52/92 | 30+ |
| **LOWER MIDDLE PLEISTOCENE** | | |
| *Region 9: East Anglia* | | |
| (*a*) *Lowestoftian glacial* | | |
| *33.* Lowestoft–Corton, Suffolk† | 62/59 | 30 |
| (*b*) *Cromerian warm period with early-Lowestoftian and late-Beestonian* | | |
| *34.* Bacton, Norfolk† | 63/33 | 30 |
| *35.* Mundesley, Norfolk† | 63/33 | 30 |
| WR. West Runton, Norfolk | 63/14 | 30− |
| (*c*) *Beestonian cold* | | |
| *36.* Beeston, Norfolk† | 63/14 | 30− |
| (*d*) *Pastonian warm and early-Beestonian* | | |
| *37.* Happisburgh, Norfolk† | 63/33 | 30 |
| *35.* Mundesley, Norfolk† | 63/33 | 30 |
| *38.* Ostend, Norfolk† | 63/33 | 30 |
| *39.* Paston, Norfolk† | 63/33 | 30 |
| (*e*) '*Lower Middle Pleistocene*' unclassified | | |
| *34.* Bacton, Norfolk | 63/33 | 30 |
| *38.* Ostend, Norfolk | 63/33 | 30 |
| *40.* Pakefield, Suffolk | 62/58 | 30 |
| **'LOWER PLEISTOCENE' UNCLASSIFIED** | | |
| *Region 9: East Anglia* | | |
| *37.* Happisburgh, Norfolk† | 63/33 | 30 |

TABLE 6. *Regional variation in maximum depressions of maximum summer temperature for the Late-Weichselian*

Data from figure 27 with dubious and uncertainly dated records ignored. In column (*b*) asterisked entries refer to pollen determinations only; figures in brackets refer to *Betula nana* and are included only if greater than those of column (*a*). Heavy type indicates reliable entries for three or more species at the figure given.

** Pollen of *Artemisia norvegica*.
– No records. '.' signifies no relevant entries.

For East Anglia and south-east England certain figures would be reduced were the calculations based on Scandinavian limiting isotherms:
$^{x}$ Reduced to 3·2 °C. $^{y}$ Reduced to 0 °C.
$^{z}$ Reduced to 0·2 or 0 °C. $^{v}$ Reduced to 1·0 °C.

| | Late-Weichselian | | | | | | | |
|---|---|---|---|---|---|---|---|---|
| | Zone I | | Zone II | | Zone III | | *Sensu lato* | |
| Region | (*a*) °C | (*b*) °C | (*a*) °C | (*b*) °C | (*a*) °C | (*b*) °C | (*a*) °C | (*b*) °C |
| Scotland | **0** | . | 1·5 | 3* (3–) | **0** | . | . | . |
| Ireland | 2+ | . | 3 to 3·5 | (3) | **3** | (3) | (III/IV 4·2*) | |
| Isle of Man | . | 0·2* | 0·2 | . | 0·2 | 2+* | **2** | . |
| North-west Britain | 0 | 3* (2–) | 1·8 | 3* (3·3) | 2·5 | 4·5* | . | . |
| Wales | 0 | 2·5** (1)* | 0 | 2·5** (1–)* | 0 | 2·5** (1–)* | . | . |
| South-west England | 1 | . | 1 to 1+ | (2–) | . | . | **1·0** | (2–) |
| Midland England | . | 2·3* | . | 2* to 2·3* | . | **2* to 3*** | 4 | (4·5) |
| North-east Britain | <0·5 | (4 to 5)* | **2·5 to 3–** | (4 to 5)* | **2·5 to 3·5** | (4 to 5)* | 3 | (4–) |
| East Anglia and London | No certainly dated and determined records | | | | 3+$^{z}$ to 4·4$^{x}$ | (5)$^{y}$ | 3+$^{v}$ to 4·4$^{x}$ | (5)$^{y}$ |
| South-east England | 3 | . | 2$^{z}$ | . | . | (5)$^{y}$ | | |

TABLE 7. *Regional variation in maximum depressions of maximum summer temperature for the Post-Glacial*

Data from figure 28 with dubious and uncertainly dated records ignored. In column (*b*) asterisked entries refer to pollen determinations only; those in brackets to *Betula nana* and are included only if greater than entries in column (*a*).

– No records. '.' Signifies no relevant entries.

For East Anglia and south-east England certain figures would be reduced were the Scandinavian data to be used as basis for calculations:
$^{x}$ Reduced to 0·2 °C. $^{y}$ Reduced to 0 °C.

| | Post glacial | | | |
|---|---|---|---|---|
| | Early (Zone IV–VI) | | Later (Zone VII onward) | |
| Region | (*a*) °C | (*b*) °C | °C | |
| Scotland | 0 | (0·3)* | 0 | |
| Ireland | 0 | . | . | (III/IV 4·2*) |
| Isle of Man | . | . | – | |
| North-west Britain | 0 | 1* (3+)* | 0 | (III/IV 3*) |
| Wales | 0 | (1·3)* | 0 | |
| South-west England | – | – | – | |
| Midland England | 0 | 2* (4·3) | 0 | (III/IV 2·5) |
| North-east Britain | 2·5 | . | 0 | |
| East Anglia | 2$^{x}$ | . | – | |
| South-east England | . | (5–)$^{y}$ | – | |

TABLE 8. *Maximum depressions of maximum summer temperature for Weichselian and Pre-Weichselian sites*

Data from figures 29–32 with dubious and uncertainly dated records ignored. In column (*b*) asterisked entry refers to pollen determination only and those in brackets to *Betula nana*, and are included only if greater than those in column (*a*). Heavy type indicates a minimum of three or more reliable entries.

– No records. '.' Signifies no relevant entries.

** *Salix polaris* 5·2 or 5·5 °C.

For East Anglia and South-east England certain figures would be reduced were calculations based on Scandinavian limiting isotherms:

[x] Reduced to 0·3 °C or less. [y] Reduced to *c*. 3·2 °C.

[z] Reduced to 4± °C.

| | Weichselian | | | | | Ipswichian | | Gipping | | Hoxnian | |
|---|---|---|---|---|---|---|---|---|---|---|---|
| | Late Mid- | | Mid- | Early | | | | | | | |
| | (*a*) | (*b*) | | (*a*) | (*b*) | (*a*) | (*b*) | (*a*) | (*b*) | (*a*) | (*b*) |
| Region | °C | °C | °C | °C | °C | °C | °C | °C | °C | °C | °C |
| Ireland | – | – | – | – | – | – | – | – | – | 3 | . |
| Midland England | – | – | **3·5 to 4+** | 2 | (4) | . | . | 2·5 | . | . | (4) |
| East Anglia and London | 4 | (5–)*[x] | **5**–6 | *c*. 4·5[y] | (5·3)[x] | 3+[x] | (5)[x] | **4** to **4·4**[z]<br>** | (5)[x] | 4·5 to 5[z]<br>** | (5+)[x] |
| South-east England | – | – | – | . | . | . | . | **4** to 5[z] | (5)[x] | – | – |

| | Lowestoftian | Cromerian | | Beestonian | Pastonian | |
|---|---|---|---|---|---|---|
| | | (*a*) | (*b*) | | (*a*) | (*b*) |
| | °C | °C | °C | °C | °C | °C |
| East Anglia and London | 5–[z] | 4+[z] to 5–[z] | (5–)[x] | 4 | 5–[z] | (5–)[x] |

TABLE 9. *Regional variation in the maximum possible depression of maximum summer temperature, based on the potentiality of the whole region (column 2) and on the current location of fossil sites (column 5)*

Maximum depressions (column 2 and 5) are based on the theoretical occurrence of the 'coldest' arctic-montane species with a firm determination found fossil in the British Isles from any period, i.e. a species limited by an isotherm of 21 °C for Highland Scotland and regionally relevant isotherms elsewhere.
* m.a.m.s. = mean annual maximum summit.
[x] Corrected on basis of Scandinavian data.
PGl = Later Post-glacial. PGe = Early Post-glacial. LW = Late-Weichselian. W = Weichselian. I = Ipswichian. G = Gipping. H = Hoxnian. L.M.P. = Lower Middle Pleistocene. BT = Bovey Tracey.

| Region | Highest possible m.a.m.s.* isotherm in region (1) °C | Maximum possible depression of summer temperature for region (2) °C | Highest m.a.m.s.* isotherm for existing sites (3) °C | Number of sites with isotherm value not more than 1 °C below highest possible (see (1))/ total no. of sites (4) | Maximum possible depression of summer temperatures based on existing site locations (5) °C |
|---|---|---|---|---|---|
| Scotland | 25·5 | 4·5 | LW 25 | 3/13 | LW 4 |
| | | | PGe 22·5 | 0/5 | PGe 1·5 |
| | | | PGl 21·5 | 0/1 | PGl 0·5 |
| Ireland | 25·5 | 4·5 | LW 25·5 | 12/18 | LW 4·5 |
| | | | PGe 25+ | 4/7 | PGe 4·2 |
| | | | PGl 24+ | 0/1 | PGl 3·2 |
| | | | H 25 | 1/3 | H 4·0 |
| Isle of Man | 23+ | 2·2 or less | LW, PG 23+ | 4/4 | LW, PG 2·2 or less |
| North-west Britain | 28 | 5·0–6·0 | LW, PGe 27·5 | 5/22<br>1/11 | LW, PGe 4·5 (5·5) |
| | | | PGl 26+ | 0/8 | PGl 3·2–4·2 |
| Wales | 28+ | 5·2 | LW 25·5 | 0/8 | LW 2·5 |
| | | | PGe 25+ | | PGe 2·2 |
| | | | PGl 25 | | PGl 2·0 |
| South-west England | 28·5 | 6·5 | (BT) 26 | 0/3 | (BT) 4·0 |
| | | | LW, PGe 25 | | LW, PGe 3·0 |
| Midland England (*a*) North | 29 | 5·5–6·0 | W 28+ | 1/1 | W 5+ |
| | | | LW, PGe 28·5 | 3/3<br>3/5 | LW, PGe 5·0–5·5 |
| | | | PGl 28+ | 1/2 | PGl 4·7–5·2 |
| (*b*) South | 30+ | 6·4–6·7 | W 30+ | 2/3 | W 6·4–6·7 |
| | | | G 29·5 | 2/2 | G 6·0–5·7 |
| | | | H 29 | 0/1 | H 5·2–5·5 |
| | | | LW, PGe 29·5 | 2/2<br>1/2 | LW, PGe 6·0–5·7 |
| North-east Britain (*a*) North | 26·5 | 3·5–4·5 | LW, PGe 25·5 | 2/6<br>1/4 | LW, PGe 2·5–3·5 |
| (*b*) South | 30 | 6·5–7·0 | LW 29+ | 1/7 | LW 5·7–6·2 |
| | | | PG 27·5 | 0/3 | PG 4·0–4·5 |
| East Anglia and London | 30·5 | 6·5 (5·5)[x] | W 30·5 | 13/13 | W 6·5 |
| | | | I, G, H 30+ | 8/8 | 6·2 |
| | | | LMP 30 | 8/8 | |
| | | | LW, PGe 30+ | 3/3<br>1/3 | |
| South-east England | 30+ | 6·2 | G 30+ | 1/1 | G 6·2 |
| | | | LW, PGe 30 | 1/3<br>2/3 | LW, PGe 6·0 |

# THE CAMBRIDGE POLLEN REFERENCE COLLECTION

by R. Andrew

*Subdepartment of Quaternary Research, Botany School, University of Cambridge*

## INTRODUCTION

The Cambridge pollen reference collection has been built up over the last twenty-two years from a handful of basic pollen slides into a collection comprising over 1700 British species, as well as a collection of small representative groups of European, Mediterranean, North and South American and Canadian species. and with modest collections of moss spores and Malayan fern spores. The British collection possesses several examples of most species, obtained from different localities and prepared by different methods. It is therefore a particularly valuable collection.

The purpose of this essay is to produce an account of the British collection and to describe some details of the curating aspects and to set out a useful list of identifiable pollen types. It is not intended to provide a key but rather a practical guide for the use of students new to pollen identification. It does not pretend to cover the whole field of pollen identification but simply to outline a working list of taxa of the British Flora which may be readily identifiable from reference collections.

## SOURCE AND PREPARATION OF REFERENCE MATERIAL

The type material has been acquired from various sources: from collectors in the field, from the Herbarium of the Botany School, from the Botanic Gardens, from abroad, and from other collections. We are fortunate in possessing an extensive collection of reference slides made up in silicone oil under the supervision of Dr A. G. Smith of Belfast.

With material obtained from herbarium sources it is usual to keep a permanent record connecting the reference slide to the parent herbarium sheet, so that it is possible to recover the original plant from which the material was obtained. This is particularly important where the taxonomy of a particular group is uncertain or undergoing revision.

To suit the requirements of the many research students who have worked in the Subdepartment over the years, reference material has been prepared in various ways, since it is advisable to match the reference slide consulted to the method of preparation of the fossil slide.

Professor Godwin's is the basic method used and consists in boiling for 5 to 10 minutes the anthers or whole small flowers of fresh or herbarium specimens in an acetolysis mixture of 9:1 glacial acetic acid and concentrated sulphuric acid. I find that a second acetolysis process using 9:1 acetic anhydride and concentrated sulphuric acid clears excessive debris, sharpens up the structure of the pollen grains and fills them out without undue swelling. Other methods of preparation may be found in the literature listed in the Bibliography, notably in Faegri & Iversen (1964).

Permanent preparations are made up in glycerine jelly stained with safranin and sealed. Experiments have been made with several types of seal: DPX mountant, paraffin wax, varnish and shellac. All are more or less satisfactory. Sometimes paraffin wax tends to invade the preparation and so spoil the slide. DPX mountant has the advantage of being easily removed to enable a dried-up slide to be reconstituted with fresh safranin glycerine jelly.

## SYSTEMATIC LIST

The list which forms the important part of this paper is based on details of identification contained in an unpublished classified index which I have compiled during the work of building-up the reference collection. This list shows the degree of specific identification which can be attained within families and genera.

The arrangement of the taxa is simple. On the left are the families; the second column repeats the name of the family if no more detailed identification can be made, the third gives groups of genera identifiable, the fourth gives identifiable genera, and the last identifiable species. Notes on particular problems are added in certain families.

This working list contains only those species which are normally found in fossil slides. There are many more which could be identified but which have not yet been found by us. There are also a number of families, such as Cruciferae, Umbelliferae or Labiatae, the species of which could be identified with further detailed study. But groups which hybridize freely such as *Salix*, *Epilobium* or *Rumex* or groups which have a critical taxonomy, such as *Rosa*, or polymorphic forms such as *Linum* still present a problem. It may be remarked that there is very little literature on the identification of British pollen taxa; perhaps the following list may indicate the particular gaps in our knowledge and may stimulate the necessary work on pollen morphology of particular genera important for the elucidation of ecological and vegetational history.

The nomenclature is that of Dandy's list of *British Vascular Plants* (1958):

| Family | Family type | Groups of genera | Genera | Species |
|---|---|---|---|---|
| Lycopodiaceae | | | *Lycopodium* | *selago* |
| | | | *L.* | *inundatum* |
| | | | *L.* | *annotinum* |
| | | | *L.* | *clavatum* |
| | | | *L.* | *alpinum* |
| Selaginellaceae | | | *Selaginella* | *selaginoides* |
| Isoëtaceae | | | *Isoëtes* | *lacustris* |
| | | | *I.* | *echinospora* |
| | | | *I.* | *hystrix* |
| Equisetaceae | | | *Equisetum* | |
| Osmundaceae | | | *Osmunda* | *regalis* |
| [Other *Osmunda* species (e.g. *O. claytoniana* and *O. cinnemomea*) are found fossil and can be distinguished from *O. regalis*.] | | | | |
| Hymenophyllaceae | | | *Trichomanes* | *speciosum* |
| | | | *Hymenophyllum* | |
| Dennstaedtiaceae | | | *Pteridium* | *aquilinum* |
| Adiantaceae | | | *Cryptogramma* | *crispa* |
| | | | *Adiantum* | *capillus-veneris* |
| Blechnaceae | Filicales | | | |
| Aspleniaceae | Filicales | | | |
| Athyriaceae | Filicales | | | |
| Aspidiaceae | Filicales | | | |
| Thelypteridaceae | Filicales | | *Thelypteris* | *dryopteris* |
| Polypodiaceae | | | *Polypodium* | *vulgare* |
| Marsileaceae | | | *Pilularia* | *globulifera* |
| Azollaceae | | | *Azolla* | |
| Ophioglossaceae | | | *Botrychium* | |
| | | | *Ophioglossum* | *vulgatum* |
| Pinaceae | | | *Pinus* | *sylvestris* |
| [*Picea*, *Abies*, *Sequoia*, *Tsuga*, *Pinus haploxylon*, *Sciadopitys* are found fossil and can be identified.] | | | | |
| Cupressaceae | | | *Juniperus* | *communis* |
| Taxaceae | | | *Taxus* | *baccata* |
| Ranunculaceae | | *Ranunculus*, *Anemone* } type | *Caltha* | *palustris* |
| | | | *Trollius* | *europaeus* |
| | | | *Thalictrum* | |
| Nymphaeaceae | | | *Nymphaea* | |
| | | | *Nuphar* | |
| Cruciferae | Cruciferae | | | |
| Polygalaceae | | | *Polygala* | |
| Guttiferae | | | *Hypericum* | |
| Cistaceae | | | *Helianthemum* | |
| Caryophyllaceae | Caryophyllaceae | *Dianthus*, *Silene*, *Lychnis* } type | *Agrostemma* | *githago* |
| | | *Cerastium*, *Stellaria*, *Minuartia*, *Arenaria*, *Sagina* } type | *Stellaria* | *holostea* |
| | | | *Spergula* | *arvensis* |
| Illecebraceae | | | *Scleranthus* | *annuus* |
| | | | *S.* | *perennis* |
| Portulacaceae | | | *Montia* | |

| Family | Family type | Groups of genera | Genera | Species |
|---|---|---|---|---|
| Chenopodiaceae | Chenopodiaceae | | | |
| Tiliaceae | | | *Tilia* | *platyphyllos* |
| | | | *T.* | *cordata* |
| Malvaceae | | | *Malva* | |
| | | | *Althaea* | *officinalis* |
| Linaceae | | *Linum bienne*, *L. usitatissimum* } type | | |
| | | | *Linum* | *anglicum* |
| | | | *L.* | *catharticum* |

[The pollen grains of *Linum* can vary within the species in relation to the style form.]

| Family | Family type | Groups of genera | Genera | Species |
|---|---|---|---|---|
| Geraniaceae | | | *Geranium* | |
| | | | *Erodium* | |
| Balsaminaceae | | | *Impatiens* | *noli-tangere* |
| | | | *I.* | *parviflora* |
| Aceraceae | | | *Acer* | *campestre* |
| Hippocastanaceae | | | *Aesculus* | *hippocastanum* |
| Aquifoliaceae | | | *Ilex* | *aquifolium* |
| Celastraceae | | | *Euonymus* | *europaeus* |
| Buxaceae | | | *Buxus* | *sempervirens* |
| Rhamnaceae | | | *Rhamnus* | *catharticus* |
| | | | *Frangula* | *alnus* |
| Vitaceae | | | *Vitis* | *vinifera* |
| Leguminosae | Leguminosae | | *Sarothamnus* | *scoparius* |
| | | | *Ononis* | |
| | | | *Trifolium* | |
| | | | *Lotus* | |
| | | | *Astragalus* | |
| | | | *Onobrychis* | *viciifolia* |
| | | | *Vicia* | |
| | | | *Lathyrus* | |
| Rosaceae | | | *Filipendula* | *ulmaria* |
| | | | *Rubus* | *chamaemorus* |
| | | *Potentilla* type | | |

[*Potentilla* type includes *Geum*, *Fragaria*, *Potentilla*, *Sibbaldia*, and *Comarum*.]

| Family | Family type | Groups of genera | Genera | Species |
|---|---|---|---|---|
| | | | *Dryas* | *octopetala* |
| | | | *Alchemilla* | |
| | | | *Sanguisorba* | *officinalis* |
| | | | *Poterium* | *sanguisorba* |
| | | | *Prunus* | |
| | | | *Crataegus* | |
| | | | *Sorbus* | |
| Crassulaceae | | | *Sedum* | *rosea* |
| Saxifragaceae | | | *Saxifraga* | *stellaris* |
| | | | *S.* | *nivalis* |
| | | | *S.* | *hypnoides* |
| | | | *S.* | *aizoides* |
| | | | *S.* | *oppositifolia* |
| Droseraceae | | | *Drosera* | *rotundifolia* |
| | | | *D.* | *anglica* |
| | | | *D.* | *intermedia* |
| Lythraceae | | | *Lythrum* | *salicaria* |
| | | | *Peplis* | *portula* |
| Eleagnaceae | | | *Hippophaë* | *rhamnoides* |
| Onagraceae | | | *Epilobium* | |
| | | | *Circaea* | |
| Haloragaceae | | | *Myriophyllum* | *verticillatum* |
| | | | *M.* | *spicatum* |
| | | | *M.* | *alterniflorum* |
| Hippuridaceae | | | *Hippuris* | *vulgaris* |
| Callitrichaceae | | | *Callitriche* | |
| Loranthaceae | | | *Viscum* | *album* |
| Cornaceae | | | *Thelycrania* | *sanguinea* |
| Araliaceae | | | *Hedera* | *helix* |
| Umbelliferae | Umbelliferae | | *Hydrocotyle* | *vulgaris* |
| | | | *Apium* | |
| | | | *Heracleum* | *sphondylium* |
| Euphorbiaceae | | | *Mercurialis* | *perennis* |
| | | | *M.* | *annua* |
| | | | *Euphorbia* | |

| Family | Family type | Groups of genera | Genera | Species |
|---|---|---|---|---|
| Polygonaceae | | | *Polygonum* | *aviculare* |
| | | | *P.* | *viviparum* |
| | | | *P.* | *bistorta* |
| | | | *P.* | *amphibium* |
| | | *Polygonum persicaria* type | | |
| [*Polygonum persicaria* type includes *Polygonum persicaria*, *P. lapathifolium*, *P. nodosum*, *P. hydropiper*, *P. mite*, *P. minus*.] | | | | |
| | | | *P.* | *convolvulus* |
| | | | *Fagopyrum* | *esculentum* |
| | | | *Koenigia* | *islandica* |
| | | | *Rumex* | |
| | | | *R.* | *acetosella* |
| | | | *R.* | *acetosa* |
| | | | *R.* | *hydrolapathum* |
| Urticaceae | Urticaceae | | *Urtica* | *urens* |
| | | | *U.* | *dioica* |
| Cannabiaceae | | | *Humulus* | *lupulus* |
| [*Cannabis sativa* is also found fossil and can be distinguished from *Humulus* when in good condition.] | | | | |
| Ulmaceae | | | *Ulmus* | |
| Juglandaceae | | | *Juglans* | *regia* |
| [*Pterocarya*, *Platycarya*, *Engelhardtia*, *Carya* are also found fossil and can be identified.] | | | | |
| Myricaceae | | | *Myrica* | *gale* |
| Betulaceae | | | *Betula* | |
| | | | *B.* | *nana* |
| | | | *Alnus* | *glutinosa* |
| [*Betula* species including *B. pubescens* ssp. *odorata* an d*B. tortuosa* are not always easy to distinguish from one another. At various times statistical, size (Birks, 1968) and pore/grain ratio methods (Walker, 1955) have been used but the problem has not yet been really reliably resolved for use on fossil grains.] | | | | |
| Corylaceae | | | *Carpinus* | *betulus* |
| | | | *Corylus* | *avellana* |
| Fagaceae | | | *Fagus* | *sylvatica* |
| | | | *Quercus* | |
| [*Quercus ilex* may be found fossil and identified when in good condition (Van Campo & Elhai, 1956).] | | | | |
| Salicaceae | | | *Populus* | |
| | | | *Salix* | |
| | | | *S.* | *herbacea* |
| Ericaceae | | | *Rhododendron* | *ponticum* |
| | | | *Andromeda* | *polifolia* |
| | | | *Arctostaphylos* | *uva-ursi* |
| | | | *Calluna* | *vulgaris* |
| | | | *Erica* | |
| | | | *E.* | *terminalis* |
| | | | *Vaccinium* | |
| Diapensiaceae | | | *Diapensia* | *lapponica* |
| Empetraceae | | | *Empetrum* | *nigrum* |
| | | | *E.* | *hermaphroditum* |
| Plumbaginaceae | | | *Armeria* | *maritima* |
| Primulaceae | | | *Lysimachia* | |
| Oleaceae | | | *Fraxinus* | *excelsior* |
| Gentianaceae | Gentianaceae | | *Gentiana* | |
| | | | *G.* | *verna* |
| Menyanthaceae | | | *Menyanthes* | *trifoliata* |
| Polemoniaceae | | | *Polemonium* | *caeruleum* |
| Boraginaceae | Boraginaceae | | *Cynoglossum* | |
| | | | *Symphytum* | *officinale* |
| Convolvulaceae | | | *Convolvulus* | *arvensis* |
| | | | *Calystegia* | |
| | | | *Cuscuta* | |
| Solanaceae | | | *Solanum* | *dulcamara* |
| | | | *S.* | *nigrum* |
| Scrophulariaceae | | | *Veronica* | |
| | | | *Euphrasia* | |
| | | | *Pedicularis* | |
| | | | *Melampyrum* | |
| Lentibulariaceae | | | *Pinguicula* | |
| | | | *Utricularia* | |
| Labiatae | | Labiatae M type | *Mentha* | |
| | | | *Lycopus* | *europaeus* |
| | | Labiatae L type | *Lamium* | |
| Plantaginaceae | | | *Plantago* | *major* |

| Family | Family type | Groups of genera | Genera | Species |
|---|---|---|---|---|
| Plantaginaceae (*cont.*) | | | *Plantago* | *media* |
| | | | *P.* | *lanceolata* |
| | | | *P.* | *maritima* |
| | | | *P.* | *coronopus* |
| | | | *Littorella* | *uniflora* |
| Campanulaceae | | | *Campanula* | |
| | | | *Jasione* | *montana* |
| Rubiaceae | Rubiaceae | | | |
| Caprifoliaceae | | | *Sambucus* | |
| | | | *Viburnum* | *lantana* |
| | | | *V.* | *opulus* |
| | | | *Linnaea* | *borealis* |
| | | | *Lonicera* | *periclymenum* |
| Adoxaceae | | | *Adoxa* | *moschatellina* |
| Valerianaceae | | | *Valerianella* | |
| | | | *Valeriana* | *officinalis* |
| | | | *V.* | *dioica* |
| Dipsacaceae | | | *Dipsacus* | |
| | | | *Knautia* | *arvensis* |
| | | | *Scabiosa* | *columbaria* |
| | | | *Succisa* | *pratensis* |
| Compositae | Compositae 'Tub.' | *Senecio* type | *Bidens* | *tripartita* |
| | | | *Ambrosia* | |
| [*Senecio* type includes Dandy genera 501 to 525.] | | | | |
| | | *Matricaria* type | *Artemisia* | |
| | | | *A.* | *norvegica* |
| [*Matricaria* type includes Dandy genera 526 to 535.] | | | | |
| | | *Arctium* type | | |
| | | | *Saussurea* | *alpina* |
| [*Arctium* type includes Dandy genera 537, 538, 542, 543 and 545.] | | | | |
| | | *Cirsium* type | | |
| [*Cirsium* type includes Dandy genera 539 to 541.] | | | | |
| | | | *Centaurea* | *scabiosa* |
| | | | *C.* | *cyanus* |
| | | | *C.* | *nigra* |
| | Compositae 'Lig.' | *Taraxacum* type | | |
| [*Taraxacum* type includes Dandy genera 546 to 560.] | | | | |
| Alismataceae | | | *Alisma* | *plantago-aquatica* |
| | | | *Sagittaria* | *sagittifolia* |
| Butomaceae | | | *Butomus* | *umbellatus* |
| Hydrocharitaceae | | | *Hydrocharis* | *morsus-ranae* |
| | | | *Stratiotes* | *aloides* |
| Scheuchzeriaceae | | | *Scheuchzeria* | *palustris* |
| Juncaginaceae | | | *Triglochin* | |
| Potamogetonaceae | | | *Potamogeton* | |
| Liliaceae | | | *Tofieldia* | *pusilla* |
| | | | *Narthecium* | *ossifragum* |
| | | | *Endymion* | *non-scriptus* |
| Amaryllidaceae | | | *Allium* | |
| Iridaceae | | | *Iris* | *pseudacorus* |
| Dioscoreaceae | | | *Tamus* | *communis* |
| Orchidaceae | | *Listera* type | | |
| [*Listera* type includes those genera of Orchidaceae which produce distinctive pollen tetrads.] | | | | |
| Lemnaceae | | | *Lemna* | |
| Sparganiaceae | | | *Sparganium* | *erectum* |
| | | | *S.* | *minimum* |
| Typhaceae | | | *Typha* | *latifolia* |
| | | | *T.* | *angustifolia* |
| Cyperaceae | Cyperaceae | | *Rhynchospora* | *alba* |
| | | | *Cladium* | *mariscus* |
| Gramineae | Gramineae | | | |
| [Pollen grains of Gramineae are not easy to determine but useful studies have been made, under phase contrast, of cereal pollen enabling *Secale cereale* and *Triticum* to be identified with some certainty (Faegri & Iversen, 1964; Grohne, 1957).] | | | | |

### Some notes on other pollen and spore types and microfossils encountered on fossil slides

Pollen grains found fossil but which have no link with British families and are therefore excluded from the list are *Ephedra*, *Brasenia*, *Liquidambar*, *Nyssa*, *Rhus* and other so-called 'Tertiary types'. Other finds in fossil slides include moss spores: some are identifiable to genera such as *Anthoceros* and *Sphagnum*. Carboniferous spores, Jurassic pollen grains such as *Classopollis*, *Caytonanthus*, and spores of *Schizaceae* are frequently encountered in inorganic sediments. Fungus spores are sometimes found in great quantity: a few can be identified such as the rust spores of *Ustilago* and teleutospores of *Tilletia sphagni*. Fungus fruiting bodies of Microthyriaceae also occur (Cookson, 1947; Godwin & Andrew, 1951). Also found in micro-fossil material are the massulae and glochidia of *Azolla* (West, 1953), *Salvinia* spores, colonies of *Pediastrum*, *Botryococcus*, *Tetraploa*, testaceous rhizopods, diatoms, *Crassosphaera* (Cookson & Manum, 1960), Hystricosphaeridae (Churchill & Sargeant, 1962), fragments of conifer wood, animal remains, small seeds such as *Juncus*, plant tissues such as *Sphagnum* leaves, leaf hairs of *Hippophaë*, hairs of *Anthrenus* larvae, algal aplanospores (Churchill, 1960), leaf spines of *Ceratophyllum*, and rhizoid gemmae of mosses (Whitehouse, 1966).

## SUMMARY

The pollen reference collection at Cambridge is briefly described and methods of preparation of reference slides are outlined. A systematic list of taxa considered identifiable is given and reference is made to publications useful for identification of British pollen and spores.

## BIBLIOGRAPHY

The following bibliography includes publications referred to in the text, and also publications containing illustrations of pollen grains and spores which have been particularly useful in problems of identification.

Andersen, S. T. (1961). Vegetation and its environment in Denmark in the Early Weichselian Glacial. *Danm. Geol. Unders.* (II), **75**.

Bassett, I. J. and Terasmae, J. (1962). Ragweeds, Ambrosia species, in Canada and their history in postglacial time. *Can. J. Bot.* **40**, 141–50.

Berglund, B. E. (1966). Late-Quaternary vegetation in Eastern Blekinge, south-eastern Sweden. I. Late-glacial time. *Op. bot. Soc. bot. Lund*, **12**, no. 1.

Bertsch, A. (1960). Untersuchungen an rezenten und fossilen pollen von Juniperus. *Flora*, **150**, 503–13.

Beug, H.-J. (1957). Untersuchungen zur spätglazialen und frühpostglazialen Floren und Vegetationsgeschichte einiger Mittelgebirge. *Flora*, **154**, 167–211.

Beug, H.-J. (1960). Beiträge zur postglazialen Floren und Vegetationsgeschichte in Süddalmatien: Der See 'Malo Jezero' auf Mljet. *Flora*, **150**, 632–56.

Beug, H.-J. (1961). *Leitfaden der Pollenbestimmung für Mitteleuropa und angrenzende Gebiete. Lieferung 1.* Stuttgart: Fischer Verlag.

Beug, H.-J. (1964). Untersuchungen zur spät- und postglazialen Vegetationsgeschichte in Gardaseegebiet unter besonderer Berücksichtigung der Mediterranean Arten. *Flora*, **154**, 401–44.

Birks, H. J. B. (1968). The identification of *Betula nana* pollen. *New Phytol.* **67**, 304–14.

Chambers, T. C. & Godwin, H. (1961). The fine structure of the pollen wall of *Tilia platyphyllos*. *New Phytol.* **60**, 393–9.

Chanda, S. (1962). Pollen morphology of some Scandinavian Caryophyllaceae. *Grana Palynol.* **3**, 67–89.

Churchill, D. M. (1960). Living and fossil unicellular algae and aplanospores. *Nature, Lond.* **186**, 493–4.

Churchill, D. M. & Sarjeant, W. A. S. (1962). Freshwater microplankton from Flandrian (Holocene) peats of south-western Australia. *Grana Palynol.* **3**, 29–53.

Cookson, I. C. (1947). Fossil fungi from Tertiary deposits in the southern hemisphere, Pt 1. *Proc. Linn. Soc. N.S.W.* **72**, 207–14.

Cookson, I. C. & Manum, S. (1960). On *Crassosphaera*, a new genus of micro-fossils from Mesozoic and Tertiary deposits. *Nytt Mag. Bot.* **8**, 5–9.

Dandy, J. E. (1958). *List of British Vascular Plants.* London.

Erdtman, G. (1943). *An introduction to pollen analysis.* Waltham, Mass.: Chronica Botanica.

Erdtman, G. (1957). *Pollen and Spore Morphology/Plant taxonomy. Gymnospermae, Pteridophyta, Bryophyta.* Stockholm: Almquist & Wiksell.

Erdtman, G., Berglund, B. & Praglowski, J. (1961). An introduction to a Scandinavian pollen flora. *Grana Palynol.* **2**, no. 3, 3–92.

Erdtman, G. (1966). *Pollen morphology and plant taxonomy. Angiosperms.* New York: Hafner.

Faegri, K. & Iversen, J. (1964). *Textbook of pollen analysis.* (2nd edn.) Oxford: Blackwell.

Godwin, H. (1956). *History of the British Flora.* Cambridge University Press.

Godwin, H. & Andrew, R. (1951). A fungal fruit body common in post-glacial peat deposits. *New Phytol.* **50**, 179–83.

Grohne, U. (1957). Die bedeutung des phasenkontrastverfahrens für die pollenanalyse, dargelegt am beispiel der Gramineenpollen vom Getreidetyp. *Photographie Forsch.* **7**, 237–48.

Hedberg, O. (1946). Pollen morphology in the genus Polygonum L S. Lat. and its taxonomical significance. *Sv. Bot. Tidskr.* **40**, 371–404.

Hyde, H. A. & Adams, K. F. (1958). *An atlas of airborne pollen grains*. London: Macmillan.

Iversen, J. (1954). (Ed.) Studies in vegetational history in honour of Knud Jessen. *Danm. Geol. Unders.* (II), **80**.

Jessen, K., Andersen, S. T. & Farrington, A. (1959). The interglacial deposit near Gort, Co. Galway, Ireland. *Proc. R. Ir. Acad.* **60** B, 1–77.

Knox, E. M. (1951). Spore morphology in British ferns. *Tr. Bot. Soc. Edinb.* **35**, 437–49.

Koelbloed, K. K. & Kroeze, J. M. (1965). Hauwmossen (Anthoceros) als cultuurbegeleiders. *Boor en Spade*, **14**, 104–9.

Kremp, G. (1949). Pollenanalytische untersuchung des Miozänen Braunkohlenlagers von Konin an der Warthe. *Palaeontographica*, **90**, B, 53–93.

Kuprianova, L. A. (1965). *The palynology of the Amentiferae*. Moscow: Academy of Sciences of the U.S.S.R., The Komarov Botanical Institute.

Lang, G. (1951). Nachweis von Ephedra in südwestdeutschen Spätglazial. *Naturwissenschaften*, **14**, 334–5.

Lang, G. (1952). Späteiszeitliche pflanzenreste in südwestdeutschland. *Beitr. Naturk, Forsch. Südw. Dtl*, **11**, 89–110.

Oldfield, F. (1959). The pollen morphology of some of the West European Ericales. *Pollen spores*, **1**, 19–48.

Perring, F. H. & Walters, S. M. (1962). *Atlas of the British Flora*, Botanical Society of the British Isles. London: Nelson.

Reitsma, T. (1966). Pollen morphology of some European Rosaceae. *Acta Bot. Neerl.* **15**, 290–307.

Rudolph, K. (1935). Microfloristische untersuchung Tertiärer ablagerungen im Nördlichen Böhmen. *Beih. bot. Zbl.* **54** B, 244–328.

Sorsa, P. (1964). Studies on the spore morphology of Fennoscandian fern species. *Ann. Bot. Fenn.* **1**, 179–201.

Thiergart, F. (1940). Die Mikropaläontologie als pollenanalyse im dienst der Braunkohlenforschung. *Schr. Geb. Brennstoffgeol.* **13**, 1–60.

Van Campo, M. & Elhai, H. (1956). Etude comparative des pollens de quelques Chênes. *Bull. Soc. bot. Fr.* **103**, 254–60.

Van der Spoel-Walvius, M. R. (1963). Les charactéristiques de l'exine chez quelques espèces de Quercus. *Acta Bot. Neerl.* **12**, 525–32.

Walker, D. (1955). Studies in the Post-glacial history of British vegetation. XIV. Skelsmergh Tarn and Kentmere, Westmorland. *New Phytol.* **54**, 222–54.

Watts, W. A. (1959). Interglacial deposits at Kilbeg and Newtown, Co. Waterford. *Proc. R. Ir. Acad.* **60** B, 79–134.

Welten, M. (1957). Über das glaziale und spätglaziale Vorkommen von Ephedra am nordwestlichen Alpenrand. *Ber. Schweiz. Bot. Ges.* **67**, 33–54.

West, R. G. (1953). The occurrence of Azolla in British interglacial deposits. *New Phytol.* **52**, 267–72.

Whitehouse, H. L. K. (1966). The occurrence of tubers in European mosses. *Trans. Brit. bryol. Soc.* **1**, 103–16.

Wodehouse, P. P. (1935). *Pollen Grains*. New York: McGraw Hill.

Zagwijn, W. H. (1960). Aspects of the Pliocene and Early Pleistocene vegetation in the Netherlands. *Med. Geol. Sticht.* series C III, **1**, no. 5.

Zagwijn, W. H. (1963). Pollen analytic investigations in the Tiglian of the Netherlands. *Med. Geol. Sticht.* New Series **16**, 49–71.

Zoller, H. (1960). Pollenanalytische untersuchungen zür vegetationsgeschichte der insubrischen Schweiz. *Denkschr. Schweiz. Naturf. Ges.* **83**, no. 2.

# THE STUDY OF PLANT MACROFOSSILS IN BRITISH QUATERNARY DEPOSITS

by C. A. Dickson

*Subdepartment of Quaternary Research, Botany School, University of Cambridge*

## INTRODUCTION AND HISTORICAL SURVEY

The identification of plants by their macroscopic remains was one of the first branches of Quaternary studies. In 1862, Heer published full descriptions and drawings of fossil leaves of *Betula nana* L. and *Salix* species from Bovey Tracey, Devon. Later, in a short paper published in 1873, Nathorst demonstrated the former presence of arctic floras from the same site and other deposits in southern England and in Europe. His discovery of *Salix polaris* Wahlenb. and other plants from the Cromer Forest Bed (still exposed in the sea cliffs of north-eastern Norfolk) inspired Reid to investigate these rich deposits in detail. After examining sites ranging from the early Quaternary ones at Cromer to Roman Silchester, Reid summarized the evidence in *The Origin of the British Flora* (Reid, 1899), incorporating the work by Andersson and Nathorst in Sweden and by Weber in Germany. Early Danish work on macroscopic remains was compiled by Hartz (1909). The high quality of the photographs of fruits and seeds from the rich flora of the Cromer Forest Bed, illustrated by Reid (1908), demonstrate the criteria by which the species were recognized. The Reids' knowledge of the fruits and seeds of the British flora is shown by his comments in their paper (p. 206) 'Almost every species that can be distinguished by other characters can be distinguished also by the seed alone; and often the seed or fruit, though apparently undescribed, gives better specific characters than the whole of the rest of the plant. The cases where there is doubt are usually seeds of plants belonging to closely allied species, which give almost as much difficulty when we compare the whole plant. Thus we do not pretend to be able to distinguish the different fruticose *Rubi* by their stones, nor can we find satisfactory specific characters in the carpels of *Rosa*, or of the Batrachian *Ranunculi*'. The great body of taxonomic work since the Reids' time serves to validate his assessment, if a certain justifiable optimism is granted. The Reids, with Chandler, continued as the British experts in this field adding especially to our knowledge of the rich floras of the last glaciation, e.g. Barnwell Station (1921) and Lea Valley, summarized by Mrs Reid in her 1949 paper.

In the nineteen thirties, the identification of Quaternary plants gained a new impetus from a greater understanding of Quaternary stratigraphy and chronology resulting from the development of pollen analysis, especially by Professor and Mrs Godwin in England and Wales and Jessen in Ireland.

It is noteworthy that, apart from Heer's pioneer study, few detailed descriptions were given until the paper by Clifford (1936), who took pains to justify his identifications not only with photographs but with full written accounts.

Improved techniques have increased the scope of identification and minute seeds and spores are now recovered more frequently than in the past. By 1956, Godwin, in his compilation of British records in *The History of the British Flora* (p. 351), revealed that over five hundred species of the present British vascular plant flora had been recorded, largely from macroscopic fossils. The illustrations and descriptions of macrofossils in his book show the diversity of plant remains which can be identified. Many of these are from the rich Late-glacial floras of Cornwall (Conolly, Godwin & Megaw, 1950) and the Lea Valley, Essex (Allison, Godwin & Warren, 1952) together with those from Ireland (Jessen, 1949). Since Godwin's book was published many sites have yielded rich macro-floras. Most notable are the inter-glacial sites such as the Hoxnian (penultimate inter-glacial) at Gort, Co. Galway, Ireland (Jessen, Andersen & Farrington, 1959) and the Ipswichian (last inter-glacial) at Bobbitshole, Suffolk (West, 1957) both having plants no longer native in Britain. Especially noteworthy floras of the last glaciation include those from Sidgwick Avenue, Cambridge

(Lambert, Pearson & Sparks, 1963), Upton Warren, Worcestershire (Coope, Shotton & Strachan, 1961) and Colney Heath, Hertfordshire (Godwin, 1964).

## COLLECTION AND PREPARATION OF REFERENCE MATERIAL

Ideally a reference collection should comprise only specimens which have been vouched for by a taxonomist. As succinctly stated by Conolly (1961): 'The adequacy of the taxonomic determination is the basis upon which ecological deduction rests. Thus type material must represent the full range of morphological variability and geographical source.' Species which produce fertile hybrids present special problems and a wide enough range may show a complete series linking fruits, seeds and leaves of two or three species, as in *Betula*. Diagnostic features are often found in capsules, pods, persistent calyces, immature fruits and inflorescences; examples of these, buds and catkin scales, leaves and leaf bases, sheathing stipules and spines have all been found fossil and should be included in a reference collection. It is important that fruits and seeds are ripe; unripe achenes are often greenish and rather deflated around the funicle; Gramineae caryopses should be plump and unwrinkled; and Caryophyllaceae seeds of many genera are blackish when fully ripe.

If the collection is made from a herbarium the sheets should be labelled with a code number and this recorded with the collected reference material. Taxonomic changes or amended identifications can then be passed on to the Quaternary worker.

Cardboard or plastic trays, the size of a microscope slide, with glass or cellophane covers and each containing one large or perhaps two or more smaller cavities can be specially made by laboratory suppliers. The boxes enable a number of species to be quickly compared and also permit the fossils to be placed alongside for direct comparison. If the reference seeds are to be affixed, gum tragacanth is useful since the seeds can be easily removed, if desired, by moistening the gum.

In many cases, in the course of time, the fossils have lost soft fleshy parts such as thin-walled epidermal and hypodermal layers, and hairs and spines may have disappeared. Seeds with delicate testas usually collapse as the living contents decay. Even the more durable ribs, borders, style bases and stalks of fruits may be missing in less well-preserved material. Some fruits and seeds can be reduced to a flattened embryo (when this has a cuticular wall), or consist of pieces of testa, pericarp or perianth; though fragmentary, such fossils may still be identified.

In order to simulate the fossil state, fruits and seeds may be gently heated in a dilute sodium hydroxide solution (about 5 %) for a few minutes; the concentration and time should be varied according to the durability of the specimen. After washing, the softened parts may then be gently eased away. Care must be taken initially to avoid overheating; for instance, the fruitstones of *Potamogeton* species may be reduced to little more than embryos and the difference between the species then disappear. It may, however, be useful to prolong the treatment for some specimens to test whether underlying layers have a distinctive appearance. After suitable washing these specimens may be kept alongside the untreated ones. With some species the delicate thin walls of seeds become ruptured by the alkali treatment, but if, after washing, the contents are removed, the testa can be mounted in gum chloral. Generally the testa is sufficiently translucent for the cell pattern to be examined using high-power microscopy.

If it is desirable to remove the opaque contents but keep the testa intact, e.g. in order to photograph the details of cell pattern of the reference seeds, the more lengthy method described by Körber-Grohne (1964) may be employed. The seeds are first wetted, then put into dilute sulphuric acid (5 to 7 drops of concentrated sulphuric in 1 cm of water). The digestion of the starch may take four to eight weeks, but it can be hastened by raising the temperature to 30 °C, though this may damage delicate seeds. The mixture should be shaken and changed occasionally to hasten the process. When the seeds have become softened, the acid is washed off and absolute alcohol is added; this is kept at boiling-point (in some cases for a matter of hours) and is changed and renewed at intervals. When the contents are fully dispersed the seeds are washed and mounted in a permanent medium.

If the reference collection covers the British flora, the majority of Quaternary fossil plant fragments should be identifiable. Nevertheless there may be some fossils of species extinct in Britain for which reference to the European flora should suffice for identification; these, for the most part, will have recognizable family or generic characters.

## COLLECTION AND TREATMENT OF SAMPLES

A cleaned exposure of sediment is the ideal situation for collecting large samples for macroscopic plant studies; in the absence of a vertical face a wide diameter borer, such as the piston type described by Cushing & Wright (1965) should be used. The Hiller type of borer, used primarily for pollen samples, often yields too little material for a detailed macro-analysis. Ideally the cores should be taken back intact to the laboratory, wrapped in thin plastic sheeting and aluminium foil, sealed with drafting tape to minimize drying out. When the cores are sampled, cut surfaces should be individually scraped with a spatula to remove contaminants (Watts & Winter, 1966). If the cores are divided in the field, contiguous samples, each 5 or 10 cm long, should be taken; or shorter segments if a change of lithology necessitates it. Strong polythene bags, secured with a wired label, are useful for keeping these samples moist for some time. A rot-proof label, e.g. white celluloid or plastic, should be used inside to indicate the top of the sample. Any minor lithological changes can then be seen in the comfort of the laboratory.

Breaking open a sample along the bedding planes often reveals leaves and leafy stems of mosses which might well be fragmented with subsequent sieving. With the exception of moss stems and small tough leaves, such as those of Ericales and dwarf Salices, it is usually very difficult to remove leaves intact and they are best drawn or photographed before removal is attempted.

Some muds, silts and unhumified peat will break up in water, but in general samples containing humic matter need stronger treatment to separate the component parts. Dilute hydrochloric acid may be used for calcareous samples; effervescence can be checked with a little alcohol. Nitric acid has the advantage of causing fruits and seeds to rise to the surface (Godwin, 1956, p. 7) but is more noxious. Watts (1959) has used dry samples in 10% nitric acid. The resulting froth is transferred to a flask which is filled to the brim with water. Watts notes that a fair sample of the identifiable macrofossils remain floating whilst other vegetable debris sinks. By this method, followed by decanting and sieving to separate the different size ranges, Watts extracted 20,000 seeds from cores totalling 174 cm of 2·5 cm diameter.

Dilute NaOH may be best for certain peats, especially if they have dried out. Gentle heating may be required to fragment dry hard peat and lignite. Whatever method is used, the sediment should be left until no lumps remain. Any seeds which have floated to the top can be removed, and a small amount transferred to a 100-mesh sieve (100 meshes to the inch, each mesh measuring about 150$\mu$ in diameter), washed with a spray of water and stirred or agitated until the washing water becomes clear. Coarse detritus may necessitate a coarser sieve (about 35 meshes to the inch) above the fine one, and then the material is examined in two fractions. If a peat is mouldered and structureless, subsequent sieving to remove the humic solutes may be difficult; constant gentle stirring with gloved fingers whilst running water through the sieve helps the washing process. Practically all this material will probably disappear and only the most resistant plant remains will be found. Alkali should always be used as dilute as possible, as it does soften plant remains.

A small amount of the debris on the sieve is tipped or spooned on to a dish with a little water. A Perspex trough has been found useful, with a grid etched on the base so that the contents can be systematically examined using a low-power binocular microscope. The coarse fraction may well prove the most rewarding in terms of identifiable plant remains, but small seeds and megaspores are often found floating when the fine fraction is closely examined.

Identifiable remains can be removed with a paint brush; a brush with a few hairs collects small seeds quickly. Round porcelain or plastic paint palettes, with 3 to 4 divisions, are useful for segregating the remains, and they can be covered with a Petri dish. A little formalin added to the water prevents fungal growth. The fruits and seeds of many species become distorted on drying or adhere to the bottom of the dish. Glycerine can be added to prevent them drying but is difficult to remove completely and its presence will obscure the cell pattern. The cell pattern is shown most clearly when the surface is drying. The plant remains will stay 'damp dry' if placed on damp filter paper whilst they are being examined.

Delicate structures, especially translucent leaves and seeds, capsule valves and leafy stems of mosses are best preserved on microscope slides; cavity slides are useful for thicker mounts. Gum chloral is recommended as a mounting medium by Godwin (1956, p. 8) who gives the method of preparation. This gum helps to clear plant material which can be mounted direct from water or glycerine; it is always possible to remove a specimen at a later date by dissolving the gum in water.

Larger specimens, after identification, can be preserved in small specimen tubes; some achenes and fruitstones can be kept dry, while others are prevented from cracking and distortion by adding a mixture of glycerine, alcohol and formalin. Although tubes take up less storage room, they are less easy to locate and re-examination of a specimen is liable to lead to damage in the course of removal from the tube. Impregnating the fruits or seeds with paraffin wax as described by Reid & Reid (1908) was successfully used by early workers and their macrofossils stored in shallow boxes sealed with coverslips, are still in a good state of preservation after 45 years. However, this method is laborious when large numbers of specimens are involved. Labelling should of course be permanent. A thin coat of colourless varnish over Indian ink (on glass) will ensure this.

## SPERMATOPHYTA

### Flowers

Flowers and anthers are occasionally found in unhumified peat and other sediments. When fossil anthers containing pollen are squashed and mounted in glycerine, the liberated pollen can usually be identified. Flowers of *Myriophyllum alterniflorum* DC have been thus identified (Conolly, Godwin & Megaw, 1950). The same site produced an immature perianth of *Rumex aquaticus* L. with stamens and a nutlet attached. Florets of Gramineae have been recorded by Duigan (1955). Each consisted of a lemma, palea, stamens and pistil and were tentatively identified by C. E. Hubbard as flowers of the two subspecies of *Festuca ovina* L. which differ in the length of their anthers (Hubbard, 1954). Lemmas especially have characters on which generic and even specific identification can be made, as shown by the illustrations and descriptions of florets by Hubbard (*op. cit.*). An outstanding example of flower preservation was found in a Middle- and Late-Weichselian site in Poland (Klimaszewski, Szafer, Szafran & Urbanski, 1939), which yielded quantities of flowering shoots of *Dryas octopetala* L. Photographs of petals, sepals, stipules and leafy shoots show the venation and perfect state of preservation. The excellent condition of this material can be attributed to its rapid burial, probably *in situ*, in 'plastic solifluction loams'.

### Fruits and seeds

Aids to identification. The most useful reference book is probably the *Atlas and Keys of Fruits and Seeds Occurring in the Quaternary Deposits of the U.S.S.R.* (Katz, Katz & Kipiani, 1965). It has descriptions and illustrations of about 1000 European species and many of the fossil localities are given. The cell patterns have been accurately drawn in and some enlarged details of cells in surface and cross sections are illustrated; drawings of fossil organs are often included as well as photographs of some of the fossil fruits and seeds. There is a general key, and family ones for those genera most commonly represented fossil, e.g. Cyperaceae and Potamogetonaceae. A brief summary of the scope of the book is given in English.

The atlas by Beijerinck (1947) deals exclusively with fruits and seeds of the Netherlands flora. Although it includes most species found in lowland Britain, the absence of northern and montane species is an obvious disadvantage for the examination of glacial floras. There are brief descriptions of the fruits and seeds and a key.

Bertsch's *Früchte und Samen* (1941) is based on the German flora and so includes certain European species not found in the above-mentioned books. Sections and details of cell patterns are given for some species. Keys are an important feature of the book, and genera are keyed in a detailed manner.

There is such a diversity of species and of fruiting parts liable to be found in the Quaternary that no single comprehensive book or key exists at present. Keys made for modern seeds are often of limited help since they normally use appendages, textures and colours which may be lacking or modified in the fossil state. However, for certain families and genera, useful keys have been constructed with reference to the fossil condition. Many genera and species which have been inadequately illustrated in seed atlases have been monographed or described by taxonomists elsewhere. Some of this literature is cited in the present chapter.

Betulaceae. The fruit of *Betula*, flattened with two stigmas and translucent wings, and the three-lobed female catkin scales, are among the most easily recognizable fruiting organs found fossil. Specific identification is based on the relation between the breadth of the wing and that of the achene and the height of the wing in relation to the stigmas; the female catkin scales are identified by the shape of the lateral lobes (Clapham, Tutin & Warburg, 1962). Further descriptions are given by Tutin *et al.* (1964).

Detailed biometric studies on the species living in Poland have been made by Białobrzeska and Truchanowiczówna (1960). The species studied were *Betula pubescens* Ehrh., *B. carpatica* Willd., *B. tortuosa* Ledeb., and *B. verrucosa* Ehrh. in Sect. albae, and *B. humilis* Schrank and *B. nana* L. in Sect. nanae. The measurements were made with a view to facilitating the determination of *Betula* species in Quaternary deposits. There are measurements of achene, complete fruit and catkin scale, and photographs are given of both average and atypical fruits and scales showing the extremes of the various characters. It was concluded that all the above-mentioned species can be distinguished on both the fruits and catkin scales.

Fossil achenes without wings were specifically identified on the several characters of the nutlets although it was pointed out that their sample of eighty-one nutlets gave a more reliable result than the sample of seven nutlets. One achene with wings of Sect. nanae proved to have some characters of both *B. humilis* and *B. nana*. The identification of hybrids is not discussed, though all the species are interfertile (Tutin *et al.*, 1964).

In Britain, where so much birch forest is secondary and mixed, it is most important that well-authenticated reference material is obtained, and hybrids must be included in the collection. Fossils of *B. pubescens* × *B. verrucosa* and *B. nana* × *B. pubescens* have been reported (e.g. Franks & Pennington, 1961) and a continuous sequence of both fruit and catkin scales between two or more species is not unusual.

CARYOPHYLLACEAE. The seeds are usually reniform with sinuous cell margins and varying degrees of ornamentation. The colour and condition may help in the identification of a fossil specimen. Seeds of *Cerastium* are usually yellow or brown and are often fragile. Those of *Arenaria ciliata* agg., *Dianthus*, *Herniaria*, *Lychnis*, *Silene* and *Stellaria* are usually blackish. Some tend to retain their shape and fragment rather than flatten, e.g. *Arenaria*, *Herniaria*, *Lychnis*, *Moehringia* and *Spergularia*.

Some species show dimorphism of the seed coat. The tuberculate and 'armadillo' types of *Silene maritima* With. and *S. vulgaris* (Moench) Garcke described by Marsden-Jones & Turrill (1957) are well-known. Similar dimorphic tendencies have been noticed in other species, e.g. *Arenaria serpyllifolia* L., and the cell shape is variable in species of the *A. ciliata* aggregate.

The two subspecies of *Stellaria nemorum* L. are separable on seed characters (Green, 1954; Clapham *et al.*, 1962); those of subsp. *glochidispermum* Murb. have seeds with marginal papillae which are long and cylindrical with barbed caps.

An example of introgressive hybridization is found in *Cerastium alpinum* L. × *C. arcticum* Lange (Hultèn, 1956). Although the seeds of the hybrid forms are not described, it seems likely that a range of size and ornamentation exists within the hybrid complex.

A wide range of seed size is found in *Cerastium fontanum* Baumg. (incl. *C. holosteoides* Fr.) ranging from 0·4–1·0 mm in Britain. Jalas & Sell (1967) describe those of subsp. *scoticum* Jalas & Sell as being 0·8–1·0 mm with tubercles *c.* 50 $\mu$ high × 125 $\mu$ wide and of subsp. *triviale* (Murb.) Jalas, 0·4–0·9 mm, tubercles 15–40 × 15–40 $\mu$. Although the seeds of the different varieties within subsp. *triviale* have not yet been fully described it seems probable that the different ecotypes will have their own ranges of seed sizes.

Subspecific variation in seed size is also found in *Arenaria serpyllifolia* agg., described by Perring & Sell (1967). *A. serpyllifolia* subsp. *leptoclados* (Reichenb.) Nyman (= *A. leptoclados* (Rchb.) Guss.), has seeds 0·4 × 0·4 mm; subsp. *serpyllifolia*, seeds 0·5 × 0·4 mm; and subsp. *macrocarpa* (Lloyd) Perring & Sell, seeds exceeding 0·6 × 0·4 mm. The last named subsp. is confined to coastal dunes and fossil seeds have been recognized from a Late-glacial site on the Isle of Man (Kirkmichael, unpublished).

As is to be expected, only the thick-walled indehiscent fruits of *Scleranthus*, and not the thin-walled seeds, have been found fossil. The ribbed fruits of *S. annuus* (L.) s.l., and *S. perennis* L. would seem to be separable in the fossil state on the relative stoutness of the ribs of the sepals; a fruit, tentatively identified as *S. annuus*, is illustrated in West *et al.* (1964).

The normally indehiscent fruits of *Herniaria* have seeds with a tough but brittle testa and these are found separately (West *et al.*, 1964). The black, shiny seeds resemble small seeds of species of *Chenopodium*; they are 0·5–0·8 mm in diameter, but their less regular shape and more prominent radicle and keel distinguish them from seeds of that genus.

CHENOPODIACEAE. It is well known that seeds of *Chenopodium* species have diagnostic sculpturing on the testa and this is illustrated, for all species now present in Britain, by Clapham *et al.* (1962). Guinet (1959) gives detailed descriptions of the seeds, much enlarged details of cell patterns for some species, and for all species a section of the seed from the margin to the centre showing the pattern variation thereon. A key includes all species native in Britain excepting *C. botryodes* Sm. (whose seeds are inseparable from those of *C. rubrum* L.); this key gives seed characters which are equally useful for determining fossil seeds.

The specific identification of seeds of *Atriplex* is particularly difficult. Seeds of *A. patula* L. and the smaller seeds of *A. hastata* L. are inseparable on general morphology; both have elongated cells on the radicle and the surface of the rest of the testa is almost smooth. But seeds of *A. hastata* may be as small as 1·0 mm in diameter whereas those of *A. patula* are usually 1·4–2·0 mm. However some seeds of *A. hastata*, especially those from maritime habitats, are 2 to 3 mm in diameter, with a reticulate, rugose surface (Tutin *et al.*, 1964). These seeds seem inseparable from those of *A. longipes* Drejer and *A. glabriuscula* Edmondst. and indeed all these taxa are interfertile.

COMPOSITAE. The characteristic pappus of Composite achenes is usually lost in the fossil state but the broad or narrow open apex which bears the pappus often remains. Well-preserved achenes often permit specific identification. In some species the achenes of the disc and ray florets differ, as illustrated by Ross-Craig (1960–3).

The pericarp of the blackish five-sided achenes of *Eupatorium cannabinum* L. is extremely resistant to decay; and even small fragments which may well be mistaken for carbonized material can be identified by the diagnostic cell pattern of wavy rows of perforations.

The flattened narrowly transparent wings of the achenes of *Achillea millefolium* L. and *A. ptarmica* L. become rather delicate in the fossil state and the achenes may be found without wings; however, the diagnostic thin-walled cells, 10 to 15 times longer than broad, with oblique but rounded ends, still enable generic identification.

*Carduus* and *Cirsium* species have similar achenes but in both the treated and fossil states a transverse rugose surface becomes apparent in *Carduus* which is not visible in *Cirsium.*

CRUCIFERAE. Recent seeds of Cruciferae are characterized by the radicle, the form of which is clearly visible along most of the length of the seed, which is about equal to that of the cotyledons. Well-preserved fossil seeds may show similar characters, but in poorly-preserved ones the only distinguishing feature may be a blackened hilar area on a flattened oval to oblong seed. In the fossil it is rarely possible to use the taxonomic character of the relative position of the cotyledons to the radicle. A number of characters are usually needed to identify a fossil: the shape of the radicle and base of the cotyledons, the shape and extent of the wing (if present), the overall shape and size of the seed, and the cell pattern observed with a high-power microscope.

Terminology of the seed morphology, including surface ornamentation, of the Cruciferae of north-eastern North America is given by Murley (1951), and about 40 of the species she describes occur in Britain. The seeds are carefully drawn and show the surface ornamentation in a more detailed manner than in the seed atlases referred to previously. Some species are shown to have seeds which are highly variable in shape and size; this variability is related to the position of the seed within the fruit.

Seeds of *Brassica* species have been studied by Berggren (1962) and she has illustrated sections and surface patterns by photographs; it appears that seeds of the three species *B. campestris* L., *B. nigra* (L.) Koch, and *B. oleracea* L. are distinct whereas those of the hybrids are variable.

Two species which appear to have identical seeds until the cells are examined with a high-power microscope are *Diplotaxis muralis* (L.) DC. and *D. tenuifolia* (L.) DC. Measuring 1–1·3 × 0·6–0·9 mm, the seeds are more or less round in cross-section, though they may flatten with the small round hilum in the same plane as the small pointed radicle. The large thin-walled surface cells do not seem to preserve, but the small polygonal cells of the next cell layer are seen in both treated and fossil seeds. In *D. muralis* they average 20 $\mu$ diameter and in *D. tenuifolia* about 14 $\mu$. The latter species has been found fossil (Plate 1 *a, b*) from various glacial sites though it has until recently been misidentified, sometimes as *Draba.*

Some species have a testa of several layers of more or less translucent cells but not all are preserved in the fossil. This seems to be the case with *Cardamine pratensis* L. where up to four layers can be determined in recent seeds but only one or two generally remain in the fossil state.

An instance of a fertile hybrid with seeds intermediate in size and markings between those of the parents is found in *Rorippa microphylla* (Boenn.) Hyland. × *R. nasturtium-aquaticum* (L.) Hayek (Clapham *et al.*, 1962). Fossil seeds have been recovered together with those of both parents (Sparks & Lambert, 1961).

CYPERACEAE. The usually biconvex or trigonous nuts of the Cyperaceae often have a cell pattern of square to polygonal cells with raised margins.

The cell pattern alone distinguishes nuts of *Eleocharis palustris* (L.) Roem. & Schult. and *E. uniglumis* (Link) Schult. (Clapham *et al.*, 1962). Even when this distinctive coat is lost in the fossil, the swollen style base and biconvex shape enable the nut to be identified to the *E. palustris* aggregate species in the British flora.

The biconvex nuts of British *Schoenoplectus* and *Scirpus* spp. may be distinguished from the biconvex ones of *Carex* by the fact that they are almost flat on one face whereas those of *Carex* are equally convex on both faces; in addition fossil nuts of *Scirpus* and *Schoenoplectus* are usually black in the fossil state whereas those of *Carex* often retain their original colour.

An example of two morphologically similar species with strikingly different nuts is found in *Isolepis* spp. The distinctive longitudinally-ribbed trigonous nut with narrow transversely elongated cells of *I. setacea* (L.) R. Br. is quite distinct among the Cyperaceae and easily separated from the nut of *I. cernua* (Vahl) Roem. which has papillose cells.

Nuts of *Cladium mariscus* (L.) Pohl are readily identified in the fossil state; not only has the small, 1·2–2·0 mm, urn-like nut three grooves and three projecting flanges, but its remarkable carbonized appearance without a clear cell pattern distinguishes it from all other members of the Cyperaceae.

In the fossil state, nuts of *Carex* are often found without their utricles. However, variation in shape, colour and surface texture allows specific identification of well-preserved nuts for most species. An excellent paper by Nilsson & Hjelmquist (1967) gives a key and drawings for the determination of both living and fossil nuts. The paper, which covers some seventy species from southern Scandinavia emphasizes the salient features separating the nuts of closely allied species; about three-quarters of British carices are included.

Nuts of *C. aquatilis* Wahl. are smooth, strongly lustrous and lack papillae; Nilsson and Hjelmquist distinguish them from the other species of the section *Acutae* Christ. which are generally papillose and have broad bases. Tentative identifications have been made of fossil nuts of *C. aquatilis* (e.g. Simpson & West, 1958).

Although nuts with utricles bearing the ill-defined nerves characteristic of this section are sometimes recovered fossil it does not seem easy to separate them on utricle characters (Jermy, 1967).

The thickness of the utricle is not the deciding factor in preservation. The thin translucent inflated utricles of *C. rostrata* Stokes, with about 12–20 narrow even-spaced nerves, are often well-preserved. On the other hand the thick utricles of *C. paniculata* L. are frequently reduced to the thickened corky base; in their often blackened state with obscure ribs, they are then difficult to identify, unless the ovate biconvex nut with its rather truncate apex is seen.

Some closely allied species may be separable on the character of the utricle; *C. vulpina* L. is distinguished from *C. otrubae* Podp. by the minute papillae on the utricle. Fruits of the Section *Flavae* Christ. have been separated on the length of the utricles (Davies, 1953).

ERICACEAE. The surprisingly few fossil records of seeds of *Erica* species and *Calluna vulgaris* (L.) Hull may be in part due to their small size. The seeds are mostly *c.* 0·5 mm in length and oval, and differ mainly in cell pattern, best seen on mounted seeds using a high-power microscope.

The rather elongate cells of *Calluna* seeds are shown clearly by Katz *et al.* (1965). Pitting is only present in the cell margin and is usually lost or indistinct in the fossil state. The densely pitted more or less isodiametric cells of *E. tetralix* L., *E. mackaiana* Bab. and *E. ciliaris* L. are photographed by Watts (1959). It is interesting to note that the fossil seed tentatively identified by Watts (1959) as *E. ciliaris* from a Gortian (=Hoxnian) interglacial has more strongly sinuous cells than the present-day seeds of that species.

Evidence of a probably extinct variety of a species no longer represented in Britain, *Erica scoparia* L. var. *macrosperma* n. var. foss., comes from the same deposit (Watts, 1959), and not only other sites, of Hoxnian age in Ireland (Jessen *et al.*, 1959) but also the Shetlands (Moar, N. 1962. The History of the Late-Weichselian and Flandrian Vegetation in Scotland, Thesis, Cambridge University.)

Seeds of *Daboecia cantabrica* (Huds.) C. Koch are also described and illustrated by Watts (1959), who points out that the small tubercules are completely lost in the fossil state leaving a delicately pitted surface.

GRAMINEAE. Despite the fact that the uncarbonized fruits of the Gramineae can be separated from those of other plants in the fossil condition few published records have appeared so far. Fossil caryopses are usually flattened, having lost their endosperm, and are often translucent. The hilum, ranging from a small round thickened area near the base to a long narrow line extending the length of the caryopsis, may be the only distinct feature, though sometimes the scutellum (seat of the embryo) is preserved. Some caryopses are transparent with no apparent cell structure, e.g. *Anthoxanthum odoratum* L., but these sometimes may be identified by the size and shape of the hilum and its lateral or ventral position, together with the overall size and shape of the caryopsis. Often the cell pattern is quite clearly preserved, and its examination with a high-power microscope usually enables at least generic identification.

Usually reference material must be treated to remove the endosperm since the cell pattern is obscure in the living state. In order to simulate the fossil, starch and fats are removed with sulphuric acid followed by absolute alcohol, as described on p. 234. This method was first given by Körber-Grohne (1964) and her admirable large photographs and drawings of treated caryopses show how successful this can be; a key in English is included. Since the caryopses were studied for a specific investigation many species, especially terrestrial ones, are unmentioned and about half the genera present in Britain are omitted. Nevertheless the paper shows clearly the possibilities of identification and suggests that specific identification is possible in many cases.

The grass fruits most commonly recognized so far in Quaternary deposits are those of the aquatic genus *Glyceria*. The fossil fruits are very tough, dark and opaque with a long hilum, and the bifid style base is often still present. The cell pattern is only seen clearly if the cells of the pericarp have begun to disintegrate (Körber-Grohne, 1964). However, the caryopses can usually be specifically identified on their shape and size; Borrill (1956) describes caryopses of the following three species: *G. declinata* Bréb., oblong-elliptic 1·75–2·25 mm; *G. fluitans* (L.) R. Br., oblong-elliptic 2·5–3·0 mm; and *G. plicata* Fr., obovate 1·5–2·5 mm. Obovate caryopses of *G. maxima* (Hartm.) Holmberg are illustrated by Hubbard (1954), length 1·2–2·0 mm.

The uncarbonized grains of cereals are also described by Körber-Grohne (1964). They fall into the size range of both large and medium caryopses of wild grasses, but the cell pattern is shown to be different for the four genera, *Avena*, *Hordeum*, *Secale* and *Triticum*.

Helbaek (1958, plate III) has separated fragments of 'integument' of *Triticum* and *Secale* by the hair structure. He also states (in Brothwell & Higgs, 1963) that hairs of grains of *Avena* and *Hordeum* have diagnostic value.

The majority of Quaternary cereal records are of carbonized grains and spikelets or their impressions in pottery, and an impressive list of these is given, with illustrations, in Godwin, 1956. Most of the identifications were made by Jessen & Helbaek (1944) and Helbaek (1953). The identification of carbonized material is difficult, as Helbaek emphasized (in Brothwell & Higgs, 1963): 'The technique of identifying carbonized plant remains is based upon the same principles as in the case of other types of material, viz. comparison with fresh homologous plant parts. The examiner must, however, be intimately acquainted with the specific changes of shape, size and proportions caused by heat in order to visualize the original appearance of the deformed and often mutilated carbonized matter, and to put a name to it.'

The carbonized flowers of *Avena sativa* L. and the *A. strigosa* group have been distinguished (Jessen & Helbaek, 1944). As described on p. 236 uncarbonized flowers of Gramineae are occasionally found and may permit specific identification.

JUNCACEAE. Seeds of *Juncus* species, though fairly distinct as a genus, have proved difficult in some cases to identify specifically. Their small size, 0·3–0·9 mm, and tendency to collapse on drying make them difficult to handle. Since the fossil seeds are often pale yellow and tend to float during the extraction, they may be overlooked. The seeds are ovate to ellipsoidal, and, in some species, asymmetrical, with a darker top and bottom and usually rather regular quadrate to hexagonal cells, which are often transversely orientated. The seeds are best examined on damp filter paper to preserve their shape. However, it may be necessary to mount them for a high-power examination of the cell pattern. Some species seem inseparable when recent untreated seeds are compared, e.g. *J. acutiflorus* Hoffm. and *J. articulatus* L. However, Körber-Grohne (1964) has shown that the nineteen species she studied (including these) can be distinguished by high-power microscopy. She gives a detailed key including an English translation, with drawings and photographs of recent and fossil seeds. The preparation of reference seeds to remove their contents for comparison with the fossil ones with a high-power microscope is described above on p. 238. The diagnostic outer coats of some species may be absent from badly preserved fossils and then only a membrane of large approximately isodiametric cells remains (Körber-Grohne, 1964, plates I and IV). This is known to occur in *J. acutiflorus*, *J. articulatus*, *J. bufonius* L., *J. compressus* Jacq., *J. gerardi* Lois., *J. mutabilis* Lam. (*J. pygmaeus* Rich.) and *J. subnodulosus* Schrank. It is not then generally possible to make a specific identification. A key has also been made by Watts (1959) using untreated reference seeds; it is for British species excluding those restricted to salt marshes and montane habitats. Körber-Grohne's key includes all lowland species native in Britain excepting *J. acutus* L., the outer transparent coat of which extends to form a short appendage at both ends, and may be removed by heating in water for a few minutes. This reveals oval seeds 0·75–0·85 mm long with a cell layer of rectangular cells with raised margins. The two species of *Juncus* which are a certain indication of saline conditions in Britain, *J. gerardii* Lois. and *J. maritimus* Lam., have easily recognizable seeds and are sometimes the only seeds found in estuarine deposits.

PAPAVERACEAE. The reniform, large reticulate, seeds of the British species of *Papaver* are clearly separable and are figured in Beijerinck (1947) and Bertsch (1941). Seeds of a non-British species of *Papaver*, which seem referable to Section Scapiflora Reichenb. (Conolly, 1957), are occasionally found in Late- or Full-glacial deposits. However, the seeds within this section (treatment of which as regards species and subspecies varies according to the authority) are very similar.

POLYGONACEAE. In the fossil state the fruits of *Polygonum* species are usually found without the perianth. They resemble somewhat the nutlets of the Cyperaceae, but the trigonous or biconvex nutlets, 2–6 mm long, are easily separable from those of the Cyperaceae by their cell pattern. Nutlets of *Polygonum* species either have raised papillae in fairly regular longitudinal rows or appear quite smooth, but in all cases the cell walls, though obscure, are strongly sinuous; nutlets of *Rumex* have similarly smooth surfaces.

The pyriform fruits of *Koenigia islandica* L. have longitudinally elongated cells with sinuous raised margins giving the appearance of many ridges (plate 1 *c*, *d*). Eight measured nutlets ranged from 1·25–1·45 × 0·75–0·85 mm.

Fossil fruits of *Polygonum aviculare* L. are usually referred to the aggregate species, but biometric studies by Styles (1962) have shown that fruits of *P. boreale* (Lange) Small and *P. arenastrum* Bor. are separable on size ranges from those of *P. aviculare* and *P. rurivagum* Jord. ex Bor.

The bulbils (fruit is rarely produced) of *P. viviparum* L., illustrated by Katz *et al.* (1965), are identifiable in the fossil state; the scarious, grooved pedicel contrasts with the thicker-walled black bulbous part which has small polygonal cells with raised margins.

Fruit polymorphism is found in *Polygonum persicaria* L.; biconvex, trigonous and even occasional tetragonal fruits can grow on the same plant (Timson, 1965). A key to the nuts of the Section Persicaria is given by Timson (1964).

The generally triquetrous nutlets of *Rumex* species are not usually specifically identifiable without the perianth segments; however the large number of specific identifications in Godwin (1956) show that the latter are often preserved. *Rumex acetosella* agg. has small, 0·8–1·5 mm, trigonous nutlets which may become translucent and show the thick walled strongly sinuous cells, illustrated in Bertsch (1941). It is fortunate that the nutlet is diagnostic because in *R. tenuifolius* (Wallr.) Löve and *R. acetosella* L. the perianth is not persistent.

PORTULACACEAE. Non-sinuous cell margins distinguish the seeds of *Montia fontana* L. from those of the Caryophyllaceae. The cell pattern is variable (Clapham *et al.*, 1962). Despite the occurrence of intermediates, four subspecies are recognized on the basis of the sculpturing of the testa. Illustrations are given by Walters (1953) and a greater range of variation is shown within each subspecies by Moore (1963).

POTAMOGETONACEAE. The removal of the fleshy exocarp from reference fruits of *Potamogeton* species is essential before comparison is possible with the fossil fruit-stones. Very useful illustrations, and a key and graph of size ranges of the fruit-stones, are given by Jessen (1955). Almost all the *Potamogeton* species in Britain at the present time are illustrated, but *P. panormitanus* is synonymous with *P. pusillus* L. and his *pusillus* with *P. berchtoldii* Fieb. Other illustrations showing a greater range of variation within the species are found in Katz *et al.* (1965). The relation between the dorsal lid and the style base, the shape of the ventral margin and the possible presence of spines or warts are important characters, as are the presence or absence of a central depression and a keel. However, badly worn stones may lack style base, stalk, keel and warts, and the shape of the embryo may be more obvious as the outer layers of the stone erode, thus forming a central depression in species which normally have convex sides. As stated already (p. 234) it is worth treating the reference fruit for varying lengths of time to provide a range of modification of the fruit-stones. Some species are still recognizable when badly preserved and almost all well-preserved fruits can be specifically identified.

Resting buds or turions are occasionally found fossil. Those of *P. pectinatus* L. have been found with fruit of the species and are figured by West (1956).

RANUNCULACEAE. Achenes of the genus *Ranunculus* are, for the most part, laterally compressed and asymmetric. The style, stalk and border often disappear in the fossil state, but with the exception of the subgenus *Batrachium* (DC.) A. Gray, specific identification is usually possible if the parenchyma is still present.

In the section Ranunculus the separation of the fossil achenes of the three species often proves difficult because of great similarity in shape and size. Reference fruits treated to remove the thin-walled epidermal cells show isodiametrical parenchyma cells forming pits with clear margins. Reid (1949) differentiates between the cell pattern of two of the species as follows: 'In *R. acris* L. the pits are of fairly uniform size and distribution, measuring from 25–30$\mu$ across the hollow, with dividing ridges about 12·5$\mu$ across; whilst in *R. repens* L. corresponding measurements are 30–50$\mu$ across the hollows, and 25$\mu$ the width of the dividing ridges. Also towards the margin the pits are smaller and more crowded in *R. repens*, being there comparable in size and distribution with those of *R. acris*.' The author has found that achenes of *R. repens* generally have rather larger pits in the centre 50–75$\mu$ and those of *R. acris* may be up to 37$\mu$ across. Measurements across pits of a few achenes of *R. bulbosus* L. showed them to be 35–50$\mu$. Helbaek (1958) figures crystals in this coarse parenchyma which are present in every cell of *R. acris* but more dispersed and irregular in *R. repens*, but this was in exceptionally well-preserved material and the crystals are easily lost. Badly preserved achenes may have this parenchyma eroded away to reveal the elongate sclerenchyma, but achenes of *R. repens* may still be tentatively identified by the thicker achene (than either *R. acris* or *R. bulbosus*) with its markedly bevelled margin.

Achenes of *R. flammula* L. and *R. lingua* L. are corroded in a similar manner, but differences in size and shape distinguish these two species from those of the Section Ranunculus.

Achenes of the subgenus *Batrachium* are characterized by the presence of transverse ridges on the achenes. The British species are about 1·0–2·2 mm long without the beak. Since beak, wings and hairs are not preserved in the fossils, only differences in size and shape are left to separate the species. It is possible that achenes of certain species could be separated by biometric measurements, for instance those of *R. baudotii* Godr. and *R. circinatus* Sibth. are generally smaller than the achenes of other species in Britain.

Achenes of *R. sceleratus* L., 1·0–1·3 mm long, are also transversely ridged, but they have a marginal area of thin-walled cells forming a broad swollen keel. Also the longitudinal section shows markedly swollen ends to the sclereids, thus differentiating them from those of the subgenus *Batrachium* (Cook, 1963).

Achenes of *R. hyperboreus* Rottb. have a similar structure and size range but differ from *R. sceleratus* in the parenchyma cells. These either cover the achene completely or leave small irregular areas through which the underlying elongated cells are sometimes seen, but do not form obvious transverse ridges (plate 2*a*). This species, no longer present in the British flora, has been found in Quaternary deposits including those of the last glaciation.

SAXIFRAGACEAE. The small, 0·3 to 1·3 mm long, fusiform to oblong seeds of British species of *Saxifraga* are often difficult to distinguish specifically on recent material, because of the similarity of their cell patterns. Most species have tuberculate or papillate surfaces. Fossil seeds tend to become translucent and the tubercules may be partially collapsed and this condition often only allows tentative identification.

In the section Sedoides a continuous overlapping series of ornamentation is found between seeds of the four species. Webb (1950) gives photographs of seeds of the notoriously variable *S. rosacea* Moench; the ornamentation of the seeds varies in different populations between the fine tuberculate pattern, found also in *S. caespitosa* L. and *S. hartii* D. A. Webb, and the coarsely papillose type found in *S. hypnoides* L. The intraspecific hybrids produce seed intermediate between that of the two parents.

However, some species have seeds which are fairly distinctive (as shown in the drawings by Ross-Craig, 1957). A fossil seed of *S. hirculus* L., a species lacking ornamentation and with straight or slightly sinuous cell margins, has been illustrated (Godwin, 1964).

The black, shiny seeds of *Chrysosplenium*, 0·5–0·6 mm long, are quite distinct; they are keeled and obovoid, and keep their shape in the fossil state. The surface is highly refractive and smooth in *C. alternifolium* L., but finely warted in *C. oppositifolium* L. Photographs of fossil seeds of *C. oppositifolium* are given in Godwin (1962), and those of *C. alternifolium* in Katz *et al.* (1965).

SPARGANIACEAE. The fruits of *Sparganium erectum* L. are distinguished from those of other British *Sparganium* species by the presence of 6–10 longitudinal ribs, and fruits of the four subspecies in Britain, separable on fruit characters, are described and figured by Cook (1961). The ribs are usually well preserved on the fossil fruit-stones and sometimes the subspecies are identifiable.

The other three species in Britain, *S. emersum* Rehm., *S. angustifolium* Michx. and *S. minimum* Wallr. have smooth fruits which, though distinguishable on size and shape in recent material (Clapham *et al.*, 1962), have fruit-stones which are not so easily differentiated in either the living or fossil states. Those of *S. emersum* are the longest, up to *c.* 3·0 mm; those of *S. angustifolium*, *c.* 2·0–2·5 mm, have up to 7 longitudinal grooves, and those of *S. minimum* are similar but generally have fewer less distinct grooves. Katz *et al.* (1965) figure rather variable fossils of these three species. Recent immature fruits, treated with sodium hydroxide to remove the outer coat, may dissolve and leave just the embryo, the resistant outer coat of which is sometimes recovered fossil. When fossil fruit-stones are dried the outer cells oxidize and may be removed until only the embryo remains. A continuous size sequence of fruit-stones was found in a single layer of a Late-glacial deposit at Kirkmichael (Dickson, unpublished); the larger strongly-grooved ones are *S. angustifolium* but the smaller ones have eroded surfaces lacking grooves. The erosion possible in fossil stones suggests that in some cases they may not be identified beyond the Subgenus. The identification problem is heightened by *S. emersum* and *S. angustifolium* forming fertile introgressive hybrids (Cook, 1961).

UMBELLIFERAE. The taxonomic value of fruits of the Umbelliferae is well known and the conspectus in Clapham *et al.* (1962) includes fruiting characters. The separated fruits or mericarps are generally compressed on one side; they usually show 6 or more resin canals (vittae), diagnostic for the family, and 5 or 9 ribs. Most illustrations include cross-sections of fruit which are useful in showing the position of the vittae in relation to the ribs. Vittae are often more clearly seen on treated reference fruits and fossils; they are very dark and may be transversely segmented. Well-preserved fossil fruits are specifically identifiable and many records are given in Godwin (1956).

There are, however, species in which the vittae are absent or deep seated and here the family resemblance, especially in the fossils, may be obscure. The fruit of *Conium maculatum* L. lacking vittae, has five prominent ridges which may disappear (A. P. Conolly, personal communication). A distinctive cell pattern is then revealed with longitudinal thick walls about 100 $\mu$ apart and thin transverse cell walls 15–30 $\mu$ apart (plate 2*b*). The fossil fruits retain the deep longitudinal groove down the contact face, and well-preserved fruits may resemble the grain of a grass or cereal.

The fossil fruits of *Berula erecta* (Huds.) Coville retain the flattened contact face but the ribs may be reduced to give slightly angular fruit which can erode in a similar way to the fruit-stones of *Sparganium*, with the resulting formation of smaller and less angular fruits. Only in very eroded fruits are the many deep-seated vittae seen.

VIOLACEAE. Seeds of the British species of *Viola* are distinctive as a genus. They are 1·2–3·0 × 0·7–2·2 mm, pointed ovoid, with a large round hilum, up to 0·8 mm across, at the broad base; the apex may be acute or subobtuse. The elaiosome, which obscures the shape of the apex, can be softened by heating in water or even sucking the seed briefly, and it is then easily removed. The high gloss which tends to obscure the cell pattern can be removed by very brief alkali treatment. However, the thin-walled outer cells are also easily lost, both in treated reference and fossil seeds; this leaves visible the longitudinally elongated thick-walled sclerenchymatous cells with narrow lumina which seem common to all species. The seed easily splits lengthwise and half or quarter seeds are frequently found.

In the British flora seeds of *V. hirta* L. and *V. odorata* L. are separable on their large size, 2·5–2·9 × 1·7–2·0 mm; the two species also differ in the shape of the apex, that of *V. hirta* being more obtuse than in *V. odorata*. Seeds of *V. reichenbachiana* Jord. ex Bor. measure 2–2·4 × 1·3–1·5 mm and those of *V. riviniana* Rchb. 1·7–2·0 × 1·1–1·4 mm; seeds of about 2·0 mm long seem inseparable in these two species. Many of the small seeds of *Viola* spp. (1·3–2·0 × 0·7–1·4 mm) are very similar. Only complete seeds with a well-preserved cell pattern are worth attempting to give a further identification, unless, as is the case with subgenus *Melanium*, the apex is particularly asymmetrical. Seeds of *V. palustris* L. (1·5–1·9 × 1·0–1·2 mm) appear distinct from other species in having a clear punctate cell pattern, and in this species incomplete seeds are recognizable (plate 2*c*).

### BUD AND CATKIN SCALES

Some genera have bud and catkin scales which are readily identified with the use of a low-power microscope, whilst others require examination of the cells of the epidermal layers. Gentle bleaching may be necessary for both recent and fossil scales to reveal the cell pattern. Sometimes these scales contribute the only recognizable macrofossils of a tree or shrub, e.g. *Populus*, where the identification is of particular value since the pollen is difficult to determine. The presence of male catkin scales may reveal the over-representation of the relevant pollen-type.

The bud scales of *Picea*, *Taxus*, *Ulmus*, *Betula*, *Carpinus*, *Quercus*, *Populus* and *Salix* have all been recovered from a last interglacial site on the continent by Rabien (1953), who describes and figures the epidermal cells. The scales were recovered in sufficient quantity for a diagram to be constructed showing their relative frequencies from zones *e–i*.

Rabien describes the bud scales of *Juniperus* and *Taxus* as readily distinguished by their distinctive gymnospermous stomata with five or six subsidiary cells surrounding the guard cells of the stomata. The stomata of *Juniperus* are more numerous than those of *Taxus*. Part of a *Taxus* scale is figured by Jessen *et al.* (1959). Scales of *Abies*, *Picea* and *Pinus* each have two resin ducts, but differences in the epidermal, hypodermal and parenchymatous cells distinguish the three genera microscopically (Rabien, 1953).

The three-lobed female catkin scales of *Betula* are well known and may permit specific identification (Clapham *et al.* 1962). Biometrical studies have been made and are described together with those for the fruit in an earlier part of this chapter. The male scales are much smaller, *c.* 1 mm long; they are peltate and rather delicate in the fossil state in contrast to the similar shaped but woody male scales of *Alnus*.

The single, entire, thimble-like bud scales of *Salix* species are distinctive; though often flattened they can be readily identified by the small apical notch and the longitudinally elongated cell pattern which gives the impression of faint narrow ridges (plate 2*d*). Epidermal fragments can be identified by the usually very short hairs which are frequent among the epidermal cells. The scales, which range from about 1–10 mm in length, are commonly found fossil; since they vary in size within an individual shrub or tree but are otherwise very similar, specific identification does not seem possible.

The distinctive feature of the outer bud scales of *Populus* species are the broad parallel bands of parenchyma (the intervening vascular tissue is small and insignificant). These parenchymatous 'ribs' are usually preserved at least as well as the thick-walled nearly isodiametric cells of the epidermis and are clearly visible (Jessen *et al.* 1959). The catkin scales of *Populus tremula* L. appear distinctive; they are deeply laciniate and in spite of their delicate appearance have been found as fossils (Pennington, 1947).

### Leaves, stems and rhizomes

Despite the enormous variation in leaves, even within individual plants, diagnostic features are often presented by leaf shape and size, character of the leaf margin, venation, stomata and epidermal cells. Because of the diversity of shape and size within many species a comprehensive selection of both young and mature leaves is needed for reference. For instance prophylls of *Salix* will differ from older leaves, and leaves of *Quercus* produced in the late summer may be unusually long and narrow. Leaves of hybrids must be included.

Treatment with a caustic alkali, useful for clearing a leaf to show the venation, may disorganize the epidermal cells. Thin leaves heated in water for a few minutes may show the venation clearly without further preparation. Gum chloral is a useful mountant for treated leaves.

In some cases it may be necessary to compare the epidermal cells and stomata of recent and fossil leaves. For this purpose reference leaves should be treated to remove all but the cuticle of the upper and lower epidermis. Stace (1965) gives a method for this using a modification of Schulze's solution, and describes in detail the characters which serve to distinguish species and the descriptive nomenclature used.

As already stated, leaves often show up clearly when the sample is split along the bedding planes; but since it is often difficult to remove a fossil leaf intact from the sediment it is advisable to illustrate it *in situ*. To remove a leaf from calcareous sediment dilute hydrochloric acid can be used to disintegrate the surrounding matrix. But a series of preliminary trials using acids of varying concentrations should be undertaken to avoid damaging the leaf tissue (Richardson, 1960*b*). Fossil leaves, however extracted, must be treated with great care to avoid damaging them. It may be possible to float a leaf onto a glass slide (preferably smeared with a little egg albumen) and there irrigate it with water after chemical treatment; or, if a leaf adheres to its substratum, the edge of a large coverslip slid under the leaf may remove it intact, and by inverting the coverslip on to a slide final clearing of the leaf may be possible with a soft paint brush. If the outline of the leaf is insufficiently diagnostic for identification the leaf may be bleached, but this also is a matter for cautious experiment.

Some families or genera have leaves which in their gross morphology are diagnostic of a particular taxon, and some which have been found fossil are described here. From the presence of associated bud scales, wood or fruiting organs, others may be identified which have no such obvious characteristics.

Leaves of conifers frequently show rows of stomata parallel to the long axis. Shape of the leaf apex, and positions of stomata and resin canals (if present) are important characters for identification. Dallimore & Jackson (1966) in their comprehensive survey illustrate a number of cross-sections of leaves. Sections showing stomatal distribution of some European species are given by Katz *et al.* (1965), and Ferré's (1952) illustrations include detailed anatomy of juvenile and mature leaves. Florin (1931) describes the cuticle of almost every known species and identification of epidermal fragments with well-preserved stomata is shown to be possible.

The following descriptions may help in the identification of fossil leaves of those conifers known to be present in the British Quaternary.

Leaves of *Abies alba* Mill. alone have emarginate apices, those of *Picea abies* (L.) Karst., *Pinus sylvestris* L. and *Taxus baccata* L. are acute and leaves of *Juniperus communis* L. s.l. have a spiny pointed apex. Resin canals can be seen from sections: *A. alba* and *P. abies* have two, *P. sylvestris* has several, *J. communis* one and *T. baccata*, none. *A. alba* has stomata on one surface only (in two bands), *P. abies* on all four sides, *P. sylvestris* on both sides, *J. communis* in a band on the concave surface only and *T. baccata* on either side of the prominent midrib, also on one surface only.

Other diagnostic features are: the circular and adpressed leaf bases of *Abies*, the peg-like decurrent leaf bases which remain on the twigs of *Picea*, and the two leaves on the short shoots of *P. sylvestris* (and other two needle pines), only the bases of which are sometimes preserved.

There may be modifications, as Jessen *et al.* (1959) pointed out. For instance in *A. alba* leaves in the floral region often have a few short rows of stomata on the usually stomata-free surface, and shade leaves of *P. abies* may be somewhat flattened rather than rhomboidal in section.

All these species have been identified and figured from the Gort (Hoxnian) Interglacial (Jessen *et al.* 1959).

Although *Larix* is not known in the British Quaternary, leaf fragments have been identified by their stomatal structure from an Early Weichselian site in Denmark (Andersen, 1961).

Other species which may be present, especially in pre-Flandrian deposits, may show characters which are not as clear cut generically as those mentioned above. For instance other species of *Abies* have leaves which are acute, obtuse or rounded. One should be aware of the possibility that *Tsuga* and other coniferous genera may be found in early Quaternary deposits.

The simple or lobed leaves of *Saxifraga* species are often characterized by the presence of simple or glandular hairs. The broad petiole is often as long as the blade. Tertiary nerves are rare. Leaves of the northwestern European species of section Sedoides are described by Webb (1950). Fossil leaves referable to this section have been illustrated in Jessen & Farrington (1938). Probably the most distinctive leaves of a *Saxifraga* in the British flora are those of *S. oppositifolia* L. with a single lime-secreting gland in the thickened apex; *S. aizoides* L., the only other British species with an apical gland, has oblong-linear leaves quite distinct from the small broad leaves of *S. oppositifolia*.

The distinct peltate scale-like hairs of the Eleagnaceae, represented solely in Britain by *Hippophaë rhamnoides* L. have been recorded and illustrated by West (1956); they are about 0·6 mm in diameter with radiating long narrow cells. The leaves have not been recorded from the British Quaternary but are illustrated by Kneblová (1958) from travertine deposits of Eemian age in Czechoslovakia.

Broken segments of the rather fragile leaves of *Myriophyllum* species have been described and illustrated (Pennington, 1947). However, the usually pinnate or pectinate bracts of the flowering shoots seem more resistant and have been specifically identified; Jessen (1949) has illustrated those of *M. spicatum* var. *squarrosum* Laestad.

The leaves of *Polygonum viviparum* L. are very variable, linear-lanceolate to subrotund. The prominent venation on the inrolled margin with the generally ill-defined major lateral veins, together with the broad flattened midrib, characterize the species. The leaves have been found associated with the thin sheathing stipules (ochreae) with straight, widely spaced almost parallel veins, in a full-glacial deposit at Broome, Norfolk (Dickson, unpublished).

The presence of glands may be a useful diagnostic feature because these are often still evident in fossil leaves; the sessile glands of *Myrica gale* L. distinguish the leaves of this species from other oblanceolate leaves.

Leaves of the three species of *Betula* in Britain can be separately identified (Clapham *et al.* 1962). Some leaves of *B. nana* × *B. pubescens* seem distinguishable on characters intermediate between the two parents. A probable hybrid from a Zone II deposit which contained fruits of undoubted hybrid origin is drawn by Franks & Pennington (1961).

Although leaves of deciduous trees in Britain are well described in floras, the diagnostic character of leaf scars is not so well known. Those of *Alnus*, *Fraxinus* and *Tilia* have been identified by Kelly and the latter two illustrated (Kelly & Osborne, 1965). The scars seem more resistant to decay than the twigs themselves and have been found in isolation from them.

Present-day species of *Salix* are notoriously difficult to identify in view of their frequent hybridization, and fossil leaves are no exception. Leaves of *S. herbacea* L. with few crenate indentations may, when poorly preserved, resemble those of *S. polaris* Wahl. However, Tralau & Zagwijn (1962) have pointed out that in *S. herbacea* a tertiary nervule can always be found running into the indentation, whereas in *S. polaris* the nervules are recurved before reaching the margin. It must be noted, however, that in northern Fennoscandia the two species hybridize at the present time. Although fossil leaves of *S. herbacea* hybrids have not been recorded as yet, present-day hybrids always reveal the influence of *S. herbacea* in the leaf shape (Tutin *et al.* 1964). Photographs of fossil leaves of *Salix herbacea*, *S. polaris* and *S. reticulata* L., from various European sites are given by Tralau (1963).

Many members of the Ericaceae have leaves with a revolute margin, some strongly so; since they are mostly thick or coriaceous they usually preserve well and specific identifications are often possible. The margins of the very small, *c.* 2 mm long, leaves of *Calluna vulgaris* (L.) Hull are so strongly revolute that the undersides are not seen. But it is seldom that individual leaves are found fossil, usually shoots with several overlapping dark grey or blackened leaves resembling a plaited structure are recovered. The small, entire leaves of *Erica* spp. are slightly to strongly revolute; Jessen (1949) illustrates fossil leaves of *E. cinerea* L., *E. mackaiana* and *E. tetralix*. Some species have glandular hairs (Clapham *et al.*, 1962), which may persist in the fossils. Carbonized leaves of *E. tetralix* have been found without their hairs, but they have been matched with carbonized recent leaves of that species.

The tough resistant leaf spines of *Stratiotes aloides* L. with a main pointed 'horny' cell and subsidiary basal cells have been found fossil and may be distinguished from those of the extinct *S. intermedius* Hartz; both are illustrated in Katz *et al.* (1965). Those of *Najas* species may be simple, of one horny cell only 50–150 $\mu$ long or, as in *N. marina* L., have several basal cells, the whole measuring about 250–400 $\mu$. The leaf spines of *Ceratophyllum* may only be about 40 $\mu$ long and are therefore more likely to be recorded from pollen preparations (photograph in Kelly, 1964).

Even the thin translucent leaves of *Potamogeton* species have been found fossil, though it may only be the apex which is recovered. Although the shape of the leaves of some species is extremely variable, the venation of the apical portion may be sufficiently distinct for a specific identification. Raunkiaer (1899) has illustrated tips of leaves of seventeen of the species and two hybrids in Britain at present. Others are illustrated in Hagström (1916) and Hagerup & Petersson (1959).

Stem bases and rhizomes, especially those of the Cyperaceae and Gramineae, are often found in Quaternary deposits and characterize many of the peat types. Well-preserved parts may, for some species, be identified without the aid of a microscope.

The rhizomes of *Narthecium ossifragum* (L.) Huds. and *Scheuchzeria palustris* L. are morphologically similar but in this case microscopic examination of the transverse sections shows the anatomical structure to be different (Schumacker, 1961).

Rhizomes of *Carex limosa* L., rhizomes and stem bases of *Cladium mariscus* (L.) Pohl and stem bases of *Eriophorum angustifolium* Honck. are fully described and illustrated by Grosse-Brauckmann (1964). The highly resistant and diagnostic sclerenchymatous spindles from the leaf bases of *Eriophorum vaginatum* L. are described and illustrated by Benda & Schneekloth (1965). They are about 1–3 mm long, fusiform and often crescentic, the surface has a distinct wavy-striate cell pattern (plate 2*e*) and, in the fossil state, they are often black. Epidermal fragments of Cyperaceae may be found with their characteristic stomata, having two subsidiary cells parallel to the pore (Metcalfe & Gregory, 1964).

The leaves and culms of the Gramineae can be recognized by their equally distinctive stomata and specialized silica cells, and the lamina identified to the species level. The anatomical monograph by Metcalfe (1960) includes a method of isolating the epidermal layers and descriptions of all genera. Macroscopic identifications of vegetative parts of Gramineae include the carbonized tuberous rhizomes of *Arrhenatherum tuberosum* (Gilib.) Schultz, illustrated in Godwin (1956, pl. XIII) and the basal internodes of *Molinia caerulea* (L.) Mech., described and illustrated by Benda & Schneekloth (1965).

### Wood

Godwin (1956) pointed out the danger of identifying fossil wood by casual inspection in the field, and, indeed, with small pieces lacking bark it is almost always necessary to compare thin sections of the wood with those of a reference collection. Jane (1956, p. 188) states that wood from a root may be different from that of the bole or a branch; only after about the 50th growth ring is the structure mature in a slow-growing tree. It is therefore preferable that the reference slides should originate from more than one part of a tree.

All types of anatomical structures are photographed in Brazier & Franklin (1961). The key for British hardwood trees and shrubs by M. H. Clifford in Godwin (1956) covers twigs and roots and was prepared for the identification of fossil woods. Jane's book *The Structure of Wood* (1956) gives full descriptions of the anatomy of wood on a world-wide basis. His caricatures of salient features are particularly useful for distinguishing between similar woods as are the methods of preparing a reference collection.

The succinct descriptions and illustrations of gymnosperms given in Phillips (1948) are particularly useful. A comprehensive treatment is found in Greguss (1955).

In angiosperms the features most used in fossil wood identification are: the size, arrangement and density of the vessels (T.S.) and their type of perforation and thickening (L.S.). The width of the rays (T.S. and T.L.S.) and their type, homogeneous or heterogeneous (R.L.S.) is also important. In gymnosperms the following characters are used in identification: the ray tracheids (R.L.S.), the presence (or absence) of resin canals (T.S. and T.L.S.), the type of associated epithelial cells (T.L.S.), and the cross-field pitting of the ray parenchyma (R.L.S.). The presence of spiral thickening of the tracheids (L.S.) is characteristic of *Taxus* among native British conifers.

Natural degradation of the wood results in a partial breakdown of the cell walls and often only the primary and outermost secondary walls remain (Barghoorn, 1949). Thus scalariform perforation and spiral thickening may vanish from the vessels of angiosperms leaving long undifferentiated tubes. In gymnosperms the spiral thickening of the tracheids seems very resistant but narrow oblique slits in the tracheids in compression wood (Jane, 1956, figure 38*e*) may be mistaken for this.

Well-preserved fossil wood of a cheesy consistency can be quickly hand-sectioned with an ordinary razor blade and gum chloral is a useful permanent mountant. The transverse section should be examined first; if it is strongly ring porous, the wood is almost certainly *Fraxinus*, *Quercus* or *Ulmus* and the distribution of the vessels of the late summer wood will indicate which genus (Jane, 1956, figure 136). *Corylus*, though less strongly ring porous, is often recognizable from a well-preserved transverse section. However, radial and tangential sections are usually needed for a certain identification.

Clifford (1936) recommends soaking dry wood in a solution of potash (approx. 1 %) for periods up to one week followed by a hardening process in 70 % alcohol for a further week. Badly preserved wood which has dried out tends to split radially due to greater tangential shrinkage (the radial surface can be distinguished by the intermittent parallel lines of the rays seen with a hand lens or low-power microscope). Thus radial sections can almost always be made. Sections may sometimes be obtained from such unpromising wood by mounting it in a water or alcohol soluble wax and sectioning with a sledge microtome. Small pieces of wood a few mm in diameter can be similarly prepared and sectioned. Sometimes sections, the transverse ones especially, will be dark and compressed and a mild bleaching agent, such as sodium hypochlorite, may be used to bleach and expand them (Richardson, 1960*a*). Wood which is heavily minerally impregnated, or has otherwise proved impossible to section satisfactorily, may be treated as charcoal. Wet wood (e.g. an archaeological artifact) may be kept in its original state by immersing it over a period in a synthetic wax, polyethylene glycol (Western, in Brothwell & Higgs, 1963); thus preserved indefinitely the wood can, if necessary, be hand-sectioned at a later date.

### Charcoal

Apart from genera with a distinctive vessel structure as seen on the transverse surface, charcoals can be both difficult and time-consuming to identify. It is necessary to examine in detail vessel and ray structure in the longitudinal sections, but the irregular, highly refractive surface of the charcoal at the magnification needed (×40–120) and the problem of lighting make these features difficult to observe. It is therefore desirable to obtain flat transverse, radial and tangential longitudinal surfaces by clean breaks (made by snapping). Very small pieces a few millimetres in diameter which cannot be broken to expose a fresh face are more difficult to identify. It is possible to impregnate the charcoal with a synthetic resin and grind it with carborundum, a laborious method but the best one available at present if microphotographs are wanted (Western, 1963). Excellent transverse sections of charcoals of trees and shrubs prepared in this way are figured by Dimbleby (1967). Even pulverized charcoal may be identified to some extent on the pattern of pits (Paulssen, 1964). Wood sections can be used for comparison and charcoal for reference quickly prepared by enclosing a small piece of reference wood in a ball of clay and heating until the clay is red hot. Larger pieces can be heated in sand within an iron tube, the ends of which are plugged with clay (a method used by E. Singer, Botany Department, Leicester).

## PTERIDOPHYTA

Rhizomes, stems, leaves, sporangia, megaspores, and microspores (the latter in pollen preparations) of the Pteridophyta have been identified from Quaternary deposits.

Despite the large number of *Lycopodium* spores (all British species have been found fossil) no macroscopic remains of Lycopodiales have been recovered. Similarly the genera *Botrychium* and *Ophioglossum* are represented only by spores.

The triradiate megaspores of *Selaginella selaginoides* (L.) Link. are distinguished from those of *Isoetes* by their larger size and absence of an equatorial ridge. The surface of *S. selaginoides* may be tuberculate or almost smooth, and often grey in the fossil megaspore; although the size usually ranges from 0·6–1·2 mm diameter, abortive tetrahedral spores, 0·2–0·4 mm, are occasionally produced, and have been recorded fossil. They have been illustrated, together with normal spores, by Watts (1959). The rounded spores of *S. helvetica* (L.) Spring are about 0·4 mm diameter. They have been recorded from deposits of the last glaciation in the Netherlands (van der Vlerk & Florschütz, 1950).

Recent megaspores of the three British species of *Isoetes* can be separately identified on their surface ornamentation (Clapham *et al.* 1962). However, the white or yellowish silica coat of the megaspore may be lost leaving translucent fossil spores, often smaller than untreated reference material and with a modified surface pattern. Reference spores, heated in dilute sodium hydroxide, and fossil spores are often smaller than the recognized size range (Tutin *et al.* 1964) so these measurements are given in brackets. Megaspores of *I. lacustris* L. are (200–500 $\mu$) 530–700 $\mu$, and have separate low broad tubercles which may almost disappear in the fossil state. *I. echinospora* Durieu (200–400 $\mu$) 440–550 $\mu$, has long fragile spines which may be reduced to broken bases in the fossil, but these are distinguished by their parallel sides from those of *I. lacustris* (Watts, 1959). Reference spores of *I. hystrix* Bory are (300–400 $\mu$) 400–560 $\mu$ and have small tubercles which often become confluent especially on the basal surface; the ornamentation of the spores is one of the variable features of this species. These species and all the other European species of *Isoetes* are described, with size ranges, by Tutin *et al.* (1964).

Cones composing the sporangiophores of *Equisetum* have not been found fossil but the distinctive shiny, grooved, and often black stems and rhizomes have been specifically identified by their anatomy.

Species of Filicales have been identified in Quaternary deposits by their leaves, rachides, rhizomes and sporangia. The three families of the Filicales present in Britain, Osmundaceae, Hymenophyllaceae and Polypodiaceae have distinctive sporangia; these are figured in Hyde & Wade (1954). Their small size, 0·3–0·5 mm, suggests that they may be overlooked; on the rare occasions when spores retaining their episporal coats have been preserved inside the sporangia, specific identification has been possible (e.g. Jessen, 1949).

Identification of leaves of Filicales (which must be mature) is based on leaf shape and venation. Leaves of all British genera, most of which have dichotomous open-ended venation, are figured by Hyde & Wade (1954) and all species described, the venation being fairly constant within any one species. Leaves of *Osmunda regalis* L. and *Pteridium* have been found fossil and are illustrated in Jessen *et al.* (1959).

Megaspores of *Pilularia globulifera* L. are distinctive. Fossil ones are drawn and described by Watts (1959) as follows: 'like minute acorns about 0·6 mm long. They have a constriction about the middle dividing the spore into a smooth-surfaced "fruit" and a rough-surfaced "cupule". In damaged and corroded specimens the surface pattern and constriction tend to be lost—single specimens in this condition are probably not recognizable. *P. globuli-*

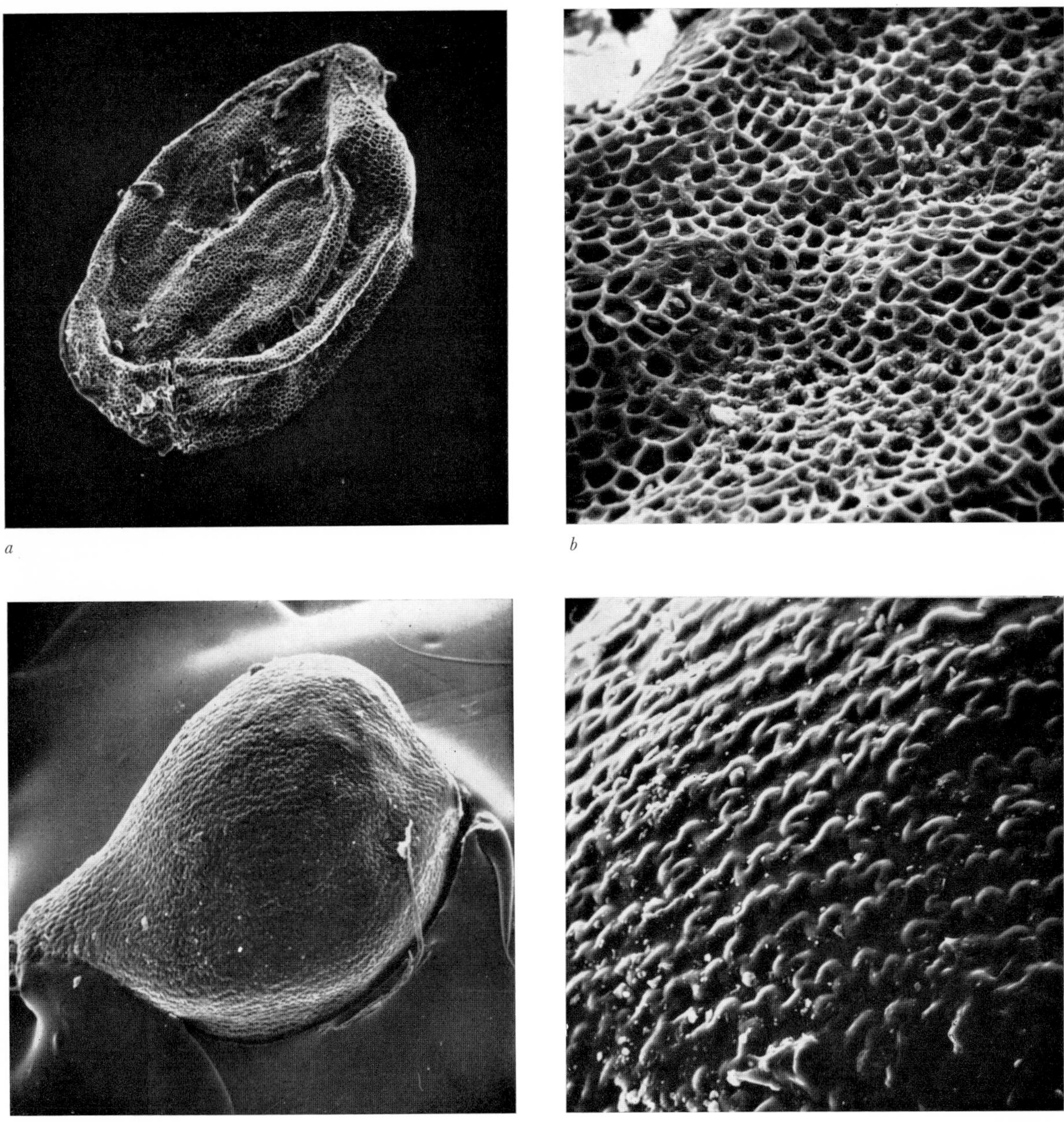

1 *a*, *b*. Seed of *Diplotaxis tenuifolia* distinguished from seed of *D. muralis* by smaller cells (average 14$\mu$ diameter), see p. 238 (Weichselian, Sidgwick Avenue, Cambridge), *a* ×52, *b* ×313. 1 *c*, *d*. Fruit of *Koenigia islandica* showing the diagnostic raised sinuous cell margins, see p. 241 (recent, Isle of Skye), *c* ×46, *d* ×264.

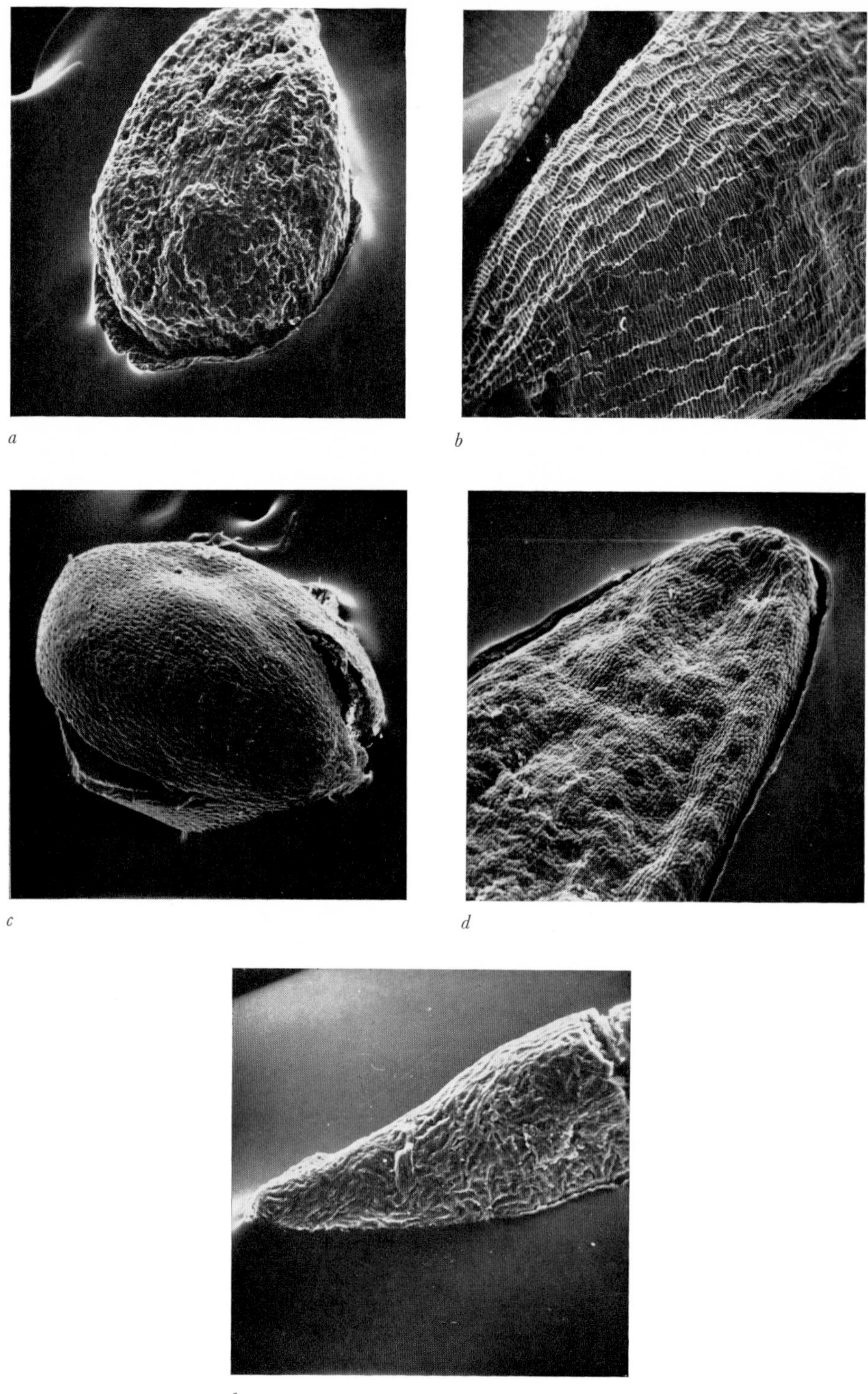

2 *a*. Fruit of *Ranunculus hyperboreus* showing irregular covering of parenchyma cells in the middle of the fruit, see p. 242 (Weichselian, Wretton, Norfolk) ×34. 2 *b*. Part of fruit of *Conium maculatum* showing longitudinal groove and diagnostic cell pattern revealed clearly when the ridges have disappeared, see p. 243 (ident. A. P. Conolly, Roman Well, Denton, Lincolnshire) ×33. 2 *c*. Seed of *Viola palustris* showing punctate cell pattern, see p. 243 (Weichselian, Kirkmichael, Isle of Man) ×32. 2 *d*. Upper part of bud scale of *Salix* sp. with apical notch and characteristic ridged surface, see p. 244 (Weichselian, Sidgwick Avenue, Cambridge) ×43. 2 *e*. Part of sclerenchymatous spindle of *Eriophorum vaginatum* showing the distinctive wavy-striate cell pattern, see p. 246 (Weichselian, Kirkmichael, Isle of Man) ×62.

*fera* is the only N. temperate species of the genus with constricted megaspores.'

*Azolla filiculoides* Lam. has been shown to be specifically identifiable on the terminal structure of the non-septate glochidia attached to the massulae which are about 0·2 mm in diameter (West, 1953). The tuberculate megaspores with three floats are 0·5 mm long; a fossil one is illustrated with the massulae attached by West (1953). It may be noted that *Azolla tegeliensis* Florsch., a fossil species, has been described from early Quaternary deposits in the Netherlands (van der Vlerk & Florschütz, 1950).

Both male and female fruiting organs of *Salvinia natans* (L.) All. have also been identified in the fossil state. West's (1957) photographs show a massula, *c.* 0·2 mm diameter, inside the coarsely reticulate microsporangium. The megaspores are about 0·6 mm long with three small valves at the top and have less coarsely reticulate sporangial walls. The wall of the megaspore is finely pitted. In both cases the reticulate wall may be absent in the fossil. The megaspores of *Salvinia* are highly resistant to decay and have been found in situations which suggest that they have been derived from earlier deposits (Dickson, unpublished).

Derived megaspores which are usually black and about 1 to 2 mm diameter with a triradiate scar and varying ornamentation have been illustrated from Quaternary deposits (Chandler, 1946; Dijkstra, 1950). They are of various pre-Quaternary ages and should thus be recorded as evidence of the presence of ancient carbon if radiocarbon dating of the sample is to be carried out.

## BRYOPHYTA

Some 210 species of mosses are known from British Quaternary deposits. This total, approaching 30% of the present flora, compares favourably with that of the flowering plants, although these have received much more attention. It is rare to find liverworts; the meagre, disproportionate total of about 30 species must result from differential preservation.

British mosses range from species about 1 mm high to those of several cm. Fossils of the larger species are easily recovered; those of diminutive species must be sought. Similarly detached capsules should not be neglected.

The texts most useful for identifying British mosses are Dixon (1924), Nyholm (1954–65) and Watson (1968). Proctor's paper (1955) is a modern guide to British species of *Sphagnum*. It can be deceptively easy to key out a species with the aid of a handbook alone; gross errors have resulted from such action. The use of properly authenticated reference material in addition to the handbooks is essential. Often the good state of preservation of moss remains allows determination to the species level.

Extraction techniques are similar to those of the flowering plants. Often it is possible to mount fossils in gum chloral on slides.

Mosses are often important constituents of peat. The peat stratigrapher discovers various mosses apart from *Sphagnum*. '*Hypnum* peat' is a term often encountered in the British and Irish literature; this is a usage of *Hypnum* in the widest sense. The genera *Acrocladium* (*Calliergon*), *Campylium*, *Drepanocladus* and *Scorpidium* (covered by *Hypnum* in Dixon's handbook), are all important peat formers. Species of *Aulacomnium*, *Camptothecium*, *Cinclidium*, *Helodium*, *Meesia*, *Paludella* and *Polytrichum* also occur.

It has long been realized that environmental inferences may be drawn from moss assemblages; some examples in the British literature are Dixon's remarks (p. 233) in Warren (1912), Proctor's (p. 143) in Sparks & West (1959) and J. H. Dickson's (p. 210) in West *et al.* (1964).

As strikingly shown by Gams (1932) the study of moss remains elucidates present distribution patterns. J. H. Dickson (1967) has postulated that the disjunct ranges of certain British mosses can be explained in the ways advanced by Godwin (1956) to account for disjunct patterns in the flowering plants.

## CHARACEAE

The oosporangia of the Characeae are frequently found in Quaternary deposits. They range from 0·2–0·9 mm in length and have five diagnostic spiral cells, originating from the basal pore and coiling up to the apex. The spiral cells of some species take up calcium carbonate from the water and are preserved as lime-shells. Other species retain only the imprint of the spirals on the brown or black oosporangia. Lime-shells can be removed with dilute acid so that the sculpturing of the ectosporestine (outer membrance of the oosporangium) can be seen. The following characters are the most important ones used by Horn af Rantzien (1959) whose monograph on the Characeae was compiled with view to the identification of fossil oosporangia.

Present-day oosporangia of the two sub-families, Nitelleae and Chareae, can be distinguished by the number of cells in the coronula, 10 and 5 respectively. But this structure, cut off from the apical spiral cells by a transverse septum, is not thought to be preserved (Groves, 1933). However a number of other characters which seem constant may be used to separate the fruiting bodies of the sub-families and to distinguish the five genera found in Britain at the present time.

Generally speaking the oosporangia of the Nitelleae (*Nitella* and *Tolypella*) are predominantly small, less than 0·5 mm long and have 6–10 spiral convolutions, whereas those of the Chareae (*Chara*, *Nitellopsis* and *Lamprothamnium*) are medium or large, 0·5–0·9 mm with 6–14 convolutions. Lime-shells are diagnostic of the Chareae though not present in all species. The Nitelleae though not developing lime-shells may have calcareous encrustations when growing in calcareous water.

*Nitella* has distinctive, laterally compressed, oosporangia which are usually prolate-spheroidal. The oosporangia of the British species range from 0·2–0·6 mm and the ectosporostinal (outer membrane) sculpturing, as seen with the use of a high-power microscope, is finely granular to reticulate, and is a useful character for specific identification.

*Tolypella* has terete oosporangia; the majority are subprolate. There are four British species, 0·3–0·5 mm with 7–9 convolutions; the sculpturing is smooth to tuberculate.

The only European species of *Nitellopsis*, *N. obtusa* (Desvaux) Groves, has comparatively large oosporangia, 0·75–0·96 mm; they are prolate-spheroidal to sub-prolate-ellipsoidal with 7–8 convolutions; the base and apex are rather truncated. At the present time the plant is confined to the Norfolk Broads and rarely fruits.

The seventeen species of *Chara* in Britain have, as a rule, prolate and ellipsoidal oosporangia with rounded or protruding apical poles and rather truncate basal ones. The granular or foveolate sculpturing does not facilitate specific determination. The size range is from 0·4–0·9 mm and there are 6–14 convolutions.

The oosporangia of the single very rare species of *Lamprothamnium* are very similar to those of *Chara* but differ in the cylindrical basal pore (in *Chara* it is always conical). The lime-shells, which are always produced, are distinctive in having little or no apical calcification.

Horn af Rantzien lists a number of characters for distinguishing species of the Characeae with or without lime-shells and suggests that it may be possible to make specific determinations. Indeed, his descriptions of oosporangia, together with photographs which include the ectosporostinal structure, do show differences between the species he has examined.

He lists specific identifications of Quaternary oosporangia which include five species of *Chara* from Flandrian peats in northern France identified and described by Froment (1954). Specimens of *Nitellopsis obtusa* from the Cromer Forest Bed have been found by Reid and identified by Groves (1933). It appears from the literature that there are many records of *Chara* and *Nitella* from Quaternary deposits but very few named species.

## PRESENTATION AND INTERPRETATION OF RESULTS

The degree of certainty that can be attached to an identification is bound to vary according to the taxonomy and the nature and condition of the fossil. The terms of reference stated by Watts & Winter (1966) are worth stressing here even though these terms are used in a similar way by most workers.

'Where a species name is used without qualification, it means that that species, and no other, matches the fossil perfectly, all reasonable alternatives considered. Where a genus or family name is followed by "sp." (species), the generic or family determination is certain, but the lower taxon is undetermined. Where a species name is followed by "type", the fossil matches not only this species but also others; here the selection of the species name is based on ecologic or geographic grounds. The use of "cf." (*confer*—compare) connotes uncertainty because of poor or fragmentary preservation, inadequate reference material or ill-defined morphology.'

It is obvious that qualified determinations for the most part severely limit phytogeographical conclusions and environmental inferences. Also it must not be overlooked, as stated by Conolly (1961), that: 'Closely allied species, or sub-species or polyploid races within an aggregate may each have different ranges and requirements so that the validity of deductions made from the present-day range will depend on the taxon represented.'

It is useful to give notes for the reader amplifying the identification in certain cases. For instance whether 'sp.' means that the species has not been matched from British material or if it is merely poorly preserved. If an identified object is of a species which has closely related taxa and has not previously been described, the distinguishing features should be stated. A full list of the terminology used for the description of plant organs is to be found in Stearn

(1966). Illustrations are often useful to supplement the descriptions. Drawing and photographic techniques are given in the *Handbook of Palaeontological Techniques* (Kummel & Raup, 1965) which includes photography of small fossils and the excellent photographs instantly obtainable with Polaroid Land film. Details of cell pattern of opaque small objects are excellently revealed by the scanning electron microscope (plate 1).

When it is possible to analyse a more or less continuous series of samples, and a sufficient number of comparable plant remains have been determined, a diagrammatic presentation of results is often helpful. Such a diagram with frequency curves or histograms often proves complementary to the pollen diagram. The results may best be shown as a percentage of the total specimens at each level (West, 1957) but can, in addition, be graphed as an absolute number from a standard volume (Watts & Winter, 1966); in either case it is useful to group the species ecologically as Watts has done. Climatic groups have also been used and Sparks & West (1959) show ecological and climatic similarities between plants and mollusca.

Although the value of the macroscopic evidence lies mostly in the detailed stratigraphical and ecological succession it often reveals, in some cases it may substantiate the rather tentative evidence from the pollen analyses. Needles of *Pinus sylvestris* L. of Zone III age in southern England gave irrefutable proof of pine in the Late-glacial when pollen alone might have been attributed to long-distance transport (Seagrief, 1959). Definitive evidence of snow patch communities has been shown by the presence of *Polytrichum norvegicum* Hedw., identified by J. H. Dickson in Kirk & Godwin (1963), and substantiated by the presence of other species in the pollen analyses, insufficiently diagnostic by themselves of chiono-philous vegetation.

Likewise the thermophilous aquatic flora of Zone *f* of the Eemian (last interglacial) is poorly represented by pollen in England (West, 1957); the macrofossil finds, which include *Najas minor* L., add to the pollen evidence.

Watts & Winter (1966) cite the characteristic thermophilous aquatic macrofossils of Zone *f* of the Eemian as an example of regional parallelism, that is vegetational change reflected in the macroflora, in response to climatic change over a wide area; regional parallelism was first described by von Post (1929) in respect to pollen analysis. Watts also gives an example from Ireland where the transition 'from a pine-oak to a fir-yew-alder pollen assemblage at Great Interglacial sites in Ireland was accompanied by a change from macrofossils indicating mesic lake and forest conditions to those of heath plants' (Watts, 1959, 1964).

The degree to which the macroflora reflects the composition of the flora in its vicinity is not clear. Obviously the macroflora from the majority of peat deposits will be of a more local origin than the water-borne flora of a lake mud. It is well known that the disseminules of some plants float for some time and are widely dispersed whereas others have little buoyancy and so their presence or absence in the sample is fortuitous. Obviously a plant which produces a large quantity of wind- or water-dispersed seeds is likely to be over represented. Seed analyses of surface lake samples in Minnesota are being studied in relation to the local flora (Watts & Winter, 1966) and the results of this work may add to the interpretation of macro-diagrams.

## SUMMARY

The literature and illustrations concerned with the identification of macroscopic plant remains of the Quaternary period are found in numerous, often obscure, publications. The aim of this essay is to indicate the most relevant books and papers; and to show the scope and limitations in the determination of plant macrofossils.

The section on fruits and seeds of selected families includes discussion of the diagnostic features of the family or genus. References are given to taxonomic or biometric studies which point to difficulties or distinctions in determining species on their fruiting parts; where relevant, modifications resulting from fossilization are discussed. The sections on other types of plant remains give the rank to which the taxa can be identified, together with the characters used for the identification or references to relevant literature. The essay concludes with a discussion of the presentation and interpretation of the data.

My thanks are due to the Cambridge Instrument Company Ltd for the use of the 'Stereoscan' Electron Microscope and to Dr P. Echlin, Botany School, Cambridge for photographic facilities; also to Miss F. G. Bell for indicating the possibilities of 'Stereoscan' photography for Quaternary macrofossils. I am indebted to Miss A. P. Conolly for her helpful suggestions and for reading the manuscript. Similarly I am most grateful to my husband, Dr J. H.

Dickson for his section on the Bryophyta, and his assistance which in many ways made this contribution possible.

## REFERENCES


Allison, J., Godwin, H. & Warren, S. H. (1952). Late-glacial deposits at Nazeing in the Lea Valley, North London. *Phil. Trans. R. Soc.* B **236**, 169–240.

Andersen, S. T. (1961). Vegetation and its environment in Denmark in the Early Weichselian Glacial (Last Glacial). *Danm. geol. Unders.* (II), **80**, 7–175.

Barghoorn, E. S. (1949). Degradation of plant remains in organic sediments. *Bot. Mus. Leafl. Harv. Univ.* pp. 1–20.

Beijerinck, W. (1947). *Zadenatlas der Nederlandsche Flora.* Wageningen: Veenman.

Benda, L. & Schneekloth, H. (1965). Das Eem-Interglacial von Köhlen, Krs. Wesermünde. *Geol. Jb.* **83**, 699–716.

Berggren, G. (1962). Reviews on the taxonomy of some species of the Genus *Brassica*, based on their seeds. *Svensk. bot. Tidskr.* **56**, 65–133.

Bertsch, K. (1941). *Früchte und Samen. Handbücher der praktischen Vorgeschichtsforschung.* I. Stuttgart: Ferdinand Enke.

Białobrzeska, M. & Truchanowiczówna, J. (1960). The variability of shape of fruits and scales of the European birches (*Betula* L.) and their determination in fossil materials. *Monographiae bot.* **9**, 2, 3–86. (Polish with English summary.)

Borrill, M. (1956). A Biosystematic Study of some *Glyceria* species in Britain. *Watsonia* **3**, 291–8.

Brazier, J. D. & Franklin, G. L. (1961). Identification of hardwoods—a microscope key. *Forest Prod. Res. Bull.* no. 46, pp. 1–96.

Chandler, M. E. J. (1921). The arctic flora of the Cam Valley at Barnwell, Cambridge. *Q. Jl geol. Soc. Lond.* **77**, 4–22.

Chandler, M. E. J. (1946). Note on some abnormally large spores formerly attributed to *Isoetes*. *Ann. Mag. nat. Hist.* 11th series, **13**, 684–9.

Clapham, A. R., Tutin, T. G. & Warburg, E. F. (1962). *Flora of the British Isles.* (2nd ed.) Cambridge University Press.

Clifford, M. H. (1936). A Mesolithic flora in the Isle of Wight. *Proc. Isle Wight nat. Hist. Archaeol. Soc.* **2**, 7, 582–94.

Conolly, A. (1957). The occurrence of seeds of Papaver sect. Scapiflora in a Scottish Late-glacial site. *Veröff. geobot. Inst. Zurich* **34**, 27–9.

Conolly, A. P. (1961). Some climatic and edaphic indications from the Late-glacial flora, in Symposium on Quaternary Ecology. *Proc. Linn. Soc. Lond.* **172**, 56–62.

Conolly, A. P., Godwin, H. & Megaw, E. M. (1950). Studies in the post-Glacial history of British vegetation. XI. Late-glacial deposits in Cornwall. *Phil. Trans. R. Soc.* B **234**, 397–469.

Cook, C. D. K. (1961). *Sparganium* in Britain. *Watsonia* **5**, 1–10.

Cook, C. D. K. (1963). Studies in *Ranunculus* Subgenus Batrachium (DC.) A. Gray. II. General morphological considerations in the taxonomy of the subgenus. *Watsonia* **5**, 294–303.

Coope, G. R., Shotton, F. W. & Strachan, I. (1961). A Late Pleistocene fauna and flora from Upton Warren, Worcestershire. *Phil. Trans. R. Soc.* B **244**, 379–421.

Cushing, E. J. & Wright, H. E., Jr. (1965). Hand-operated piston corers for lake sediments. *Ecology*, **46**, 380–4.

Dallimore, W. & Jackson, A. B. (1966). *A handbook of Coniferae and Ginkoaceae.* (4th ed. Revised G. G. Harrison.) London: Arnold.

Davies, E. W. (1953). Notes on *Carex flava* and its allies: III. The taxonomy and morphology of the British representatives. *Watsonia*, **3**, 74–9.

Dickson, J. H. (1967). The British moss flora of the Weichselian Glacial. *Rev. Palaeobot. Palynol.* **2**, 245–53.

Dijkstra, S. J. (1950). Carboniferous Megaspores in Tertiary and Quaternary Deposits of S.E. England. *Ann. Mag. nat. Hist.* 12th series, **3**, 865–77.

Dimbleby, G. W. (1967). *Plants and Archaeology.* London: Baker.

Dixon, H. N. (1924). *The student's Handbook of British Mosses.* (3rd ed.) Eastbourne: Sumfield and Day.

Duigan, S. L. (1955). Plant remains from the gravels of the Summertown–Radley Terrace near Dorchester, Oxfordshire. *Q. Jl geol. Soc. Lond.* **140**, 3, 225–38.

Ferré, Y. de (1952). Les formes de jeunesse des Abiet acees. Ontogenie–Phylogenie. *Trav. Lab. for. Toulouse*, **4**, 1.

Florin, R. (1931). Untersuchungen zur Stammesgeschichte der Coniferales und Cordaitales. I. Morphologie und Epidermisstruktur des Assimilationsorgane bei den rezenten Koniferen. *K. svenska Vetensk. Akad. Handl.* **3**, ser. 10, 1–588.

Franks, J. W. & Pennington, W. (1961). The Late-glacial and Post-glacial deposits of the Esthwaite basin, North Lancashire. *New Phytol.* **60**, 27–42.

Froment, P. (1954). 4. Charophytes Fossiles. Découverte d'oogones de *Chara* dans une tourbe flandrienne du Nord de la France. *Rapp. Communic. Huitième Congr. Intern. Bot. Paris*, Sect. 5, pp. 237–8.

Gams, H. (1932). Quaternary Distribution. In *Manual of Bryology.* (Ed. F. Verdoorn), pp. 297–322. The Hague: Martinus Nijhoff.

Godwin, H. (1956). *The History of the British Flora.* Cambridge University Press.

Godwin, H. (1962). Vegetational History of the Kentish Chalk Downs as seen at Wingham and Frogholt. *Veröff. geobot. Inst. Zürich* **37**, 83–99.

Godwin, H. (1964). Late-Weichselian conditions in south eastern Britain: organic deposits at Colney Heath, Herts. *Proc. Roy. Soc.* B **160**, 258–75.

Green, P. S. (1954). *Stellaria nemorum* L. subspecies *glochidisperma* Murbeck in Britain. *Watsonia*, **3**, 122–6.

Greguss, P. (1955). *Identification of Living Gymnosperms on the Basis of Xylotomy*. Budapest: Akad. Kiadó.

Grosse-Brauckmann, G. (1964). Einige wenig beachtete Pflanzenreste in nordwestdeutschen Torfen und die Art ihres Vorkommens. *Geol. Jb.* **81**, 621–44.

Groves, J. (1933). Charophyta. *Fossilium Catalogus*, II. *Plantae*, **19**, 1–74. Berlin.

Guinet, P. (1959). Essai d'identification des graines de Chénopodes commensaux des Cultures on cultivés en France. *J. Agric. trop. Bot. appl.* **6**, 241–66.

Hagerup, O. & Petersson, V. (1959). *Botanisk Atlas*, vol. 1, Angiosperms (transl. H. Gilbert-Carter), p. 19. Copenhagen: Munksgaard.

Hagström, J. O. (1916). Critical researches on the Potamogetons. *K. Svenska Vetensk. Akad. Handl.* **55**, 1–280.

Hartz, N. (1909). Bidrag til Danmarks Tertiaere og diluviale Flora. *Danm. geol. Unders.* (II) **20**, pp. 1–292.

Heer, O. (1862). On the fossil flora of Bovey Tracey. *Phil. Trans. R. Soc.* **152**, 1039–86.

Helbaek, H. (1953). Early Crops in Southern England. *Proc. Prehist. Soc.* **18**, 194–233.

Helbaek, H. (1958). Grauballe mandens sidste Måltid. *Kuml, Årbog f. Jysk. Ark. Selskab, Århus.* pp. 83–116. (Danish with English summary.)

Helbaek, H. (1963). In *Science in Archaeology* (ed. D. Brothwell & E. Higgs), pp. 174–94. London: Thames & Hudson.

Horn af Rantzien, H. (1959). Recent charophyte fructifications and their relations to fossil charophyte gyrogonites. *Ark. Bot.* **4**, no. 7, 165–332.

Hubbard, C. E. (1954). *Grasses*. Harmondsworth: Penguin Books.

Hultèn, E. (1956). The *Cerastium alpinum* complex. A case of World-wide Introgressive Hybridization. *Svensk bot. Tidskr.* **50**, 3, 411–95.

Hyde, H. A. & Wade, A. E. (1954). *Welsh Ferns*. (3rd ed.) Cardiff: National Museum of Wales.

Jalas, J. & Sell, P. D. (1967). Taxonomic and nomenclatural notes on the British flora. *Watsonia* **6**, 292–4.

Jane, F. W. (1956). *The Structure of Wood*. London: Black.

Jermy, A. C. (1967). *Carex* section Carex. *Proc. bot. Soc. Br. Isl.* **6**, 375–9.

Jessen, K. (1949). Studies in Late Quaternary deposits and flora history of Ireland. *Proc. R. Ir. Acad.* **52** B **6**, 85–290.

Jessen, K. (1955). Key to subfossil *Potamogeton*. *Bot. Tidsskr.* **52**, 1–7.

Jessen, K., Andersen, S. T. & Farrington, A. (1959). The Interglacial deposits near Gort, Co. Galway, Ireland. *Proc. R. Ir. Acad.* **60** B **1**, 1–77.

Jessen, K. & Farrington, A. (1938). The bogs at Ballybetagh, near Dublin, with remarks on Late-glacial conditions in Ireland. *Proc. R. Ir. Acad.* **44** B **10**, 205–60.

Jessen, K. & Helbaek, H. (1944). Cereals in Great Britain and Ireland in prehistoric and early historic time. *K. danske Vidensk. Selsk, Skr.* **3**, 2.

Katz, N. J., Katz, S. V. & Kipiani, M. G. (1965). *Atlas and Keys of Fruits and Seeds Occurring in the Quaternary deposits of the U.S.S.R.* (In Russian). Moscow: Nauka.

Kelly, M. R. (1964). The Middle Pleistocene of North Birmingham. *Phil. Trans. R. Soc.* B **247**, 533–92.

Kelly, M. & Osborne, P. J. (1965). Two faunas and floras from the alluvium at Shustoke, Warwickshire. *Proc. Linn. Soc. Lond.* **176**, 37–65.

Kirk, W. & Godwin, H. (1963). A Late-glacial site at Loch Droma, Ross and Cromarty. *Trans. R. Soc. Edinb.* **65**, no. 11, 225–49.

Klimaszewski, M., Szafer, W., Szafran, B. & Urbanski, J. (1939). The *Dryas* Flora of Krościenko on the river Dunajec. (Polish with English summary.) *Biul. pánst. Inst. geol.* **24**, 1–85.

Kneblová, V. (1958). The Interglacial Flora in Gánovce Travertines in Eastern Slovakia. *Acta biol. cracov.* **1**, 1–5.

Körber-Grohne, U. (1964). *Probleme der küstenforschung im südlichen nordseegebiet.* 7. *Bestimmungsschlüssel für subfossile Juncus — Samen und Gramineen — Früchte.* Hildersheim: August Lax.

Kummel, B. & Raup, D. (ed.) (1965). *Handbook of Palaeontological techniques*, pp. 423–68. San Francisco: Freeman.

Lambert, C. A., Pearson, R. G. & Sparks, B. W. (1963). A flora and fauna from Late Pleistocene deposits at Sidgwick Avenue, Cambridge. *Proc. Linn. Soc. Lond.* **174**, 13–29.

Marsden-Jones, E. M. & Turrill, W. B. (1957). *Bladder Campions*, p. 37. London: Ray Society.

Metcalfe, C. R. (1960). *Anatomy of the Monocotyledons:* vol. 1. *Gramineae*. Oxford: Clarendon.

Metcalfe, C. R. & Gregory, M. (1964). Comparative Anatomy of Monocotyledons. Some New Descriptive Terms for Cyperaceae with a Discussion of Variations in Leaf Form noted in the Family. Notes from the Jodrell Laboratory. London: Royal Botanic Gardens.

Moore, D. (1963). The subspecies of *Montia fontana* L. *Bot. Notiser* **116**, 1, 16–30.

Murley, M. M. (1951). Seeds of the Cruciferae of North-eastern North America. *Am. Midl. Nat.* **46**, 1, 1–81.

Nathorst, A. G. (1873). On the Distribution of Arctic plants during the Post-glacial Epoch. *J. Bot., Lond.* n.s., **2**, 225–8.

Nilsson, O. & Hjelmquist, H. (1967). Studies on the nutlet structure of South Scandinavian species of *Carex*. *Bot. Notiser.* **120**, 460–85.

Nyholm, E. (1954–1965). *Illustrated Moss Flora of Fennoscandia.* II Musci. **1** to **5**, Lund: Gleenips.

Paulssen, L. M. (1964). *Identification of Active Charcoals and Wood Charcoals.* Trondheim: Scandinavian University Books.

Pennington, W. (1947). Studies of the post-Glacial History of British vegetation. VII. Lake sediments: Pollen diagrams from the bottom deposits of the North Basin of Windermere. *Phil. Trans. R. Soc.* B **233**, 137–75.

Perring, F. H. & Sell, P. D. (1967). Taxonomic and Nomenclatural Notes on the British Flora. *Watsonia* **6**, 294–5.

Phillips, E. W. J. (1948). Identification of Softwoods by their microscopic structure. *Forest Prod. Res. Bull.* no. 22, pp. 1–56.

Post, L. von (1929). Die postarktische Geschichte der europaïschen Wälder nach den vorliegenden Pollendiagrammen. *Meddn. Stockh. Högsk. bot. Inst.* no. 16, pp. 1–27.

Proctor, M. C. F. (1955). A key to British species of *Sphagnum*. *Trans. Brit. bryol. Soc.* **2**, 552–9.

Rabien, I. (1953). Zur Bestimmung fossiler knospenschuppen. *Paläont. Z.* **27**, 1/2, 57–66.

Raunkiaer, C. (1895–9). *De Danske blomsterplanters Naturhistorie,* vol. 1, *Monocotyledons*. Copenhagen: Hos Gyl den dalske.

Reid, C. (1899). *The Origin of the British Flora.* London: Dulau.

Reid, C. & Reid, E. M. (1908). On the Pre-Glacial Flora of Britain. *J. Linn. Soc.* (Bot.) **38**, 206–27.

Reid, E. M. (1949). The Late Glacial flora of the Lea Valley. *New Phytol.* **48**, 245–52.

Richardson, F. (1960*a*). Laboratory techniques in plant anatomy: I. A method of reviving sections of compressed wood from archaeological sites and peatbogs. *Kew Bull.* **14**, 85–6.

Richardson, F. (1960*b*). Laboratory techniques in plant anatomy: II. A method of examining fossilized or semi-fossilized plant tissues impregnated with calcareous substances. *Kew Bull.* **14**, 87.

Ross-Craig, S. (1957). *Drawings of British Plants.* Part X. Saxifragaceae, Crassulaceae. London: Bell.

Ross-Craig, S. (1960–3). *Drawings of British Plants.* Parts XV–XVIII. Compositae I–IV. London: Bell.

Schumacker, R. (1961). Etude d'une Tourbe à *Scheuzeria palustris* dans les couches inférieures des dépots de la Fagne Wallonne. *Bull. Soc. r. Sci. Liège*, nos. 11–12, pp. 496–511.

Seagrief, S. C. (1959). Pollen diagrams from Southern England: Wareham, Dorset, and Nursling, Hampshire. *New Phytol.* **58**, 316–25.

Simpson, I. M. & West, R. G. (1958). On the stratigraphy and palaeobotany of a Late Pleistocene organic deposit at Chelford, Cheshire. *New Phytol.* **57**, 239–50.

Sparks, B. W. & Lambert, C. A. (1961). The Post-glacial deposits at Apethorpe, Northamptonshire. *Proc. Malac. Soc. Lond.* **34**, 6, 302–15.

Sparks, B. W. & West, R. G. (1959). The Palaeoecology of the Interglacial deposits at Histon Road, Cambridge. *Eiszeitalter Gegenw.* **10**, 123–43.

Stace, C. A. (1965). Cuticular studies as an aid to plant taxonomy. *Bull. Br. Mus. nat. Hist. bot.* **4**, 1, 1–78.

Stearn, W. T. (1966). *Botanical Latin.* London: Nelson.

Styles, B. T. (1962). The Taxonomy of *Polygonum aviculare* and its allies in Britain. *Watsonia* **5**, 177–214.

Timson, J. (1964). Fruits of *Polygonum* Section Persicaria. *Proc. bot. Soc. Br. Isl.* **5**, 381.

Timson, J. (1965). Fruit variation in *Polygonum persicaria* L. *Watsonia* **6**, 106–8.

Tralau, H. (1963). The recent and fossil distribution of some boreal and arctic montane plants in Europe. *Ark. Bot.* **5**, 3, 533–82.

Tralau, H. & Zagwijn, W. H. (1962). Fossil *Salix polaris* Wahlbg. in the Netherlands. *Acta bot. neerl.* **11**, 425–7.

Tutin, T. G. *et al.* (ed.) (1964). *Flora Europaea,* **1**. Cambridge University Press.

Van der Vlerk, I. M. & Florschütz, F. (1950). *Nederland in het Ijstijdvak.* Utrecht: W. de Haan N.V.

Walters, S. M. (1953). *Montia fontana* L. *Watsonia* **3**, 1–6.

Warren, S. H. (1912). On a Late-glacial stage in the valley of the River Lea. *Q. Jl geol. Soc. Lond.* **68**, 213–51.

Watson, E. V. (1968). *British Mosses and Liverworts.* (2nd ed.) Cambridge University Press.

Watts, W. A. (1959). Interglacial deposits at Kilbeg and Newtown, Co. Waterford. *Proc. R. Ir. Acad.* **60** B **2**, 79–134.

Watts, W. A. (1964). Interglacial deposits at Baggotstown, near Bruff, Co. Limerick. *Proc. R. Ir. Acad.* **63** B **9**, 167–89.

Watts, W. A. & Winter, T. C. (1966). Plant Macrofossils from Kirchner Marsh, Minnesota—A Paleoecological Study. *Bull. geol. Soc. Am.* **77**, 1339–60.

Webb, D. A. (1950). A revision of the Dactyloid Saxifrages of North-western Europe. *Proc. R. Ir. Acad.* **53** B **12**, 207–40.

West, R. G. (1953). The occurrence of *Azolla* in British Interglacial deposits. *New Phytol.* **52**, 267–72.

West, R. G. (1956). The Quaternary deposits at Hoxne, Suffolk. *Phil. Trans. R. Soc.* B **239**, 265–356.

West, R. G. (1957). Interglacial deposits at Bobbitshole, Ipswich. *Phil. Trans. R. Soc.* B **241**, 1–31.

West, R. G., Lambert, C. A. & Sparks, B. W. (1964). Interglacial deposits at Ilford, Essex. *Phil. Trans. R. Soc.* B **247**, 185–212.

Western, A. C. (1963). In *Science in Archaeology* (ed. D. Brothwell & E. Higgs), pp. 150–8. London: Thames & Hudson.

# AUTHOR INDEX

(*Note: Figures in bold type indicate pages on which references are listed*)

# AUTHOR INDEX

# SUBJECT INDEX

# SUBJECT INDEX

# SUBJECT INDEX

## SUBJECT INDEX